The Institute of Mathematics and its Applications Conference Series

Volumes in the previous series were published by Academic Press to whom all enquiries should be addressed. The following and all forthcoming titles are published by Oxford University Press throughout the world.

NEW SERIES

1. *Supercomputers and parallel computation* Edited by D. J. Paddon
2. *The mathematical basis of finite element methods*
   Edited by David F. Griffiths
3. *Multigrid methods for integral and differential equations*
   Edited by D. J. Paddon and H. Holstein
4. *Turbulence and diffusion in stable environments* Edited by J. C. R. Hunt
5. *Wave propagation and scattering* Edited by B. J. Uscinski
6. *The mathematics of surfaces* Edited by J. A. Gregory
7. *Numerical methods for fluid dynamics II*
   Edited by K. W. Morton and M. J. Baines
8. *Analysing conflict and its resolution* Edited by P. G. Bennett
9. *The state of the art in numerical analysis*
   Edited by A. Iserles and M. J. D. Powell
10. *Algorithms for approximation* Edited by J. C. Mason and M. G. Cox
11. *The mathematics of surfaces II* Edited by R. R. Martin
12. *Mathematics in signal processing* Edited by T. S. Durrani, J. B. Abbiss, J. E. Hudson, R. N. Madan, J. G. McWhirter, and T. A. Moore
13. *Simulation and optimization of large systems*
   Edited by Andrzej J. Osiadacz
14. *Computers in mathematical research*
   Edited by N. M. Stephens and M. P. Thorne
15. *Stably stratified flow and dense gas dispersion*
   Edited by J. S. Puttock
16. *Mathematical modelling in non-destructive testing*
   Edited by Michael Blakemore and George A. Georgiou
17. *Numerical methods for fluid dynamics III*
   Edited by K. W. Morton and M. J. Baines
18. *Mathematics in oil production*
   Edited by Sir Sam Edwards and P. R. King
19. *Mathematics in major accident risk assessment*
   Edited by R. A. Cox
20. *Cryptography and coding* Edited by Henry J. Beker and F. C. Piper
21. *Mathematics in remote sensing* Edited by S. R. Brooks
22. *Applications of matrix theory*
   Edited by M. J. C. Gover and S. Barnett
23. *The mathematics of surfaces III* Edited by D. C. Handscomb

Continued overleaf

24. *The interface of mathematics and particle physics*
Edited by D. G. Quillen, G. B. Segal, and Tsou, S. T.
25. *Computational methods in aeronautical fluid dynamics*
Edited by P. Stow
26. *Mathematics in signal processing II*
Edited by J. G. McWhirter
27. *Mathematical structures for software engineering*
Edited by Bernard de Neumann, Dan Simpson, and Gil Slater
28. *Computer modelling in the environmental sciences*
Edited by D. G. Farmer and M. J. Rycroft
29. *Statistics in medicine* Edited by F. Dunstan and J. Pickles
30. *The mathematical revolution inspired by computing*
Edited by J. H. Johnson and M. J. Loomes
31. *The mathematics of oil recovery* Edited by P. R. King
32. *Credit scoring and credit control*
Edited by L. C. Thomas, J. N. Crook, and D. B. Edelman
33. *Cryptography and coding II* Edited by Chris Mitchell
34. *The dynamics of numerics and the numerics of dynamics*
Edited by D. S. Broomhead and A. Iserles
35. *The unified computation laboratory: modelling, specifications, and tools* Edited by Charles Rattray and Robert G. Clark
36. *Mathematics in the automotive industry* Edited by James R. Smith
37. *The mathematics of control theory*
Edited by N. K. Nichols and D. H. Owens
38. *Mathematics in transport planning and control* Edited by J. D. Griffiths
39. *Computational ordinary differential equations*
Edited by J. R. Cash and I. Gladwell

The Mathematics of Control Theory

Based on the proceedings of a conference on Control Theory, organized by The Institute of Mathematics and its Applications and held at the University of Strathclyde in September 1988

Edited by

N. K. NICHOLS
Department of Mathematics
University of Reading, UK

and

D. H. OWENS
Centre for Systems and Control Engineering
University of Exeter, UK

CLARENDON PRESS · OXFORD · 1992

Oxford University Press, Walton Street, Oxford OX2 6DP

Oxford New York Toronto
Delhi Bombay Calcutta Madras Karachi
Petaling Jaya Singapore Hong Kong Tokyo
Nairobi Dar es Salaam Cape Town
Melbourne Auckland

and associated companies in
Berlin Ibadan

Oxford is a trade mark of Oxford University Press

Published in the United States
by Oxford University Press, New York

A catalogue record for this book is available from the British Library

Library of Congress Cataloging in Publication Data

The Mathematics of control theory : based on the proceedings of a conference on control theory, organized by the Institute of Mathematics and Its Applications, and held at the University of Strathclyde in September 1988 / edited by N.K. Nichols and D.H. Owens.
(The Institute of Mathematics and Its Applications conference series ; new ser., 37)
"[Papers from] the 5th IMA Conference on Control Theory"—Pref.
Includes bibliographical references.
1. Control theory—Congresses. I. Nichols, N. K. II. Owens, D. H. (David H.) III. Institute of Mathematics and Its Applications. IV. Series.
QA402.3.M3484 1992 629.8'312—dc20 92-23022
ISBN 0-19-853640-2

Printed in Great Britain by
Bookcraft (Bath) Ltd
Midsomer Norton, Avon

Preface

5th IMA Conference on Control Theory

The 5th IMA Conference on Control Theory was held at the University of Strathclyde in Glasgow in September 1988. Its aims were to attract a wide spectrum of academic researchers from the international community and industrialists with an interest in the potential of mathematical control and control algorithms for improving the performance and efficiency of their control schemes. The conference was organized under the auspices of the IMA Control Group under the chairmanship of Dr A. Zinober and secretary Dr P. A. Cooke. Local arrangements were supported by the academic staff of the University of Strathclyde and, in particular, Dr W. E. Leithead who carried the bulk of the administrative burden. Professor C. J. Harris and Dr J. E. Marshall also supported the conference in their roles as editors of the IMA Journal of Mathematical Control and Information.

The efforts of the organizers backed up by the staff of the IMA combined with the high quality of submitted papers to produce an interesting meeting attended by representatives of all of the main disciplines in control. Particular mention must be made here of Professor Gareth Knowles, whose efforts produced a special invited session in the area of robust and adaptive control, and Professors A. J. Pritchard and D. Hinrichsen, who organized an invited session on robust stabilization.

The venue for the conference was the exciting, cultural city of Glasgow in Scotland. The hosts were the control community of the University of Strathclyde. The conference did not, however, focus its eyes solely on the splendours of control theory or the opportunities to taste the nightlife of Glasgow. Indeed, the delegates were to be seen following the pipes into the hills to taste the delights of Loch Lomond at the University of Strathclyde Staff Club, Ross Priory. The details are omitted.

In its final form, the conference consisted of 33 contributed papers with special plenary contributions from Professor H. Kwakernaak, Professor Lindquist, Professor J. C. Willems, Professor C. J. Harris, Professor R. Curtain, and Dr D. J. Bell. As a whole, the programme was a fair reflection of current worldwide activity in areas of mathematical control and its applications. This volume contains those papers submitted in complete form. We have tried to group the papers together in a way which matches the original conference schedule while following a natural pattern of mathematical development and industrial needs. Where possible, papers have been grouped around a plenary talk representative of the area of study.

The first group of papers are representative of the traditional area of algebraic linear systems theory. Linear systems have played and will always play a central role in control studies due to their analytical tractability, their potential for generating physical insights into systems dynamics and control, and their demonstrated influences in the development of CAD procedures for systems design based on stability as a central requirement of systems performance. The natural relationship between linear systems, linear algebra, and polynomials is the theme moving through these papers on pole and eigenstructure assignment, polynomial systems representation and optimization. The ideas of polynomial systems and optimization are brought together in the invited paper of Kwakernaak which is concerned with the computation of minimum norm H_∞ or 'minimum sensitivity' feedback controls using an algebraic characterization of the optimizing feedback compensator.

The area of nonlinear control has been and increasingly will be a challenging area for control theorists worldwide. The challenge lies at all levels including model selection, dynamic analysis, systematic stabilization procedures, and performance enhancements. The technical detail underpinning such developments is extremely diverse and is illustrated in these proceedings by the survey paper by Bell in the area of systems invertibility and by the supporting contributions in the area of frequency response characterization and geometric analysis of singular perturbations of nonlinear systems. Of course, in control systems design, the control engineer has to contend with not only plant nonlinearities but also an increasing potential to use nonlinearities as a control mechanism. The use of switching functions is illustrated by the section on variable structure systems, and the potential for elaborate switching-type and other nonlinear controls to cope with model uncertainty is highlighted by the paper by Ryan who uses set-valued maps to characterize uncertainty and construct stabilizing control schemes.

The potential for both linear and nonlinear control algorithms to cope with plant uncertainty is an intriguing and increasingly important area of study that will preoccupy many control theorists in the foreseeable future. There is no one universally applicable paradigm for this area as the solution can be expected to depend upon the degree of uncertainty and the chosen characterization of that uncertainty. One of the extreme cases in this interpretation is that of adaptive control where plant uncertainty is 'almost' complete and control is attempted by continuously adjusting parameters in a linear control structure using an on-line, recursive (nonlinear) parameter estimation algorithm. At the other extreme lies the situation of the availability of a nominal plant model plus explicit bounds on the error in modelling the real plant by such a nominal model. Athough such a situation can be approached by an adaptive control philosophy, the robust control design methodology attempts to create a fixed parameter controller capable of stabilizing all plants close enough to the nominal system in the sense that the modelling error is less than the available error bound. The difference between adaptive and robust control also lies in their distinctive on-line and off-line nature respectively. The diversity of plant models, error characterization, and on-line and off-line priorities automatically leads to a diversity of solution procedures, some of which are reflected in the papers on adaptive control and robust adaptive control in these proceedings.

It is well known that the most mathematically tractable problems in control theory are concerned with systems of finite state dimension. There is, however, a need for an explicitly constructed theory of control of infinite-dimensional systems exemplified by models containing partial differential equations and/or delay-differential elements. The state of the art in this area is illustrated elegantly by the paper by Bontsema and Curtain and supported by papers on approximation and computational aspects of delay systems.

Applications and applications related theoretical developments are increasingly important for the credibility of mathematical control. The conference was fortunate to receive a number of excellent applications papers in the area of chemical process control and automotive engineering and to be able to include the paper by Ponnapalli *et al.* providing an overview of control structures for industrial application and an indication of present and future industrial needs and attitudes.

Control theory and concepts are finding increasing application in many fields of endeavour by combination with the ideas of information theory and the methods of computer science. An indication of the possiblities in this area is provided by the concluding paper in the proceedings by Harris on the theory of evidence in multi-sensor data fusion.

Finally, we would like to take this opportunity to thank everybody involved with the conference including the invited speakers, the authors of contributed papers, members of the national and local organizing committees, and the IMA secretariat. It is the goodwill and enthusiasm of this group that has enabled the following proceedings to be made available to the control community as a whole.

Reading N. K. N.
Exeter D. H. O.
October 1991

Contents

Contributors

Allgöwer, F.
Institut für Systemdynamik und Regelungstechnik, Universität Stuttgart, Pfaffenwaldring 9, D-7000 Stuttgart 80, Germany.

Baker, R. P. D.
Department of Mathematics, University of Strathclyde, Glasgow, G1 1XH, UK.

Banks, S. P.
Department of Control Engineering, University of Sheffield, Mappin Street, Sheffield, S1 3JD, UK.

Barnes, A.
Department of Computer Science and Applied Mathematics, Aston University, Birmingham, B4 7ET, UK.

Bartolini, G.
Dipartimento di Informatica, Sistemistica e Telematika, Via Opera Pia 11a, 16145 Genova, Italy.

Bell, D. J.
Department of Mathematics, UMIST, Manchester, M60 1QD, UK.

Bonstsema, J.
Rijksuniversiteit Groningen, Mathematisch Instituut, Postbus 800, 9700 AV Groningen, The Netherlands.

Burton, J. A.
B. S. O./Automation Technology, Kon. Wilheminalaan 5, 3503 RB Utrecht, The Netherlands.

Butt, R.
Department of Applied Mathematical Studies, School of Mathematics, University of Leeds, LS2 9JT, UK.

Celentano, G.
Dipartimento di Informatica e Sistemistica, Università degli Studi di Napoli, Via Claudio 21, 80125 Napoli, Italy.

Chanane, B.
Department of Control Engineering, University of Sheffield, Mappin Street, Sheffield, S1 3JD, UK.

Chotai, A.
Control and Systems Group, Department of Environmental Science, University of Lancaster, Lancaster, LA1 4YQ, UK.

Cook, P. A.
Control Systems Centre, UMIST, Manchester, M60 1QD, UK.

Curtain, R. F.
Rijksuniversiteit Groningen, Mathematisch Instituut, Postbus 800, 9700 AV Groningen, The Netherlands.

Davies, R.
Control Systems Centre, UMIST, Manchester, M60 1QD, UK.

Dayawansa, Wijesuriya
Department of Mathematics, Texas Tech University, Lubbock, Texas 79409-1042, USA.

Garofalo, F.
Dipartimento di Informatica e Sistemistica, Università degli Studi di Napoli, Via Claudio 21, 80125 Napoli, Italy.

Glielmo, L.
Dipartimento di Informatica e Sistemistica, Università degli Studi di Napoli, Via Claudio 21, 80125 Napoli, Italy.

Goodall, D. P.
Department of Mathematics, Coventry Polytechnic, Priory Street, Coventry, CV1 5FB, UK.

Greenough, R. D.
Department of Electronic Engineering and Applied Physics, University of Hull, Hull, HU6 7RX, UK.

Grimble, M. J.
Industrial Control Unit, Department of Electronic and Electrical Engineering, University of Strathclyde, Royal College Building, 204 George Street, Glasgow, G1 1XW, UK.

Harris, C. J.
Department of Aeronautics and Astronautics, Southampton University, Highfield, Southampton, SO9 5NH, UK.

Hayton, G. E.
Department of Electronic Engineering, Hull University, Hull, HU6 7RX, UK.

Hinrichsen, D.
Institut für Dynamische Systeme, Universität Bremen, Bibliotekstrasse, Postfach 330 440, D-2800 Bremen 33, Germany.

Ho, W. C. Daniel
Industrial Control Unit, Department of Electronic and Electrical Engineering, 204 George Street, Glasgow, G1 1XW, UK.

Hydon, P. E.
DAMTP, University of Cambridge, Silver Street, Cambridge, CB3 9EW, UK.

Ilchmann, A.
Institut für Angewandte Mathematik, Universität Hamburg, D-2000 Hamburg 13, Germany.

Johnson, M. A.
Industrial Control Unit, Department of Electronic and Electrical Engineering, University of Strathclyde, Royal College Building, 204 George Street, Glasgow, G1 1XW, UK.

Jones, R. P.
Department of Engineering, University of Warwick, Coventry, CV4 7AL, UK.

Jones, R. W.
Department of Mathematics, University of Strathclyde, Glasgow, G1 1XH, UK.

Kafai, A.
Department of Mathematical Sciences, Loughborough University of Technology, Loughborough, Leicestershire, LE11 3TU, UK.

Karcanias, N.
City University, Control Engineering Centre, Department of Electrical Electronics and Information Engineering, Northampton Square, London, EC1V 0HB, UK.

Knowles, G.
Systems Research Division, Grumann Aerospace, Bethpage, New York, USA.

Kwakernaak, Huibert
Department of Applied Mathematics, University of Twente, P.O. Box 217, 7500 AE Enschede, The Netherlands.

Logemann, H.
Institut für Dynamische Systeme, Universität Bremen, Postfach 330 440, D-2800 Bremen 33, Germany.

Marshall, J. E.
School of Mathematical Sciences, University of Bath, Bath, BA2 7AY, UK.

Martin, Clyde
Department of Mathematics, Texas Tech University, Lubbock, Texas 79409-1042, USA.

Mårtensson, Bengt
Institut für Dynamische Systeme, Universität Bremen, Postfach 330 440, D-2800 Bremen 33, Germany.

Milonidis, E.
City University, Control Engineering Centre, Department of Electrical Electronics and Information Engineering, Northampton Square, London, EC1V 0HB, UK.

Newmann, M. M.
Department of Engineering Mathematics, The Queen's University of Belfast, Northern Ireland, BT7 1NN, UK.

Nichols, N. K.
Department of Mathematics, University of Reading, Box 220, Reading, RG6 2AX, UK.

O'Reilly, J.
Industrial Control Unit, Department of Electronic and Electrical Engineering, University of Strathclyde, Royal College Building, 204 George Street, Glasgow, G1 1XW, UK.

Owens, D. H.
Centre for Systems and Control Engineering, University of Exeter, Exeter, EX4 4QF, UK.

Partington, Jonathan R.
Department of Pure Mathematics, University of Leeds, Leeds, LS2 9JT, UK.

Parvinmehr, A.
Department of Electronic Engineering and Applied Physics, University of Hull, Hull, HU6 7RX, UK.

Patton, R. J.
Department of Electronics, University of York, Heslington, York, YO1 5DD, UK.

Pritchard, A. J.
Department of Mathematics, University of Warwick, Coventry, CV4 7AL, UK.

Ponnapalli, P. V. S.
Industrial Control Unit, Department of Electronic and Electrical Engineering, University of Strathclyde, Royal College Building, 204 George Street, Glasgow, G1 1XW, UK.

Pugh, A. C.
Department of Mathematical Sciences, Loughborough University of Technology, Loughborough, Leicestershire, LE11 3TU, UK.

Raisch, J.
Institut für Systemdynamik und Regelungstechnik, Universität Stuttgart, Pfaffenwaldring 9, D-7000 Stuttgart 80, Germany.

Roberts, A. P.
Department of Engineering Mathematics, The Queen's University of Belfast, Northern Ireland, BT7 1NN, UK.

Rogers, E.
Department of Aeronautics and Astronautics, The University of Southampton, Southampton, SO9 5NH, UK.

Rubio, J. E.
Department of Applied Mathematical Studies, School of Mathematics, University of Leeds, LS2 9JT, UK.

Ryan, E. P.
School of Mathematical Sciences, University of Bath, Claverton Down, Bath, BA2 7AY, UK.

Sharkey, P. M.
Industrial Control Unit, Department of Electronic and Electrical Engineering, University of Strathclyde, Royal College Building, 204 George Street, Glasgow, G1 1XW, UK.

Spurgeon, S. K.
Department of Engineering, University of Leicester, Leicester, UK.

Townley, Stuart
Department of Mathematics, University of Exeter, Exeter, EX4 4QF, UK.

Walker A. B.
Department of Mathematics, North East London Polytechnic, London, UK.

Wang, C.
Control and Systems Group, Department of Environmental Science, University of Lancaster, Lancaster, LA1 4YQ, UK.

Wang, L.
Department of Chemical Engineering, University of Toronto, Toronto, Canada.

Warwick, Kevin
Department of Cybernetics, University of Reading, Whiteknights, Reading, RG6 2AL, UK.

Walton, K.
School of Mathematical Sciences, University of Bath, Bath, BA2 7AY, UK.

Wilkinson, A. J.
Department of Electronic Engineering and Applied Physics, University of Hull, Hull, HU6 7RX, UK.

Young, P. C.
Control and Systems Group, Department of Environmental Science, University of Lancaster, Lancaster, LA1 4YQ, UK.

Zarrop, M. B.
Control Systems Centre, UMIST, Manchester, M60 1QD, UK.

Zinober, A. S. I.
Department of Applied and Computational Mathematics, University of Sheffield, Sheffield, S10 2TN, UK.

Zolezzi, T.
Dipartimento di Matematica, Via L B Alberti 4, 16132 Genova, Italy.

Simplified polynomial solution of the standard $\mathcal{H}_\infty$ optimal control problem

H. Kwakernaak

Abstract

A new polynomial parametrization for the "standard" $\mathcal{H}_\infty$ optimal control problem leads to a quadratic matrix polynomial equation. The optimal control problem is reduced to finding a suitable solution of this equation.

1 Introduction

We study the "standard" $\mathcal{H}_\infty$ optimal control problem (Francis [4]), which may be formulated as follows. Consider a linear time-invariant plant with external input w, control input u, control error signal z and feedback output y as in Fig. 1, whose real rational transfer matrix G may be partioned as

$$\begin{pmatrix} z \\ y \end{pmatrix} = \begin{pmatrix} G_{11} & G_{12} \\ G_{21} & G_{22} \end{pmatrix} \begin{pmatrix} w \\ u \end{pmatrix}. \tag{1.1}$$

The dimensions of the blocks follow from

$$\begin{matrix} & k_1 & k_2 \\ m_1 & G_{11} & G_{12} \\ m_2 & G_{21} & G_{22} \end{matrix} . \tag{1.2}$$

It is desired to find a feedback $u = Ky$ with compensator transfer function K that stabilizes the closed-loop system and minimizes

$$\|H_K\|_\infty , \tag{1.3}$$

with H_K the closed-loop transfer matrix

$$H_K = G_{11} + G_{12}K(I - G_{22}K)^{-1}G_{21} \tag{1.4}$$

from the external input w to the control error signal z.

The emphasis in this paper is on *polynomial* solution methods, i.e., solution methods where the manipulations and computations involve polynomial matrices. Earlier work in this direction (Kwakernaak [6], [7], [8]) indicates that relatively simple algorithms may be obtained in this way, without introducing spurious dynamics. Other recent work with this feature(Doyle et. al. [3]) focuses on state space algorithms.

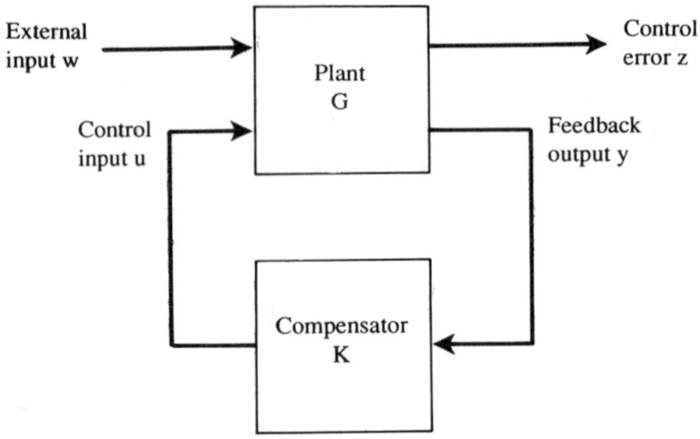

Figure 1. Standard feedback configuration.

1.1 Assumptions. Throughout, we assume the following:

1. $k_1 \geq m_2$, i.e., the dimension of the external input w is at least as great as that of the observed output y;

2. $m_1 \geq k_2$, i.e., the dimension of the control error z is at least as great as that of the control input u;

3. $$(G_{21}\ G_{22}) \text{ and } \begin{pmatrix} G_{12} \\ G_{22} \end{pmatrix} \text{ have full row rank and full column rank, respectively;}$$

4. $k_2 \geq m_2$, i.e., the dimension of the control input u is at least as great as that of the observed output y.

The assumptions (a), (b) and (c) are satisfied in many problems of practical interest. If (d) does not hold we may equivalently consider the problem of minimizing the norm $\|H_K^\top\|_\infty$ of the *transpose* $H_K^\top$ of H_K with respect to the transpose $K^\top$ of the compensator transfer function K. Transposing the problem does not affect assumptions (a) to (c). •

In Appendix A the dimensions of the most important matrices occurring in the remainder of the paper are recorded.

2 Solution method

The method of solution is based on the following lemma (Kwakernaak [8]). Here and in the sequel, if R is any real rational matrix, $R^\star$ is its adjoint defined by

$R(s) = R^{\mathsf{T}}(-s)$ for all $s \in \mathcal{C}$.

2.1 Fundamental Lemma. Define

$$Z_{K,L} = H_K^\star H_K + L^\star L, \tag{2.1}$$

with L a real rational matrix, let Φ be a power spectral density matrix (i.e., the real rational matrix Φ is strictly proper, $\Phi^\star = \Phi$ and Φ is nonnegative-definite on the imaginary axis), and consider the *auxiliary problem* of minimizing

$$\operatorname{tr} \int_{-\infty}^{\infty} \Phi(i\omega) Z_{K,L}(i\omega) d\omega \tag{2.2}$$

with respect to all stabilizing K and all strictly wide L (i.e., all L that have strictly fewer rows than columns.) Suppose that the auxiliary problem has an *equalizing* solution $K_\circ$, $L_\circ$, i. e., there exists a real constant $\lambda_\circ$ such that

$$Z_{K_\circ, L_\circ} = \lambda_\circ^2 I. \tag{2.3}$$

Then $K_\circ$ minimizes $\|H_K\|_\infty$ with respect to all stabilizing K. •

2.2 Proof. We have $\|H_{K_\circ}\|_\infty = \|Z_{K_\circ,L_\circ} - L_\circ^\star L_\circ\|_\infty = \left\|\lambda_\circ^2 I - L_\circ^\star L_\circ\right\|_\infty = \lambda_\circ^2$, where the last equality follows from the fact that $L_\circ$ is strictly wide and hence $L_\circ^\star L_\circ$ is singular. Suppose that there exists a stabilizing $K_\star$ such that $\|H_{K_\star}\|_\infty < \|H_{K_\circ}\| = \lambda_\circ^2$. Then $H_{K_\star}^\star(i\omega) H_{K_\star}(i\omega) < \lambda_\circ^2 I$ for all real ω and hence

$$\begin{aligned} \operatorname{tr} \int_{-\infty}^{\infty} \Phi(i\omega) Z_{K_\star,0}(i\omega) d\omega &= \operatorname{tr} \int_{-\infty}^{\infty} \Phi(i\omega) H_{K_\star}^\star(i\omega) H_{K_\star}(i\omega) d\omega \\ &< \int_{-\infty}^{\infty} \Phi(i\omega) \lambda_\circ^2 I = \operatorname{tr} \int_{-\infty}^{\infty} \Phi(i\omega) H_{K_0}^\star(i\omega) H_{K_0}(i\omega) d\omega, \end{aligned} \tag{2.4}$$

which contradicts the assumption that $K_\circ$, $L_\circ$ solves the auxiliary problem. Hence the hypothesis that there exists a stabilizing $K_\star$ such that $\|H_{K_\star}\|_\infty < \|H_{K_\circ}\|_\infty$ is false, and $K_\circ$ minimizes $\|H_K\|_\infty$. •

In view of the fundamental lemma we concentrate on *equalizing* compensators, i.e., compensators such that there exists a real constant λ and a real rational strictly wide L such that

$$H_K^\star H_K + L^\star L = \lambda^2 I. \tag{2.5}$$

First, in Section 3 we derive a convenient polynomial parametrization of all compensators, and, in particular, all stabilizing compensators. Next, in Section 4 this characterization is used to obtain a simple quadratic polynomial equation that is satisfied by equalizing compensators, and it is discussed how $\mathcal{H}_\infty$ optimal compensators are obtained from this equation. Section 5 presents an example.

3 Parametrization

In this section we derive convenient parametrizations of all closed-loop transfer matrices.

3.1 Theorem: Parameterization of all closed-loop transfer matrices. Suppose that

$$G = \begin{pmatrix} N_1 \\ N_2 \end{pmatrix} \begin{pmatrix} D_1 \\ D_2 \end{pmatrix}^{-1} \tag{3.1}$$

where N_1, N_2, D_1, and D_2 have m_1, m_2, k_1 and k_2 rows, respectively, is a right coprime polynomial matrix fraction representation of the transfer matrix G. Let T_1 be a square polynomial matrix of least degree such that

$$\begin{pmatrix} N_1 \\ D_1 \\ \hline N_2 \\ D_2 \end{pmatrix} T_1 = \begin{pmatrix} N_{11} & N_{12} \\ D_{11} & D_{12} \\ \hline bI & 0 \end{pmatrix}, \tag{3.2}$$

with b a scalar polynomial and I an $(m_2+k_2)\times(m_2+k_2)$ unit matrix. Then given any compensator transfer matrix $K = YX^{-1}$, the closed-loop transfer function H_K may be written as

$$H_K = PR^{-1}, \tag{3.3}$$

where the polynomial matrices P and R are given by

$$\begin{aligned} P &= \begin{pmatrix} N_{11}V & N_{12} \end{pmatrix}, \\ R &= \begin{pmatrix} D_{11}V & D_{12} \end{pmatrix}, \end{aligned} \tag{3.4}$$

and

$$V = \begin{pmatrix} X \\ Y \end{pmatrix}. \tag{3.5}$$

•

Theorem 3.1 parametrizes all closed-loop transfer matrices. Its proof is given in Appendix B. We next consider the parametrization of *stable* closed-loop systems.

3.2 Theorem: Stable closed-loop systems. Consider the parametrization of 3.1. Let $D_{12}^{\perp}$ be a full-rank polynomial matrix with as many rows as possible such that

$$D_{12}^{\perp} D_{12} = 0. \tag{3.6}$$

Suppose that the tall matrix D_{12} may be unimodularly transformed from the left to

$$\begin{pmatrix} D^c_{12} \\ 0 \end{pmatrix}, \tag{3.7}$$

with D^c_{12} square nonsingular, and that the wide matrix $D^{\perp}_{12}D_{11}$ may be unimodularly transformed from the right to

$$\begin{pmatrix} D^c_{11} & 0 \end{pmatrix}, \tag{3.8}$$

with D^c_{11} square nonsingular. Then

1. necessary and sufficient conditions for the existence of a stable closed-loop system are that D^c_{12} and D^c_{11} be stable;

2. the closed-loop poles consist of the roots of D^c_{12} and D^c_{11} together with those of V_1, where the square polynomial matrix V_1 is defined by

$$U^{-1}V = \begin{pmatrix} V_1 \\ V_2 \end{pmatrix}, \tag{3.9}$$

with U the (square) unimodular polynomial matrix such that

$$D^{\perp}_{12}D_{11}U = \begin{pmatrix} D^c_{11} & 0 \end{pmatrix}. \tag{3.10}$$

•

The *roots* of a square polynomial matrix A are the roots of the polynomial $\det(A)$. A is said to be *stable* if all its roots lie in the open left-half complex plane.
3.3 Proof. The closed-loop poles are the roots of

$$R = \begin{pmatrix} D_{11}V & D_{12} \end{pmatrix}. \tag{3.11}$$

Let U' be the (square) unimodular polynomial matrix that transforms the polynomial matrix D_{12} from the left to

$$U'D_{12} = \begin{pmatrix} D^c_{12} \\ 0 \end{pmatrix}. \tag{3.12}$$

Partitioning conformingly

$$U' = \begin{pmatrix} U_1 \\ D^{\perp}_{12} \end{pmatrix} \tag{3.13}$$

it follows that

$$U_1D_{12} = D^c_{12}, \qquad D^{\perp}_{12}D_{12} = 0. \tag{3.14}$$

Because U' is unimodular, the roots of R are the same as the roots of

$$\begin{aligned} U'R &= U' \begin{pmatrix} D_{11}V & D_{12} \end{pmatrix} = \begin{pmatrix} U_1 \\ D_{12}^{\perp} \end{pmatrix} \begin{pmatrix} D_{11}V & D_{12} \end{pmatrix} \\ &= \begin{pmatrix} U_1 D_{11} V & D_{12}^c \\ D_{12}^{\perp} D_{11} V & 0 \end{pmatrix}. \end{aligned} \tag{3.15}$$

This shows that the roots of R consist of the roots of D_{12}^c together with those of $D_{12}^{\perp} D_{11} V$. Writing

$$D_{12}^{\perp} D_{11} V = D_{12}^{\top} D_{11} U \cdot U^{-1} V = \begin{pmatrix} D_{11}^c & 0 \end{pmatrix} \begin{pmatrix} V_1 \\ V_2 \end{pmatrix} = D_{11}^c V_1, \tag{3.16}$$

it immediately follows that the roots of R consist of the roots of D_{12}^c and D_{11}^c together with those of V_1. This proves (b). The polynomial matrices D_{12}^c and D_{11}^c are fully determined by the plant and do not depend on the choice of the compensator. V_1 is fully determined by the compensator. Hence, a stable closed-loop system exists if and only if D_{12}^c and D_{11}^c are stable. This proves (a) and at the same time completes the proof of the theorem. •

It sometimes is useful to apply additional transformations to structure the parametrization further.

3.4 Theorem: Structured parametrization. Let U be the unimodular polynomial matrix of 3. Then there exists a unimodular transformation from the right

$$\begin{pmatrix} N_{11} & N_{12} \\ D_{11} & D_{12} \\ bI & 0 \end{pmatrix} \begin{pmatrix} U & 0 \\ 0 & I \end{pmatrix} \begin{pmatrix} I & 0 \\ T & I \end{pmatrix} = \left(\begin{array}{cc|c} N_{111} & N_{112} & N_{12} \\ D_{111} & 0 & D_{12} \\ \multicolumn{2}{c|}{bU} & 0 \end{array} \right) \tag{3.17}$$

such that $D_{12}^{\perp} D_{111} = D_{11}^c$. Correspondingly, any closed-loop transfer matrix may be parametrized as

$$H_K = P R^{-1}, \tag{3.18}$$

where

$$P = \begin{pmatrix} N_{111} V_1 + N_{112} V_2 & N_{12} \end{pmatrix}, \qquad R = \begin{pmatrix} D_{111} V_1 & D_{12} \end{pmatrix}, \tag{3.19}$$

with V_1 and V_2 arbitrary polynomial matrices of appropriate dimensions. The closed-loop poles are the roots of D_{11}^c and D_{12}^c together with those of the square polynomial matrix V_1. The compensator $K = Y X^{-1}$ follows from

$$\begin{pmatrix} X \\ Y \end{pmatrix} = U \begin{pmatrix} V_1 \\ V_2 \end{pmatrix}. \tag{3.20}$$

3.5 Proof. First consider the unimodular transformation

$$\begin{pmatrix} N_{11} & N_{12} \\ D_{11} & D_{12} \\ bI & 0 \end{pmatrix} \begin{pmatrix} U & 0 \\ 0 & I \end{pmatrix} = \begin{pmatrix} N_{11}U & N_{12} \\ D_{11}U & D_{12} \\ bU & 0 \end{pmatrix}, \tag{3.21}$$

and use the top two rows to parametrize the closed-loop transfer function as in 3. Then $D_{12}^{\perp}(D_{11}U) = (D_{11}^{c}\ 0)$ and the closed-loop poles consist of the roots of D_{12}^{c}, D_{11}^{c} and V_1, where the square polynomial matrix V_1 is defined by partitioning

$$V = \begin{pmatrix} V_1 \\ V_2 \end{pmatrix}. \tag{3.22}$$

Next, let $D_{11}U = (D_{111}\ D_{112})$, where D_{111} has the same number of columns as D_{11}^{c}. Since $D_{12}^{\perp}(D_{11}U) = (D_{11}^{c}\ 0)$, it follows that $D_{12}^{\perp}D_{111} = D_{11}^{c}$ and $D_{12}^{\perp}D_{112} = 0$. From the latter equality we infer that the columns of D_{112} are spanned by the columns of D_{12}. Hence we may apply a unimodular transformation from the right such that

$$\begin{pmatrix} N_{11}U & N_{12} \\ D_{11}U & D_{12} \\ bU & 0 \end{pmatrix} \begin{pmatrix} I & 0 \\ T & I \end{pmatrix} = \left(\begin{array}{cc|c} N_{111} & N_{112} & N_{12} \\ D_{111} & 0 & D_{12} \\ \multicolumn{2}{c|}{bU} & 0 \end{array}\right). \tag{3.23}$$

This results in the given parametrization, while the compensator may be found as indicated. •

From the polynomial parametrization of 3 we may derive the following rational parametrization.

3.6 Theorem: Rational parametrization. The closed-loop transfer matrix may be parametrized as

$$H_K = H_\circ + N_{112}FD_{12}^{\perp}, \tag{3.24}$$

where the rational matrices $H_\circ$ and F are given by

$$H_\circ = \begin{pmatrix} N_{111} & N_{12} \end{pmatrix} \begin{pmatrix} D_{111} & D_{12} \end{pmatrix}^{-1}, \tag{3.25}$$

$$F = V_2V_1^{-1}(D_{11}^{c})^{-1}, \tag{3.26}$$

and N_{111}, N_{112}, N_{12}, D_{111}, D_{12}, V_1 and V_2 are as in 3. If the system is stabilizable, i.e., D_{11}^{c} and D_{12}^{c} are stable, every stable closed-loop transfer matrix may be parametrized as

$$H_k = H_\circ + N_{112}FD_{12}^{\perp} \tag{3.27}$$

with F rational and stable. •

The proof follows by writing

$$R = \begin{pmatrix} D_{111}V_1 & D_{12} \end{pmatrix} = \begin{pmatrix} D_{111} & D_{12} \end{pmatrix} \begin{pmatrix} V_1 & 0 \\ 0 & I \end{pmatrix}, \tag{3.28}$$

using simple algebra, and proving that

$$\begin{pmatrix} I & 0 \end{pmatrix} \begin{pmatrix} D_{111} & D_{12} \end{pmatrix}^{-1} = (D_{11}^{c})^{-1} D_{12}^{\perp}. \tag{3.29}$$

4 The main equation

We are now ready to establish the quadratic polynomial equation that characterizes equalizing compensators.

4.1 Theorem: The Main Equation. Suppose that the closed-loop transfer function is parametrized as in 3. Let $M_{2,\lambda}$ be the spectral cofactor defined by

$$M_{2,\lambda}M_{2,\lambda}^{\star} = D_{12}^{\star}D_{12} - \frac{1}{\lambda^2}N_{12}^{\star}N_{12}, \tag{4.1}$$

and define the polynomial coefficient matrices Λ_λ and Δ_λ by the left-to-right conversion

$$M_{2,\lambda}^{-1}\left(D_{12}^{\star}D_{11} - \frac{1}{\lambda^2}N_{12}^{\star}N_{11}\right) = \Delta_\lambda\Lambda_\lambda^{-1}. \tag{4.2}$$

Suppose that the polynomial matrix W_λ satisfies the quadratic polynomial equation

$$W_\lambda^{\star}\Gamma_\lambda W_\lambda = 0, \tag{4.3}$$

where Γ_λ is the para-Hermitian matrix

$$\Gamma_\lambda = \Lambda_\lambda^{\star}\left(D_{11}^{\star}D_{11} - \frac{1}{\lambda^2}N_{11}^{\star}N_{11}\right)\Lambda_\lambda - \Delta_\lambda^{\star}\Delta_\lambda. \tag{4.4}$$

Then the compensator determined by

$$\begin{pmatrix} X_\lambda \\ Y_\lambda \end{pmatrix} = V_\lambda, \tag{4.5}$$

with

$$V_\lambda = \Lambda_\lambda W_\lambda \tag{4.6}$$

is equalizing. •

4.2 Proof. Writing the closed-loop transfer matrix as $H_K = PR^{-1}$, the equation (2.1) takes the form

$$R^{-\star}P^{\star}PR^{-1} + L^{\star}L = \lambda^2 I, \tag{4.7}$$

where $R^{-\star} := (R^{-1})^{\star} = (R^{\star})^{-1}$. We rewrite this as $P^{\star}P + R^{\star}L^{\star}LR = \lambda^2 R^{\star}R$. Since $P^{\star}P - \lambda^2 R^{\star}R$ is polynomial, so is $R^{\star}L^{\star}LR$ and hence also LR. Setting $LR = M$ with M polynomial it follows that equalizing solutions need satisfy

$$\lambda^2 R^{\star}R = P^{\star}P + M^{\star}M. \tag{4.8}$$

We partition this equation conformingly to the partitioning of 3.1. To this end, let

$$M = \lambda \begin{pmatrix} M_1^{\star} & M_2^{\star} \end{pmatrix}, \tag{4.9}$$

with M_1 and M_2 polynomial matrices to be determined, and where the factor λ is introduced for later convenience. M_2 is square with its dimension equal to the number of columns of D_{12} and N_{12}.

Substituting P and R as given by (3.4) and Mas given by (4.9) into (4.8), we find after expanding the equations that

$$V^{\star}D_{11}^{\star}D_{11}V = \frac{1}{\lambda^2}V^{\star}N_{11}^{\star}N_{11}V + M_1M_1^{\star}, \tag{4.10}$$

$$V^{\star}D_{11}^{\star}D_{12} = \frac{1}{\lambda^2}V^{\star}N_{11}^{\star}N_{12} + M_1M_2^{\star}, \tag{4.11}$$

$$D_{12}^{\star}D_{12} = \frac{1}{\lambda^2}N_{12}^{\star}N_{12} + M_2^{\star}M_2. \tag{4.12}$$

Rewriting (4.12) in the form

$$M_2M_2^{\star} = D_{12}^{\star}D_{12} - \frac{1}{\lambda^2}N_{12}^{\star}N_{12} \tag{4.13}$$

shows that M_2 may be obtained by spectral factorization. It will be proved below that M_2 need be stable, so that it is in fact a spectral cofactor. The spectral factorization exists for λ large enough. The equation (4.11) may be rearranged as

$$\left(D_{12}^{\star}D_{11} - \frac{1}{\lambda^2}N_{12}^{\star}N_{11}\right)V = M_2M_1^{\star}. \tag{4.14}$$

Multiplication on the left by M_2^{-1} results in

$$M_2^{-1}\left(D_{12}^{\star}D_{11} - \frac{1}{\lambda^2}N_{12}^{\star}N_{11}\right)V = M_1^{\star}. \tag{4.15}$$

By the left-to-right matrix polynomial conversion (4.2) this assumes the form

$$\Delta_\lambda \Lambda_\lambda^{-1} V = M_1^{\star}. \tag{4.16}$$

It follows that there exists a polynomial matrix W_λ such that

$$V = \Lambda_\lambda W_\lambda, \qquad M_1^{\star} = \Delta_\lambda W_\lambda. \tag{4.17}$$

Substitution of V and M_1 into (4.10) results in the quadratic polynomial matrix equation

$$W_\lambda^\star \Gamma_\lambda W_\lambda = 0, \tag{4.18}$$

where Γ_λ is the polynomial matrix (4.4). This equation is also valid if the "structured" parameterization of 3.4 is used, provided one takes $D_{11} = (D_{111}\ 0)$ and $N_{11} = (N_{111}\ N_{112})$.

It remains to establish that M_2 should be chosen stable. To this end, consider the case that λ is very large. Using the "structured" parametrization of 3.4 it follows from (4.2) after partitioning Δ_λ and Λ_λ that

$$M_{2,\infty}\left(\Delta_{1,\infty}\ \Delta_{2,\infty}\right) = \begin{pmatrix} D_{12}^\star D_{111} & 0 \end{pmatrix} \begin{pmatrix} \Lambda_{11,\infty} & \Lambda_{12,\infty} \\ \Lambda_{21,\infty} & \Lambda_{22,\infty} \end{pmatrix}, \tag{4.19}$$

or

$$M_{2,\infty}\Delta_{1,\infty} = D_{12}^\star D_{111}\Lambda_{11,\infty}, \tag{4.20}$$

$$M_{2,\infty}\Delta_{2,\infty} = D_{12}^\star D_{111}\Lambda_{12,\infty}. \tag{4.21}$$

From this we conclude that $\Delta_{12,\infty} = 0$ and $\Lambda_{12,\infty} = 0$, while $\Lambda_{11,\infty}$ has the same roots as $M_{2,\infty}$. From $\Lambda_{12,\infty} = 0$ it follows that $V_{1,\infty} = \Lambda_{11,\infty}W_{1,\infty}$. Because for closed-loop stability $V_{1,\infty}$ should be stable it follows that $\Lambda_{11,\infty}$ need be stable, so that also $M_{2,\infty}$ is required to be stable. Because the optimal solution is homotopic to a suitable stable equalizing solution for $\lambda = \infty$[8], we need choose $M_{2,\lambda}$ stable for all λ.

The solution of the Main Equation is by no means unique. The one we are looking for results in a stable closed-loop system and allows the construction of a spectral density matrix Φ such that the auxiliary problem is solved. From earlier work [8] it is known that the optimal solution

1. is *homotopic* to a suitable stabilizing equalizing solution for $\lambda = \infty$ (i.e., there exists a solution curve that is analytic in λ connecting the solutions), and

2. is *reduced* (i.e., such that the degree of one of the columns of W_λ drops by one compared to the solution for $\lambda = \infty$.)

Below it is shown how to construct the solution at ∞. Newton-Raphson methods have been successfully used to solve the homotopy problem numerically (Kwakernaak [8], [9]; Boekhoudt [1].) The quadratic equation (4.3) is especially amenable to solution by the Newton-Raphson method.

4.3 Theorem: Solution at Infinity. Suppose that $k_2 = m_2$. In the structured parametrization of 3.4 the desired equalizing solution is in the limit $\lambda \to \infty$ asymptotically characterized by

$$V_{2,\lambda}V_{1,\lambda}^{-1} \simeq \lambda\,(N_{112}^{\circ})^{-1}\left(D_{12}^{\perp\mathrm{co}}\right)^{-1} D_{11}^{c}, \tag{4.22}$$

where N_{112}° is an outer factor of N_{112} and $D_{12}^{\mathsf{T}\mathrm{co}}$ a co-outer factor of D_{12}^{T}. •

4.4 Proof. In the rational parametrization of 3.6 the closed-loop transfer function is given by

$$H_K = H_\circ + N_{112} F D_{12}^{\perp}, \tag{4.23}$$

where $F = V_2 V_1^{-1} \left(D_{11}^{c}\right)^{-1}$ is the "parameter." The equality $H_K^\star H_K + L^\star L = \lambda^2 I$ may in the limit $\lambda \to \infty$ only be satisfied by letting F be very large, so that $H_K^\star H_K \simeq \left(D_{12}^{\perp}\right)^\star F^\star N_{112}^\star N_{112} F D_{12}^{\perp}$. Under the assumption $k_2 = m_2$, choosing $F = \lambda (N_{112}^{0})^{-1} \left(D_{12}^{\perp co}\right)^{-1}$ - which is equivalent to (4.22) - results in $H_K^\star H_K \simeq \lambda^2 \left(D_{12}^{\perp ci}\right)^\star D_{12}^{\perp ci}$, where $D_{12}^{\perp ci}$ is a co-inner factor of D_{12}. From $H_K^\star H_K + L^\star L = \lambda^2 I$ it follows that $L^\star L \simeq \lambda^2 \left[I - \left(D_{12}^{\perp ci}\right)^\star D_{12}^{\perp ci}\right] = \lambda^2 D_{12}^{i} \left(D_{12}^{i}\right)^\star$, where D_{12}^{i} is an inner factor of D_{12}. This is satisfied if $L \simeq \lambda \left(D_{12}^{i}\right)^\star$. It may be shown that the asymptotic solution $F \simeq \lambda \left(N_{112}^{\circ}\right)^{-1} \left(D_{12}^{\perp co}\right)^{-1}$ and $L \simeq \lambda \left(D_{12}^{i}\right)^\star$ is homotopic to the optimal solution. •

The case $k_2 \neq m_2$ involves a minor modification. Theorem 4.3 allows the determination of the asymptotic solution for the "structured" parametrization. Examples indicate that structured parametrization may destroy row- or column-reducedness and thus leads to numerical ill-conditioning, so that the direct parametrization of 3.1 is more advisable. Given the asymptotic solution V_1, V_2 of 4.3, the resulting compensator X, Y follows from (3.20). The corresponding asymptotic solution for W of the Main Equation in the direct parametrization may then be obtained from

$$\Lambda_\lambda W_\lambda \simeq \begin{pmatrix} X \\ Y \end{pmatrix}. \tag{4.24}$$

5 Example

By way of illustration we consider the example of Boekhoudt and Kwakernaak[2]. This example concerns a SISO plant with disturbances and measurement noise, and is a four-block problem in the terminology of Doyle. The transfer matrix G is given by

$$G(s) = \left(\begin{array}{cc|c} \frac{s+1}{s-2} & 0 & \frac{1}{s-2} \\ 0 & 0 & 1 \\ \hline \frac{s+1}{s-2} & 1 & \frac{1}{s-2} \end{array}\right) \tag{5.1}$$

Parametrization
For the problem at hand we have $m_1 = 2$, $m_2 = 1$, $k_1 = 2$ and $k_2 = 1$. A coprime

right factorization $G = ND^{-1}$ with the correct partitioning is given by

$$N = \begin{pmatrix} N_1 \\ N_2 \end{pmatrix} = \left(\begin{array}{ccc} 1 & -1 & 0 \\ s-2 & 3 & 0 \\ \hline 1 & -1 & 1 \end{array}\right), D = \begin{pmatrix} D_1 \\ D_2 \end{pmatrix} = \left(\begin{array}{ccc} 0 & -1 & 0 \\ 0 & 0 & 1 \\ \hline s-2 & 3 & 0 \end{array}\right). \tag{5.2}$$

It follows that

$$\begin{pmatrix} N_1 \\ D_1 \\ N_2 \\ D_2 \end{pmatrix} = \left(\begin{array}{ccc} 1 & -1 & 0 \\ s-2 & 3 & 0 \\ \hline 0 & -1 & 0 \\ 0 & 0 & 1 \\ \hline 1 & -1 & 1 \\ \hline s-2 & 3 & 0 \end{array}\right). \tag{5.3}$$

By elementary column operations this may be brought into the form

$$\begin{pmatrix} N_{11} & N_{12} \\ D_{11} & D_{12} \\ N_{21} & N_{22} \\ D_{21} & D_{22} \end{pmatrix} = \left(\begin{array}{c|cc} 0 & -1/3 & s+1 \\ 0 & 1 & 0 \\ \hline 0 & -1/3 & s-2 \\ 1 & 1/3 & -s-1 \\ \hline 1 & 0 & 0 \\ \hline 0 & 1 & 0 \end{array}\right), \tag{5.4}$$

so that $b = 1$. By 3.1 the closed-loop systems are parametrized by the 2×1 matrix

$$V = \begin{pmatrix} V_1 \\ V_2 \end{pmatrix}. \tag{5.5}$$

It is easily seen that by elementary row operations the tall matrix D_{12} may be brought into the form $\begin{pmatrix} 1 \\ 0 \end{pmatrix}$, so that $D_{12}^c = 1$. Inspection shows that we may choose

$$D_{12}^{\perp} = \begin{pmatrix} s+1 & s-2 \end{pmatrix}, \tag{5.6}$$

so that

$$D_{12}^{\perp} D_{11} = \begin{pmatrix} s-2 & -1 \end{pmatrix}. \tag{5.7}$$

The unimodular matrix

$$U = \begin{pmatrix} 0 & 3 \\ -1 & 3(s-2) \end{pmatrix} \tag{5.8}$$

transforms (5.7) from the right to (1 0), so that also $D_{11}^c = 1$. Thus, the necessary and sufficient conditions of 3.2 for the existence of a stabilizing compensator are satisfied.

Application of the unimodular transformation of (3.21) results in

$$\left(\begin{array}{cc|c} 0 & -1/3 & s+1 \\ 0 & 1 & 0 \\ \hline 0 & -1/3 & s-2 \\ 1 & 1/3 & -s-1 \\ \hline 1 & 0 & 0 \\ \hline 0 & 1 & 0 \end{array}\right) \left(\begin{array}{cc|c} 0 & 3 & 0 \\ -1 & 3(s-2) & 0 \\ \hline 0 & 0 & 1 \end{array}\right) = \left(\begin{array}{cc|c} 1/3 & -s+2 & s+1 \\ -1 & 3(s-2) & 0 \\ \hline 1/3 & -s+2 & s-2 \\ -1/3 & s+1 & -s-1 \\ \hline 0 & 3 & 0 \\ \hline -1 & 3(s-2) & 0 \end{array}\right). \tag{5.9}$$

Adding the last column to the second and dividing the result by 3 results in the structured form

$$\left(\begin{array}{cc|c} N_{111} & N_{112} & N_{12} \\ D_{111} & 0 & D_{12} \\ \multicolumn{2}{c|}{bU} & 0 \end{array}\right) = \left(\begin{array}{cc|c} 1/3 & 1 & s+1 \\ -1 & s-2 & 0 \\ \hline 1/3 & 0 & s-2 \\ -1/3 & 0 & -s-1 \\ \hline 0 & 1 & 0 \\ \hline -1 & s-2 & 0 \end{array}\right). \tag{5.10}$$

The Main Equation

To set up the Main Equation we use the direct form (5.4). It follows from

$$M_{2,\lambda} M_{2,\lambda}^{\star} = D_{12}^{\star} D_{12} - \frac{1}{\lambda^2} N_{12}^{\star} N_{12} = \left(5 - 2s^2\right) - \frac{1}{\lambda^2}\left(1 - s^2\right) \tag{5.11}$$

that

$$M_{2,\lambda} = p_\lambda s + q_\lambda , \tag{5.12}$$

where

$$p_\lambda = \sqrt{2 - \frac{1}{\lambda^2}}, \qquad q_\lambda = \sqrt{5 - \frac{1}{\lambda^2}}. \tag{5.13}$$

Furthermore, since

$$D_{12}^{\star} D_{11} - \frac{1}{\lambda^2} N_{12}^{\star} N_{11} = \left(s - 1 \qquad e_\lambda s + f_\lambda\right), \tag{5.14}$$

where

$$e_\lambda = \frac{1}{3}\left(2 - \frac{1}{\lambda^2}\right), \qquad f_\lambda = \frac{1}{3}\left(1 + \frac{1}{\lambda^2}\right), \tag{5.15}$$

it follows from the left-to-right matrix fraction conversion

$$M_{2,\lambda}^{-1}\left(D_{12}^{\star} D_{11} - \frac{1}{\lambda^2} N_{12}^{\star} N_{11}\right) = \Delta_\lambda \Lambda_\lambda^{-1} \tag{5.16}$$

that we may take

$$\Delta_\lambda = \left(\begin{array}{cc} 1 & 0 \end{array}\right), \qquad \Lambda_\lambda = \left(\begin{array}{cc} c_\lambda & e_\lambda s + f_\lambda \\ d_\lambda & -s + 1 \end{array}\right), \tag{5.17}$$

where

$$c_\lambda = \frac{1}{3}\left(1+\frac{1}{\lambda^2}\right)\sqrt{2-\frac{1}{\lambda^2}} - \frac{1}{3}\left(2-\frac{1}{\lambda^2}\right)\sqrt{5-\frac{1}{\lambda^2}}, \tag{5.18}$$

$$d_\lambda = \sqrt{2-\frac{1}{\lambda^2}} + \sqrt{5-\frac{1}{\lambda^2}}. \tag{5.19}$$

To find the coefficient matrix Γ_λ of the Main Equation we need the matrix

$$D_{11}^\star D_{11} - \frac{1}{\lambda^2} N_{11}^\star N_{11} = \begin{pmatrix} 1 & 1/3 \\ 1/3 & g_\lambda \end{pmatrix}, \tag{5.20}$$

where

$$g_\lambda = \frac{1}{9}\left(2 - \frac{10}{\lambda^2}\right). \tag{5.21}$$

The matrix $\Gamma_\lambda = \Lambda_\lambda^\star \left(D_{11}^\star D_{11} - \frac{1}{\lambda^2} N_{11}^\star N_{11}\right) \Lambda_\lambda - \Delta_\lambda^\star \Delta_\lambda$ follows as

$$\Gamma_\lambda = \begin{pmatrix} \Gamma_{11,\lambda} & \Gamma_{12,\lambda} \\ \Gamma_{21,\lambda} & \Gamma_{22,\lambda} \end{pmatrix}, \tag{5.22}$$

where

$$\begin{aligned}
\Gamma_{11,\lambda} &= c_\lambda^2 + \frac{2}{3} c_\lambda d_\lambda + g_\lambda d_\lambda^2 - 1, \\
\Gamma_{12,\lambda} &= \left(c_\lambda e_\lambda - \frac{1}{3} c_\lambda - \frac{1}{3} d_\lambda e_\lambda - d_\lambda g_\lambda\right) s + \left(c_\lambda f_\lambda + \frac{1}{3} c_\lambda + \frac{1}{3} d_\lambda f_\lambda + d_\lambda g_\lambda\right), \\
\Gamma_{21,\lambda} &= -\left(c_\lambda e_\lambda - \frac{1}{3} c_\lambda + \frac{1}{3} d_\lambda e_\lambda - d_\lambda g_\lambda\right) s + \left(c_\lambda f_\lambda + \frac{1}{3} c_\lambda + \frac{1}{3} d_\lambda f_\lambda + d_\lambda g_\lambda\right), \\
\Gamma_{22,\lambda} &= -\left(e_\lambda^2 - \frac{2}{3} e_\lambda + g_\lambda\right) s^2 + \left(f_\lambda^2 + g_\lambda\right).
\end{aligned} \tag{5.23}$$

The degrees of the elements of Γ_λ are given by

$$\deg(\Gamma_\lambda) = \begin{pmatrix} 0 & 1 \\ 1 & 2 \end{pmatrix}. \tag{5.24}$$

A consistent set of equations is obtained by taking $\deg(W_{1,\lambda}) = 1$, $\deg(W_{2,\lambda}) = 0$. Expansion of the equation of the equation results in two (homogeneous) coefficient equations in the three unknown coefficients of the polynomials $W_{1,\lambda}$ and $W_{2,\lambda}$.

The optimal solution

For this simple example the optimal solution may be obtained quite easily. The reason is that for the reduced-degree solution the degrees of $W_{1,\lambda}$ and $W_{2,\lambda}$ both drop by 1. This means that for the reduced solution $W_{1,\lambda}$ is a constant, while $W_{2,\lambda}$ reduces to 0. Thus, the *reduced* solution $W_{\lambda_{\text{opt}}}$ satisfies the equation

$$W_{1,\lambda_{\text{opt}}}^\star \Gamma_{11,\lambda_{\text{opt}}} W_{1,\lambda_{\text{opt}}} = 0. \tag{5.25}$$

This implies

$$\Gamma_{11,\lambda_{\mathrm{opt}}} = 0, \tag{5.26}$$

or

$$c^2_{\lambda_{\mathrm{opt}}} + \frac{2}{3} c_{\lambda_{\mathrm{opt}}} d_{\lambda_{\mathrm{opt}}} + g_{\lambda_{\mathrm{opt}}} d^2_{\lambda_{\mathrm{opt}}} - 1 = 0. \tag{5.27}$$

This is a scalar nonlinear equation in the unknown λ_{opt}, which may easily be solved. Its root is found to be

$$\lambda_{\mathrm{opt}} = 5.3112764. \tag{5.28}$$

From

$$W_{1,\lambda_{\mathrm{opt}}} = 1, \; W_{2,\lambda_{\mathrm{opt}}} = 0, \tag{5.29}$$

it immediately follows that

$$\begin{pmatrix} X \\ Y \end{pmatrix} = V_{\lambda_{\mathrm{opt}}} = \Lambda_{\lambda_{\mathrm{opt}}} W_{\lambda_{\mathrm{opt}}} = \begin{pmatrix} c_{\lambda_{\mathrm{opt}}} \\ d_{\lambda_{\mathrm{opt}}} \end{pmatrix} = \begin{pmatrix} -0.9753198 \\ 3.6297517 \end{pmatrix}. \tag{5.30}$$

The optimal compensator hence is

$$K = YX^{-1} = -3.7216014. \tag{5.31}$$

This solution has been obtained with considerably less effort than in [2].

6 Conclusions

The quadratic polynomial equation that yields the solution to the "standard" $\mathcal{H}_\infty$ optimal control problem (the "Main Equation") is considerably simpler than the set of linear and quadratic polynomial matrix equations obtained in [2]. It looks as if the Main Equation has the minimal complexity required for the solution of the $\mathcal{H}_\infty$ optimal control problem, and also as if its complexity is comparable to that of the new state space methods that are currently being developed [3].

The Main Equation is based on a polynomial parametrization of the standard $\mathcal{H}_\infty$ optimal control problem. This parametrization and associated "structured" and rational parametrizations may have further interesting applications, and their exploration seems worthwhile.

Further work need be done on the development of numerical methods for the solution of the Main Equation. It would be helpful to have a complete understanding of the degree structure of the polynomial solution of the standard problem as developed in [7] for the multivariable mixed-sensitivity problem.

Appendix A: Dimensions

In this appendix we list the dimensions of the important matrices arising in the text.

Section 1

G_{11}	$m_1 \times k_1$
G_{12}	$m_1 \times k_2$
G_{21}	$m_2 \times k_1$
G_{22}	$m_2 \times k_2$
K	$k_2 \times m_2$
H_k	$m_1 \times k_1$

Section 2

L	$(k_1 - m_2) \times k_1$

Section 3

Y	$k_2 \times m_2$
X	$m_2 \times m_2$
P	$m_1 \times k_1$
Q	$m_2 \times k_1$
R	$k_1 \times k_1$
N_1	$m_1 \times (k_1 + k_2)$
N_2	$m_2 \times (k_1 + k_2)$
D_1	$k_1 \times (k_1 + k_2)$
D_2	$k_2 \times (k_1 + k_2)$
N_{11}	$m_1 \times (m_2 + k_2)$
N_{12}	$m_1 \times (k_1 - m_2)$
D_{11}	$k_1 \times (m_2 + k_2)$
D_{12}	$k_1 \times (k_1 - m_2)$
U	$(m_2 + k_2) \times (m_2 + k_2)$
D_{11}^c	$m_2 \times m_2$
$D_{12}^\perp$	$m_2 \times k_1$
D_{12}^c	$(k_1 - m_2) \times (k_1 - m_2)$
D_{111}	$k_1 \times k_1$
N_{111}	$m_1 \times k_1$
N_{112}	$m_1 \times k_2$
V	$(m_2 + k_2) \times m_2$
V_1	$m_2 \times m_2$
V_2	$k_2 \times m_2$
F	$k_2 \times m_2$
H_o	$m_1 \times k_1$

Section 4

M	$(k_1 - m_2) \times k_1$
M_1	$m_2 \times (k_1 - m_2)$
M_2	$(k_1 - m_2) \times (k_1 - m_2)$
Λ	$(k_2 + m_2) \times (k_2 + m_2)$
Δ	$(k_1 - m_2) \times (k_2 + m_2)$
Γ	$(k_2 + m_2) \times (k_2 + m_2)$
W	$(k_2 + m_2) \times m_2$
N°_{112}	$k_2 \times k_2$
$D^{\mathrm{T}\,\mathrm{co}}_{12}$	$m_2 \times m_2$

Appendix B: Proof of Theorem 3.1.

The real rational transfer matrix G may be represented in *left* polynomial matrix fraction form as

$$G = (D^{\ell})^{-1} N^{\ell}, \tag{B.1}$$

where the (left coprime) polynomial matrices D^{ℓ} and N^{ℓ} are partitioned conforming to the partitioning of G as

$$D^{\ell} = \begin{pmatrix} D^{\ell}_{11} & D^{\ell}_{12} \\ D^{\ell}_{21} & D^{\ell}_{22} \end{pmatrix}, \qquad N^{\ell} = \begin{pmatrix} N^{\ell}_{11} & N^{\ell}_{12} \\ N^{\ell}_{21} & N^{\ell}_{22} \end{pmatrix}. \tag{B.2}$$

It follows from (1.1) and (B.1) that

$$\begin{pmatrix} D^{\ell}_{11} & D^{\ell}_{12} \\ D^{\ell}_{21} & D^{\ell}_{22} \end{pmatrix} \begin{pmatrix} z \\ y \end{pmatrix} = \begin{pmatrix} N^{\ell}_{11} & N^{\ell}_{12} \\ N^{\ell}_{21} & N^{\ell}_{22} \end{pmatrix} \begin{pmatrix} w \\ u \end{pmatrix}. \tag{B.3}$$

By assumption the compensator has the real rational transfer function $K = YX^{-1}$ with Y and X polynomial. Defining $X^{-1}y = v$, with v an intermediate signal, it follows from $u = Ky = YX^{-1}y = Yv$ that

$$y = Xv, \qquad u = Yv. \tag{B.4}$$

Substitution into (B.3) results in

$$\begin{pmatrix} z \\ v \end{pmatrix} = \begin{pmatrix} D^{\ell}_{11} & D^{\ell}_{12}X - N^{\ell}_{12}Y \\ D^{\ell}_{21} & D^{\ell}_{22}X - N^{\ell}_{22}Y \end{pmatrix}^{-1} \begin{pmatrix} N^{\ell}_{11} \\ N^{\ell}_{21} \end{pmatrix} w. \tag{B.5}$$

Inspection shows that the transfer matrix H_K from the external input w to the control error signal z equals

$$H_K = \begin{pmatrix} I & 0 \end{pmatrix} \begin{pmatrix} D^{\ell}_{11} & D^{\ell}_{12}X - N^{\ell}_{12}Y \\ D^{\ell}_{21} & D^{\ell}_{22}X - N^{\ell}_{22}Y \end{pmatrix}^{-1} \begin{pmatrix} N^{\ell}_{11} \\ N^{\ell}_{21} \end{pmatrix}. \tag{B.6}$$

By left-to-right polynomial matrix fraction conversion we introduce polynomial matrices P, Q and R such that

$$\begin{pmatrix} D^\ell_{11} & D^\ell_{12}X - N^\ell_{12}Y \\ D^\ell_{21} & D^\ell_{22}X - N^\ell_{22}Y \end{pmatrix}^{-1} \begin{pmatrix} N^\ell_{11} \\ N^\ell_{21} \end{pmatrix} = \begin{pmatrix} P \\ Q \end{pmatrix} R^{-1}. \tag{B.7}$$

As a consequence, the closed-loop transfer matrix H_K may be expressed as the right polynomial matrix fraction

$$H_K = \begin{pmatrix} I & 0 \end{pmatrix} \begin{pmatrix} P \\ Q \end{pmatrix} R^{-1} = PR^{-1}. \tag{B.8}$$

The polynomial matrices P and R depend on the compensator $K = YX^{-1}$ through (B.7). By cross-multiplication it follows from (B.7) that for given X and Y the polynomial matrices P, Q and R may be solved from

$$\begin{pmatrix} D^\ell_{11} & D^\ell_{12}X - N^\ell_{12}Y \\ D^\ell_{21} & D^\ell_{22}X - N^\ell_{22}Y \end{pmatrix} \begin{pmatrix} P \\ Q \end{pmatrix} = \begin{pmatrix} N^\ell_{11} \\ N^\ell_{21} \end{pmatrix} R. \tag{B.9}$$

We rewrite this in the form

$$\begin{pmatrix} D^\ell_{11} \\ D^\ell_{21} \end{pmatrix} P + \begin{pmatrix} D^\ell_{12}X - N^\ell_{12}Y \\ D^\ell_{22}X - N^\ell_{22}Y \end{pmatrix} Q = \begin{pmatrix} N^\ell_{11} \\ N^\ell_{21} \end{pmatrix} R, \tag{B.10}$$

which in turn may be rearranged as

$$\begin{pmatrix} D^\ell_{11} \\ D^\ell_{21} \end{pmatrix} P + \begin{pmatrix} D^\ell_{12} \\ D^\ell_{22} \end{pmatrix} XQ - \begin{pmatrix} N^\ell_{11} \\ N^\ell_{21} \end{pmatrix} R - \begin{pmatrix} N^\ell_{12} \\ N^\ell_{22} \end{pmatrix} YQ = 0 \tag{B.11}$$

or

$$\begin{pmatrix} D^\ell & -N^\ell \end{pmatrix} \begin{pmatrix} P \\ XQ \\ R \\ YQ \end{pmatrix} = 0. \tag{B.12}$$

The left and right factorizations $G = \left(D^\ell\right)^{-1} N^\ell = ND^{-1}$ are related by the equality $D^\ell N = N^\ell D$, so that

$$\begin{pmatrix} D^\ell & -N^\ell \end{pmatrix} \begin{pmatrix} N \\ D \end{pmatrix} = 0. \tag{B.13}$$

Comparison of (B.12) and (B.13) reveals that there exists a polynomial matrix T of suitable dimensions such that

$$\begin{pmatrix} P \\ XQ \\ R \\ YQ \end{pmatrix} = \begin{pmatrix} N \\ D \end{pmatrix} T = \begin{pmatrix} N_1 \\ N_2 \\ D_1 \\ D_2 \end{pmatrix} T, \tag{B.14}$$

where we partition N and D conforming to the left hand side.

In (B.12), X and Y are the *independent* variables, and P, Q and R the *dependent* variables. Using (B.14), we should like to consider the polynomial matrix T as the independent variable. For this we need a further transformation of N and D because of the coupling introduced by Q in the relations

$$\begin{aligned} XQ &= N_2 T, \\ YQ &= D_2 T, \end{aligned} \tag{B.15}$$

implied by (B.14). To this end, let $T := T_1 T$, where T_1 is a square polynomial matrix (not necessarily unimodular but of least degree) such that

$$\begin{pmatrix} N_2 \\ D_2 \end{pmatrix} T_1 = \begin{pmatrix} bI & 0 \end{pmatrix}, \tag{B.16}$$

with I a unit matrix of suitable dimensions, and b a scalar polynomial. The matrix T_1 may be determined by first finding a unimodular matrix T_1' such that

$$\begin{pmatrix} N_2 \\ D_2 \end{pmatrix} T_1' = \begin{pmatrix} B & 0 \end{pmatrix} \tag{B.17}$$

with B square nonsingular, and then multiplying T_1' on the right by $B^{-1}b$ with $b = \det(B)$, so that $T_1 = T_1' B^{-1} b$. For this transformation to succeed we need the assumptions (a) and (c) of 1.1.

With this transformation (B.15) takes the form

$$\begin{pmatrix} X \\ Y \end{pmatrix} Q = \begin{pmatrix} bI & 0 \end{pmatrix} T. \tag{B.18}$$

Right multiplication of (B.11) by a suitable unimodular polynomial matrix brings the wide matrix Q without loss of generality into the form

$$Q := \begin{pmatrix} Q_0 & 0 \end{pmatrix}, \tag{B.19}$$

with Q_0 square nonsingular. Partitioning

$$T = \begin{pmatrix} T_{11} & T_{12} \\ T_{21} & T_{22} \end{pmatrix} \tag{B.20}$$

it follows from (B.18) with (B.19) and (B.20) that

$$\begin{pmatrix} X \\ Y \end{pmatrix} Q_0 = bT_{11}, \qquad 0 = bT_{12}, \tag{B.21}$$

so that

$$Q_0 = bI, \qquad \begin{pmatrix} X \\ Y \end{pmatrix} = T_{11}, \qquad T_{12} = 0. \tag{B.22}$$

This shows that we may write the solution of (B.14) in the form

$$\begin{aligned} P &= N_1 T_1 T, \qquad\qquad R = D_1 T_1 T, && \text{(B.23)} \\ Q &= \begin{pmatrix} bI & 0 \end{pmatrix}, && \text{(B.24)} \end{aligned}$$

where T is any polynomial matrix of the form

$$T = \begin{pmatrix} T_{11} & 0 \\ T_{21} & T_{22} \end{pmatrix}, \tag{B.25}$$

and where the role of X and Y as independent variables is now taken over by the polynomial matrices T_{11}, T_{21} and T_{22}.

The polynomial matrices T_{21} and T_{22} may be eliminated. To see this we write

$$\left(\begin{array}{c} N_1 \\ D_1 \\ \hline N_2 \\ D_2 \end{array} \right) T_1 =: \left(\begin{array}{cc} N_{11} & N_{12} \\ D_{11} & D_{12} \\ \hline bI & 0 \end{array} \right). \tag{B.26}$$

It follows that P and R may be expressed as

$$\begin{aligned} P &= \begin{pmatrix} N_{11}T_{11} + N_{12}T_{21} & N_{12}T_{22} \end{pmatrix}, && \text{(B.27)} \\ R &= \begin{pmatrix} D_{11}T_{11} + D_{12}T_{21} & D_{12}T_{22} \end{pmatrix}. && \text{(B.28)} \end{aligned}$$

The polynomial matrices P, $Q = (bI \quad 0)$ and R have a common right divisor

$$\begin{pmatrix} I & 0 \\ 0 & T_{22} \end{pmatrix}. \tag{B.29}$$

This divisor may be canceled by simply setting $T_{22} = I$, so that

$$\begin{aligned} P &= \begin{pmatrix} N_{11}T_{11} + N_{12}T_{21} & N_{12} \end{pmatrix}, && \text{(B.30)} \\ R &= \begin{pmatrix} D_{11}T_{11} + D_{12}T_{21} & D_{12} \end{pmatrix}. && \text{(B.31)} \end{aligned}$$

The terms $N_{12}T_{21}$ and $D_{12}T_{21}$, in turn, may be removed by multiplication of P, Q and R on the right by the unimodular polynomial matrix

$$\begin{pmatrix} I_{11} & 0 \\ -T_{21} & I \end{pmatrix}, \tag{B.32}$$

so that

$$\begin{aligned} P &= (N_{11}T_{11} \quad N_{12}), && \text{(B.33)} \\ R &= (D_{11}T_{11} \quad D_{12}). && \text{(B.34)} \end{aligned}$$

The proof is completed by replacing T_{11} with V.

References

[1] Boekhoudt, P., *The $\mathcal{H}_\infty$ Control Design Method: A Polynomial Approach.* Doctoral dissertation, University of Twente, 1988.

[2] Boekhoudt, P., and H. Kwakernaak, "A polynomial approach to the $\mathcal{H}_\infty$ control design method." *Proc. 1988 American Control Conference*, Atlanta, Georgia, p. 1219-1224. IEEE, Piscataway, N.J., U.S.A., 1988.

[3] Doyle, J., K. Glover, P. Khargonekar and B. Francis, "State-space solution to standard $\mathcal{H}_\epsilon$ and $\mathcal{H}_\infty$ control problems. "*Proc. 1988 American Control Conference*, Atlanta, Georgia, p. 1691-1696. IEEE, Piscataway, N.J., U.S.A., 1988.

[4] Francis, B.A., A COURSE IN $\mathcal{H}_\infty$ Control Theory (corrected printing). LNCIS vol. 88, Springer-Verlag, Heidelberg, 1987.

[5] Ježek, J., and V. Kučera, "Efficient algorithm for matrix spectral factorization." *Automatica*, vol. 21, pp. 663-669 (1985).

[6] Kwakernaak, H., "Minimax frequency domain performance and robustness optimization of linear feedback systems." *IEEE Trans. Aut. Control*, vol. AC-30, pp. 994-1004(1985).

[7] Kwakernaak, H., "A polynomial approach to minimax frequency domain optimization of multivariable systems." *Int. J. Control*, vol. 44, pp. 117-156 (1986).

[8] Kwakernaak, H., "A polynomial approach to $\mathcal{H}_\infty$ optimization of control systems." R.F. Curtain, Ed., *Modeling, Robustness and Sensitivity Reduction in Control Systems*, Springer-Verlag, Berlin, 1987.

Polynomial optimization

A. P. Roberts M. M. Newmann

Abstract

The polynomial method is used to optimize the stationary stochastic feedback control of linear multivariable plants. Both discrete–time and continuous–time systems are considered but it is assumed that the plant is stable. With this proviso, optimization is achieved in each case with two spectral factorizations and the solution of only one diophantine equation.

1 Introduction

The polynomial method of optimization was pioneered by Kučera during the 1970's. His many publications culminated with his famous book [2] on discrete linear control. A remarkable feature is the purely algebraic nature of the work using polynomials in the delay operator d. However, with continuous–time systems it is necessary to take Laplace transforms to obtain polynomials in the transform variables [3]. Transforms can also be used in discrete–time and Grimble and Johnson [1] used the Z transform to produce polynomials in z^{-1}. Since transforms must be used for continuous–time, transforms are also used for discrete–time here. however, the authors find it more convenient to use the Δ transform which is the Z transform with z^{-1} replaced by δ .

The methods of solution given by Kučera[2,3] are more powerful than those presented here because they produce the optimal control even when the plant is unstable. Two diophantine equations were introduced (sometimes more [4]) and he proved that their solution gave the desired results. Unfortunately, neither their function nor their derivation were immediately apparent. in contrast, the one diophantine equation here is used simply to split an expression into the sum of its stable and unstable parts. Thus, although the method given here is much more restrictive, when it can be used its derivation is relatively simple. Furthermore, it forms a useful introduction to the polynomial method of optimization for both discrete–time and continuous–time systems.

The block diagram in fig. 1 indicates the type of feedback control system to be considered. it serves equally well for discrete–time and continuous–time cases. The lower–case letters correspond to the transforms of vector variables and upper–case to transfer matrices, the elements of the latter being constant rational functions of the appropriate transform variable. The variables ξ, η and ζ correspond to stationary independent white–noise vectors with zero means and identity spectral matrices. S, S_d, S_n and S_r are known stable transfer matrices and the aim is to find a controller transfer matrix R to minimize a quadratic cost which will be defined later. •

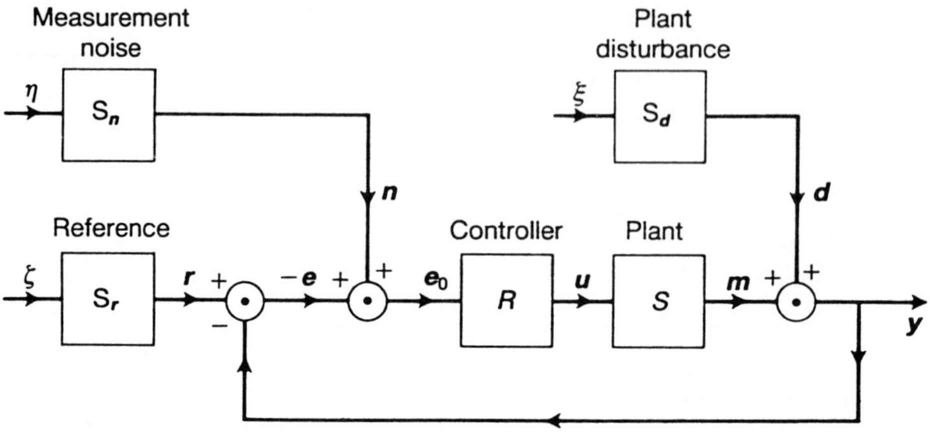

Figure 1. Stochastic feedback control system

The S matrices are written in terms of polynomial matrices:

$$S = A^{-1}B, \; S_d = A^{-1}C_d, \; S_n = A^{-1}C_n, \; S_r = A^{-1}C_r \tag{1.1}$$

where A is the least common left denominator of S, S_d, S_n, and S_r, and the matrices A and $[B, C_d, C_n, C_r]$ are left–coprime. Also, $[C_d, C_n, C_r]$ must have full row rank. Furthermore, S may be written in terms of left–coprime and right–coprime fractions of polynomial matrices, i.e.

$$S = A_1^{-1}B_1 = B_2A_2^{-1} \tag{1.2}$$

It should also be noted that the matrix

$$\begin{bmatrix} \Phi A_2 \\ \Psi B_2 \end{bmatrix}$$

must have full column rank, where Φ and Ψ are weighting matrices in cost functions (see later).

The controller transfer matrix can be written as a left or right matrix fraction, i.e.

$$R = M_1^{-1}N_1 = N_2M_2^{-1} \tag{1.3}$$

where M_1, N_1, M_2 and N_2 are not polynomials but are stable transfer matrices. Kučera[2] showed that for satisfactory feedback control, the following relations must hold between the matrices in (1.2) and (1.3):

$$\left.\begin{array}{c} A_1M_2 + B_1N_2 = I, \\ M_1A_2 + N_1B_2 = I, \\ A_2M_1 + N_2B_1 = I, \\ M_2A_1 + B_2N_1 = I, \\ A_2N_1 = N_2A_1, \\ B_2M_1 = M_2B_1. \end{array}\right\} \quad (1.4)$$

This applies in continuous–time as well as in discrete–time.

Using (1.4), the transforms of the control input u and the tracking error e may be written as

$$u = N_2A_1(r - d + n) \quad (1.5)$$

and

$$-e = r - y = M_2A_1(r - d + n) - n. \quad (1.6)$$

All of the above has been included in the introduction because it applies equally well in both discrete–time and continuous–time. Now the method of optimization will be sketched for the discrete–time case in Chapter 2 and for continuous–time in Chapter 3. Detailed proofs are given in another paper [5]. Two simple examples, one for discrete–time and one for continuous–time, are given in Chapter 4 to illustrate the solution procedure.

2 Discrete–time control

The quadratic cost function involving control input and tracking error may be written as

$$\nu = \frac{1}{2\pi j}\oint \mathrm{tr}\,(\Phi W_u + \Psi W_e)\,\frac{d\delta}{\delta}, \quad (2.1)$$

where W_u and W_e are the spectral matrices of the control input and tracking error vectors, respectively. Φ and Ψ are constant non–negative definite weighting matrices. $\oint$ indicates anticlockwise integration around the unit circle $|\delta| = 1$.

The spectral matrices of d, n and r are

$$W_d = A^{-1}C_dC_{d^*}A_*^{-1}, W_n = A^{-1}C_nC_{n^*}A_*^{-1}, W_r = A^{-1}C_rC_{r^*}A_*^{-1}, \quad (2.2)$$

respectively, where the star replaces δ by δ^{-1} and transposes the matrix. Since the white–noise sources are independent, the spectral matrix of $r - d + n$ is

$$W_{r-d+n} = A^{-1}(C_dC_{d^*} + C_nC_{n^*} + C_rC_{r^*})A_*^{-1} = A^{-1}DD_*A_*^{-1}. \quad (2.3)$$

Here, use has been made of the spectral factorization

$$DD_* = C_dC_{d^*} + C_nC_{n^*} + C_rC_{r^*} \quad (2.4)$$

in which D is Hurwitz.

On using (2.2) and (2.3), it follows from (1.5) and (1.6) that

$$W_u = N_2 A_1 A^{-1} D D_* A_*^{-1} A_{1*} N_{2*} \tag{2.5}$$

and

$$\begin{aligned} W_e &= M_2 A_1 A^{-1} D D_* A_*^{-1} A_{1*} M_{2*} + A^{-1} C_n C_{n*} A_*^{-1} \\ &\quad - M_2 A_1 A^{-1} C_n C_{n*} A_*^{-1} - A^{-1} C_n C_{n*} A_*^{-1} A_{1*} M_{2*} \end{aligned} \tag{2.6}$$

After substituting (2.5) and (2.6) in (2.1), a considerable amount of manipulation is required which includes using (1.4) to replace M_2 and N_2 by terms in N_1 and completing the square with respect to N_1 terms.

The following substitutions are also required. The Hurwitz polynomial E emerges from the spectral factorization

$$E_* E = A_{2*} \Phi A_2 + B_{2*} \Psi B_2 \tag{2.7}$$

Then

$$D_3 A_3^{-1} / p(\delta) = \Psi A^{-1} (D - C_n C_{n*} D_*^{-1}), \tag{2.8}$$

where D_3 and A_3 are right–coprime polynomial matrices and the polynomial $p(\delta)$ is the least common multiple of the unstable denominator terms in the elements of the rational matrix given by the right–hand side of (2.8).

Also

$$B_3 = \delta^h B_2 \text{ and } E_3 = \delta^h p_* E, \tag{2.9}$$

where $h = \deg p$ and

$$\overline{B}_3 = \delta^g B_{3*} \text{ and} \overline{E}_3 = \delta^g E_{3*}, \tag{2.10}$$

where $g = \max\{\deg B_3, \deg E_3\}$.

The appropriate manipulations produce

$$\nu = \nu_1 + \nu_3 \tag{2.11}$$

where

$$\begin{aligned} \nu_1 = \frac{1}{2\pi j} \oint \Big[&\operatorname{tr} \left(D_* A_*^{-1} \Psi A^{-1} D - A_{3*}^{-1} D_{3*} B_3 E_3^{-1} E_{3*}^{-1} B_{3*} D_3 A_3^{-1} \right) \\ &- \operatorname{tr} \left(C_{n*} A_*^{-1} \Psi A^{-1} C_n \right) \Big] \frac{d\delta}{\delta} \end{aligned} \tag{2.12}$$

and

$$\nu_3 = \frac{1}{2\pi j} \oint \operatorname{tr} \left[\left(E N_1 A^{-1} D - \overline{E}_3^{-1} \overline{B}_3 D_3 A_3^{-1} \right)_* \left(E N_1 A^{-1} D - \overline{E}_3^{-1} \overline{B}_3 D_3 A_3^{-1} \right) \right] \frac{d\delta}{\delta} \tag{2.13}$$

The controller does not affect ν_1, so N_1 must be chosen to minimize ν_3.

A diophantine equation is now used to split the term $\overline{E}_3^{-1}\overline{B}_3 D_3 A_3^{-1}$ into the sum of stable and unstable parts. By finding polynomial matrices Y and Z to satisfy the diophantine equation

$$\overline{E}_3 Y + Z A_3 = \overline{B}_3 D_3, \tag{2.14}$$

it follows that

$$\overline{E}_3^{-1}\overline{B}_3 D_3 A_3^{-1} = Y A_3^{-1} + \overline{E}_3^{-1} Z. \tag{2.15}$$

Notice that YA_3^{-1} has no poles inside $|\delta| = 1$. In solving (2.14), it is also necessary to find the solution with minimum–degree with respect to Z.

Equation (2.13) can now be written as

$$\nu_3 = \frac{1}{2\pi j}\oint \operatorname{tr}\left[\left(\overline{E}_3^{-1}Z + V\right)_*\left(\overline{E}_3^{-1}Z + V\right)\right]\frac{d\delta}{\delta}, \tag{2.16}$$

where

$$V = YA_3^{-1} - EN_1A^{-1}D. \tag{2.17}$$

Since N_1 and A^{-1} are stable, V has no poles inside $|\delta| = 1$.

Consider the cross term in (2.16) which involves $\oint \operatorname{tr} Z_*\overline{E}_{3*}^{-1}V\frac{d\delta}{\delta}$. Since Z is the minimum degree solution of (2.14), $Z_*\overline{E}_{3*}^{-1}V/\delta$ has no pole at $\delta = 0$. Also $\operatorname{tr} Z_*\overline{E}_{3*}^{-1}V$ has no poles inside $|\delta| = 1.$, Hence, by Cauchy's theorem, the integral is zero. It them follows [5] that the other cross term is also zero. Consequently,

$$\nu_3 = \frac{1}{2\pi j}\oint \operatorname{tr} Z_*\overline{E}_{3*}^{-1}\overline{E}_3^{-1}Z\frac{d\delta}{\delta} + \frac{1}{2\pi j}\oint \operatorname{tr} V_*V\frac{d\delta}{\delta} \tag{2.18}$$

The first term in (2.18) is unaffected by choice of N_1. The second term takes its minimum value of zero if

$$V = YA_3^{-1} - EN_1A^{-1}D = 0, \tag{2.19}$$

i.e.

$$N_1 = E^{-1}YA_3^{-1}D^{-1}A. \tag{2.20}$$

Substituting from the second equation in (1.4) gives

$$M_1 = (I - N_1B_2)\,A_2^{-1} \tag{2.21}$$

The optimum transfer matrix for the controller is then $R = M_1^{-1}N_1$.

Finally the minimum cost is

$$\nu_{min} = \nu_1 + \nu_2, \tag{2.22}$$

where ν_1 is given by (2.12) and

$$\nu_2 = \frac{1}{2\pi j}\oint \operatorname{tr} Z_*\overline{E}_{3*}^{-1}\overline{E}_3^{-1}Z\frac{d\delta}{\delta}. \tag{2.23}$$

3 Continuous–time control

The Δ transform used in discrete–time is now replaced by the Laplace transform. The quadratic cost function becomes

$$\nu = \frac{1}{2\pi j}\int_{-j\infty}^{j\infty} \mathrm{tr}(\Phi W_u + \Psi W_e) ds \qquad (3.1)$$

Spectral matrices W_u and W_e are formed as before but now the effect of the star symbol is to replace s by $-s$ and transpose the matrix. With this change, the spectral factors D and E are again formed as in (2.4) and (2.7) but now Hurwitz means that the polynomial has no roots in the right–half of the s–plane.

As far as other substitutions are concerned, (2.8) is replaced by

$$D_3 A_3^{-1}/p(s) = \Psi A^{-1}(D - C_n C_{n*} D_*^{-1}), \qquad (3.2)$$

where the polynomial $p(s)$ is the least common multiple of the unstable denominator terms in the elements of the rational matrix given by the right–hand side of (3.2). (2.9) is replaced by

$$E_2 = p_* E. \qquad (3.3)$$

With these changes, ν is manipulated as before to give $\nu = \nu_1 + \nu_3$, where

$$\begin{aligned}\nu_1 &= \frac{1}{2\pi j}\int_{-j\infty}^{j\infty} \left[\mathrm{tr}\left(D_* A_*^{-1}\Psi A^{-1} D - A_{3*}^{-1} D_{3*} B_2 E_2^{-1} E_{2*}^{-1} B_{2*} D_3 A_3^{-1}\right)\right. \\ &- \left.\mathrm{tr}\left(C_{n*} A_*^{-1}\Psi A^{-1} C_n\right)\right] ds \qquad (3.4)\end{aligned}$$

and

$$\nu_3 = \frac{1}{2\pi j}\int_{-j\infty}^{j\infty} \mathrm{tr}\left[\left(E N_1 A^{-1} D - E_{2*}^{-1} B_{2*} D_3 A_3^{-1}\right)_* \left(E N_1 A^{-1} D - E_{2*}^{-1} B_{2*} D_3 A_3^{-1}\right)\right] ds \qquad (3.5)$$

Again, the controller dose not affect ν_1, so N_1 must be chosen to minimize ν_3.

A diophantine equation is used to split the term $E_{2*}^{-1} B_{2*} D_3 A_3^{-1}$ into the sum of stable and unstable parts. By finding polynomial matrices Y and Z to satisfy the diophantine equation

$$E_{2*} Y + Z A_3 = B_{2*} D_3 \qquad (3.6)$$

it follows that

$$E_{2*}^{-1} B_{2*} D_3 A_3^{-1} = Y A_3^{-1} + E_{2*}^{-1} Z \qquad (3.7)$$

The term $Y A_3^{-1}$ has no poles in the right–half s-plane and $E_{2*}^{-1} Z$ has no poles in the left–half s-plane.

Equation (3.5) can now be written as

$$\nu_3 = \frac{1}{2\pi j}\int_{-j\infty}^{j\infty} \mathrm{tr}\left[\left(E_{2*}^{-1} Z + V\right)_* \left(E_{2*}^{-1} Z + V\right)\right] ds, \qquad (3.8)$$

where

$$V = YA_3^{-1} - EN_1A^{-1}D. \tag{3.9}$$

In solving (3.6), the solution must be found with minimum degree with respect to Z because this ensures that $E_{2*}^{-1}Z$ is strictly proper. For a feasible solution to the optimization, ν_3 must be finite, which implies that $E_{2*}^{-1}Z + V$ is strictly proper. Thus V must be strictly proper.

It follows that $s\mathrm{tr}V_*D_{2*}^{-1}Z$ is strictly proper and consequently

$$\int_\Gamma \mathrm{tr}V_*E_{2*}^{-1}Zds = 0, \tag{3.10}$$

where Γ is the infinite semicircle on the left–half of the s-plane. $\mathrm{tr}V_*E_{2*}^{-1}Z$ has no poles in the left half–plane, so by Cauchy's theorem,

$$\int_{-j\infty}^{j\infty} \mathrm{tr}V_*E_{2*}^{-1}Zds = 0. \tag{3.11}$$

The other cross term in ν_3 is also zero so

$$\nu_3 = \frac{1}{2\pi j}\int_{-j\infty}^{j\infty} \mathrm{tr}Z_*E_2^{-1}E_{2*}^{-1}Zds + \frac{1}{2\pi j}\int_{-j\infty}^{j\infty} \mathrm{tr}V_*Vds \tag{3.12}$$

The first term in (3.12) is unaffected by choice of N_1. The second term takes its minimum value of zero if

$$V = YA_3^{-1} - EN_1A^{-1}D = 0, \tag{3.13}$$

i.e.

$$N_1 = E^{-1}YA_3^{-1}D^{-1}A. \tag{3.14}$$

As before,

$$M_1 = (I - N_1B_2)A_2^{-1} \tag{3.15}$$

and

$$R = M_1^{-1}N_1 \tag{3.16}$$

Finally, the minimum cost is

$$\nu_{min} = \nu_1 + \nu_2, \tag{3.17}$$

where ν_1 is given by (3.4) and

$$\nu_2 = \frac{1}{2\pi j}\int_{-j\infty}^{j\infty} \mathrm{tr}Z_*E_2^{-1}E_{2*}^{-1}Zds. \tag{3.18}$$

4 Examples

4.1 Discrete–time

Let

$$S = \begin{bmatrix} \delta/(1-0.5\delta) \\ \delta^2 \end{bmatrix}, \tag{4.1}$$

$$S_d = \begin{bmatrix} 1/(1-0.5\delta) \\ 0 \end{bmatrix}, \tag{4.2}$$

$$S_n = \begin{bmatrix} 0 \\ 1/(1-0.5\delta) \end{bmatrix}, \tag{4.3}$$

$$S_r = \begin{bmatrix} 0 \\ 1/(1-0.9\delta) \end{bmatrix}, \tag{4.4}$$

$$\Phi = 1, \tag{4.5}$$

$$\Psi = \begin{bmatrix} 1 & 0 \\ 0 & 1 \end{bmatrix}. \tag{4.6}$$

From (1.1),

$$A = \begin{bmatrix} 1-0.5\delta & 0 \\ 0 & (1-0.5\delta)(1-0.9\delta) \end{bmatrix}, \tag{4.7}$$

$$B = \begin{bmatrix} \delta \\ (1-0.5\delta)(1-0.9\delta)\delta^2 \end{bmatrix}, \tag{4.8}$$

$$C_d = \begin{bmatrix} 1 \\ 0 \end{bmatrix}, \tag{4.9}$$

$$C_n = \begin{bmatrix} 0 \\ 1-0.9\delta \end{bmatrix}, \tag{4.10}$$

$$C_r = \begin{bmatrix} 0 \\ 1-0.5\delta \end{bmatrix} \tag{4.11}$$

and (1.2) gives $A_2 = 1 - 0.5\delta$ and

$$B_2 = \begin{bmatrix} \delta \\ (1-0.5\delta)\delta^2 \end{bmatrix}. \tag{4.12}$$

From (2.4)

$$D = \begin{bmatrix} 1 & 0 \\ 0 & 1.4653 - 0.9554\delta \end{bmatrix}. \tag{4.13}$$

From (2.7), $E = 1.785 - 0.5602\delta$ and from (2.8)

$$\frac{D_3 A_3^{-1}}{p(\delta)} = \begin{bmatrix} 1/(1-0.5\delta) & 0 \\ 0 & 0.5233(1-2\delta)/(1-0.9\delta)(1-1.5337\delta) \end{bmatrix}. \tag{4.14}$$

with $p(\delta) = 1 - 1.5337\delta$ and $h = \deg p = 1$, leaving

$$A_3 = \begin{bmatrix} 1-0.5\delta & 0 \\ 0 & 1-0.9\delta \end{bmatrix}, \tag{4.15}$$

$$D_3 = \begin{bmatrix} 1-1.5337\delta & 0 \\ 0 & 0.5233(1-2\delta) \end{bmatrix}, \tag{4.16}$$

Now

$$B_3 = \begin{bmatrix} \delta^2 \\ (1-0.5\delta)\delta^3 \end{bmatrix}, \tag{4.17}$$

and $E_3 = -2.7377(1-0.652\delta)(1-0.3138\delta)$ from (2.9), with $g = \max\{\deg B_3, \deg E_3\} = \max\{4, 2\} = 4$; so

$$\overline{B}_3 = [\delta^2, -0.5(1-2\delta)], \tag{4.18}$$
$$\overline{E}_3 = -0.5601\delta^2 + 2.6439\delta^3 - 2.7375\delta^4. \tag{4.19}$$

Now $\overline{B}_3 D_3 = [c_1, c_2]$, where

$$c_1 = \delta^2 - 1.5337\delta^3, \tag{4.20}$$
$$c_2 = -0.2617 + 1.0466\delta - 1.0466\delta^2. \tag{4.21}$$

The diophantine equation $\overline{E}_3 Y + Z A_3 = \overline{B}_3 D_3$ can be written as

$$e_3[y_1, y_2] + [z_1, z_2] \begin{bmatrix} 1-0.5\delta & 0 \\ 0 & 1-0.9\delta \end{bmatrix} = [c_1, c_2], \tag{4.22}$$

where $e_3 = \overline{E}_3$. This corresponds to two independent scalar equations

$$e_3 y_1 + (1-0.5\delta) z_1 = c_1, \tag{4.23}$$
$$e_3 y_2 + (1-0.9\delta) z_2 = c_2. \tag{4.24}$$

The minimum degree solution with respect to Z gives

$$Y = [0.3323, 0.3160] \tag{4.25}$$

Then, from (2.20) and (2.21),

$$N_1 = \frac{1}{1-0.3138\delta}\left[0.1862, \frac{0.1208(1-0.5\delta)}{1-0.652\delta}\right], \tag{4.26}$$
$$M_1 = \frac{(1-0.6152\delta)(1-0.0368\delta+0.0982\delta^2)}{(1-0.3138\delta)(1-0.652\delta)}. \tag{4.27}$$

Hence, the optimal transfer matrix for the controller is

$$R = M_1^{-1} N_1 = \frac{[0.1862(1-0.652\delta), 0.1208(1-0.5\delta)]}{(1-0.6152\delta)(1-0.0368\delta+0.0982\delta^2)} \tag{4.28}$$

4.2 Continuous–time

Let

$$S = \begin{bmatrix} 1/(1+s) \\ (1-s)/(1+s) \end{bmatrix}, \tag{4.29}$$

$$S_d = \begin{bmatrix} 1/(1+s) \\ 0 \end{bmatrix}, \tag{4.30}$$

$$S_n = \begin{bmatrix} 0 \\ 1/(1+0.1s) \end{bmatrix}, \tag{4.31}$$

$$S_r = \begin{bmatrix} 0 \\ 1/(1+s) \end{bmatrix}, \tag{4.32}$$

$$\Phi = 1, \tag{4.33}$$

$$\Psi = \begin{bmatrix} 1 & 0 \\ 0 & 1 \end{bmatrix}. \tag{4.34}$$

From (1.1),

$$A = \begin{bmatrix} 1+s & 0 \\ 0 & (1+0.1s)(1+s) \end{bmatrix}, \tag{4.35}$$

$$B = \begin{bmatrix} 1 \\ (1+0.1s)(1-s) \end{bmatrix}, \tag{4.36}$$

$$C_d = \begin{bmatrix} 1 \\ 0 \end{bmatrix}, \tag{4.37}$$

$$C_n = \begin{bmatrix} 0 \\ 1+s \end{bmatrix}, \tag{4.38}$$

$$C_r = \begin{bmatrix} 0 \\ 1+0.1s \end{bmatrix} \tag{4.39}$$

and (1.2) gives $A_2 = 1 + s$ and

$$B_2 = \begin{bmatrix} 1 \\ 1-s \end{bmatrix}. \tag{4.40}$$

From (2.4),

$$D = \begin{bmatrix} 1 & 0 \\ 0 & 1.4142 + 1.005s \end{bmatrix} \tag{4.41}$$

and $E = 1.7321 + 1.4142s$ from (2.7). Then (3.2) gives

$$\frac{D_3 A_3^{-1}}{p(s)} = \begin{bmatrix} 1/(1+s) & 0 \\ 0 & (1-0.1s)/(1+s)(1.4142 - 1.005s) \end{bmatrix}, \tag{4.42}$$

with $p(s) = 1.4142 - 1.005s$, leaving

$$A_3 = \begin{bmatrix} 1+s & 0 \\ 0 & 1+s \end{bmatrix} \tag{4.43}$$

and

$$D_3 = \begin{bmatrix} 1.4142 - 1.005s & 0 \\ 0 & 1 - 0.1s \end{bmatrix}. \tag{4.44}$$

From (3.3),

$$E_2 = (1.4142 + 1.005s)(1.7321 + 1.4142s) \tag{4.45}$$

and

$$\begin{aligned} E_{2*} &= (1.4142 - 1.005s)(1.7321 - 1.4142s) \\ &= 2.4495 - 3.7408s + 1.4213s^2. \end{aligned} \tag{4.46}$$

Also, $B_{2*} = [1, 1+s]$ and $B_{2*}D_3 = [c_1, c_2]$, where

$$c_1 = 1.4142 - 1.005s \tag{4.47}$$

and

$$c_2 = 1 + 0.9s - 0.1s^2. \tag{4.48}$$

The diophantine equation (3.6) $E_{2*}Y + ZA_3 = B_{2*}D_3$ can be written as

$$e_2[y_1, y_2] + [z_1, z_2]\begin{bmatrix} 1+s & 0 \\ 0 & 1+s \end{bmatrix} = [c_1, c_2], \tag{4.49}$$

where $e_2 = E_{2*}$. This corresponds to two independent scalar equations

$$e_2y_1 + (1+s)z_1 = c_1 \tag{4.50}$$
$$e_2y_2 + (1+s)z_2 = c_2. \tag{4.51}$$

The minimum degree solution with respect to Z gives

$$Y = [0.3178, 0]. \tag{4.52}$$

Then from (3.14) and (3.15),

$$N_1 = \frac{[0.3178, 0]}{1.7321 + 1.4142s} \text{ and } M_1 = \frac{1.4142}{1.7321 + 1.4142s}. \tag{4.53}$$

Hence, the optimal transfer matrix for the controller is

$$R = M_1^{-1}N_1 = [0.2247, 0]. \tag{4.54}$$

References

[1] Grimble, M.J. and Johnson, M.A. ,"Optimal Control and Stochastic Estimation" Wiley, Chichester, 1988.

[2] Kučera, V., "Discrete Linear Control" Wiley, Chichester, 1979.

[3] Kučera, V., Stochastic Multivariable Control: a polynomial Equation Approach, IEEE Trans. Auto. Control, AC–25, 913–919, 1980.

[4] Roberts, A.P., Polynomial Optimization of Stochastic Feedback Control for Unstable plants, Int. J. Control, 45, 1953–1961, 1987.

[5] Roberts, A.P. and Newmann, M.M., Polynomial Optimization of Stochastic Feedback Control for Stable Plants, IMA of Math. Control and Inf., 5, 243–257, 1988.

A general transformation for polynomial systems

G.E. Hayton A.B. Walker A.C. Pugh

Abstract

The transformation of "full equivalence" between general polynomial matrices of possibly different dimensions is described and its properties summarised. The transformation is then restricted to provide a relationship between general polynomial system matrices which preserves both the finite and the infinite frequency structure of the system. Finally, connections with existing system transformations are investigated.

1 Introduction

Given a certain description of a system it is often advantageous for analysis and design purposes to consider an alternative representation of the system model. Thus for example in LCR network theory one realisation of a given admittance matrix may utilise fewer reactive elements than another which achieves the same purpose. Hence a question of fundamental importance in systems theory is when are two system descriptions equivalent. That is, when do two system representations actually describe the same physical system.

For state-space descriptions of a system the transformation of system similarity (Rosenbrock, 1970), obtained by a change of basis in the state-space, is fundamental and preserves the essential dynamics of the system. In the more general setting of polynomial system matrix descriptions the transformation of strict system equivalence (Rosenbrock, 1970; see also Fuhrmann, 1977, Pugh and Shelton, 1978) plays an analogous role and is seen to preserve the order of the system, the transfer function matrix and various poles and zeros associated with the system at finite frequencies.

More recently Verghese(1978) has shown that if a dynamical system is formed as a result of component failure or switching then it becomes important to consider not only the finite frequency behaviour as the theory had done hitherto, but also the impulsive solutions that may arise in response to the unconstrained nature of the input. Such impulsive solutions are associated with significant dynamical structure attributable to the point at infinity. Rosenbrock and Pugh (1974) have also shown the importance of infinite frequency considerations when a system is represented as a set of subsystems. Unfortunately the usual transformation of "strict system equivalence" is not suited in this respect since it generally destroys infinite frequency characteristics. Consequently in recent years the dynamic behaviour of linear, time-invariant systems at infinite frequency and transformations which preserve such behaviour have become of interest.

The simplest systems which exhibit infinite frequency behaviour are the so-called generalised state-space systems. Transformations of such systems which preserve both finite and infinite frequency dynamics have been explored in depth by a number of authors (Rosenbrock, 1974,Verghese, 1978, Pugh, Fretwell and Hayton, 1983, Anderson, Coppel and Cullen, 1985). These ideas have been extended to more general polynomial matrices by Anderson, Coppel and Cullen (1985), Hayton, Pugh and Fretwell (1987), and Hayton, Walker and Pugh (1987). The approach adopted by Anderson et al (1985) is to give two separate transformations, one preserving only finite (called 'system equivalence') the other only infinite frequency structure (called 'system equivalence at infinity'). Two polynomial system descriptions are then said to be strongly system equivalent if they are both system equivalent and system equivalent at infinity. It is asserted that the appropriate setting for studying the behaviour at infinity of a linear system is through the use of causal, rational matrices, since in the polynomial setting the transforming(polynomial) matrices all have a pole structure at infinity which can cause undesirable pole/zero cancellation.

An alternative approach to this problem is described here. This sets out a single polynomial matrix transformation which simultaneously preserves both the finite and infinity frequency structure. It is shown that the development depends largely on a natural extension of the idea of infinite elementary divisors of a matrix pencil to matrix polynomials. In the proposed matrix transformation a difficulty noted by Anderson, et. al.(1985) is resolved by the insistence that a certain McMillan degree condition be satisfied.

The restriction of the above matrix transformation to the linear systems context is next considered. The resulting system matrix transformation is termed full system equivalence. Several results concerning invariants of this transformation are noted. Further the detailed connection between full system equivalence and the strong system equivalence notion of Anderson et al(1985) is presented.

2 Infinite zeros of a polynomial matrix

Infinite frequency structure has been defined by a number of authors including Verghese(1978) and Pugh and Ratcliffe (1970) who make use of the bilinear transformation $s = 1/w$ to map the point at infinity in the s plane to the origin of the w plane. Infinite poles(resp. zeros) of the original rational matrix are the defined as poles (reps. zeros) at $w = 0$ of the transformed matrix. An alternative characterisation of infinite poles and zeros may be given via the following natural extension of the concept of infinite elementary divisors of a matrix pencil to matrix polynomials

Definition 1 *Consider the $m \times l$ matrix polynomial of degree n:*

$$P(s) \triangleq P_0 + P_1 s + \cdots + P_n s^n \tag{2.1}$$

where $P_i, i = 1, 2, \ldots, n$ $(P_n \neq 0)$ *are constant matrices. The dual of* $P(s)$, *denoted* $DP(\hat{s})$, *is*

$$DP(\hat{s}) = P_0\hat{s}^n + P_1\hat{s}^{n-1} + \cdots + P_n \tag{2.2}$$

and $P(s)$ *is said to have an* infinite elementary divisor (i.e.d.) of degree q *whenever* $DP(\hat{s})$ *has a finite elementary divisor of the form* $\hat{s}^q$.

The relationship between the i.e.d.s and the infinite pole/zero structure of the matrix is summarised in the following (Hayton et. al., 1988):

Theorem 1 *Let* $P(s)$ *be an* $m \times l$ *matrix polynomial of degree* n. *Then*

1. $P(s)$ *has an i.e.d. of degree* $q(< n)$ *iff it has an infinite pole of multiplicity* $n - q$.
2. $P(s)$ *has an i.e.d. of degree* $q(> n)$ *iff it has an infinite zero of multiplicity* $q - n$.
3. *i.e.d.s of degree* $q = n$ *have no dynamical significance.*

3 Transformation of polynomial matrices

A number of transformations which preserve certain properties of a polynomial matrix have been proposed. Those of direct relevance to the current argument are now listed together with a brief summary of their properties. Let $\mathcal{P}(m, l)$ be the class of $(r + m) \times (r + l)$ polynomial matrices where l and m are fixed integers and r ranges over all integers greater than $max(-m, -l)$ and let $\mathcal{P}_o(m, l)$ be the set of matrix pencils. Let $P_i(s) \in \mathcal{P}(m, l), i = 1, 2$, and suppose $\exists$ polynomial $M(s)$, $N(s)$ of appropriate dimensions s.t.

$$\begin{pmatrix} M(s) & P_2(s) \end{pmatrix} \begin{pmatrix} P_1(s) \\ -N(s) \end{pmatrix} = 0 \tag{3.1}$$

Definition 2 *If* $M(s)$, $N(s)$ *are constant, square and invertible then the matrices* $P_i(s)$ *related as in (3.1) are said to be* strictly equivalent *(s.e.).*

s.e. preserves the finite and infinite zeros, and the infinite poles (and thus the MacMillan degree) of the $P_i(s)$ it relates, note however that it only relates matrices of the same dimensions (Gantmacher, 1959).

Definition 3 *If* $M(s)$, $N(s)$ *are constant,* $P_i(s) \in \mathcal{P}_o(m, l)$, $i = 1, 2$ *and*

$$\begin{bmatrix} M & P_2(s) \end{bmatrix}, \begin{bmatrix} P_1(s) \\ -N \end{bmatrix} \quad \begin{matrix} \textit{have full rank } \forall s \\ \textit{and} \\ \textit{no infinite zeros} \end{matrix}$$

then $P_i(s)$ *related as in (3.1) are said to be* completely equivalent. *(c.e.)*

c.e. preserves the finite and infinite zeros of matrix pencils and allows pencils of different dimensions to be related. It is an equivalence relation for regular pencils (Hayton & Pugh, 1985) and is a natural generalisation of s.e. and similarity. Regular pencils which are c.e. can also be related by operations of s.e followed by trivial inflation or deflation.

Definition 4 *If $M(s)$, $N(s)$ are square and unimodular then $P_i(s)$ related as in (3.1) are said to be* unimodular equivalent *(u.e.).*

u.e. is another classical transformation relating only matrices of the same dimension it may, however, be generalised to:

Definition 5 *If $[M(s)\ P_2(s)]$, $\begin{bmatrix} P_1(s) \\ -N(s) \end{bmatrix}$ have full rank $\forall s$ then $P_i(s)$ related as in (3.1) are said to be* extended unimodular equivalent *(e.u.e.).*

e.u.e. allows matrices of different dimensions to be related and both e.u.e. and u.e. preserve finite zero structure, they do not, however, preserve infinite zero structure (Pugh & Shelton, 1978).

Definition 6 *If $[M(s)\ P_2(s)]$, $\begin{bmatrix} P_1(s) \\ -N(s) \end{bmatrix}$ have full rank $\forall s$ and no infinite zeros*

$$\left.\begin{aligned} \delta\,[\,M(s) \quad P_2(s)\,] &= \delta[P_2(s)] \\ \delta\begin{bmatrix} P_1(s) \\ -N(s) \end{bmatrix} &= \delta[P_1(s)] \end{aligned}\right\} \tag{3.2}$$

where $\delta(\cdot)$ indicates MacMillan degree then $P_i(s)$ related as in (3.1), are said to be fully equivalent *(f.e.).*

Note first the relationship between f.e. and the other transformations listed above:

Lemma 1 *(1) s.e. and c.e. are special cases of full equivalence.*
(2) f.e. is a special case of e.u.e.

Theorem 2 *If $P_1(s)$, $P_2(s) \in \mathcal{P}(m,l)$ are related by f.e. then they possess identical finite and infinite zero structure.*

Proof The original result is due to Ratcliffe (1982) although at this time the reason for the MacMillan degree condition was obscure. The characterisation of infinite frequency structure given in Def. 1 above results in a different proof (Hayton et. al. 1988) which is summarised as follows:

From Lemma 1 $P_1(s)$, $P_2(s)$ are e.u.e and hence have the same finite zeros. The dual of (3.1) is then constructed and it is shown that iff the MacMillan degree condition holds, this dual can be reduced by the removal of certain divisors to a second transformation of e.u.e which makes transparent the preservation of the infinite zeros of $P_1(s)$, $P_2(s)$. The MacMillan degree condition ensures that removal of the common divisors affects only information about infinite pole structure, an analogous condition for finite frequency is not needed since *polynomial* matrices do not have finite poles.

Another result of interest is the following:

Lemma 2 *Any polynomial matrix is f.e. to a matrix pencil having identical zero structure at finite and infinite frequencies.*

Proof (Walker, Hayton & Pugh, 1988).

The above lead to a proof of:

Theorem 3 *f.e. is an equivalence relation on the set of nonsingular matrices in* $\mathcal{P}(m,m)$.

Proof (Walker, Hayton & Pugh, 1988).

4 Transformations of linear systems

The dynamic behaviour of a linear, time–invariant invariant system is encapsulated in the system matrix

$$P(s) = \begin{bmatrix} T(s) & U(s) \\ -V(s) & W(s) \end{bmatrix} \tag{4.1}$$

together with $G(s) = V(s)T^{-1}(s)U(s) + W(s)$, the transfer function matrix. Note that $P(s)$ as in (4.1) form that subset of $\mathcal{P}(m,l)$ such that $r \geq 0$ and $|T(s)| \neq 0$ this will be denoted $\mathcal{P}_s(m,l)$. Various submatrices of $P(s)$ and $G(s)$ define the finite pole and zero structure of the system (Rosenbrock, 1970), however, in order to provide consistent definitions of pole and zeros at both finite and infinite frequency it is necessary to consider the normalised system matrix (Verghese & Kailath, 1981)

$$\overline{P}(s) = \begin{bmatrix} \overline{T}(s) & \overline{U} \\ \overline{V} & 0 \end{bmatrix} = \left[\begin{array}{ccc|c} T(s) & U(s) & 0 & 0 \\ -V(s) & W(s) & I & 0 \\ 0 & I & 0 & I \\ \hline 0 & 0 & -I & 0 \end{array}\right] \tag{4.2}$$

Note that $\overline{P}(s) \in \mathcal{P}(m,l)$ and $\overline{G}(s) = \overline{V}\overline{T}^{-1}(s)\overline{U} = G(s)$.

By suitably restricting a matrix transformation a system transformation may be obtained. The nessary restrictions arise from the need to preserve the transfer function matrix and may be described in general terms by:

$$\begin{bmatrix} M(s) & 0 \\ X(s) & I \end{bmatrix} P_1'(s) = P_2'(s) \begin{bmatrix} N(s) & Y(s) \\ 0 & I \end{bmatrix} \tag{4.3}$$

Definition 7 *If $P_i'(s) = P_i(s)$ of (4.1) $M(s), N(s), X(s), Y(s)$ are polynomial and $M(s)T_1(s) = T_2(s)N(s)$ transformation of e.u.e. then systems related as in (4.3) are said to be* extended strict system equivalent *(e.s.s.e.), (Pugh & Shelton, 1978). E.s.s.e. preserves the finite pole and zero structure of the systems it relates, but not the infinite frequency structure (Pugh & Shelton, 1978). It is a natural extension of Rosenbrock's strict system equivalence (s.s.e.) in which the relationship between the $T_i(s)$ is of u.e. and is seen to be identical to the transformation of polynomial system equivalence proposed by Anderson and Coppel (1985).*

Coppel and Cullen (1985) give a number of characterisations of a separate transformation which preserves infinite frequency structure, among these is:

Definition 8 *If $P_i'(s) = \overline{P}_i(s)$ in (4.2), $M(s), N(s), X(s), Y(s)$ are causal, rational matrices and $[M(s)\ \overline{T}_2(s)]$, $\begin{bmatrix} \overline{T}_1(s) \\ -N(s) \end{bmatrix}$ have causal right and left inverses respectively then systems related as in (4.3) are* system equivalent at infinity.*(s.e.∞)*

S.e.∞ merely preserves the infinite frequency structure of the system. To preserve the total pole and zero structure definitions 7 and 8 are combined to give (Coppel & Cullen, 1985):

Definition 9 *$P_1(s)$, $P_2(s) \in \mathcal{P}_s(m,l)$ are* Strongly system equivalent *(str.s.e.) if they are simultaneously e.s.s.e. and s.e.∞.*

As an alternative approach using the matrix transformation of f.e. a single transformation preserving both finite and infinite frequency structure can be defined in terms of the normalised system matrices by:

Definition 10 *If $P_i'(s) = \overline{P}_i(s)$ in (4.2), $X(s) = 0$, $Y(s) = 0$ and the transformation $M(s)\overline{T}_1(s) = \overline{T}_2(s)N(s)$ is f.e. then the systems realed in (4.3) are* normally full system equivalent*(n.f.s.e.).*

N.f.s.e. is readily shown to induce transformations of f.e. on the various submatrices of P(s) which define the pole and zero structure of the system and thus it preserves all the finite and infinite zeros of the systems it relates (Hayton,Walker & Pugh, 1988).

A second definition of system equivalence based on f,e., but acting on the more usual Rosenbrock system matrices of (4.1) as opposed to the normalised form of (4.2) is given by:

Definition 11 *If $P_i'(s) = \overline{P}_i(s)$ in (4.1) and (4.3) is a f.e. transformation the systems are* fully system equivalent *(f.s.e.).*

5 Unification of the transformations

We first demonstrate that f.s.e. and n.f.s.e. are equivalent notions:

Theorem 4 *$P_i(s) \in \mathcal{P}_s(s)$, $i = 1,2$ and f.s.e. iff the associated normal system matrices $\overline{P}_i(s)$ are n.f.s.e.*

Proof Suppose $P_i(s)$ are n.f.s.e. then (4.3) may be written as:

$$\left[\begin{array}{cccc|cccc} M_{11}(s) & M_{12}(s) & M_{13}(s) & 0 & T_2(s) & U_2(s) & 0 & 0 \\ M_{21}(s) & M_{22}(s) & M_{23}(s) & 0 & -V_2(s) & W_2(s) & I & 0 \\ M_{31}(s) & M_{32}(s) & M_{33}(s) & 0 & 0 & I & 0 & -I \\ 0 & 0 & 0 & I & 0 & 0 & -I & 0 \end{array}\right] \cdot \left[\begin{array}{cccc} T_1(s) & U_1(s) & 0 & 0 \\ -V_1(s) & W_1(s) & I & 0 \\ 0 & I & 0 & -I \\ 0 & 0 & -I & 0 \\ \hline -N_{11}(s) & -N_{12}(s) & -N_{13}(s) & 0 \\ -N_{21}(s) & -N_{22}(s) & -N_{23}(s) & 0 \\ -N_{31}(s) & -N_{32}(s) & -N_{33}(s) & 0 \\ 0 & 0 & 0 & -I \end{array}\right] = 0 \tag{5.1}$$

Examination of (block) elements (3,3), (4,i), and (i,4), $i = 1,2,3$ in (5.1) shows that it may be simplified to give:

$$\left[\begin{array}{ccc|ccc} M_{11}(s) & M_{12}(s) & 0 & T_2(s) & U_2(s) & 0 \\ M_{21}(s) & M_{22}(s) & 0 & -V_2(s) & W_2(s) & I \\ M_{31}(s) & M_{32}(s) & I & 0 & I & 0 \end{array}\right] \cdot \left[\begin{array}{ccc} T_1(s) & U_1(s) & 0 \\ -V_1(s) & W_1(s) & I \\ 0 & I & 0 \\ \hline -N_{11}(s) & -N_{12}(s) & -N_{13}(s) \\ -N_{31}(s) & -N_{32}(s) & -N_{33}(s) \\ 0 & 0 & I \end{array}\right] = 0 \tag{5.2}$$

(5.2) is a transformation of f.e. between $\overline{T}_1(s)$ and $\overline{T}_1(s)$ and by considering the degrees of various minors of the matrices it may be shown that $M_{31}(s)$, $M_{32}(s)$ and $N_{13}(s)$ must be constant in order that the McM illan degree condition holds. Postmultiplying the first matrix in (5.2) by:

$$Q = \begin{bmatrix} I & 0 & 0 & 0 & 0 & 0 \\ 0 & I & 0 & 0 & 0 & 0 \\ 0 & 0 & I & 0 & 0 & 0 \\ 0 & -N_{13} & 0 & I & 0 & 0 \\ -M_{31} & -M_{32} & 0 & 0 & I & 0 \\ 0 & 0 & 0 & 0 & 0 & I \end{bmatrix} \tag{5.3}$$

and simultaneously premultipling the second matrix by Q^{-1} gives:

$$\left[\begin{array}{ccc|ccc} M'_{11}(s) & M'_{12}(s) & 0 & T_2(s) & U_2(s) & 0 \\ M'_{21}(s) & M'_{22}(s) & 0 & -V_2(s) & W_2(s) & I \\ 0 & 0 & I & 0 & I & 0 \end{array}\right] \left[\begin{array}{ccc} T_1(s) & U_1(s) & 0 \\ -V_1(s) & W_1(s) & I \\ 0 & I & 0 \\ \hline -N'_{11}(s) & -N'_{12}(s) & 0 \\ -N'_{31}(s) & -N'_{32}(s) & 0 \\ 0 & 0 & I \end{array}\right] = 0 \tag{5.4}$$

Since Q is constant and nonsingular the left (respectively right) hand matrix in (5.4) is s.e. to the left (respectively right) hand matrix in (5.3). Hence (5.4) is also a transformation of f.e. between $\overline{T}_1(s)$ and $\overline{T}_2(s)$. Consideration of individual (block) elements in (5.4) allows further reduction to:

$$\left[\begin{array}{cc|cc} M'_{11}(s) & 0 & T_2(s) & U_2(s) \\ M'_{21}(s) & I & -V_2(s) & W_2(s) \end{array}\right] \left[\begin{array}{cc} T_1(s) & U_1(s) \\ -V_1(s) & W_1(s) \\ -N'_{11}(s) & -N'_{12}(s) \\ \hline 0 & I \end{array}\right] = 0 \tag{5.5}$$

which is a transformation of f.e. between $P_1(s)$ and $P_2(s)$ showing the systems to be f.s.e.

For the converse, given that the systems are f.s.e. a transformation of f.e. of the form (5.4) may be constructed between $\overline{T}_1(s)$ and $\overline{T}_2(s)$ and this is readily expanded to (5.3).

Using the results above it is clear that str.s.e. and f.s.e. both preserve the total pole and zero structure, finite and infinite, of the system they relate. In order to show that they are in fact identical notions of equivalence we first provide a new characterisation of s.e.∞ involving polynomial rather than rational transforming matrices. This requires:

Theorem 5 *A rational matrix $G(s)$ has a causal inverse iff it has full normal rank and no infinite zeros.*

Proof Let $G(s)$ be an $m \times l$ rational matrix, $m \leq l$. There exist rational, bicausal $B_i(s)$, $i = 1, 2$, such that:

$$B_1(s)G(s)B_2(s) = [\text{diag.}(s^{q_1}, \cdots, s^{q_r}, 0, \cdots, 0)|0] \tag{5.6}$$

The right hand side of (5.6) is the Smith MacMillan form at infinity of $G(s)$ and shows the infinite pole/zero structure of the matrix (Vardulakis & Karcanias, 1983). If $G(s)$ has full normal rank $r = m$ and

$$G'(s) = B_2(s)\begin{bmatrix} I \\ 0 \end{bmatrix} \text{diag.}(s^{-q_1}, \cdots, s^{-q_m})B_1(s) \tag{5.7}$$

$G'(s)$ is readily shown to be a right inverse of $G(s)$. Further, is $G(s)$ has no infinite zeros $q_i \geq 0\ \forall i$ and hence $G'(s)$ is causal. The converse result is obtained in a similar way.

Theorem 4 leads immediately to the result:

Lemma 3 *$P_i(s) \in \mathcal{P}(m,1)$ are s.e.∞ if the associated normalised system matrices $\overline{P}_i(s)$ are related as in (4.3) with $M(s)$, $N(s)$ polynomial, $X(s) = 0$, $Y(s) = 0$ and $[M(s)\ \overline{T}_2(s)] \begin{bmatrix} \overline{T}_1(s) \\ -N(s) \end{bmatrix}$ have no infinite zeros and respective MacMillan degrees $\delta(\overline{T}_2(s))$, $\delta(\overline{T}_1(s))$.*

Proof Suppose $P_1(s)$, $P_2(s)$ are s.e.∞ One possible choice is $X(s) = 0$, $Y(s) = 0$. Now consider

$$[M(s)\ \overline{T}_2(s)] \begin{bmatrix} \overline{T}_1(s) \\ -N(s) \end{bmatrix} = 0 \tag{5.8}$$

From Theorem 4 since each of the two compound matrices in (5.8) has a causal inverse, neither has infinite zeros. Also, since $M(s)$ is causal and $\overline{T}_2(s)$ polynomial, $\delta[M(s)\ \overline{T}_2(s)]$ is equal to the number of finite poles in $M(s)$ (i.e. the order of $M(s)$) plus the number of infinite poles in $\overline{T}_2(s)$ (i.e. $\delta(\overline{T}_2(s))$). For any proper, rational matrix $C(s)$ the matrices in (5.8) can be transformed by bicausal operations to:

$$[M(s) - \overline{T}_2(s)C(s)\ \overline{T}_2(s)] \begin{bmatrix} \overline{T}_1(s) \\ -N(s) - C(s)\overline{T}_1(s) \end{bmatrix} = 0 \tag{5.9}$$

Choosing $C(s) = (\overline{T}_2^{-1}(s)M(s))_{sp} = (N(s)\overline{T}_1^{-1}(s))_{sp}$ where $(.)_{sp}$ denotes the strictly proper part of the matrix ensures that $M(s) = M(s) - \overline{T}_2(s)C(s)$ and $N(s) = -N(s) - C(s)\overline{T}_1(s)$ are polynomial and have no finite poles. Since the operations were bicausal the infinite frequency structures of the matrices are unaltered and hence $[M(s)\ \overline{T}_2(s)]$ has no infinite zeros and MacMillan degree equal to $\delta(\overline{T}_2(s))$ Similar remarks may be made concerning the second compound matrix in (5.9) completing this part of the proof. For the converse a causal transformation is generated from the polynomial one by reversing the above argument.

The following theorem is the main result and demonstrates that f.s.e. and str.s.e. are equivalent notions.

Theorem 6 *$P_i(s) \in \mathcal{P}_s(m,l)$, $i = 1,2$ are str.s.e. if they are f.s.e.*

Proof f.s.e is clearly a special case of e.s.s.e. thus satisfying the first part of Def. 9. Further, from Theorem 4 the systems are n.f.s.e. and thus ∃ a relationship satisfying the conditions of Lemma 3, hence showing that they are s.e.∞.

For the converse assume $P_1(s)$, $P2(s)$ are str.s.e. with associated normalised forms $\overline{P}_1(s)$, $\overline{P}_2(s)$ as in (4.2) then $P_1(s)$, $P_2(s)$ are e.s.s.e. This relationship may be extended and manipulated to give a transformation of e.u.e. between $\overline{T}_1(s)$ and $\overline{T}_2(s)$ of the form:

$$[R(s) - \overline{T}_2(s)Z_p(s)|\overline{T}_2(s)] \begin{bmatrix} \overline{T}_1(s) \\ -Q(s) - Z_p(s)\overline{T}_1(s) \end{bmatrix} = 0 \tag{5.10}$$

where $R(s)$, $Q(s)$ are compound polynomial matrices formed from the transforming matrices of Def.7 and $Z_p(s)$ is any polynomial matrix. $P_1(s)$, $P_2(s)$ are also s.e.∞ Thus using Lemma 3 together with further bicausal transformations $\exists$ the following relationship between $\overline{T}_1(s)$ and $\overline{T}_2(s)$:

$$[\overline{M}(s) + \overline{T}_2(s)Z_c(s)|\overline{T}_2(s)] \begin{bmatrix} \overline{T}_1(s) \\ -\overline{N}(s) - Z_c(s)\overline{T}_1(s) \end{bmatrix} = 0 \tag{5.11}$$

with $\overline{M}(s)$, $\overline{N}(s)$ polynomial as in the lemma and $Z_c(s)$ any causal matrix. Since (5.11) is obtained from (5.8) via bicausal transformations the compound matrices in (5.11) have no infinite zero, and $\delta(\overline{T}_2(s))$, $\delta(\overline{T}_1(s))$ infinite poles respectively. Defining:

$$D(s) = \overline{T}_2^{-1}(s)\left(\overline{M}(s) - r(s)\right) = \left(\overline{N}(s) - Q(s)\right)\overline{T}_1^{-1}(s) \tag{5.12}$$

and

$$Z_p(s) = D[(s)]_p \text{ –the polynomial part of } D(s) \tag{5.13}$$

$$Z_c(s) = D[(s)]_{sp} \tag{5.14}$$

makes (5.10) and (5.11) identical transformation of f.e. between $\overline{T}_1(s)$ and $\overline{T}_2(s)$. Hence $P_1(s)$, $P_2(s)$ are n.f.s.e. and, as noted above, they f.s.e. which completes the proof.

References

[1] Anderson, B.D.O., Coppel, W.A. & Cullen; D.J., 1985,J. Aust. Math. Soc., B27, 393-411.

[2] Coppel,W.A. & Cullen,D.J.; 1985, J. Aust. Math. Soc., B27, 223-237.

[3] Furhmann, P. A.; 1977, Int. J. Con. 25, 5-10.

[4] Gantmacher, F.r.; 1959, " The Theory of Matrices ", Chelsea Pub. Co.

[5] Hayton, G.E. & Pugh, A.c.; 1985, Proc. M.T.N.S. pp. 323-337

[6] Hayton, G.E.; Pugh, A.C. & Fretwell, P.; 1988 Int. J. Control 47 53

[7] Hayton, G.E., Walker, A. & Pugh, A.C.; 1987, Proc. M.T.N.S., Phoenix, U.S.A.

[8] Pugh, A.C., Fretwell, P. & Hayton, G.E.; 1983, Proc. ACC, San Francisco USA, 633-7.

[9] Pugh, A.C. & Ratclife, P.A.; 1979, Int. J. Control, 30,213.

[10] Pugh, A.C.; Shelton, A.K.; 1978, Int. J. Control, 27, 657

[11] Ratcliffe, P.A.; 1982, Ph.D. thesis, Plymouth Polytechnic U.K.

[12] Rosenbrock, H.H,; 1970 "State Space and Multivariable Theory", Nelson, London, U.K.

[13] Rosenbrock, H.H.; 1974, Int. J. Control, 20, 191-202.

[14] Vardulakis, A.I.G.; & Karcanias N.; 1983, Int. J. Control, 35, 779-809.

[15] Verghese, G.C. & Kailath, T.; 1981, I.E.E.E. Trans. A.C. 26 235-243.

[16] Walker, A.B.; Hayton, G.E. & Pugh, A.C.; 1988, "Linear Systems Requirements of Matrix Transformations", Proc. IMA Conf. on Applications of Matrices, University of Bradford, UK.

Some local feedback results for linear systems

A.C. Pugh A. Kafai G.E. Hayton

Abstract

A local theory for linear systems is presented and in particular a local state realisation is derived, In the context of this model the concept of "local" state feedback is considered and certain of its effects are characterised.

1 Introduction

A large part of Linear Systems Theory works through the medium of matrix fraction descriptions (M.F.D.'s), and frequently the conditions on the M.F.D. appear excessive for the analysis that is being carried out. The relative primeness condition insisted on by the M.F.D. is of course a full rank requirement which must be satisfied at every frequency in the finite complex plane. Often however the investigation is only concerned with one specific frequency and in such cases much of the information carried by the M.F.D. is not utilised. A prime example in this respect occurs in the study of the infinite frequency behaviour of a linear system with transfer function matrix $G(s)$. The usual approach to such problems is to perform the bilinear transformation ($s \rightarrow 1/\omega$), and to study the resulting rational matrix $G(1/\omega)$ at the specific frequency $\omega = 0$. To this end a M.F.D. of $G(1/\omega)$ is utilised since it is well-known that the denominator matrix in the M.F.D. describes the pole structure of $G(1/\omega)$ and that the numerator matrix describes the zero structure. While to a large extent the M.F.D. of $G(1/\omega)$ can be suggested by the available M.F.D. of $G(s)$, it is often difficult to verify completely the required relative primeness condition although it is usually simple to verify the full rank requirement at certain frequencies. The question thus arises as to which parts of the complete information carried by the M.F.D. are necessary for the investigation in hand.

The proposed paper takes its lead from the above problem and provides a highly localised approach to linear systems theory. Specifically an analysis of the system structure at a single frequency $s = s_0$ is presented to the exclusion of all other frequencies$\in \mathcal{C} \cup \{\infty\}$. As such the paper follows the matrix theory ideas of Van Dooren et al. (1979) although the development presented is more in the spirit of Vardulakis et al. (1982). Pernebo (1978; 1981a; 1981b) has in certain respects adopted a rather more general approach, as will be indicated.

The relevant results from matrix theory and linear systems theory are presented. Among the latter is an interesting and particularly useful local state space form realisation of a given transfer function matrix. This realisation enables a result for proper systems, concerning the invariance of the infinite zero structure under state feedback (originally due to Vardulakis(1980)), to be generalised to non-proper systems. The approach also reveals a surprising property of the infinite pole structure. The final part of the paper examines local state feedback (in the context of the above local state-space realisation) and will determine its invariants.

2 Preliminaries

Let F be a field (usually assumed to be $\mathcal{R}$, or $\mathcal{C}$) and let $F[s]$ denote the ring of polynomials and $F(s)$ the field of rational functions in the indeterminate s with coefficients taken from F.

Definition 1 *A rational function $g(s) \in F(s)$ is said to be PROPER AT s_0 if*

$$\lim_{s \to s_0} g(s) \tag{2.1}$$

exists. Additionally $g(s)$ is called STRICTLY PROPER AT s_0 if this limit is zero, and BIPROPER AT s_0 if it is not zero.

If $g(s) \in F(s)(g(s) \not\equiv 0)$, it follows immediately that $g(s)$ may be written uniquely as

$$g(s) = (s - s_0)^k g^*(s) \tag{2.2}$$

where $g^*(s)$ is biproper at s_0 and k is some integer. It therefore follows that $g(s)$ has a pole at s_0 if $k < 0$ while $g(s)$ has a zero at s_0 if $k > 0$. Thus in these terms $g(s)$ is proper at s_0 iff it has no poles at s_0, and it is biproper at s_0 iff additionally it has no zeros at s_0. The integer k as determined by the factorisation (2.2) will be called the DEGREE OF $g(s)$ AT s_0.

The above ideas are extended to the case of rational matrices in the following way.

Definition 2 *$G(s) \in F^{m \times l}(s)$ is said to be PROPER AT s_0 iff*

$$\lim_{s \to s_0} G(s) \tag{2.3}$$

exists. Additionally $G(s)$ is said to be STRICTLY PROPER AT s_0 if this matrix limit is zero, and BIPROPER AT s_0 if it is nonsingular.

Definition 3 *Two $m \times l$ rational matrices $G_1(s)$, $G_2(s)$ are said to e EQUIVALENT AT s_0 iff $\exists$ rational matrices $M(s)$, $N(s)$ both biproper at s_0 such that*

$$G_2(s) = M(s)G_1(s)N(s) \tag{2.4}$$

It is possible to define the same equivalence in terms of the following elementary operations:
(E.O.) (i) interchange any two rows (resp. columns)
(E.O.) (ii) multiply any row (resp. column) by a nonzero rational functions which

is biproper at s_0
(E.O.) (iii) add to any row (resp. column) a multiple by any rational function which is proper at s_0, of any other row (resp. column).

The effect that these operations have on a given matrix may be represented by pre–(post–) multiplication of the given matrix by an appropriate rational matrix, biproper at s_0. Conversely multiplication of the given matrices by a rational matrix which is biproper at s_0 can be represented by a sequence of elementary operations performed on that matrix. thus it can be said that two rational matrices are equivalent at s_0 in case one can be obtained from the other by a sequence of the above elementary operations.

Under the above notion of equivalence there is a canonical form for any given rational matrix. This was originally noted by Verghese (1978) and is described as follows:

Theorem 1 *An $m \times l$ rational matrix $G(s)$ of rank ρ is equivalent at s_0 to a matrix $M_{s_0}(G)$ where*

$$M_{s_0}(G) = \begin{cases} (Q(s), Q_{m,l-m}), & m < l \\ Q(s), & m = l \\ \begin{pmatrix} Q(s) \\ Q_{m-l,m} \end{pmatrix}, & m > l \end{cases} \tag{2.5}$$

and

$$Q(s) = diag[(s-s_0)^{q_1}, \cdots, (s-s_0)^{q_\rho}, 0, \cdots, 0] \tag{2.6}$$

with

$$q_1 \leq \cdots \leq q_k \leq 0 \leq q_{k+1} \leq \cdots \leq q_\rho \tag{2.7}$$

The matrix $M_{s_0}(G)$ as defined in Theorem 1 will be called the SMITH – MACMILLAN FORM OF $G(s)$ AT s_0.

Clearly from (2.6) it can be seen that those elements $(s - s_0)^{q_i}$ for $q_i < 0$ determine the pole structure of $G(s)$ ats_0 , while those elements $(s - s_0)^{q_i}$ for which $q_i > 0$ describe the zero structure at s_0. Thus the essential invariants of the transformation "equivalence at s_0" are those detailing the pole and zero structure at s_0. As can be seen the pole/zero structure at no other frequency (including the point at infinity) is preserved.

The systems theory transformation arising from the notion of equivalence at s_0 is as follows. Let

$$P_i(s) = \left(\begin{array}{c|c} T_i(s) & U_i(s) \\ \hline -V_i(s) & W_i(s) \end{array} \right) \quad (i = 1, 2) \tag{2.8}$$

be two $(p+n) \times (p+l)$ rational systems matrices. The ORDER OF $P_i(s)$ AT s_0 will be defined as the degree at s_0 of the rational function $|T_i(s)|$.

Definition 4 *$P_1(s)$, $P_2(s)$ are said to be SYSTEM EQUIVALENT AT s_0 if $\exists$ rational matrices $M(s)$, $N(s)$ which are biproper at s_0 and rational matrices $X(s)$,*

$Y(s)$ which are proper at s_0, such that

$$\left(\begin{array}{c|c} M(s) & 0 \\ \hline X(s) & I \end{array}\right)\left(\begin{array}{c|c} T_1(s) & U_1(s) \\ \hline -V_1(s) & W_1(s) \end{array}\right)\left(\begin{array}{c|c} N(s) & Y(s) \\ \hline 0 & I \end{array}\right) = \left(\begin{array}{c|c} T_2(s) & U_2(s) \\ \hline -V_2(s) & W_2(s) \end{array}\right) . \quad (2.9)$$

It is noted that an alternative phrasing of this definition is possible, in terms of elementary operations if the elementary operations (E.O.) (i)-(iii) are restricted in the same manner as those of unimodular equivalence are used to generate the elementary operations of strict system equivalence (Rosenbrock,1970). As can be seen from the following result, equivalence at s_0 is a system transformation which in terms of generality lies somewhere between the polynomially based strict system equivalence and the transformation of system equivalence (ibid) which is based on general rational functions. the important point concerning Definition 4 is that the transforming matrices should contain no elements possessing poles at s_0. Thus under this transformation it will be possible to remove decoupling zeros from the system matrix provided these are not located at s_0. it then follows that the essential invariants of the transformation of system equivalence at s_0 are those summarised by the order at s_0 and the system transfer function matrix (see Pugh et al., 1988).

3 Local structure results for linear systems

The above ideas relate to general rational system matrices although the main context in which they will be applied will be that of rational system matrices which are proper at s_0. These matrices in relation to the ring of rational functions which are proper at s_0, are analogous to the polynomial system matrices of conventional linear systems theory. A result fundamental to much of what follows concerns the existence and structure of (local) state space realisations of such system matrices. These are described as follows.

Theorem 2 *Let*

$$P(s) = \left(\begin{array}{c|c} T(s) & U(s) \\ \hline -V(s) & W(s) \end{array}\right) \quad (3.1)$$

be a $(p+m) \times (p+l)$ rational system matrix which is proper at s_0 and of order n at s_0 $(p \geq n)$. If the associated transfer function $G(s)$ has Laurent expansion about s_0

$$G(s) = G_{s_0}(s) + Q(s) \quad (3.2)$$

where $G_{s_0}(s)$ denotes the Principal Part of this expansion, then $P(s)$ is system equivalent at s_0 to a rational system matrix of the form

$$P'(s) = \left(\begin{array}{cc|c} I_{p-n} & 0 & 0 \\ 0 & sI_n - A & B \\ \hline 0 & -C & Q(s) \end{array}\right) \quad (3.3)$$

where A, B, C are constant matrices, and A has eigenvalues only at s_0.

Proof. Multiply each row in the first block row of $P(s)$ of (3.1) by its least common denominator. Operating similarly with the first block column and noting that these operations are permitted under equivalence at s_0, a matrix of the form (3.1) results in which $T(s), U(s), V(s)$ are polynomial and $W(s)$ is unchanged and therefore still proper at s_0. With $P(s)$ in such a form it may then by conventional strict system equivalence be brought to the trivially extended state–space form

$$P_1(s) = \left(\begin{array}{ccc|c} I_{p-n_1} & 0 & 0 & 0 \\ 0 & sI_{n_1-n} - A_1 & 0 & B_1 \\ 0 & 0 & sI_n - A_2 & B_2 \\ \hline 0 & -C_1 & -C_2 & D(s) \end{array}\right) \tag{3.4}$$

where $diag(A_1, A_2)$ is in some canonical form (say, first natural normal form), with A_1 having no eigenvalues at s_0 while A_2 has eigenvalues only at s_0 ,and where $D(s)$ remains proper at s_0.

Now $sI_{n_1-n} - A_1$ is biproper at s_0, and so by system equivalence at s_0, $P(s)$ may be reduced successively to the forms

$$\left(\begin{array}{ccc|c} I_{p-n_1} & 0 & 0 & 0 \\ 0 & I_{n_1-n} & 0 & B_1 \\ 0 & 0 & sI_n - A_2 & B_2 \\ \hline 0 & -C_1 & -C_2 & D(s) \end{array}\right) \rightarrow \left(\begin{array}{ccc} I_{p-n} & 0 & 0 \\ 0 & sI_n - A_2 & B_2 \\ 0 & -C_2 & R(s) \end{array}\right) \tag{3.5}$$

where

$$R(s) \triangleq D(s) + C_1(sI_{n_1-n} - A_1)^{-1}B_1 \tag{3.6}$$

is proper at s_0.

Now (3.5) and the original $P(s)$ are seen to be system equivalent at s_0 and so they have the same transfer function matrix,

$$\text{i.e. } G(s) \equiv C_2\,(sI_n - A_2)^{-1}\,B_2 + R(s). \tag{3.7}$$

However the first matrix on the R.H.S., of (3.7) has only poles at s_0 while $R(s)$ is proper at s_0, hence by the uniqueness of the Laurent expansion it follows that

$$C_2\,(sI_n - A_2)^{-1}\,B_2 \equiv G_{s_0}(s) \tag{3.8}$$

and more specifically that

$$R(s) \equiv Q(s) \tag{3.9}$$

which establishes the theorem.

In conventional linear system theory strict system equivalence is the basic relationship between least order system matrices giving rise to the same transfer function matrix. In the present context system equivalence at s_0 can be shown to play the same role as follows.

Definition 5 *Two rational matrices $T(s)$, $U(s)$ of respective dimensions $p \times p$ and $p \times l$ and which are proper at s_0 are said to be relatively left prime at s_0 iff*

$$RANK[T(s_0)\ U(s_0)] = p. \tag{3.10}$$

Relative right primeness at s_0 is similarly defined.

Definition 6 *The $(p+m) \times (p+l)$ rational system matrix*

$$P(s) = \left(\begin{array}{c|c} T(s) & U(s) \\ \hline -V(s) & W(s) \end{array} \right) \tag{3.11}$$

which is proper at s_0 is said to have LEAST ORDER AT s_0 iff $T(s)$, $U(s)$(resp.$T(s)$, $-V(s))^T$) are relatively left (resp. right) prime at s_0.

Theorem 3 *Let $P_1(s), P_2(s)$ be two $(p+m)\times(p+l)$ rational system matrices which are proper at s_0 and have least order at s_0. Then $P_1(s), P_2(s)$ gives rise to same transfer function matrix $G(s)$ iff they are system equivalent at s_0.*

Proof. See Pugh et al. (1988)

As expected the above results lead in a natural way to the basic structure theorem relating the structure of the transfer function matrix at s_0 to that of the system matrix.

Theorem 4 *Let the $(p+m) \times (p+l)$ rational system matrix $P(s)$ of (3.11) be proper at s_0 and have least order at s_0. Suppose $P(s)$ gives rise to the transfer function matrix $G(s)$, and that $G(s)$ has the Smith–McMillan form at s_0 described in Theorem 1. Then the Smith-McMillan form at s_0 of $P(s)$ is*

$$M_{s_0}(P) = \begin{pmatrix} I_p & 0_{p,m} \\ 0_{m,p} & diag(\phi_i) \end{pmatrix} \quad (l=m) \tag{3.12}$$

where

$$\phi_i(s) \triangleq \begin{cases} 1, & if\ q_i \le 0,\ i \le \rho \\ (s-s_0)^{q_i}, & if\ q_i > 0,\ i \le \rho \\ 0, & if\ \rho < i \le m \end{cases} \tag{3.13}$$

(with obvious changes if $l \neq m$). Further the Smith–McMillan form at s_0 of $T(s)$ is

$$M_{s_0}(T) \triangleq \begin{pmatrix} I_{p-m} & 0_{p-m,m} \\ 0_{m,p-m} & diag(\psi_{m-i+1}) \end{pmatrix} \tag{3.14}$$

where

$$\psi_i(s) \triangleq \begin{cases} (s-s_0)^{-q_i}, & if\ q_i < 0, i \le \rho \\ 1, & if\ q_i \ge 0\ or\ \rho < i < m \end{cases} \tag{3.15}$$

Proof. See again Pugh et al. (1988).

The above result thus shows that system matrices with least order at s_0 describe the structure at s_0 of the associated transfer function matrix in a completely irredundant manner.

4 Some properties of state and local state feedback

Let $G(s)$ be a general $m \times l$ rational matrix and let $P(s)$ be a conventional state space realisation of $G(s)$. Thus

$$P(s) = \left(\begin{array}{c|c} sI_n - A & B \\ \hline -C & D(s) \end{array} \right) \tag{4.1}$$

where $D(s)$ is an $m \times l$ polynomial matrix and A, B, C are constant matrices of the appropriate dimensions, and

$$G(s) = C\,(sI_n - A)^{-1}\,B + D(s). \tag{4.2}$$

Consider the state feedback control law

$$u = Fx + H\nu \tag{4.3}$$

where F is an $l \times n$ constant matrix and H an $l \times l$ constant nonsingular matrix. Let $G_F(s)$ denote the closed-loop transfer function matrix corresponding to (4.3) and (4.1), then

$$G_F(s) = (C + D(s)_F)(sI - A - BF)^{-1}BH + D(s)H. \tag{4.4}$$

We are interested in determining the effect that the state feedback controller (4.3) has on the infinite frequency structure of $G(s)$. In this respect, recall that the infinite zero (resp. pole) structure of $G(s)$ and $G_F(s)$ is defined as the finite zero (resp. pole) structure of $G(\frac{1}{\omega})$ and $G_F(\frac{1}{\omega})$ at $w = 0$. (Verghese, 1978; Pugh and Ratcliffe, 1979). Now

$$G\left(\frac{1}{\omega}\right) = C\left(\frac{1}{\omega}I_n - A\right)^{-1} B + D\left(\frac{1}{\omega}\right) \tag{4.5}$$

where $D(\frac{1}{\omega})$ is a rational matrix in ω which possesses only poles at $\omega = 0$ (since $D(s)$ is polynomial). Let therefore

$$D\left(\frac{1}{\omega}\right) = C_1\,(\omega I - A_1)^{-1}\,B_1 + D_1 \tag{4.6}$$

be a conventional state space realisation of $D(\frac{1}{\omega})$ having least order at $\omega = 0$. Note that D_1 is a constant matrix. It then follows from (4.6) and (4.5) that

$$G\left(\frac{1}{\omega}\right) = C\,(I - A\omega)^{-1}\,B\omega + C_1\,(\omega I - A_1)^{-1}\,B_1 + D_1. \tag{4.7}$$

Notice from (4.7) that $G(\frac{1}{\omega})$ may be viewed as the parallel connection of the two transfer function matrices $C(I - A\omega)^{-1}B\omega$ and $D(\frac{1}{\omega})$, thus a polynomial system matrix realisation of $G(\frac{1}{\omega})$ may be written as

$$P'(\omega) = \left(\begin{array}{cc|c} I - A\omega & 0 & B\omega \\ 0 & \omega I - A_1 & B_1 \\ \hline -C & -C_1 & D_1 \end{array}\right). \tag{4.8}$$

Now since the realisation (4.6) has least order at $w = 0$, it is clear that $P'(\omega)$ is a realisation of $G(\frac{1}{\omega})$ with least order at $w = 0$. It will not be necessary to establish that (4.8) has least order in the conventional sense in view of the local theory established previously.

Consider now the infinite frequency structure of the feedback system (4.4). Now

$$G_F\left(\frac{1}{\omega}\right) = \left(C + D\left(\frac{1}{\omega}\right)F\right)(I - A\omega - BF\omega)^{-1}BF\omega + D\left(\frac{1}{\omega}\right)H \tag{4.9}$$

and using (4.6) this become

$$\begin{aligned} G_F\left(\frac{1}{\omega}\right) = {} & \left(C + C_1(\omega I - A_1)^{-1}B_1F + D_1F\right)(I - A\omega - BF\omega)^{-1}BH\omega \\ & + \left(C_1(\omega I - A_1)^{-1}B_1 + D_1\right)H. \end{aligned}$$

After some manipulation (4.9) may be written as

$$\begin{aligned} G_F\left(\frac{1}{\omega}\right) = {} & \left(C_1(\omega I - A_1)^{-1}B_1 + D_1\right)\{F(I - A\omega - BF\omega)^{-1}BH\omega + H\} \\ & + C(I - A\omega - BF\omega)^{-1}BH\omega. \end{aligned} \tag{4.10}$$

From (4.10) it can be seen that $G_F\left(\frac{1}{\omega}\right)$ may be viewed as the parallel connection of two transfer functions, the first being $C(I - A\omega - BF\omega)^{-1}BH\omega$ the second being the serial connection of two other transfer function matrices. This observation is an aid in the construction of the following realisation of $G_F\left(\frac{1}{\omega}\right)$. From (4.10)

$$G_F\left(\frac{1}{\omega}\right) = \begin{pmatrix} C & C_1 & D_1F \end{pmatrix}$$

$$\begin{pmatrix} (I - A\omega - BF\omega)^{-1} & 0 & 0 \\ 0 & (\omega I - A_1)^{-1} & (\omega I - A_1)^{-1}B_1F(I - A\omega - BF\omega)^{-1} \\ 0 & 0 & (I - A\omega - BF\omega)^{-1} \end{pmatrix}$$

$$\times \begin{pmatrix} BH\omega \\ B_1H \\ BH\omega \end{pmatrix} + D_1H.$$

i.e.

$$G_F\left(\frac{1}{\omega}\right) = \begin{pmatrix} C & C_1 & D_1F \end{pmatrix} \begin{pmatrix} I - A\omega - BF\omega & 0 & 0 \\ 0 & \omega I - A_1 & -B_1F \\ 0 & 0 & I - A\omega - BF\omega \end{pmatrix}^{-1} \begin{pmatrix} BH\omega \\ B_1H \\ BH\omega \end{pmatrix} + D_1H. \quad (4.11)$$

Hence a polynomial system matrix realisation of $G_F(\frac{1}{\omega})$ is seen to be

$$P_F'(\omega) = \left(\begin{array}{ccc|c} I - A\omega - BF\omega & 0 & 0 & BH\omega \\ 0 & \omega I - A_1 & -B_1F & B_1H \\ 0 & 0 & I - A\omega - BF\omega & BH\omega \\ \hline -C & -C_1 & -D_1F & D_1 \end{array}\right). \quad (4.12)$$

Because of the complicated nature of (4.12) it is not possible to establish that $P_F'(\omega)$ has least orde in the usual sense. Recall however that is only the infinite frequency structure of $G_F(s)$ or what is equivalent, the structure of $G_F\left(\frac{1}{\omega}\right)$ at $\omega = 0$, that is of interest. In this respect and in view of the theorem developed above it will be sufficient to verify that (4.12) has least order at $\omega = 0$. in fact this follows quite readily from the form of (4.12) and the least order property of (4.6). We are now able to establish the following result.

Theorem 5 *The infinite frequency structure of $G(s)$ is invariant under state variable feedback.*

Proof. Since (4.12) has least order at $\omega = 0$ it follows from Theorem 4 that the infinite zero structure of $G_F(s)$ is the zero structure of $P_F'(\omega)$ at $\omega = 0$. Now $P_F'(\omega)$ may be written as

$$P_F'(\omega) = L(\omega)P_E(\omega)R(\omega) \quad (4.13)$$

where

$$L(\omega) = \begin{pmatrix} I - A\omega - BF\omega & I & 0 & 0 \\ 0 & 0 & I & 0 \\ 0 & I & 0 & 0 \\ -C & 0 & 0 & I \end{pmatrix}$$

$$R(\omega) = \begin{pmatrix} I & 0 & -I & 0 \\ 0 & 0 & I & 0 \\ 0 & I & 0 & 0 \\ 0 & 0 & -F & H \end{pmatrix} \quad (4.14)$$

$$P_E(\omega) = \begin{pmatrix} I & 0 & 0 & 0 \\ 0 & I - A\omega & 0 & B\omega \\ 0 & 0 & \omega I - A_1 & B_1 \\ 0 & -C & -C_1 & D_1 \end{pmatrix} \quad (4.15)$$

Clearly $L(\omega)$ and $R(\omega)$ are biproper at $\omega = 0$ and so from (4.13) it can be seen that $P'_F(\omega)$ is equivalent at $\omega = 0$ to $P_F(\omega)$. However note that $P_E(\omega)$ is merely a trivial expansion of $P'(\omega)$. Hence $P'_F(\omega)$ and $P'(\omega)$ have the same Smith–McMillan form at $\omega = 0$ modulo a trivial expansion. Now $P'(\omega)$ has least order at $\omega = 0$ and gives rise to $G(\frac{1}{\omega})$, hence from Theorem 4 the zero structure of $P'(\omega)$ at $\omega = 0$ is the zero structure of $G\left(\frac{1}{\omega}\right)$ at $\omega = 0$.

Thus the zero structures of $G_F\left(\frac{1}{\omega}\right)$ and$G\left(\frac{1}{\omega}\right)$ at $\omega = 0$ are identical i.e. the infinite zero structures of $G_F(s)$ and $G(s)$ are identical.

Consider now the infinite pole structure of $G_F(s)$ or equivalently the pole structure of $G_F\left(\frac{1}{\omega}\right)$ at $\omega = 0$. Since $P'_F(\omega)$ is a realisation of $G_F\left(\frac{1}{\omega}\right)$ with least order at $\omega = 0$, it follows from Theorem 4 that this zero structure is given by the zero structure at $\omega = 0$ of

$$T_F(\omega) \triangleq \begin{pmatrix} I - A\omega - BF\omega & 0 & 0 \\ 0 & \omega I - A_1 & -B_1 F \\ 0 & 0 & I - A\omega - BF\omega \end{pmatrix}. \tag{4.16}$$

This can be written as

$$T_F(\omega) = L'(\omega) T_E(\omega) R'(\omega) \tag{4.17}$$

where

$$L'(\omega) = \begin{pmatrix} I - A\omega - BF\omega & 0 & 0 \\ 0 & I & -B_1 F \\ 0 & 0 & I - A\omega - BF\omega \end{pmatrix} \tag{4.18}$$

$$R'(\omega) = \begin{pmatrix} (I - A\omega)^{-1} & 0 & 0 \\ 0 & I & 0 \\ 0 & 0 & I \end{pmatrix} \tag{4.19}$$

$$T_E(\omega) = \begin{pmatrix} I - A\omega & 0 & 0 \\ 0 & \omega I - A_1 & 0 \\ 0 & 0 & I \end{pmatrix} \tag{4.20}$$

Clearly $L'(\omega)$ and $R'(\omega)$ are biproper at $\omega = 0$ and so from (4.17), $T_F(\omega)$ and $T_E(\omega)$ are equivalent at $\omega = 0$. However from (4.19) and (4.18) $T_E(\omega)$ is simply a trivial expansion of

$$T(\omega) \triangleq \begin{pmatrix} I - A\omega & 0 \\ 0 & \omega I - A_1 \end{pmatrix}. \tag{4.21}$$

Hence $T_F(\omega)$ and $T(\omega)$ have the same Smith-McMillan form at $\omega = 0$ which establishes the invariance of the infinite pole structure of $G(s)$.

The invariance of the infinite zero structure under state feedback was first established by Vardulakis (1980), but only in the case of proper transfer function matrices. Ferreira (1982) considered the extension of this result to the non-proper case, however there was an added assumption that the transfer function matrix should have full normal rank. The above result thus represents the complete extension of the Vardulakis result to the case of general rational transfer function matrices.

The invariance of the infinite pole structure under state feedback is altogether more surprising, since it is well-known that this structure is not invariant under constant output feedback. this occurrence can however be explained to some extent by the fact that the states used in the control scheme (4.3) is determined from entirely finite frequency considerations and as such can only be expected to change that part of the system to which it relates. This observation leads naturally to the interesting question of what remains invariant if a state feedback controller is implemented on the local state pace realisation (3.3). In this case the" state" contains only information relating to the frequency s_0 and if the above observation is correct then only the structure of $G(s)$ at s_0 should change under this action. Consider therefore the local state space realisation of $G(s)$ given by

$$P(s) = \begin{pmatrix} sI - A & B \\ -C & Q(s) \end{pmatrix} \tag{4.22}$$

Where A has eigenvalues only at s_0 while as in (3.3)$Q(s)$ has no poles ats_0. Assume that (4.22) has least orde at s_0 so that the pole structure of $G(s)$ at s_0 is realised by (4.22) in the non-redundant manner of Theorem 4.

Consider now the " local" state feedback controller

$$u = Fx + H\nu \tag{4.23}$$

where the term "state" is used in the context of (4.22). Thus this notion of state contains information relating to s_0 alone and no other frequency. Applying (4.23) gives rise to the transfer function matrix

$$G_F(s) = (C + Q(s)F)(sI - A - BF)^{-1}BH + Q(s)H \tag{4.24}$$

i.e.

$$G_F(s) = C(sI - A - BF)^{-1}BH + Q(s)\{F(sI - A - BF)^{-1}BH + H\}. \tag{4.25}$$

In order to quantify the effect of the feedback (4.23) on the structure of $G(s)$ it is necessary to realise $Q(s)$ in the least order form

$$Q(s) = C_1(sI - A_1)^{-1}B_1 + D_1(s) \tag{4.26}$$

where $D_1(s)$ is polynomial, and A_1 has no eigenvalues at s_0. The reason for using this form is that it brings the remaining finite frequency pole structure of $G(s)$ into view. What is hidden by (4.26) and the subsequent analysis is the infinite pole structure, (as described by $D_1(s)$) which is known to be invariant from the above.

Substituting (4.26) into (4.25) gives

$$\begin{aligned} G_F(s) &= C(sI-A-BF)^{-1}BH \\ &\quad +\{C_1(sI-A_1)^{-1}B_1+D_1(s)\}\{F(sI-A-BF)^{-1}BH+H\} \\ &= \begin{pmatrix} C & C_1 & D_1(s)F \end{pmatrix} \end{aligned}$$

$$\begin{pmatrix} (sI-A-BF)^{-1} & 0 & 0 \\ 0 & (sI-A_1)^{-1} & (sI-A_1)^{-1}B_1F(sI-A-BF)^{-1} \\ 0 & 0 & (sI-A-BF)^{-1} \end{pmatrix}$$

$$\times \begin{pmatrix} BH \\ B_1H \\ BH \end{pmatrix} + D_1(s)H.$$

Hence a polynomial system matrix realisation of $G_F(s)$ is

$$P_F(s) = \left(\begin{array}{ccc|c} sI-A-BF & 0 & 0 & BH \\ 0 & sI-A_1 & -B_1F & B_1H \\ 0 & 0 & sI-A-BF & BH \\ \hline -C & -C_1 & -D_1(s)F & D_1(s)H \end{array}\right)$$

removing the input decoupling zeros in an obvious way gives the polynomial system matrix realisation.

The reason for this is that it enables the remaining finite frequency pole structure of $G(s)$ to be brought into view. What will remain hidden in (4.26) and the subsequent analysis is the infinite pole structure (as described by $D_1(s)$ which is known to be invariant from the above. Substituting (4.26) into (4.22) yields the following state space realisation of the open loop $G(s)$

$$P'(s) = \left(\begin{array}{cc|c} sI-A & 0 & B \\ 0 & sI-A_1 & B_1 \\ \hline -C & -C_1 & D_1(s) \end{array}\right) \tag{4.27}$$

while substituting (4.26) into (4.25) gives a

$$\begin{aligned} G_F(s) &= C(sI-A-BF)^{-1}BH \\ &\quad +\{C_1(sI-A_1)^{-1}B_1+D_1(s)\}\{F(sI-A-BF)^{-1}BH+H\} \\ &= \begin{pmatrix} C & C_1 & D_1(s)F \end{pmatrix} \end{aligned}$$

$$\begin{pmatrix} (sI-A-BF)^{-1} & 0 & 0 \\ 0 & (sI-A_1)^{-1} & (sI-A_1)^{-1}B_1F(sI-A-BF)^{-1} \\ 0 & 0 & (sI-A-BF)^{-1} \end{pmatrix}$$

$$\times \begin{pmatrix} BH \\ B_1 H \\ BH \end{pmatrix} + D_1(s)H. \tag{4.28}$$

Hence a polynomial system matrix realisation of $G_F(s)$ is

$$P_F(s) = \left(\begin{array}{ccc|c} sI - A - BF & 0 & 0 & BH \\ 0 & sI - A_1 & -B_1 F & B_1 H \\ 0 & 0 & sI - A - BF & BH \\ \hline -C & -C_1 & -D_1(s)F & D_1(s)H \end{array}\right) \tag{4.29}$$

This clearly does not have least order and so removing the input decoupling zeros in an obvious way gives

$$P'_F(s) = \left(\begin{array}{ccc|c} I & 0 & 0 & 0 \\ 0 & sI - A_1 & -B_1 F & B_1 H \\ 0 & 0 & sI - A - BF & BH \\ \hline -C & -C_1 & -C - D_1(s)F & D_1(s)H \end{array}\right). \tag{4.30}$$

Now $P'_F(s)$ has no input decoupling zeros but may possess output decoupling zeros (which is a well–know feature of state feedback). In fact this only occurs if the feedback is such as to produce zeros in $sI - A - BF$ which coincide with those of $sI - A_1$. In any event (4.30) represents a polynomial realisation (having no input decoupling zeros) of $G_F(s)$ which has th same (conventional) order as the least order realisation (4.25) of $G(s)$. Thus the zeros of

$$\begin{pmatrix} I & 0 & 0 \\ 0 & sI - A_1 & -B_1 F \\ 0 & 0 & sI - A - BF \end{pmatrix} \tag{4.31}$$

may be taken as the closed–loop locations of he open-loop poles under the local state feedback(4.23).

Now it is immediately clear from (4.31) that the zeros of $sI - A_1$ (representing the finite pole structure of $G(s)$ away from s_0) will always be present as zeros of (4.31) i.e. as poles of $G_F(s)$. in this respect it can be seen therefore that the local feedback (4.23) does not move any of the poles of $G(s)$ which are not located at s_0. The true effect of such feedback is thus to move only those poles of $G(s)$ which are located at s_0 . This then proves

Theorem 6 *The local state feedback (4.23) moves only those poles of the system which are located at s_0 and leaves all other poles unchanged.*

In that the local state is determined entirely from information about the system gathered only at the frequency s_0 , the result represents a restatement of the classical resut of poles invariant assignment (Porter & Crossley, 1972). That is, in order to change a single pole (of a system with distinct poles) and leave all other poles invariant then a feedback based on the left eigenvector corresponding to this eigenvalue of th plant matrix should be used.

5 Conclusions

A local approach to linear systems theory has been presented and local analogues of many standard results have been noted. Of particular interest are the local state space realisation and a local version of Rosenbrock's structure theorem. The main reason for the development of the theory is seen in the study of the infinite frequency structure of an open-loop transfer function matrix under state feedback. The theory also raises interesting questions about the effects of local state feedback and some preliminary results have been established.

References

[1] P. M. G. Ferreira. Invariance of infinite zeros under feedback. *Int. J. Control*, 1982.

[2] L. Pernebo. *Algebraic control theory for linear multivariable systems.* PhD thesis, Lund Inst. Technol, 1978.

[3] L. Pernebo. An algebraic theory for the design of controllers for linear multivariable systems–part i: Structure matrices and feed forward design. *IEEE Trans. Aut. Control.*, 1981a.

[4] L. Pernebo. An algebraic theory for the design of controllers for linear multivariable systems–part ii: Feedback realization and feedback design. *IEEE Trans. Aut. Control.*, 1981b.

[5] B. Porter and T. R. Corssley. *Model control: Theory and applications.* Taylor Francis, London, 1972.

[6] A. C. Pugh, A. Kafai, and G. E. Hayton. Local results in linear systems theory and implications. *Int. J. Control*, 1988. to appear.

[7] A. C. Pugh and P. A. Ratcliffe. On the zeros and poles of a rational matrix. *Int. J. Control*, 1979.

[8] H. H. Rosenbrock. *State space and multivariable theory.* Nelson–Wiley, London–New York, 1970.

[9] A. J. J. Van Der Weiden and O. H. Bosgra. The determination of structural properties of a linear multivariable system by operations of system similarity–2. nonproper systems. *Int. J. Control*, 1980.

[10] A. I. G. Vardulakis. On infinite zeros. *Int. J. Control*, 1980.

[11] A. I. G. Vardulakis, D. N. J. Limebeer and N. Karcanias. Structure and smith–mcmillan form of a rational matrix at infinity. *Int. J. Control*, 1982.

[12] G. C. Verghese. *Infinite frequency behaviour in generalised dynamical systems.* PhD thesis, Stanford University, Stanford, CA, 1978.

Stability analysis for discrete non-unit memory linear multipass processes

E. Rogers and D.H. Owens

Abstract

A new set of stability conditions for discrete non-unit memory linear multipass processes is derived and used to develop computationally feasible stability tests. This results in a systematic procedure which tests these conditions in a particular order with termination if the one just tested does not hold. Finally, the use of this procedure within the area of control systems design is discussed.

1 Introduction

The essential characteristic of a multipass process is the presence of a repetitive, or recursive, action with interaction between successive passes. Industrial examples include long-wall coal cutting and the rolling of metal strip. To introduce a formal definition, first suppose that the pass length α is constant. Then, Owens (1977), a multipass process is a dynamic system where the output vector $Y_k(t)$, $0 \leq t \leq \alpha < +\infty$, ($t$ ='distance' variable) generated during the kth pass acts as a forcing function on, and hence contributes to, the dynamics of the new output vector, or pass profile, $Y_{k+1}(t)$, $0 \leq t \leq \alpha + \infty$, $k \geq 0$.

In the most general case, a multipass process has nonlinear dynamics and a pass length which varies from pass to pass. Suppose, however, that the dynamics are linear and the pass length is constant. Then a large number of practical examples is known, Edwards and Owens (1982), to satisfy these conditions. Further, Owens (1977) has developed an abstract representation for this case which, in effect, regards the output on any pass as a point in a suitably chosen function space. This abstract representation includes as a special case the class known, Edwards and Owens (1982), as discrete non-unit memory linear multipass processes which are the subject of the work reported here.

Using elementary techniques from functional analysis, Owens (1977) has developed a stability theory for this abstract representation. In particular, a definition of stability which has a well defined physical meaning has been introduced and necessary and sufficient conditions for its existence developed. Further, the results of interpreting these conditions for a wide range of special cases have been reported by Owens (1977), Edwards and Owens (1982), and Rogers (1987). For the purposes of this paper, however, the most important of these are those for discrete non-unit memory linear multipass processes which can, for example, be found in Edwards and Owens (1982).

Consider, therefore, the problem of testing the reported conditions for stability of a discrete non-unit memory linear multipass process. Then it is immediately apparent that testing one of these conditions in its current form is not a computationally feasible proposition; a fact which prompts the search for equivalent conditions which are computationally feasible to test and one which forms the basis of this paper.

The main body of this paper begins in section 2 with the introduction of the relevant general results in terms of the abstract representation. In section 3 the state-space model of a discrete non-unit memory linear multipass process is introduced and the results of interpreting the general stability conditions for this special case are presented. Further, the computational feasibility of testing these conditions is discussed and an equivalent set derived to serve as the basis for the development of computationally feasible stability tests.

Section 4 considers the development of such tests. In particular, the theory of circle positivity of polynomial matrices is used to develop tests which are, in effect, well known tests from standard linear systems theory. Finally, the use of these new tests in the important area, Rogers (1987), of control systems design for multipass processes will be discussed with particular emphasis on computer aided design aspects.

2 Background results

Consider a unit memory multipass process with linear dynamics and a constant pass length. Then the unique features of such a process are, for example, summarised in chapter 2 of Rogers (1987). Suppose also that the pass profile on pass k is regarded as a point in a suitably chosen Banach space E_α, i.e.

$$Y_k \in E_\alpha, \; k \geq 0 \tag{2.1}$$

in which case the following definition, Owens (1977), provides an abstract representation, or model, for such processes.

Definition 2.1 *A linear multipass process $S(E_\alpha, W_\alpha, L_\alpha)$ of constant pass length $\alpha > 0$ consists of a Banach space E_α a linear subspace W_α of E_α and a bounded linear operator L_α of E_α into itself. The system dynamics are described by linear recursion relations of the form*

$$Y_{k+1} = L_\alpha Y_k + b_{k+1}, \; k \geq 0 \tag{2.2}$$

where $Y_k \in E_\alpha$ is the pass profile on pass k and $b_{k+1} \in W_\alpha$, $k \geq 0$. Here the term $L_\alpha Y_k$ represents the contribution from the kth pass to the $k+1$th pass profile and b_{k+1} represents known initial conditions, disturbances and control input effects.

$S(E_\alpha, W_\alpha, L_\alpha)$ includes all known examples as special cases and the details associated with a representative cross section can be found in chapter 2 of Rogers (1987).

To introduce the relevant elements of Owens (1977) stability theory for $S(E_\alpha, W_\alpha, L_\alpha)$ first note that this theory consists of two distinct concepts, termed asymptotic stability and stability along the pass. Asymptotic stability is a necessary condition for stability along the pass and it is known, Rogers (1987), that this stronger latter condition is required in practical applications. In order to state the necessary and sufficient conditions for stability along the pass, first introduce the following definition.

Definition 2.2 *Suppose that α_0 is a nominal value of pass length of interest. Then a collection of models of $S(E_\alpha, W_\alpha, L_\alpha)$ with pass lengths in the range $\alpha \geq \alpha_0$ is termed an extended linear multipass process and is denoted by $S(E_\alpha, W_\alpha, L_\alpha)_{\alpha \geq \alpha_0}$. Suppose also that $\|\cdot\|_\alpha$ denotes the norm in E_α. Then the required result is the following.*

Theorem 2.1 *$S(E_\alpha, W_\alpha, L_\alpha)_{\alpha \geq \alpha_0}$ is stable along the pass if, and only if,*

$$r_\infty = sup_{\alpha \geq \alpha_0} r(L_\alpha) < 1 \tag{2.3}$$

where $r(L_\alpha)$ denotes the spectral radius of L_α and

$$M_0 \triangleq sup_{\alpha \geq \alpha_0} sup_{|z| \geq \lambda} \|(zI - L_\alpha)^{-1}\|_\alpha < +\infty \tag{2.4}$$

for some real number λ in the range $r_\infty < \lambda < 1$.

Proof 1 *See the proof of theorem 3.1 in Owens (1977).*

Note 2.1 *(2.3) is equivalent to asymptotic stability $\alpha \geq \alpha_0$*

3 Stability of discrete processes

Multipass processes also exist, Rogers (1987), where the current pass profile is a function of the independent inputs to that pass and a finite number, $M(> 1)$ of previous pass profiles. Such processes are referred to as non–unit memory of memory length M. Consider also the case of linear dynamics and a constant pass length. Then such processes can be written in the form of definition 2.1, and hence theorem 2.1 is applicable, by use of a product space formulation as detailed in Rogers (1987).

In the remainder of this paper, attention is exclusively directed towards the sub–class known as discrete non–unit memory linear multipass processes whose defining state–space model has the form

$$X_{k+1}(p+1) = \Phi X_{k+1}(p) + \Delta U_{k+1}(p) + \sum_{j=1}^{M} \Delta_{j-1} Y_{k+1-j}(p)$$

$$Y_{k+1}(p) = CX_{k+1}(p) + D_0U_{k+1}(p) + \sum_{j=1}^{M} D_j Y_{k+1-j}(p)$$
$$X_{k+1}(p) \in R^n,\ Y_{k+1} \in R^m,\ U_{k+1}(p) \in R^l$$
$$0 \le p \le \alpha < +\infty,\ X_{k+1}(0) = d_{k+1},\ k \ge 0 \tag{3.1}$$

Suppose also that $\Delta_{j-1} = 0,\ D_j = 0,\ 1 \le j \le M$, the subscript $k+1$ is dropped, and the concept of a pass length is irrelevant. Then (3.1) reduces to the well known state–space model from standard discrete linear systems theory which in the multipass framework is termed, Rogers (1987), the derived conventional linear system. Further, it will be shown in section 4 that well established tests from the stability theory of this derived system can be used to test for multipass stability.

Application of theorem 2.1 to (3.1) now gives the following result.

Theorem 3.1 *Suppose that*

1. *the pair* $\{C, \Phi\}$ *is observable*
2. *the pair* $\{\Phi, \sum_{j=1}^{M} \Delta_{j-1}\eta^{j-1}\}$ *is controllable at all but a finite number of points* $\eta_1, \eta_2, \cdots, \eta_q$ *in the complex plane, and*
3. $|z_1 I_n - \Phi - \sum_{j=1}^{M} \Delta_{j-1}\eta_i^{j-1}\ P(\eta_i^{-1})^{-1}C|$ *has no roots on the unit circle in the complex plane,* $1 \le i \le q$, *where*

$$P(\eta) = \eta I_m - D_1 - D_2\eta^{-1} - \cdots - D_M\eta^{1-M} \tag{3.2}$$

Then the extended linear multipass process $S(E_\alpha, W_\alpha, L_\alpha)_{\alpha \ge \alpha_0}$ *generated by discrete models of the form (3.1) with* $\alpha \ge \alpha_0$ *is stable along the pass if, and only if,*

1. $$r_\infty = sup\ \{|z| : \ g(z) = 0\} < 1 \tag{3.3}$$

   where

   $$g(z) = |z^M I_m - D_1 z^{M-1} - D_2 z^{M-2} - \cdots - D_M| \tag{3.4}$$

   and

2. *there exists real numbers* $\varepsilon > 0$ *and* $r_\infty < \lambda < 1$ *such that*

   $$|z_1 I_n - \Phi - \sum_{j=1}^{M} \Delta_{j-1} z^{-j} Q(z)^{-1} C| \ne 0 \tag{3.5}$$

   for all complex numbers z_1, z *satisfying* $|z_1| \ge 1 - \varepsilon$ *and* $|z| \ge \lambda$ *where*

   $$Q(z) = I_m - D_1 z^{-1} - D_2 z^{-2} - \cdots - D_M z^{-M} \tag{3.6}$$

Proof 2 *See the proof of theorem 3.5.5 in Edwards and Owens (1982).*

Consider now the testing of (a) and (b) of theorem 3.1 for a given example. Then it follows immediately that (a) is, in effect, the standard stability result for the derived conventional linear system. Hence it can be tested by applying any one of numerous tests which avoid the need to compute the roots of $g(z)$. In the case of (b) this condition is equivalent to the requirement that all eigenvalues of the matrix $\Phi + \sum_{j=1}^{M} \Delta_{j-1} z^{-j} Q(z)^{-1} C$ have modulus strictly less than unity for all $|z| \geq \lambda$ A fact which leads immediately to the conclusion that testing this condition in its current form is not a computationally feasible proposition.

The following is the major result of this paper and, see section 4, leads to the development of computationally feasible stability tests. Its effective action is to replace (b) of theorem 3.1 by two equivalent conditions.

Theorem 3.2 *With the assumptions of theorem 3.1, the extended linear multipass process $S(E_\alpha, W_\alpha, L_\alpha)_{\alpha \geq \alpha_0}$ is stable along the pass if, and only if,*

1. *(3.3) of theorem 3.1 holds,*
2. $$\rho(z_1) = |z_1 I_n - \Phi| \neq 0, \ \forall |z_1| \geq 1 \tag{3.7}$$
   and
3. $$g(z_1, z) \neq 0, \forall |z_1| = 1, \ |z| \geq 1 \tag{3.8}$$
   where
   $$g(z_1, z) = |z^M I_m - G_1(z_1) z^{M-1} - G_2(z_1) z^{M-2} - \cdots - G_M(z_1)| \tag{3.9}$$
   and
   $$G_j(z_1) = C(z_1 I_n - \Phi)^{-1} \Delta_{j-1} + D_j, \ 1 \leq j \leq M \tag{3.10}$$

Proof 3 *This consists of showing that (3.7) and (3.8) are, together, equivalent to (b) of theorem 3.1.*

To proceed with this task, first let $|z| \to +\infty$ in (3.5) and hence (3.7). Further, with (3.7) established, note that

$$|z_1 I_n - \Phi - \sum_{j=1}^{M} \Delta_{j-1} z^{-j} Q(z)^{-1} C| = \frac{\rho(z_1)}{g(z)} g(z_1, z) \tag{3.11}$$

and hence (b) of theorem 3.1 reduces to

$$g(z_1, z) \neq 0, \forall |z_1| \geq 1 - \varepsilon, \ |z| \geq \lambda \tag{3.12}$$

Necessity now follows immediately on considering the case of $|z_1| = 1$ since (3.12) then implies that all solutions of $g(z_1, z) = 0$ have modulus strictly less than λ.

Conversely, suppose that (a)-(c) above hold and therefore $r_\infty < 1$ by (a). Further, consider the unit circle in the z_1 plane and let $z_j(z_1)$, $1 \leq j \leq N$, $N = mM$, *denote the jth root of $g(z_1, z)$. In which case it is possible to choose a real scalar $r_\infty < \lambda < 1$ such that all eigenvalues of Φ have modulus strictly less than λ and*

$$sup_{|z_1|=1} \max_{1 \leq j \leq N} |z_j(z_1)| < \lambda \tag{3.13}$$

Suppose also that z_1 traverses the unit circle in a clockwise manner. Then application of standard encirclement theorems immediately yields $|z_1 I_n - \Phi - \sum_{j=1}^{M} \Delta_{j-1} z^{-j} Q(z)^{-1} C| \neq 0$ for all complex numbers z_1, z satisfying $|z_1| \geq 1$, $|z| \geq \lambda$.

The final step in this proof is to demonstrate the existence of a suitable $\varepsilon > 0$. To accomplish this, first note again that $|z_1 I_n - \Phi - \sum_{j=1}^{M} \Delta_{j-1} z^{-j} Q(z)^{-1} C| \rightarrow \rho(z_1)$ as $|z| \rightarrow +\infty$. Hence (b) of theorem 3.1 can be replaced by the requirement that all eigenvalues of $\Phi + \sum_{j=1}^{M} \Delta_{j-1} z^{-j} Q(z)^{-1} C$ have modulus strictly less than unity for all z lying in some compact set $\lambda \leq |z| \leq R$ with R 'large'. The existence of a suitable $\varepsilon > 0$ now follows by continuity and the consequence existence of a finite covering of this set by open balls.

Theorems 3.1 and 3.2 invoke assumptions concerning the existence of uncontrollable oscillatory modes and, as yet, the precise reasons for these assumptions are not fully understood. Hence it will be necessary to assume for the remainder of this paper that these assumptions always hold. This may not always be the case, however, and hence this area is one to which further research effort could profitably be directed; a subject for which some promising preliminary results and observations can be found in Rogers (1987).

4 Stability tests

The purpose of this section is to develop computationally feasible tests for the conditions of theorem 3.2. In particular, the theory of circle positivity, Siljak (1975), is exploited to produce tests which are, in effect, well known tests from conventional linear systems theory.

Consider first, therefore, (a) of theorem 3.2 and note again that this is a necessary condition for stability along the pass and hence for (b) and (c) of this result. Consequently no further tests for the latter are required if this condition does not hole. Further, it was concluded in section 3 that if this condition is expressed in the form (3.3) then it can be tested using any one of numerous well known tests from the stability theory of the derived conventional linear system of (3.1).

Suppose, therefore, that (a) of theorem 3.2 holds and consider the condition listed under (b). Then this condition is necessary for stability along the pass and consequently no further tests are required if it does not hold. Further, the stability theory of the derived conventional linear system of (3.1) immediately indicates that it can be tested using any one of numerous well established tests.

At this stage, suppose that (a) and (b) of theorem 3.2 hold. Then the particular example under consideration is stable along the pass if, and only if, (c) holds. To develop a test for this condition first write $g(z_1, z) = 0$ in the form

$$z^N + c_{N-1}(z_1)z^{N-1} + \cdots + c_1(z_1)z + c_0(z_1) = 0 \tag{4.1}$$

where the coefficients $c_0(z_1), c_1(z_1), \cdots, c_{N-1}(z_1)$ are rational functions in z_1. Further, let $a_N(z_1)$ denote the least common denominator of the $c_j(z_1)$, $j = 0, 1, \cdots, N-1$. Then (4.1) can be written as

$$a_N(z_1)z^N + a_{N-1}(z_1)z^{N-1} + \cdots + a_1(z_1)z + a_0(z_1) = 0 \tag{4.2}$$

where the coefficients $a_0(z_1)$, $a_1(z_1), \cdots, a_N(z_1)$ are real polynomials in z_1. From which it follows immediately that this condition holds provided the, assumed irreducible, polynomial

$$\rho(z) = a_N(z_1)z^N + a_{N-1}(z_1)z^{N-1} + \cdots + a_1(z_1)z + a_0(z_1) \tag{4.3}$$

satisfies

$$\rho(z) \neq 0, \ \forall z_1 \in C = \{z_1 : |z_1| = 1\}, \ |z_1| \geq 1 \tag{4.4}$$

In order to test (4.4), first suppose that the $N \times N$ Schur–Cohn matrix $H = \{h_{ij}\}$ is constructed for $\rho(z)$ where

$$h_{ij} = \sum_{k=1}^{i} a_{N-i-k}\overline{a}_{N-j-k} - \overline{a}_{i-k}a_{j-k}, \ i \leq j \tag{4.5}$$

and is hence a polynomial in z_1 and/or its complex conjugate. Then it can be shown, Siljak (1975), that (4.4) holds if, and only if, the Hermitian polynomial matrix $H(z_1) \equiv H$ is positive definite $\forall z_1 \in C$. Such a matrix is said to be circle positive and the following result gives necessary and sufficient conditions for the existence of this property.

Theorem 4.1 *Then $N \times N$ Hermitian polynomial matrix $H(z_1)$ is circle positive if, and only if,*

1.

$$H(1) > 0 \tag{4.6}$$

and

2.

$$|H(z_1)| > 0, \ \forall z_1 \in C \tag{4.7}$$

Proof 4 *See the proof of theorem 2 in Siljak (1975).*

In order to test the conditions of theorem 4.1, first set $z_1 = 1$ in $H(z_1)$. Then $H(1)$ is simply a real constant coefficient symmetric matrix. Hence (4.6) can be tested by applying any one of numerous, Gantmacher (1959), sets of criteria for positive definiteness of real symmetric matrices. For (4.7), first note that $\overline{z}_1 = z_1^{-1}$

on C and hence, since the determinant of a hermitian matrix is real, $f(z_1) \triangleq |H(z_1)|$, $z_1 \in C$, must have the form

$$f(z_1) = \sum_{j=0}^{q} c_j(z_1^j + z_1^{-j}) \tag{4.8}$$

where the coefficients $c_0, c_1, \cdots, c_q$ are real scalars. Consequently (4.7) holds if, and only if,

$$f(z_1) > 0, \ \forall \ z_1 \in C \tag{4.9}$$

Further, this new condition can be expressed in terms of the roots of $f(z_1)$ by means of the following easily proven result.

Lemma 4.1 *The polynomial $f(z_1)$ satisfies (4.9) if, and only if, it has no roots on C and $f(z_1) > 0$ for some $z_1 \in C$.*

Note 4.1 *In the trivial case of $q = 0$, lemma 4.1 reduces to $f(z_1) \equiv c_0 > 0$ and no further analysis is required.*

The location of the roots of $f(z_1)$ with respect to C can now be determined as follows. First construct the polynomial $W(z_1)$ as

$$W(z_1) = z_1^q f(z_1) = \sum_{j=0}^{q} c_j(z_1^{q+j} + z_1^{q-j}) \tag{4.10}$$

and note that $W(z_1)$ has the same number of roots on C as $f(z_1)$ and of the same multiplicities. Hence the following serves as an alternative to lemma 4.1.

Lemma 4.2 *The polynomial $f(z_1)$ satisfies (4.9) if, and only if, $W(z_1)$ has no roots on C and $W(z_1) > 0$ for some $z_1 \in C$.*

Note 4.2 *$W(z_1) > 0$ for some $z_1 \in C$ can be replaced by either $W(1) > 0$ or $W(-1) > 0$, i.e.*

$$\sum_{j=0}^{q} c_j > 0 \tag{4.11}$$

or

$$(-1)^q \sum_{j=0}^{q} (-1)^j c_j > 0 \tag{4.12}$$

In particular, (4.11) and (4.12) are simple necessary conditions for (4.9) which should be tested before proceeding further.

The polynomial, $W(z_1)$ is an example of a self–inversive, or reciprocal, polynomial, Jury (1974), where this term is used to denote the fact that the reciprocals of roots inside C are also roots. Further, let $V(W)$ denote the number of roots of

$W(z_1)$ inside C. Then, by the self-inversive property, $W(z_1)$ also has $V(W)$ roots outside C and, since its degree is $2q$,

$$P(W) = 2(q - V(W)) \tag{4.13}$$

roots on C. Hence no roots of $W(z_1)$ on C requires that

$$V(W) = q \tag{4.14}$$

Finally, $V(W)$ can be computed using the Jury/Marden table in the manner described by Rogers (1987).

In summary, therefore, it has been shown that conditions (a) - (c) of theorem 3.2 for stability along the pass of (3.1) can be tested using tests which are, in effect, well known tests from discrete conventional linear systems theory. Further, it has been concluded that these conditions should be tested in the order of (a) followed by (b) followed by (c) with termination if the one just tested does not hold. This fact leads immediately to the following systematic procedure for determining the stability along the pass characteristics of (3.1).

step 1 Test the necessary asymptotic stability condition in the form of (3.3)-(3.4) by applying an appropriate test from the stability theory of the derived conventional linear system (3,1) where, in the context of this paper, the preferred test is the July/Marden table.

step 2 Test condition (b) by applying the July/Marden table to $\rho(z_1) = |z_1 I_n - \Phi|$.

step 3 Construct $p(z)$ of (4.3) and hence the Schur-Cohn matrix $H(z_1)$ of (4.5). Test if $H(1)$ is positive definite and stop if this is not the case.

step 4 Construct $W(z_1) = z_1^q f(z_1) = z_1^q |H(z_1)|$ of (4.10) and stop if (4.11) and (4.12) do not hold for this polynomial. Then use the July/Marden table to test (4.14).

At this stage, it is appropriate to consider the above test procedure in terms of its use in the important area, Rogers (1987), of control systems design and, in particular, CACSD (Computer Aided Control Systems Design) for (3.1). In which context, note that this procedure does not provide measures of relative stability or performance indicators. Further, it is clear that the constituent tests, in particular those for condition (c), are not suitable for software implementation. Hence this procedure is best suited to low order synthesis problems which, by analogy with the conventional linear systems case, can be expected to play a significant role in the initial development of this important area. Theorem 3.2 can, however, still be used to develop tests suitable for a CACSD environment by first reformulating conditions $(a) - (c)$ as eigenvalue problems. Complete details of these tests can be found in Rogers and Owens (1988b).

5 Conclusions

Motivated by the absence of computationally feasible tests, a new set of stability conditions for discrete non–unit memory linear multipass processes has been developed. Two of the resulting three conditions are expressed in terms of the fundamental result from the stability theory of discrete standard or, within the multipass context, conventional linear systems. In the case of the third, the theory of circle positivity of polynomial matrices has been used to show that this condition con also be tested using discrete conventional linear systems tests. This has led to a systematic computationally feasible procedure which tests these conditions in a particular order with termination if the one just tested does not hold.

The use of this procedure within a CACSD environment has been discussed. This has led to the conclusion that it is best suited to low order synthesis problems which are expected to play a significant role as benchmarks in the initial development of this important area.

References

[1] Edwards, J.B. and Owens, D.H. 1982. Analysis and Control of Multipass Processes. New York: Wiley.

[2] Jury, E.I. 1974. Inners and the Stability of Dynamic Systems. New York: Wiley.

[3] Gantmacher, F.R. 1959. Theory of Matrices Vol. I. New York: Chelsea.

[4] Owens, D.H. 1977. Stability of Linear Multipass Processes. Proc. I.E.E., 124, (11), pp1079-1082.

[5] Rogers, E. 1987. Feedback and Stability Theory for Linear Multipass Processes. Research Report, The Queen's University of Belfast, Department of Electrical and Electronic Engineering.

[6] Rogers, E. and Owens, D.H. 1988a. Poles and a Characteristic Polynomial for Discrete Linear Multipass Processes. In preparation.

[7] Rogers, E. and Owens, D.H. 1988a. Eigenvalue-based Stability Tests for Discrete Linear Multipass Processes. In preparation.

[8] Siljak, D.D. 1975. Stability Criteria for Two-Variable Polynomials. Proc. I.E.E.E. Trans. on Circuits and Systems., Vol. CAS-22, No. 3, pp185-189.

Total finite settling time stabilization for discrete–time single–input single–output systems

N. Karcanias E. Milonidis

Abstract

The problem of *Total Finite Settling Time Stabilization (TESTS)* for discrete–time, single–input, single–output, linear, time–invariant systems is investigated in this paper. The approach for its solution is purely algebraic and it reduces to the solution of a polynomial Diophantine equation

All the *Finite Settling Time (FST)* controllers are parameterized, the subfamily of the causal FST controllers is obtained and the necessary and sufficient conditions for the existence of stable FST compensators are derived. The controllers guarantee *internal* as well as *external* FST properties. The minimal design problem is considered and the parameterization of the family of the FST controllers according to their McMillan degree is obtained.

1 Introduction

The *Finite Settling Time Problem (FSTP)* and specifically the dead–beat response–ie the forcing of the state or output vector from any initial state to the origin in minimum time–is unique in discrete–time systems. [1]–[4]

Most of the work carried out until now has treated the case of state and output dead–beat regulation via state feedback, or in the case of inaccessible states via a combination of an observer and feedback of the state estimates. The main aim has been to shift all the poles (or almost all in the case of output dead–beat) of the plant to the origin and simultaneously to obtain an optimal eigenstructure. As the optimal solution is not unique specific types of dead–beat controllers are designed based on techniques varying from procedures on selection of independent vectors from a certain vector space to the solution of discrete Riccati equations. [3], [5]–[16]

The output dead–beat control particularly is also investigated under the influence of the geometric approach and minimal–time minimal–order dead–beat regulators have been proposed through state feedback or function observers. [17], [18]

V. Kucera has pioneered in state and output time–optimal control of both single–variable and multivariable discrete–time systems using polynomial algebra [19]–[24]. Many other researchers followed this approach [25], [26], with the most recent work of Y. Zhao and H. Kimura [27], where the problem of robustness of the dead–beat regulators is addressed and solved.

The present work does not directly deal with the dead–beat control, but with the rather more general problem of *Total Finite Settling Time Stabilization*. That is, all internal and external variables (signals) of the system are required to settle to a new steady–state after finite time from the application of a step change to its input and for every initial condition (see also [28]). The dead–beat control then, is a special problem of the FSTP.

A basic control scheme comprised by a precompensator and a unity feedback is used as it is the core design to a variety of control strategies [29], [30]. The powerful and elegant polynomial approach is employed for the solution of the FSTP and the whole design is reduced to the solution of a polynomial Diophantine equation which guarantees not only *internal stability* but also *internal (state) FST.*

A simple parameterization of the class of all FST as well causal FST controllers is derived and the necessary and sufficient conditions for the existence of stable FST controllers are obtained. The minimal design problem is defined and solved through the parameterization of the FST controllers according to their McMillan degree and the minimum order compensator is obtained as a solution to a standard linear algebra problem.

Finally, necessary and sufficient conditions are developed for the tracking and disturbance rejection of a class of inputs and disturbances respectively.

Through out this paper $\mathcal{R}$ is the field of real numbers, $\mathcal{R}[d]$ is the ring of polynomials with indeterminate d and $\mathcal{D}$ denotes the closed unit disc. The symbol $\mathcal{R}^{m\times n}[d]$ denotes the set of $m \times n$ polynomial matrices, $\mathcal{R}(d)$ is the field of rational functions in d and d stands for the delay operator $d = z^{-1}$.

2 Definition of the problem–preliminary results

Consider the closed–loop system of Fig. (2.1) where g_p, g_c are the transfer functions of the plant and the controller respectively and u_1 , u_2 are the externally applied inputs. Then, the problem of finite settling time is formulated as follows:

Definition 2.1 *The system of Fig. (2.1) exhibits a* Finite Settling Time *(FST) response, if for a step change in either u_1 or u_2 or both, all the signals e_1, e_2, y_1, y_2 settle to a new steady state value in finite time. The values of the finite settling time and of the steady state are left free.*

Lemma 2.1 [2] *A system characterized by a transfer function $g(d)$ exhibits an FST response iff $g(d)$ is a polynomial in d.*

Lemma 2.2 *Consider the closed–loop system of Fig. (2.1). $H(p,c)$ denote the transfer function between $\underline{e} = [e_1, e_2]^{\mathsf{T}}$ and $\underline{u} = [u_1, u_2]^{\mathsf{T}}$ and $W(p,c)$ the transfer function from $\underline{u}$ to $\underline{y} = [y_1, y_2]^{\mathsf{T}}$. Then [30]:*

$$H(p,c) \in \mathcal{R}^{2\times 2}[d] \text{ iff } W(p,c) \in \mathcal{R}^{2\times 2}[d] \tag{2.1}$$

where

$$H(p,c) = \begin{bmatrix} 1/(1+g_p g_c) & -g_p/(1+g_p g_c) \\ g_c/(1+g_p g_c) & 1/(1+g_p g_c) \end{bmatrix} \tag{2.2}$$

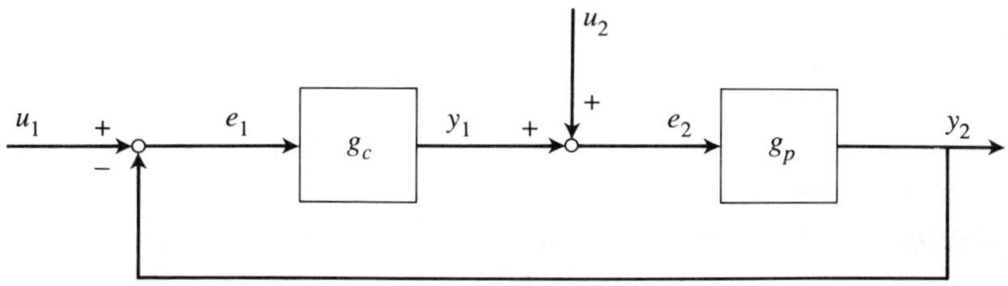

Figure 2.1. Basic control scheme.

Lemma 2.3 *The system of Fig. (2.1) exhibits an FST response iff* $H(p,c) \in \mathcal{R}^{2\times2}[d]$.

If the state-space representations of the plant and the controller are stabilizable and detectable, and the closed-loop system of Fig. (2.1) is well-posed $(1+g_p g_c \neq 0)$, then $H(p,c) \in \mathcal{R}^{2\times2}[d]$ implies that the closed-loop system is *internally stable* [30]. If in addition, the plant and the controller are controllable and observable, the overall system exhibits also *internal FST response* (all its states settle to a new steady-state in finite time for every initial state vector).

3 Parameterization of the FST controllers–causality

The solution to the FSTP and parameterization of all the controllers that guarantee total finite settling time response is examined in this section. The subfamily of all causal FST compensators is also derived and parameterized.

Theorem 3.1 *Let* $g_p = n_p/d_p$, $g_c = n_c/d_c$ *where* n_p, d_p, n_c, $d_c \in \mathcal{R}[d]$, n_p, d_p *are coprime and* n_c, d_c *are coprime, Then, the solution of the FSTP exists iff*

$$\delta(p,c) = d_p d_c + n_p n_c = c \in \mathcal{R} - \{0\} \tag{3.1}$$

Moreover, the family of all FST controllers is given by:

$$W(g_p) = \{(n_c, d_c) : n_c = \overline{n}_c + t d_p, d_c = \overline{d}_c - t n_p, t \in \mathcal{R}[d] \text{ and } \overline{d}_c - t n_c \neq 0\} \tag{3.2}$$

where $\overline{d}_c$, $\overline{n}_c$ *is a particular solution pair of Eqn. (3.1).*

Proof The proof of Th. (3.1) is similar to that given by Vidyasagar over the ring of proper and stable rational functions [30] and is omitted.

Remark 3.1 The FST controller pairs (n_c, d_c) are coprime for all the choices of the free parameter $t(d)$.

Remark 3.2 If the plant and controller are controllable and observable (ie g_p, g_c are true representations of them), $\delta(p,c) = c$ implies that the closed-loop system of Fig. (2.1):

1. is internally stable (in fact for this, g_p, g_c need only be stabilizable and detectable).
2. exhibits internal FST response because all its eigenvalues are shifted to the origin.

The stabilizing nature of the solution of the FSTP suggests that the controllers defined by the FSTP may be referred to as *Total Finite Settling Time Stabilizing* (TESTS) controllers, or simply FSTS controllers; the problem of defining them (FSTP), may also be referred to as *Total Finite Settling Time Stabilization Problem* (TESTSP), or simply FSTSP.

Remark 3.3 If the system of Fig. (2.1) is a sampled–data system, the condition that e_2 settles in finite time guarantees that y_2 settles in finite time not only at the sampling points but also in–between.

The subfamily of all causal FST controllers is parameterized next.

Corollary 3.1 *Suppose that* $g_p = n_p/d_p$ *and* $g_c = n_c/d_c \in W(g_p)$. *Then, the FSTS controller* g_c *is causal for*

1. every $t(d) \in \mathcal{R}[d]$ *if* $n_p(0) = 0$
2. almost every $t(d) \in \mathcal{R}[d]$ *if* $n_p(0) \neq 0$

Proof The controller g_c is causal iff $d_c(0) = \bar{d}_c(0) - t(0)n_p(0) \neq 0$[32]. Therefore:

1. if $n_p(0) = 0$ then $\bar{d}_c(0) \neq 0$ in order Eqn. (3.1) to be satisfied for $d = 0$. So, $d_c(0) = \bar{d}_c(0)$ for every $t(d) \in \mathcal{R}[d]$.
2. if $n_p(0) \neq 0$ then $d_c(0) = \bar{d}_c(0) - t(0)n_p(0) \neq 0$ implies that $t(0) \neq \bar{d}_c(0)/n_p(0)$.

Hence, if $W_c(g_p)$ denotes the family of all causal FST controllers, we have:

1. $W_c(g_p) = W(g_p)$ if $n_p(0) = 0$
2. $W_c(g_p) = \{(n_c, d_c) : n_c = \bar{n}_c + td_p,\ d_c = \bar{d}_c - tn_p,\ t \in \mathcal{R}[d], t(0) \neq \bar{d}_c(0)/n_p(0), \bar{d}_c - tn_p \neq 0\}$ if $n_p(0) \neq 0$.

Remark 3.4 If the plant possesses a delay all the controllers that belong to the $W(g_p)$ family are causal.

4 Strong FSTS

The problem of solving the FSTS using a *stable* compensator is examined in this section and will be defined as *strong* FSTSP. The necessary and sufficient conditions for the existence of stable FSTS compensators are derived. They turn out to be the same as those for strong stabilization [30], [31] although in the FSTS case the domains of stability of system and controller respectively, are different.

Theorem 4.1 *Suppose $g_p = n_p/d_p, g_c = n_c/d_c$ where $n_p, d_p, n_c, d_c \in \mathcal{R}[d]$, n_p, d_p are coprime and n_c, d_c are coprime. There exists a stable FSTS controller iff $d_p(d)$ has the same sign at all real zeros σ_i of $n_p(d)$ inside the closed interval $[-1, 1]$.*

Proof The proof of Th. (4.1) is based on disc algebra theory. It is similar to that given by Vidyasagar for strong stabilization [30] and it is presented in Apx. (A.1).

Corollary 4.1 *There always exists a* stable *FSTS controller iff the number of poles of $g_p(d)$ inside any interval of successive zeros of $g_p(d)$ inside the unit disc is even.*

5 Algebraic computation of the family $W(g_p)$

According to Th. (3.1) the computation of the family $W(g_p)$ requires only the computation of a particular solution $(\overline{n}_c, \overline{d}_c)$. This problem can be reduced to a standard linear algebra problem. Let

$$g_p = \frac{b_0 + b_1 d + \cdots + b_m d^m}{1 + a_1 d + \cdots + a_n d^n} = \frac{n_p}{d_p} \tag{5.1}$$

be the transfer function of a causal plant and

$$g_c = \frac{c_0 + c_1 d + \cdots + c_\mu d^\mu}{f_0 + f_1 d + \cdots + f_\nu d^\nu} = \frac{n_c}{d_c} \tag{5.2}$$

the transfer function of the controller.

For FSTS, n_c, d_c is a solution to the Diophantine Eqn. (3.1) with n_p, d_p comprime in $\mathcal{R}[d]$ and n_c, d_c coprime in $\mathcal{R}[d]$. Eqn. (3.1) implies:

$$\deg(n_p n_c) = \deg(d_p d_c) \text{ ie} \tag{5.3}$$

$$\deg(n_p) + \deg(n_c) = \deg(d_p) + \deg(d_c) \text{ ie} \tag{5.4}$$

$$m + \mu = n + \nu \tag{5.5}$$

Using Toeplitz matrices for polynomial multiplication, Eqn. (3.1) with $c = 1$, becomes:

$$\underbrace{\begin{bmatrix} b_0 & 0 & \cdots & 0 \\ b_1 & b_0 & \ddots & \vdots \\ \vdots & & \ddots & 0 \\ \vdots & & & b_0 \\ b_m & & & \vdots \\ 0 & b_m & & \vdots \\ \vdots & & \ddots & \vdots \\ 0 & \cdots & 0 & b_m \end{bmatrix}}_{\triangleq T_\mu(n_p)} \underbrace{\begin{bmatrix} c_0 \\ \vdots \\ c_\mu \end{bmatrix}}_{\triangleq \underline{c}_\mu} + \underbrace{\begin{bmatrix} 1 & 0 & \cdots & 0 \\ a_1 & 1 & & \vdots \\ \vdots & & \ddots & 0 \\ \vdots & & & 1 \\ a_n & & & \vdots \\ 0 & a_n & & \vdots \\ \vdots & & & \vdots \\ 0 & \cdots & 0 & a_n \end{bmatrix}}_{\triangleq T_\nu(d_p)} \underbrace{\begin{bmatrix} f_0 \\ \vdots \\ f_\nu \end{bmatrix}}_{\triangleq \underline{f}_\nu} = \underbrace{\begin{bmatrix} 1 \\ \vdots \\ 0 \end{bmatrix}}_{\triangleq \underline{e}^1} \quad (5.6)$$

or in compact form:

$$T_{\mu,\nu}^{m,n}(n_p, d_p) \begin{bmatrix} \underline{c}_\mu \\ \underline{f}_\nu \end{bmatrix} = \underline{e}^1 \quad (5.7)$$

where

$$T_{\mu,\nu}^{m,n}(n_p, d_p) = [T_\mu(n_p), T_\nu(d_p)] \quad (5.8)$$

Therefore the solution to the FSTSP reduces to the solution of Eqns. (5.7).

Proposition 5.1 *There always exists an FSTS controller $\tilde{g}_c = \tilde{n}_c/\tilde{d}_c$ with $deg(\tilde{n}_c) = n-1$, $deg(\tilde{d}_c) = m-1$, and the parameter vectors $\underline{c}_{n-1}$, $\underline{f}_{m-1}$, of $\tilde{n}_c$, $\tilde{d}_c$ respectively, given by:*

$$\begin{bmatrix} \underline{c}_{n-1} \\ \underline{f}_{m-1} \end{bmatrix} = \{T_{n-1,m-1}^{m,n}(n_p, d_p)\}^{-1} \underline{e}^1 \quad (5.9)$$

The compensator $\tilde{g}_c = \tilde{n}_c/\tilde{d}_c$ is unique and will be referred to as the prime *FSTS controller.*

Proof $T_{n-1,m-1}^{m,n}(n_p, d_p) \in \mathcal{R}^{(m+n)\times(m+n)}$ and for the proof of Pr. (5.1) it is sufficient to show that $T_{n-1,m-1}^{m,n}$ has full rank.

Consider $T_{n-1,m-1}^{m,n}(n_p, d_p)$ as in Eqn. (5.8). By reordering the two blocks it becomes $[T_{m-1}(d_p), T_{n-1}(n_p)]$ and then reordering the columns of $T_{m-1}(d_p)$ in a cyclic manner leads to $[T_{m-1}^0(d_p), T_{n-1}(n_p)]$. Transposing the last matrix and reordering its columns in a cyclic manner results to the matrix:

$$S(n_p, d_p) = \begin{bmatrix} a_n & a_{n-1} & \cdots & \cdots & 1 & 0 & \cdots & 0 \\ 0 & a_n & & & & 1 & \ddots & \vdots \\ \vdots & & \ddots & & & & \ddots & 0 \\ 0 & & 0 & a_n & a_{n-1} & & & 1 \\ 0 & \cdots & 0 & b_m & b_{m-1} & \cdots & \cdots & b_0 \\ \vdots & & & & & & & 0 \\ 0 & b_m & & & & & & \vdots \\ b_m & b_{m-1} & \cdots & \cdots & b_0 & 0 & \cdots & 0 \end{bmatrix} \tag{5.10}$$

However, $S(n_p, d_p)$ is the Sylvester resultant of the coprime polynomials $n_p(d), d_p(d)$ and therefore $\det S(n_p, d_p) \neq 0$. But $S(n_p, d_p)$ and $T^{m,n}_{n-1,m-1}(n_p, d_p)$ have the same rank by conctruction and the result is established.

Remark 5.1 If the plant has a non zero deadtime the *prime* FSTS compensator is generically causal.

6 Parameterization of the FSTS controllers through their McMillan degree

The McMillan degree of a causal $g(d) \in \mathcal{R}(d)$ is a well defined notion. This definition may be extended to a general $g(d)$ in the following way:

Definition 6.1 *Let $g(d) = n_g/d_g \in \mathcal{R}(d)$ with n_g, d_g coprime in $\mathcal{R}[d]$ and $deg(n_g) = m$, $deg(d_g) = n$. The* extended *McMillan degree is defined as the integer $\delta^*_M(g) = max(m, n)$.*

If $g(d)$ is causal then δ^*_M coincides with the standard McMillan degree δ_M and indicates the total number of poles (finite) of the $\tilde{g}(z) = g(z^{-1}) \in \mathcal{R}_{pr}(z)$ transfer function. If $g(d)$ is not causal then $\tilde{g}(z) \in \mathcal{R}(z)$ and it is nonproper; in this case δ^*_M denotes the total number of poles of $\tilde{g}(z)$, finite and infinite. Clearly δ^*_M is a measure of complexity of $g(d)$.

The extended notion δ^*_M of the McMillan degre will be used in the study of the McMillan degree properties of the family $W(g_p)$. The importance of δ^*_M in the study of properties of the $W(g_p)$ family is demonstrated by the following result.

Remark 6.1 Let $W(g_p)$ be the family of FSTS compensators. The following properties hold true.

1. The relationship R_M on $W(g_p)$ defined for every $g_1, g_2 \in W(g_p)$ by:

$$g_1 R_M g_2 \Longleftrightarrow \delta^*_M(g_1) = \delta^*_{M(g_2)} \tag{6.1}$$

is an equivalence relation.

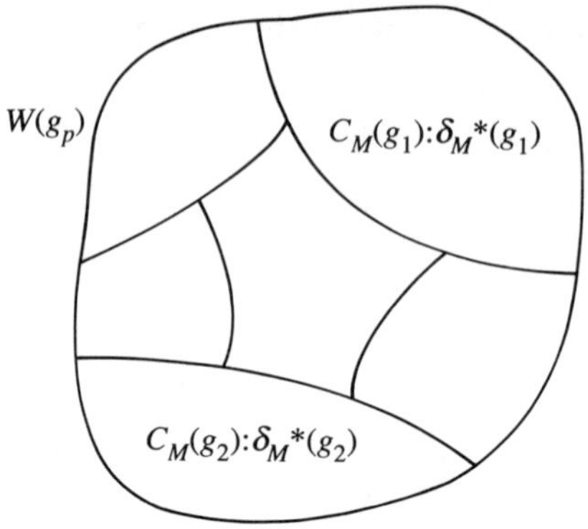

Figure 2. Partitioning of $W(g_p)$ under $\mathcal{R}_{\mathcal{M}}$.

2. Let $C_M(g)$ denote the equivalence class of $g \in W(g_p)$ under R_M. The family C_M^* of all equivalent classes $C_M(g)$ forms a partition of $W(g)$.

Note that each equivalence class is parameterized by a distinct number which is the extended McMillan degree of the equivalence class. A number of important questions that arise in this context are considered next. First define:

Definition 6.2 *Let $W(g_p)$ be the FSTS controller family, C_M^* the family of all equivalence classes $C_M(g_i)$ $g_i \in W(g_i)$ with $\delta_M^*(g_i)$ extended McMillan degree.*

1. *Define $i = \delta_M^*(g_i)$ as the* McMillan index *of $C_M^*(g_i)$ and $C_M^*(g_i)$ will be referred to as the i–*subfamily.
2. *The set of indices $I_M = \{i : i = \delta_M^*(g_i)$ index of $C_M(g_i)$ subfamily, $\forall C_M(g_i) \in C_M^*(g_p)\}$ will be called the* McMillan index *set of $W(g_p)$.*

The problems examined next are.

1. *Minimal Design Problem (MDP)*: Define the minimal McMillan degree of all FSTS causal compensators.
2. *Parameterization Problem (I)(PP(I))*: Define the McMillan index set I_M of $W(g_p)$.
3. *Parameterization Problem (II) (PP(II))*: For every $i \in I_M$ define a parametric expression of R_M equivalence class $C_M(g_i)$, where $\delta_M^* = i$. Examine the existence of causal compensators in the i–subfamily.

The solution of the above problems is examined next.

Theorem 6.1 *Let $g_p(d) = n_p(d)/d_p(d)$ be a discrete causal d–transfer function with n_p, d_p coprime in $\mathcal{R}[d]$ and $deg(n_p) = m$,* $\deg(d_p) = n$. *The prime FSTS compensator $\tilde{g}_c(d) = \tilde{n}_c(d)/\tilde{d}_c(d)$ is the* minimal *extended McMillan degree compensator* *of the $W(g_p)$ family with*

$$\delta_m^* \stackrel{\Delta}{=} \delta_M^*(\tilde{g}_c) = \max(m,n) - 1 \tag{6.2}$$

Proof Every FSTS controller can be parameterized as follows:

$$n_c = \tilde{n}_c + td_p \tag{6.3}$$
$$d_c = \tilde{d}_c - tn_p \tag{6.4}$$

where $t \in \mathcal{R}[d]$ and $\tilde{n}_c$, $\tilde{d}_c$ is the prime FSTS controller. But

$$\deg(\tilde{n}_c) = n - 1 \tag{6.5}$$

and

$$\deg(td_p) = \deg(t) + \deg(d_p) = \deg(t) + n \tag{6.6}$$

Therefore

$$\deg(n_c) \geq \deg(\tilde{n}_c). \tag{6.7}$$

$$\deg(\tilde{n}_c) = m - 1 \tag{6.8}$$

and

$$\deg(tn_p) = \deg(t) + \deg(n_p) = \deg(t) + m \tag{6.9}$$

Therefore

$$\deg(d_c) \geq \deg(\tilde{d}_c) \tag{6.10}$$

where the equality holds for $n_c = \tilde{n}_c$ and $d_c = \tilde{d}_c$. Then, according to the definition of the extend McMillan degree and of δ_m^*, δ_m^* is

$$\delta_m^* = \min \delta_M^*(g_c) = \delta_M^*(\tilde{g}_c) = \max(m,n) - 1 \tag{6.11}$$

Remark 6.2 The parameters f_{m-1}, c_{n-1} as defined by Eqn. (5.9) may be zero and thus $\delta_M^*(\tilde{g}_c) \leq \max(m,n) - 1$. The generic value of $\delta_M^*(\tilde{g}_c)$ is $\delta_M^*(\tilde{g}_c) = \max(m,n) - 1$.

The minimal extended McMillan degree of the family $W(g_p)$ will be denoted by $\delta_m^*(g_p)$ and will be referred to as the *McMillan characteristic* of $W(g_p)$. Using the parameterization for $W(g_p)$

$$n_c = \tilde{n}_c + td_p \tag{6.12}$$
$$d_c = \tilde{d}_c - tn_p \tag{6.13}$$

where $t \in \mathcal{R}[d]$, the minimal design problem and the parameterization problems (I) and (II) may be examined.

Corollary 6.1 *(MDP): The minimal McMillan degree δ_{Mmin} of the FSTS compensators is given by*

1. $\delta_{Mmin} = \delta^*_m = \max(m,n) - 1$ *if the prime FSTS compensator is causal.*
2. $\delta_{Mmin} = \max(m,n)$ *if the prime FSTS compensator is non causal.*

Proof

1. If the prime FSTS compensator is causal then the minimal McMillan degree compensator is unique (modulo $c \in \mathcal{R}$); that is, the prime FSTS compensator. Hence:
$$\delta_{Mmin} = \delta^*_m = \max(m,n) - 1 \tag{6.14}$$
2. If the prime FSTS compensator is not causal then the minimal McMillan degree compensators belong to the subfamily
$$n_c = \tilde{n}_c + cd_p \tag{6.15}$$
$$d_c = \tilde{d}_c - cn_p \tag{6.16}$$
where $c \in \mathcal{R}$ and $c \neq \tilde{d}_c(0)/n_p(0)$. Therefore $\deg(n_c) = n, \deg(d_c) = m$ and
$$\delta_{Mmin} = \delta^*_m = \max(m,n) \tag{6.17}$$

Corollary 6.2 *(PP(I)): The McMillan index set I_M of $W(g_p)$ is defined by*
$$I_M = \{\delta^*_m(g_p);\ \max(m,n) + k,\ k = 0,1,2,\cdots\} \tag{6.18}$$

Proof For $t(d) = 0$ in Eqns. (6.1), the solution is the prime one and therefore the first McMillan index is $\delta^*_M(\tilde{g}_c) = \delta^*_m(g_p)$. If $\deg(t(d)) = k$ then, $\delta^*_M(\tilde{g}_c) = \max(m+k, n+k)$ or $\delta^*_M(\tilde{g}_c) = \max(m,n) + k$.

Having specified I_M, the FSTS subfamilies of $W(g_p)$ with given McMillan degree may be parameterized as follows:

Corollary 6.3 *(PP(II)): The δ^*_m-subfamily of $W(g_p)$ has a single element, the prime FSTS pair $(\tilde{n}_c, \tilde{d}_c)$. The δ_j-subfamily, where $\delta_j = \max(m,n) + j$, $j = 0,1,2,\cdots$ is parameterized by*
$$n_c = \tilde{n}_c + t(d)d_p \tag{6.19}$$
$$d_c = \tilde{d}_c - t(d)n_p \tag{6.20}$$
$t(d) \in \mathcal{R}[d], \deg(t(d)) = j$.

Moreover, if

1. $n_p(0) = 0$ *every FSTS controller within the δ_j-subfamily is causal.*
2. $n_p(0) \neq 0$ *every FSTS controller with $t(0) \neq d_c/n_p(0)$ within the δ_j-subfamily is causal.*

Proof For any $t(d) \in \mathcal{R}[d]$ with $\deg(t(d)) = j$, $\delta^*_M(g_c) = max(m,n) + j$ and thus by fixing the degree of $t(d)$ we remain within the family of the given $\delta_j$index. For any other choice of degree of $t(d)$, ie $\deg(t(d)) = k \neq j$, the corresponding controller g_c will have $\delta^*_M(g_c) = max(m,n) + k \neq max(m,n) + j$ and it will belong to a different subfamily. Invoking Cor. (3.1), the existence of causal FSTS compensators in the δ_j–subfamily follows immediately.

7 FSTS compensators for tracking

The problem of tracking is one of the important problems in control system design where the output has to follow a particular set of inputs.

In the case of FSTS, it is required (*Fig.*(2.1)) that the output y_2 tracks the input $u_1 = r$ in finite time. In this sense, most of the tracking dead–beat controllers can be considered as time–optimum FSTS and tracking controllers. [24], [33]–[35]

Theorem 7.1 *Let $g_p = n_p/d_p, n_p, d_p \in \mathcal{R}[d], n_p, d_p$ coprime, the plant transfer function and $g_c = n_c/d_c, n_c, d_c \in \mathcal{R}[d], n_c, d_c$ coprime, an FSTS compensator. Suppose also that the input $u_1 = r = n_r/d_r, n_r, d_r \in \mathcal{R}, n_r, d_r$ coprime, belongs to a prespecified class of signals. Then, the output y_2 tracks the input u_1 in finite time iff $d_r|d_c d_p$ (see Fig. (2.1)).*

Proof The transfer function between e_1 and u_1 (Fig. (2.1)) is:

$$e_1(d) = \frac{d_c d_p}{n_c n_p + d_c d_p} r(d) \tag{7.1}$$

and for FSTS $n_c n_p + d_c d_p = c \in \mathcal{R} - \{0\}$. Hence, $e_1(d)$ becomes:

$$e_1(d) = \frac{d_c d_p n_r}{c d_r} \tag{7.2}$$

For finite tracking $e_1(d)$ has to be polynomial in $\mathcal{R}$ and this is possible iff $d_r|d_c d_p$.

Remark 7.1 If the set of inputs contains "unstable" signals, then the input e_2 to the plant will become "unstable" and this will lead to saturation. If the requirement that e_2 tends to zero in finite time will be imposed to Th. (7.1), then for finite tracking d_r has to divide d_p.

8 FSTS compensators for disturbance rejection

Another problem usually encountered in control system design is that of disturbance rejection, ie every non desirable input signal has to vanish at the output.

Let $u_2 = d$ in Fig. (2.1) is the disturbance signal to be rejected by the system. Then y_2 has to reach a zero steady–state after finite time, ie $y_2(d)$ has to be a polynomial in $\mathcal{R}$.

Theorem 8.1 *Let $g_c = n_c/d_c \in W(g_p)$ with $n_c, d_c \in \mathcal{R}[d]$, and n_c, d_c coprime. The output y_2 of Fig. (2.1) rejects the output $u_2 = d = n_d/d_d$ in finite time iff $d_d|n_p d_c$.*

Proof According to Fig. (2.1)

$$y_2(d) = \frac{n_p d_c}{n_c n_p + d_c d_p} u_2 \tag{8.1}$$

For FSTS, $n_c n_p + d_c d_p = c \in \mathcal{R} - \{0\}$ and y_2 becomes:

$$y_2(d) = \frac{n_p d_c n_d}{c d_d} \tag{8.2}$$

Therefore, for finite disturbance rejection $y_2(d)$ has to be polynomial in $\mathcal{R}$, that is d_d has to divide $n_p d_c$.

9 Conclusions

An algebraic approach for the study of *Total Finite Settling Time Stabilization* of single–input single–output discrete–time systems has been developed and it has led to a Kucera, Youla–Bongiorno type parameterization of FSTS controllers. The subfamily of causal FSTS controllers has been parametrized in terms of a simple condition on the design parameter and the minimum McMillan degree FSTS controller has been defined; its computation is reduced to the solution of a simple Toeplitz type linear equation. The nature of the FSTS problem allows the parameterization of the whole family of FSTS controllers according to McMillan degree. The conditions on strong stabilization [30], [31], have been shown to be also valid in the present context where the stability domains of controller and system are different.

The advantage of the approach is that the general family of FSTS controllers may be computed without resorting to the solution of polynomial Diophantine equations. The additional conditions for FSTS tracking and FSTS disturbance rejection may be imposed as additional constraints on the design parameter. the approach may be readily extended to the multivariable case. The parameterization of multivariable controllers according to McMillan degree is a subject under investigation at the moment.

APPENDIX A.1

Proof of Theorem (4.1) "ONLY IF". Suppose that there is a stable FSTS controller. Then:

$$d_c d_p + n_c n_p = 1 \tag{A.1}$$

and d_c is a polynomial unit in the disc algebra A_s, ie $d_c(d) \neq 0 \ \forall d \in \mathcal{D}$.

Therefore, $d_c(\sigma_i)$ has the same sign $\forall \sigma_i : n_p(\sigma_i) = 0, \ \sigma_i \in [-1, 1]$. Hence, Eqn (A.1) becomes for $d = \sigma_i$:

$$d_c(\sigma_i) d_p(\sigma_i) = 1 \tag{A.2}$$

which means that $d_p(\sigma_i)$ has the same sign as $d_c(\sigma_i)$, ie $d_p(d)$ does not change sign at the real zeros of n_p inside the unit disc $\mathcal{D}$.

"IF". The proof of this part is constructive. Let $d_p(d)$ does not change sign at the real zeros of n_p inside the unit disc $\mathcal{D}$ and $\overline{n}_c$, $\overline{d}_c$ be a particular solution of Eqn. (A.1). Also:

$$n_p = n_p^+ n_p^- \tag{A.3}$$

where $n_p^+ = \prod_{i=1}^{l}(d - z_i)^{m_i} \in \mathcal{R}[d]$ contains all the "unstable" zeros of n_p and $n_p^- \in \mathcal{R}[d]$ contains all the "stable" zeros of n_p.

Then, $\overline{d}_c(\sigma_i)d_p(\sigma_i) = 1$ ie $\overline{d}_c(d)$ does not change sign $\forall \sigma_i \in [-1, 1]$ and $n_p(\sigma_i) = 0$. Therefore, there is a unit h in A_s (not necessarily polynomial) that either h or $-h$ possesses a logarithm (depending on whether $\overline{d}_c(\sigma_i) > 0$ or $d_c(\sigma_i) < 0$ respectively) and which interpolates $\overline{d}_c(d)$ and its derivatives at the "unstable" zeros of n_p, ie:

$$h^{(j)}(z_i) = \overline{d}_c^{(j)}(z_i) \; i = 1, \cdots, l \; j = 1, \cdots, m_i - 1 \tag{A.4}$$

Then, $n_p^+ | (h - \overline{d}_c)$ in A_s and let $c = (h - \overline{d}_c)/n_p^+ = c_1 \overline{n}_p \; c, c_i \in A_s$ (because $\overline{n}_p$ is a polynomial unit in A_s). Since polynomials are dense in A_s, there exists a pol ynomial c_2 such that

$$\|c_1 - c_2\| \leq 1/\|h^{-1}\| \; \|n_p^+\| \; \|n_p^-\| \tag{A.5}$$

Now define $f = \overline{d}_c + c_2 n_p^+ n_p^- \in \mathcal{R}[d]$. Then,

$$\|h - f\| = \|(c_1 - c_2)n_p^+ n_p^-\| \leq \|n_p^+\| \; \|n_p^-\|/\|h^{-1}\|\|\|n_p^+\| \; \|n_p^-\| \tag{A.6}$$

ie

$$\|h - f\| \leq 1/\|h^{-1}\| \tag{A.7}$$

Hence, f is a polynomial unit in A_s, ie $f(d) \in \mathcal{R}[d]$ and $f(d) \neq 0 \; \forall d \in \mathcal{D}$. If $d_c = f$ and $t = -c_2$ then:

$$d_c = \overline{d}_c - t n_p, \; n_c = \overline{n}_c + t d_p \tag{A.8}$$

define a *stable* FSTS controller.

References

[1] Bergen, A.R. and J.R.Ragazzini (1954). Sampled–Data Techniques for Feedback Control Systems, **AIEE** Trans. 73(II), 236–247.

[2] Iserman, R. (1981). Digital Control Systems, Springer Verlang.

[3] Kalman, R.E. (1960). On the General Theory of Control Systems, 1st IFAC Congr. Aut. Control, Moscow, 4 481–492.

[4] Kuo, C.B. (1980). Digital Control Systems, Holt Rinenhart Winston.

[5] Ackerman, J. (1968). Zeitoptimale Mehrafach–Adtastregelsysteme, IFAC Symp, Multivariable Control Systems Band 1, Dusseldorf.

[6] Akashi, H., H. Imai and M.Adachi (1978). On time optimal control of linear discrete–time systems by geometric approach, Int.. J. Control, 28, 853–867.

[7] Jordan, D. and J. Korn (1980). Deadbeat Algorithms for Multivariable Process Control, **IEEE** trans. **AC** -25, 486-491.

[8] Kucera, V. (1971). The Structure and Properties of Time-Optimal Discrete Linear Control, **IEEE** Trans. AC-16, 375-377.

[9] Kucera, V. (1972). State Space Approach to Discrete Linear Control, Kybernetika, 8, 233-251.

[10] Leden, B. (1976) Dead-Beat Control and the Riccati Equation, **IEEE** Trans. AC-21, 791-792

[11] Lewis, F.L. (1982). A General Riccati Equation Solution to the Deadbeat Control Problem, **IEEE** Trans. AC-27, 763-766.

[12] Mullis, C.T. (1972). Time optimal Discrete Regulator Gains, **IEEE** Trans. **AC** -17, 265-266. .

[13] O'Reilly, J. (1980). The deadbeat control of linear multivariable systems with inaccessible states, Int..J.Control, 31, 645-654.

[14] O'Reilly, J. (1981). Rhe Discrete linear Time Invariant Time-Optimal Control Problem-An Overview, Automatica, 17, 363-370.

[15] Pachter, M. (1977). An Explicit Pole-Assigning Feedback formula with Application to Dead-beat Feedback construction in Discrete linear Systems, **IEEE** Trans. AC-22, 263-265.

[16] Sebakhy, O.A. and T.M. Abdel-Moneim (1980). Design of Optimal Dead-Beat Controllers, **IEEE** Trans. **AC** -25, 604-606.

[17] Kimura, H. and Y. Tanaka (1981). Minimal-Time Minimal-Order Deadbeat Regulator with Internal Stability, **IEEE** Trans. **AC** -26, 1276-1282.

[18] Leden, B. (1977). Output dead-beat control-a geometric approach, Int.J.Control, 26, 493-507.

[19] Kucera, V. (1973a). Algebraic Theory of Discrete Optimal Control for Single-variable Systems I, Kybernetika, 9,94-107.

[20] Kucera, V. (1973b). Algebraic Theory of Discrete Optimal Control for Single-variable Systems II, Kybernetika, 9, 206-221.

[21] Kucera, V. (1973c). Algebraic Theory of Discrete optimal Control for Single-variable Systems III, Kybernetika, 9, 291-312.

[22] Kucera, V. (1980a). Polynomial Design of Deadbeat Control Laws, Kybernetika, 16, 198-203.

[23] Kucera, V. (1980b). Observer-based Deadbeat Controllers:A Polynomial Design, Kybernetika, 16, 431-441.

[24] Kucera, V. (1980c). A dead-beat servo problem, Int.. J, Control, 32, 107–113

[25] Eichstaedt, B. (1982). Multivariable Closed-loop Deadbeat Control: a Polynomial-matrix Approach, Automatica, 18, 589–593.

[26] Wolovich, W.A. (1983). Deadbeat error control of discrete multivariable systems, Int.J.Control, 37, 567–582.

[27] Zhao, Y. and H. Kimura (1988). Multivariable dead-beat control with robustness, Int..J.Control, 47, 229–255.

[28] Lin, W.S., T.S. Kuo and G.J. Thaler (1983). Finite-time control of linear discrete-time systems, Int..J.Control, 37, 855–860.

[29] Kucera, V. (1979). Discrete Linear Control, John Wiley.

[30] Vidyasagar, M. (1985). Control System Synthesis: A Factorization Approach, MIT Press.

[31] Youla, D.C., J.J.Bongiorno and C.N. Lu(1974). Single-Loop Feedback Stabilization for Linear Multivariable Dynamical Plants, Automatica, 10, 159–173.

[32] Wolovich, W.A and H. Elliot (1983). Discrete models for Linear multivariable systems, Int..J.Control, 38, 337–357.

[33] Kucera, V. and M. Sebek (1984). On Deadbeat Controllers. IEEE Trans. Ac–29, 719–722.

[34] Wang, S.S. and B.S. Chen (1986). Simultaneous deadbeat tracking controller synthesis, Int..J.Control, 44, 1579–1586.

[35] Zafiriou, E. and M. Morari (1985). Digital controller for SISO systems: a review and a new algorithm, Int..J.Control, 42, 855–876.

Conditions for arbitrary pole assignability by input/output state–variable feedback

A. Chotai C. Wang P. C. Young

Abstract

This paper presents the main theoretical results which underpin a new method of state variable feedback pole–assignment control for discrete–time systems. The method depends upon the specification of a non–minimal state space (NMSS) representation of the system in which the state vector is composed only of the present and past input–output variables, together with an "integral of error" state, which ensures type–1 servomechanism performance. The results are presented for single input-single output (SISO) systems here for brevity.

1 Introduction

The idea of designing control systems which ensures that the closed loop poles are assigned at, or near, specified desirable locations in the complex plane is appealing to most control engineers. Indeed, it can be argued that the idea is inherent in many conventional methods of control system design. But the most powerful general method of pole–assignment is based on the concept of state variable feedback (SVF), since it is well known that a linear SVF control law can, in theory, arbitrarily assign the closed loop poles, provided the dynamical system is controllable (Wonham, 1967). One obvious drawback of SVF system design, which has almost certainly restricted its practical application, is the apparent need to measure all of the state variables which characterise the state vector. This requirement can be obviated by the use of a state–reconstructor (observer) or its stochastic equivalent, an optimal state estimator (Kalman filter), but these can add considerably to the complexity if the resultant design. In this paper, we consider an alternative discrete–time approach based on the definition of a special non–minimal state space (NMSS) model (see Young et al., 1988; Wang, 1988; Wang and Young, 1988a,b and Wang et al. 1988), in which the state variables are composed only of the present and past values of the output; the past values of the input; and, in the servomechanism form, the integral–of–error between the current sampled output and the reference input. In this paper we will investigate the controllability of the NMSS model for both the regulator and servomechanism cases; and develop an algorithm for calculating the feedback gains for the servomechanism case.

2 The regulator systems

Consider a single–input, single–output (SISO), discrete–time system represented in the following transfer function form,

$$y_k = \frac{B(z^{-1})}{A(z^{-1})} u_k \tag{2.1}$$

where

$$A(z^{-1}) = 1 + a_1 z^{-1} + \cdots + a_n z^{-n} \tag{2.2}$$

and

$$B(z^{-1}) = b_1 z^{-1} + \cdots + b_m z^{-m} \tag{2.3}$$

In the above polynomials, z^{-i} is the backward shift operator, i.e. $z^{-i} y_k = y_{k-i}$, a_n and b_m are non–zero; but some of the initial b_i's can be zero to accommodate any pure time delay.

The non–minimal state vector (NMSV) in this case is defined by the following vector (Wang and Young, 1988a),

$$\underline{x}_k^{\mathrm{T}} = [y_k \; y_{k-1} \; \cdots \; y_{k-n+1} u_{k-1} \; \cdots \; u_{k-m+1}] \tag{2.4}$$

With this state vector the following non–minimal state space representation can be directly obtained:

$$\underline{x}_{k+1} = F \underline{x}_k + \underline{g} u_k \tag{2.5}$$

$$y_k = \underline{h} \underline{x}_k \tag{2.6}$$

where

$$F = \begin{bmatrix} -a_1 & -a_2 & \cdots & -a_{n-1} & -a_n & b_2 & b_3 & \cdots & b_{m-1} & b_m \\ 1 & 0 & \cdots & 0 & 0 & 0 & 0 & \cdots & 0 & 0 \\ 0 & 1 & \cdots & 0 & 0 & 0 & 0 & \cdots & 0 & 0 \\ \vdots & \vdots & & \vdots & \vdots & \vdots & \vdots & & \vdots & \vdots \\ 0 & 0 & & 1 & 0 & 0 & 0 & & 0 & 0 \\ 0 & 0 & \cdots & 0 & 0 & 0 & 0 & \cdots & 0 & 0 \\ 0 & 0 & \cdots & 0 & 1 & 0 & 0 & \cdots & 0 & 0 \\ \vdots & \vdots & \cdots & \vdots & \vdots & \vdots & \vdots & \cdots & \vdots & \vdots \\ 0 & 0 & \cdots & 0 & 0 & 0 & 0 & \cdots & 1 & 0 \end{bmatrix} \tag{2.7}$$

$$\underline{g} = [b_1, 0 \; \cdots \; 0 \; 1 \; 0 \; \cdots \; 0]^{\mathrm{T}} \tag{2.8}$$

$$\underline{h} = [1 \; 0 \; \cdots \; 0] \tag{2.9}$$

It is obvious that the state variables of the NMSS system consist of the present values of y_k and the past values of y_k and u_k; and that we have perfect access to these variables at the $j-th$ step after observation is initiated, where $j = max(n-1, m-1)$.

Our objective here is to design a feedback control law

$$u_k = \underline{k}^{\mathrm{T}} \underline{x}_k$$

such that the closed loop poles (the roots of $\det(zI - F + \underline{g}\underline{k}^{\mathrm{T}}) = 0$) are at pre–assigned positions in the complex z plane. Since pole assignabililty in this sense is equivalent to the controllability of the state space representation, hence our first task is to assess the conditions under which the representation (2.5) is controllable. These conditions are provided by the following theorem.

Theorem 1 *(Wang and Young, 1988a) Given a single–input single–output, discrete–time system described by equation (1), the non–minimal state space representation (2.5), as defined by the pair* $[F, \underline{g}]$*, is completely controllable if and only if the polynomials* $A(z^{-1})$ *and* $B(z^{-1})$ *are coprime.*

Remarks

1. The proof is based on a PBH test (Kailath, 1981) and the details are given in Wang and Young (1988).
2. The coprimeness condition is equivalent to requiring that there are no pole–zero cancellations between the $A(z^{-1})$ and $B(z^{-1})$ polynomials.

3 Pole–assignment and servomechanism by NMSV feedback

The pole–assignment design problem based on this NMSV feedback is discussed in much detail in Wang and Young (1988a) and Wang (1988). Here we will consider the case with servomechanism performance, in addition to the pole–assignment, by augmenting the NMSV defined in the previous section by an integral–of–error term.

Consider again the system (2.1), as in Young et al. (1988), the non–minimal state variable is defined as,

$$\underline{x}_k^{\mathrm{T}} = [y_k \; y_{k-1} \; \cdots \; y_{k-n+1} \; u_{k-1} \; \cdots \; u_{k-m+1} \; x_{n+m,k}] \tag{3.1}$$

where $x_{n+m,k}$ is an "integral of error" variable which is introduced to ensure type–1 servomechanism behaviour and is defined as follows,

$$x_{n+m,k} = x_{n+m,k-1} + y_{dk} - y_k \tag{3.2}$$

where y_{dk} is the desired output level.

The NMSS system can be represented by the following discrete–time state equations

$$\underline{x}_k = F\underline{x}_{k-1} + \underline{g}u_{k-1} + \underline{d}y_{dk} \tag{3.3}$$

$$y_k = \underline{h}\underline{x}_k \tag{3.4}$$

where

$$\underline{g}^{\mathrm{T}} = [b_1\ 0\ \cdots\ 0\ 1\ 0\ \cdots\ 0\ -b_1] \tag{3.5}$$
$$\underline{h} = [1\ 0\ \cdots\ 0] \tag{3.6}$$
$$\underline{d}^{\mathrm{T}} = [0\ 0\ \cdots\ 0\ 1] \tag{3.7}$$

and

$$F = \begin{bmatrix} -a_1 & -a_2 & \cdots & -a_{n-1} & -a_n & b_2 & b_3 & \cdots & b_{m-1} & b_m & 0 \\ 1 & 0 & \cdots & 0 & 0 & 0 & 0 & \cdots & 0 & 0 & 0 \\ 0 & 1 & \cdots & 0 & 0 & 0 & 0 & \cdots & 0 & 0 & 0 \\ \vdots & \vdots & & \vdots & \vdots & \vdots & & \vdots & \vdots & \vdots & \\ 0 & 0 & \cdots & 1 & 0 & 0 & 0 & \cdots & 0 & 0 & 0 \\ 0 & 0 & \cdots & 0 & 0 & 0 & 0 & \cdots & 0 & 0 & 0 \\ 0 & 0 & \cdots & 0 & 0 & 1 & 0 & \cdots & 0 & 0 & 0 \\ 0 & 0 & \cdots & 0 & 0 & 0 & 1 & \cdots & 0 & 0 & 0 \\ \vdots & \vdots & & \vdots & \vdots & \vdots & & \vdots & \vdots & \vdots & \\ 0 & 0 & \cdots & 0 & 0 & 0 & 0 & \cdots & 1 & 0 & 0 \\ a_1 & a_2 & \cdots & a_{n-1} & a_n & -b_2 & -b_3 & \cdots & -b_{m-1} & -b_m & 1 \end{bmatrix} \tag{3.8}$$

As discussed by Young and Willems (1972), the servomechanism requirement is automatically accommodated by the introduction of the integral–of–error variable (Wang, 1988). Thus we will restrict discussion to the pole assignment problem. The solution to this problem can be stated in terms of the following theorem (Wang, 1988).

Theorem 2 *Given a single–input, single–output, linear discrete–time system (2.1), the non–minimal state space representation (3.3) is completely controllable if and only if*

1. *The polynomials $A(z^{-1})$ and $B(z^{-1})$ are coprime, and*
2. $b_1 + b_2 + \cdots + b_m \neq 0$

Proof The detailed proof of the theorem is given by Wang (1988). Since it follows the same scheme as the case for pole–assignment only (Wang and Young, 1988a), we will merely outline the main points of the proof, concentrating on aspects which differ from the case without integral action. In the usual manner, the state space system defined by $(F, \underline{g})$ will not be controllable if there exists a non–zero row vector $\underline{q}$ (a left eigenvector of F) which is orthogonal to $\underline{g}$, i.e. if we can find a $\underline{q} = [q_1 \cdots\ q_n q_{n-1} \cdots\ q_{n+m}] \neq 0$ such that the following two equations hold simultaneously

$$\underline{q}F = u\underline{q} \tag{3.9}$$
$$\underline{q}\underline{g} = 0 \tag{3.10}$$

where u is an eigenvalue of F. The characteristic polynomial, $S(u)$, of F can be calculated easily and is given by

$$S(u) = (u^n + a_1 u^{n-1} + \cdots + a_{n-1}u + a_n)(u-1)(u)^{m-1} \tag{3.11}$$

Now, suppose that for u, which is an eigenvalue of F, there exists a row vector $\underline{q}$ such that Eqn(3.9) and (3.10) are fulfilled. It is then possible to expand these equations to yield the following scalar equations:

$$\begin{array}{lr}
-a_1 q_1 + a_1 q_{n+m} + q_2 = uq_1 & (1) \\
-a_2 q_1 + a_2 q_{n+m} + q_3 = uq_2 & (2) \\
\cdots & \\
-a_{n-1} q_1 + a_{n-1} q_{n+m} + q_n = uq_{n-1} & (n-1) \\
-a_n q_1 + a_n q_{n+m} = uq_n & (n) \\
b_2 q_1 - b_2 q_{n+m} + q_{n+2} = uq_{n+1} & (n+1) \\
\cdots & \\
b_{m-1} q_1 - b_{m-1} q_{n+m-1} = uq_{n+m-2} & (n+m-2) \\
b_m q_1 - b_m q_{n+m} = uq_{n+m-1} & (n+m-1) \\
q_{n+m} = uq_{n+m} & (n+m)
\end{array} \tag{3.12}$$

and

$$b_1 q_1 - b_1 q_{n+m} + q_{n+1} = 0 \tag{3.13}$$

We can now note the following:

When $u = 0$. From the assumption that $a_n \neq 0$, we can see that $u = 0$ is not a root of equation $u^n + a_1 u^{n-1} + \cdots + a_n = 0$, i.e. $u = 0$ is not a pole of the original system representation (2.1). In this case, we have that $(q_1 - q_{n+m}) = 0$ from Eq(n) of (3.12) and $q_{n+m} = 0$ from $Eq(n+m)$. These relationships imply that the whole vector $\underline{q}$ is a zero vector from the equations in (3.12), i.e. the modes specified by $u = 0$ are controllable.

When $u = u_1 \neq 0$ **but** $A(u_1) = u_1^n + a_1 u_1^{n-1} + \cdots + a_n = 0$. Following the same manipulations as by Wang and Young (1988a), we can obtain the following relationships:

$$(q_1 - q_{n+m})B(u_1) = 0 \tag{3.14}$$

where

$$B(u) = b_1 u^{m-1} + \cdots + b_m \tag{3.15}$$

and

$$(-q_1 + q_{n+m})(a_1 u_1^{n-1} + \cdots + a_n) = q_1 u_1^n \tag{3.16}$$

or

$$(-q_1 + q_{n+m})A(u_1) = q_{n+m} u^n \tag{3.17}$$

Now if $A(u)$ and $B(u)$ are not coprime, i.e. there is a u_1, such that $A(u_1) = 0$ and $B(u_1) = 0$ hold, then we can always find a $q_1 \neq 0$, so that all of the equations in (3.12) and (3.13) hold. Conversely, if $A(u)$ and $B(u)$ are coprime, then, $B(u_1) \neq 0$, which implies that $q_1 - q_{n+m} = 0$ and Eqn(3.17) gives $q_{n+m} = 0$. It follows that $\underline{q} = 0$, so that the modes specified by the roots of $A(u) = 0$ are controllable if and only if $A(u)$ and $B(u)$ are coprime.

In comparison with the case without integral action, the NMSS representation here has one more mode specified by $u = 1$. If $b_1 + \cdots + b_m \neq 0$, then equations (3.14) and (3.17) will have a unique zero solution for q_1 and q_{n+m} which will allow the whole q vector to be zero, and the mode specified by $u = 1$ will be controllable. On the other hand, if $b_1 + \cdots + b_m = 0$, then equation (3.14) will possess non–zero solutions for $q_1 = q_{n+m}$, which will allow equation (3.17) to have non–zero solutions for $q_1 = q_{n+m}$. As a result, there will exist a non–zero vector $\underline{q}$ such that equations (3.12) and (3.13) hold. In this case, therefore, the mode specified by $u = 1$ is uncontrollable.

Remarks

1. The conditions in theorem 2 have obvious physical interpretation:–Condition (1) simply means that the original system representation (2.1) should contain no pole–zero cancellations and is a quite obvious requirement. This is exactly the controllability condition for the NMSS representation without the integral action, as discussed by Wang and Young (1988).

   Condition (2) simply means that $u = 1$ should not be a zero of the original transfer function representation. From the block diagram (Fig. 1) of the closed loop system, we can consider the plant transfer function together with the delayed input and output feedback as an inner loop, and the integral feedback as an outer loop of the system. Considering the "closed loop" transfer function associated with the inner loop, we see that it retains $B(z^{-1})$ as its numerator. Hence, if $z = 1$ is a zero of $B(z^{-1})$, i.e. if $b_1 + \cdots + b_m = 0$, then the pole introduced into the system by the integral action at $z = 1$ will be cancelled out by this zero. As a result, the system will contain a mode at $u = 1$, which is not controllable.

2. In the NMSS representations discussed above, if the output vector is assumed to be $\underline{h} = [1\ 0\ \cdots\ 0]$, then the representation is technically unobservable, as can be ascertained from the normal observability criterion. However, if we allow for the storage of the past values of the input and the output variables, as we must do in digital systems, then the system is trivially "observable", since all of the NMSS variables are available for measurement at an arbitrary k–th time instant after a maximum of $(n-1, m-1)$ steps from the initiation of the system.

4 The computational algorithm

Many algorithms have been proposed for calculating the control gains in state variable feedback pole assignment (see e.g. Munro, 1979; Kailath, 1980). however, the special form of the NMSS representation allows us to compute the gains in a particularly simple manner. This NMSS algorithm can be developed by either following the scheme utilised by Wang and Young (1988a), using state–space analysis; or by approaching the problem from the standpoint of block diagram manipulation (Young et al., 1987; Wang, 1988). The former method provides greater insight into the nature of the algorithm but the latter is rather simple in mathematical terms. Both types of analysis naturally yield the same computational algorithm.

Suppose the desired closed loop characteristic polynomial (CLCP) of the system is defined as

$$S_c(u) = u^{n+m} + d_1 u^{n+m-1} + d_2 u^{n+m-2} + \cdots + d_{n+m-1}u + d_{n+m} \tag{4.1}$$

If we now define the state variable feedback (SVF) control law as

$$\begin{aligned} u_k &= -\underline{k}^{\mathrm{T}}\underline{x}_k \\ &= -f_0 y_k - f_1 y_{k-1} - \cdots - f_{n-1} y_{k-n+1} \\ &\quad -g_1 u_{k-1} - \cdots - g_{m-1} u_{k-m+1} + k_I x_{n+m,k} \end{aligned} \tag{4.2}$$

where the state feedback control gain vector $\underline{k}^{\mathrm{T}}$ is defined in the form,

$$\underline{k}^{\mathrm{T}} = [f_0 \; f_1 \; \cdots \; f_{n-1} g_1 \cdots \; q_{m-1} - k_I] \tag{4.3}$$

then the closed loop system is depicted in the block diagram form of Fig. 1.

Using either of the approaches above, it is straightforward to show (Young et al., 1987; Wang and Young, 1988a; Wang, 1988) that the vector $\underline{k}^{\mathrm{T}}$ which ensures that CLCP conforms with the desired form in (4.1) can be obtained by solution of the following linear algebraic equations

$$S(n,m)\underline{k} = \underline{S}_c - \underline{f} \tag{4.4}$$

where

$$S(n,m) =$$

$$\begin{bmatrix}
b_1 & 0 & \cdots & 0 & 1 & 0 & \cdots & 0 & b_1 \\
b_2-b_1 & b_1 & & \vdots & a_1-1 & 1 & & \vdots & b_2 \\
\vdots & b_2-b_1 & & \vdots & \vdots & a_1-1 & & 0 & \vdots \\
\vdots & \vdots & & 0 & \vdots & \vdots & & 1 & \vdots \\
b_m-b_{m-1} & \vdots & & b_1 & \vdots & \vdots & & a_1-1 & b_m \\
-b_m & b_m-b_{m-1} & & b_2-b_1 & a_n-a_{n-1} & \vdots & & \vdots & 0 \\
0 & -b_m & & \vdots & -a_n & a_n-a_{n-1} & & \vdots & \vdots \\
\vdots & 0 & & \vdots & 0 & -a_n & & \vdots & \vdots \\
\vdots & \vdots & & \vdots & \vdots & 0 & & \vdots & \vdots \\
\vdots & \vdots & & b_m-b_{m-1} & \vdots & 0 & a_n-a_{n-1} & \vdots & \\
0 & 0 & \cdots & -b_m & 0 & 0 & \cdots & -a_n & 0
\end{bmatrix}$$ (4.

$$\underline{S}_c = [d_1 \; d_2 \; \cdots \; d_{n+m-1} \; d_{n+m}]^{\mathrm{T}} \tag{4.6}$$

and

$$\underline{f} = [a_1 - 1 \;\; a_2 - a_1 \;\; \cdots \;\; a_n - a_{n-1} \;\; -a_n \;\; \cdots \;\; 0]^{\mathrm{T}} \tag{4.7}$$

As might be expected from the controllability conditions in Theorem 1, it is not difficult to prove the nonsingularity of $S(n,m)$, provided the conditions of Theorem 2 are satisfied. In particular, if we add all of the rows below each row in the matrix to that row, then $S(n,m)$ becomes,

$$\begin{bmatrix}
0 & 0 & \cdots & 0 & 0 & 0 & \cdots & 0 & b_1+\cdots+b_m \\
-b_1 & 0 & \cdots & 0 & -1 & 0 & \cdots & 0 & 0 \\
-b_2 & -b_1 & & 0 & -a_1 & -1 & & 0 & 0 \\
\vdots & -b_2 & & \vdots & \vdots & -a_1 & & \vdots & \vdots \\
\vdots & \vdots & & \vdots & \vdots & \vdots & & -1 & \vdots \\
-b_{m-1} & \vdots & & -b_1 & \vdots & \vdots & & -a_1 & \vdots \\
-b_m & -b_{m-1} & & \vdots & -a_{n-1} & \vdots & & \vdots & \vdots \\
0 & -b_m & & \vdots & -a_n & -a_{n-1} & & \vdots & \vdots \\
\vdots & 0 & & \vdots & 0 & -a_n & & \vdots & \vdots \\
\vdots & \vdots & & -b_{m-1} & \vdots & \vdots & & -a_{n-1} & \vdots \\
0 & 0 & \cdots & -b_m & 0 & 0 & \cdots & -a_n & 0
\end{bmatrix} \tag{4.8}$$

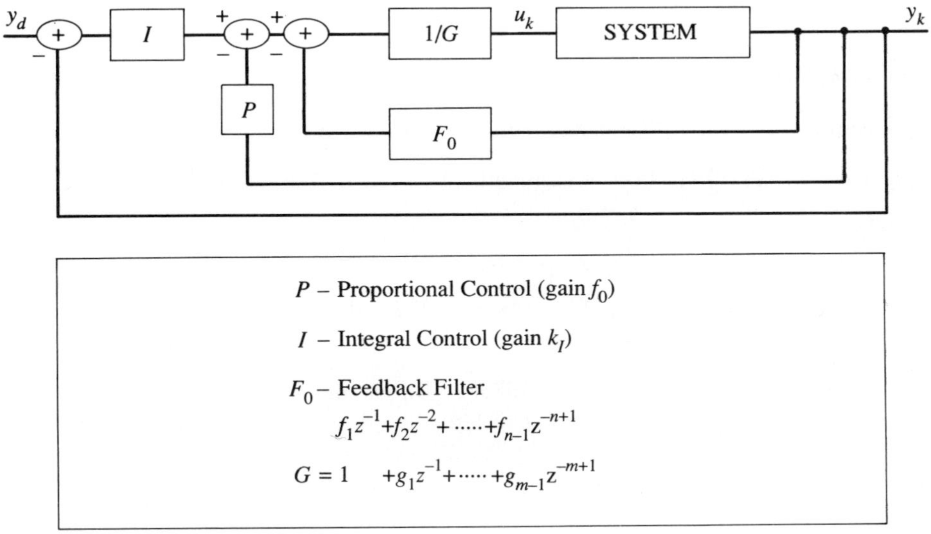

Figure 1. The input–Output Feedback, SISO, Servomechanism System.

It can be seen that the lower left block matrix is the well known Sylvester resultant matrix, which is nonsingular if the polynomials $A(u)$ and $B(u)$ are coprime. Consequently if, in addition, $b_1 + \cdots + b_m \neq 0$, then the matrix $S(n, m)$ is nonsingular.

It is clear that this algorithm is straightforward to implement on a digital computer, so that the feedback gains can be easily calculated for any specified discrete closed loop characteristic polynomial.

5 Conclusion and remarks

In this paper, we have considered a new method for direct digital pole–assignment design for discrete–time systems, based on the feedback of a non–minimal state vector defined in terms only of the output variable and the stored past sampled values of the input and output variables. Although single–input, single–output (SISO) systems are discussed in this paper, the results can be extended to multivariable systems (see e.g. Wang, 1988; Wang and Young, 1988b; Wang et al., 1988c). Moreover, additional multivariable concepts such as optimality and dynamic decoupling can be considered within this NMSS context (Wang, 1988; Wang et al., 1990). The practical implications of this approach to control system design are considerable (Young et al., 1988). In particular, it helps to unify existing methods of digital self–adaptive control and lays the foundations for the design of True Digital Control (TDC) systems (Young, 1988).

References

[1] Kailath, T. (1980) "Linear systems", Prentice–Hall, Englewood N.J.

[2] Munro, N. (1979) "Pole assignment" Proc. IEE, 126, 549–554.

[3] Wang, C.L. (1988) "New Methods for the Direct Digital Control for Discrete–Time Systems". PhDThesis, Lancaster University.

[4] Wang, C.L. and Young, P.C. (1988a) "Direct digital control by input–output, State variable feedback: theoretical back–ground", Int. J. control, 47, 97–109.

[5] Wang, C.L. and Young, P.C. (1988b) "Direct digital pole–assignment control system design for multivariable systems based on input–output state variable feedback" I.E.E. Conf. Publ. No 285, 212-217.

[6] Wang, C.L., Young, P.C. and Chotai, A. (1988c) " CAD program for direct digital pole–assignment system design", IFAC Conf. Proc. CACDS 88, 98-104.

[7] Wang, C.L., Young, P.C. and Chotai, A. (1990) "True Digital Control (TDC) non–minimal state variable feedback pole assignment", Report No. TR/85(1990) CRES, Lancaster University.

[8] Wonham, M.W. (1967) "On pole assignment in the multivariable controllable linear systems", IEEETrans. Auto. Control, AC–12, 660–665.

[9] Young, P.C. (1988) Recursive estimation, forecasting and adaptive control, Chapter appears in "Control and Dynamic systems Vol. XXXII" (C.T. Leondes ed.), Acad. Press: San Diego, 119-168.

[10] Young, P.C., Behzadi, M.A., Wang, C.L. and Chotai, A (1988) "Direct digital and adaptive control by input–output, state variable feedback pole assignment" Int. J. Control, 46, 1867–1881.

[11] Young, P.C. and Willems, J.C, (1972) "An approach to the linear multivariable servomechanism problem" Int. J. Control, 15, 961–979.

Optimal pole placement control for discrete–time systems

K. Warwick

Abstract

This paper is concerned with univariate, stochastic systems under pole placement control, within a discrete–time environment. When posed in a state–space framework the design procedure involves firstly state vector reconstruction by means of an optimal observer, and secondly the calculation of state feedback parameters which ensure that the closed–loop eigenvalues (poles) are placed in positions which can be arbitrarily selected.

The linear stochastic system models employed are initially based on a CARMA form, however when considered in the state–space, a reconstructible canonical description is used. The final controller forms which result can thus be viewed both from a polynomial and a state–space approach, with the two different methods producing identical end products only if certain conditions are met.

It is shown that by coupling an optimal state estimate with the required state feedback parameters, so the state–space pole placement controller can be regarded as equivalent to that obtained in the normally encountered, polynomial based pole placement controller. It is also shown how the optimal innovations model estimate is not that employed in the normally encountered pole placement controller, a point which questions use of the phrase 'optimal state estimate', when applied to the commonly encountered innovations model.

1 Introduction

In order to control a univariate, discrete–time system there are two distinct ways in which a solution can be found to the pole placement control problem when applied to linear stochastic systems. The output to input feedback control law can be solved by means of a polynomial description [13], in order to directly select the poles of the closed–loop transfer function. The second method however, makes use of a state–space description and employs a certain amount of Kalman filtering theory [4], combined with state feedback parameter selection, to solve for the pole placement controller[8].

It appears on the surface that the two different approaches to the pole placement control problem, polynomial and state–space, should produce a final controller form which is identical,

This is because once the requirement for arbitrarily placing poles has been made, it does not matter how the controller is obtained, there should be only one solution for that particular problem. If a polynomial based controller is found which truly results in arbitrary placement of closed–loop poles, then the problem which remains is to find a state–space based controller which is identical to that found in the polynomial case.

Many attempts have been made to connect the polynomial and state–space routes[5], [7], [8], although problems remained in that it was necessary to employ a non–optimal observer, based on an innovations model, in order to show equivalence. This would normally lead to the conclusion that if an optimal observer was employed, a 'better' pole placement controller could be obtained –this was however not the case. It is shown in this paper that the common error was in the automatic use of an innovations form of state–space model [3], despite the fact that a large number of alternative candidate models exist.

Both polynomial[13] and state–space [8] pole–placement methods have been employed within a self–tuning control framework, in which the controller parameters are calculated/updated in real–time, making use of recursively estimated parameters within a model of the system to be controlled. A common solution for the state–space based approaches is to employ an innovations form of state–space model [10] . In order to find identical pole placement control actions for both polynomial and state–space approaches to self–tuning control, the same equivalence problems occur, if an innovations model is used.

It is shown in this paper that a pole placement based state–space controller can be found, which is equivalent to the polynomial pole placement control considered, and which links an optimal interest is the use of a noise free measurement state–space system description, rather than an innovations form, and this answers many of the questions of equivalence posed by the employment of an innovations model.

There are an infinite number of different ways in which system models, of the type considered in the following section, can be posed in a state–space representation [6]. It is therefore most peculiar that by far the most commonly encountered form is the innovations model, this being particularly surprising since the optimal observer from this model, does not lead to controller equivalence with the commonly encountered polynomial based controllers.

2 System model

The univariate, stochastic system to be controlled is modelled by the discrete time equation

$$Ay(t) = Bu(t-1) + Ce(t) \tag{2.1}$$

where the polynomials A, B and C are defined by:

$$\begin{aligned} A &= 1 + a_1 z^{-1} + a_2 z^{-2} + \cdots + a_n z^{-n} \\ B &= b_0 + b_1 z^{-1} + b_2 z^{-2} + \cdots + b_n z^{-n} \\ C &= 1 + c_1 z^{-1} + c_2 z^{-2} + \cdots + c_n z^{-n} \end{aligned} \tag{2.2}$$

In which z^{-1} is the unit backward shift operator such that

$$z^{-i}y(t) \equiv y(t-i). \tag{2.3}$$

Also $\{y(t) : t \in \tau\}$ and $\{u(t) : t \in \tau\}$ are the system output and input sequences respectively whilst $\{e(t) : t \in \tau\}$ is a white noise sequence having a zero mean value and a finite variance. The input to output transport delay is restricted to one sample period in equation (2.1), which means that by definition $b_0 \neq 0$. This restriction is only for explanation purposes, in practice the transport delay can in fact be any integer number of sample periods.

As a further simplification to ease explanation, the polynomials A, B and C are all considered to be of order n in (2.2). In reality this means that higher order terms of no more than two of the polynomials A, B, C are zero. An assumption is $(A, B) = 1$, and of the pair $(C, A) = 1$.

The results given in this paper are of great relevance to adaptive control techniques, such as self-tuning control, however it is considered that the polynomial coefficients in (2.2) are known, time invariant values. The extension to the time variant/adaptive case can therefore be treated as an extrapolation if desired.

3 Pole placement controller

The polynomial based, pole placement control problem has been treated in great depth elsewhere [13], [14] and is therefore only looked at superficially in this section. The overall control objective aimed for is the arbitrary placement of poles of the closed loop transfer function. In its general sense the closed loop transfer function can be regarded as being an input/output function or a disturbance/output function or indeed both if the poles are identical. Here, only the disturbance/output transfer function is considered, the inclusion of a set-point, reference input signal being thought of as a simple extension of this study.

Let the control input signal be specified as:

$$Du(t) = Gy(t) \tag{3.1}$$

where

$$\begin{aligned} D &= 1 + d_1 z^{-1} + d_2 z^{-2} + \cdots + d_{n+1} z^{-(n+1)} \\ G &= g_0 + g_1 z^{-1} + g_2 z^{-2} + \cdots + g_n z^{-n} \end{aligned} \tag{3.2}$$

Then, by substitution of the control input $u(t)$ from (3.1), into the system equation (2.1), it follows that:

$$[AD - BG]y(t) = CDe(t) \tag{3.3}$$

For the pole placement control method, the required/specified closed loop pole polynomial is defined as:

$$T = 1 + t_1 z^{-1} + t_2 z^{-2} + \cdots + t_{n+1} z^{-(n+1)} \tag{3.4}$$

The highest index $(n+1)$ is a fairly arbitrary, but maximum, value, the aim being that by selecting t_1, t_2, $\cdots$ etc.; the operator thereby chooses the closed loop poles. If it is, for example, required that all poles must lie at the origin of the z–plane then $T = 1$.

The control polynomials, D and G, can be found from the equation:

$$AD - BG = CT \tag{3.5}$$

which is derived from the closed loop equation (3.3). It is worth noting that only the polynomials D and G are unknown in (3.5), the system polynomials A, B, C and the selected pole polynomial, T all being previously obtained values. As long as D and G are found to satisfy (3.5), it follows from (3.3) that the final closed loop equation is given by:

$$Ty(t) = De(t) \tag{3.6}$$

such that T specifies the poles relating the disturbance $e(t)$ to the output $y(t)$.

It is however necessary that a unique solution is obtained for D and G in equation (3.5), and in fact by equating like powers of z a total of $2n+1$ equations are obtained from which $2n+2$ parameters $(d_1 \cdots,\ d_{n+1},\ g_0,\ g_1,\ \cdots,\ g_n)$ must be found.

Clearly a modification must be carried out to equate the number of equations with the number of parameters. The most commonly encountered ploy [13] is to set $d_{n+1} = g_n = t_{n+1} = 0$, thereby reducing the number of equations to 2n which is then the same as the number of unknown parameters.

It is shown later that forcing the highest order terms in D and G to be zero, in order to find a unique solution, is effectively the method carried out when using a pole placement design based on an optimal state estimate from a noise free measurement state–space model. The employment of an optimal state estimate from an innovations model produces an alternative approach, as can be seen in Section 6.

4 State–space approach

Although an infinite number of state–space reconstructions of the system model (2.1) can be, in theory, found, there is only one structure which will directly result in an optimal state estimate which is that, effectively, employed in the pole placement controller considered in the previous section. This particular state–space formulation [10], [11], contains a measured output signal which is noise free, along with a state vector, at time instant t, which is directly affected by the disturbance, at the time instant t.

The noise free measurement state–space representation, of the system model in equation (2.1), is given by

$$\begin{aligned} \underline{x}(t+1) &= P\underline{x}(t) + Qu(t) + Se(t+1) \\ y(t) &= H\underline{x}(t) \end{aligned} \tag{4.1}$$

in which $\underline{x}(t)$ is the vector of state variables at time instant t, and the matrices P, Q, S and H are defined as

$$P = \begin{bmatrix} 0 & \cdots & 0 & 0 \\ 1 & \ddots & \vdots & -a_n \\ \vdots & \ddots & 0 & \vdots \\ 0 & 0 & 1 & -a_1 \end{bmatrix}$$

$$Q = \begin{bmatrix} b_n \\ \vdots \\ b_0 \end{bmatrix}$$

$$S = \begin{bmatrix} c_n \\ \vdots \\ c_1 \\ 1 \end{bmatrix}$$

$$H^{\top} = \begin{bmatrix} 0 \\ \vdots \\ 0 \\ 1 \end{bmatrix} \tag{4.2}$$

remembering that the transport delay is considered to be unity. The dimension of P is $(n+1) \times (n+1)$, whereas that of Q, $H^{\top}$ and S is $(n+1) \times 1$, and $H^{\top}$ denotes the transpose of H.

It is a straightforward procedure to check that the state–space description of (4.1) is a representation of the original system model (2.1), [12] and hence this is not laboriously shown here.

The actual control input to be applied is obtained as a combination of state feedback parameters and a representation of the state vector. However, the state vector itself is not directly available from (4.1) and an estimation must be found, the estimated state vector being used in the feedback loop in place of the actual, but unknown, value.

If $\hat{\underline{x}}(t)$ denotes the estimated state vector at time instant t, a full order observer for $\underline{x}(t)$ can be written as:

$$\hat{\underline{x}}(t+1) = [I - KH]\{P\hat{\underline{x}}(t) + Qu(t)\} + Ky(t+1) \tag{4.3}$$

in which the gain K is selected to make $\hat{\underline{x}}(t)$ an optimal estimate of $\underline{x}(t)$.

Optimality in the state estimate is considered in terms of the reconstruction error, $\Delta(t)$, which is the difference between the true and estimated state vectors, i.e.

$$\Delta(t) = \underline{x}(t) - \hat{\underline{x}}(t) \tag{4.4}$$

It is shown in [12] that the state estimate is optimal when $K = S$, which is a solution to the steady–state Kalman filter problem. With this choice of gain, under steady–state conditions

$$E[\Delta(\infty)] = 0 \tag{4.5}$$

in which $E[\cdot]$ denotes the expected value. Also, the state estimate $\hat{\underline{x}}(t)$ is the minimum mean square linear estimator of the state vector $\underline{x}(t)$, in the sense of minimization of the mean square reconstruction error, i.e. the error covariance.

The optimal state estimate of $\underline{x}(t)$ is therefore found to be:

$$\hat{\underline{x}}(t+1) = [I - z^{-1}\overline{P}]^{-1}\{[I - SH]Qu(t) + Sy(t+1)\} \tag{4.6}$$

where

$$\overline{P} = P - SHP \tag{4.7}$$

It is worth noting, and may appear to be rather obvious, that as $\hat{\underline{x}}(t)$ is the optimal estimate of $\underline{x}(t)$, *any* other estimate *must* be a non–optimal solution! So if it is required to obtain a state–space based feedback controller which is equivalent to a polynomial pole placement feedback controller, then an optimal pole placement controller is obtained when the optimal estimate $\hat{\underline{x}}(t)$ in (4.6) is employed in the feedback loop.

5 State–space controller

Let us assume, for the moment, that we actually have knowledge of the state vector $\underline{x}(t)$, this can then be used to form the control input $u(t)$ from:

$$u(t) = F\underline{x}(t) \tag{5.1}$$

where F is a $1 \times (n+1)$ row vector of state feedback parameters, which we wish to find. Note, that as with the polynomial form of control input (3.1), only regulation capabilities are being examined, the addition of an external reference signal leads to only a minor extension of this work.

By substituting for the input $u(t)$ in (4.1) it follows that

$$[I - z^{-1}\overline{F}]\underline{x}(t+1) = Se(t+1) \tag{5.2}$$

where

$$\overline{F} = P + QF. \tag{5.3}$$

Considering the output equation in (4.1), we have that the disturbance/output closed loop equation is given by

$$y(t) = H[I - z^{-1}\overline{F}]^{-1}Se(t) \tag{5.4}$$

The poles of the closed loop system can then be placed, which is the control objective, in positions dictated by the specified polynomial T, defined in (3.4). This is done by choosing the parameters in the row vector F such that

$$\det[I - z^{-1}\overline{F}] = T \tag{5.5}$$

where det[·] denotes the determinant of that function. It then follows that the numerator

$$H\text{adj}[I - z^{-1}\overline{F}]S = D \tag{5.6}$$

and this polynomial is of order n, which can be seen by simple induction, and is shown in more detail in [9]. It is highlighted directly that (5.6) is true iff (5.5) is true.

But the control input applied has the form.

$$u(t) = F\hat{\underline{x}}(t) \tag{5.7}$$

where F is obtained from (5.5) and $\hat{\underline{x}}(t)$ from (4.6). Substituting for $\hat{\underline{x}}(t)$ from (4.6), it follows that

$$\{I - z^{-1}F[I - z^{-1}\overline{P}]^{-1}[I - SH]Q\}u(t) = F[I - z^{-1}\overline{P}]^{-1}Sy(t) \tag{5.8}$$

and it is easy but extremely laborious to verify that:

$$\det[I - z^{-1}\overline{P}] = C \tag{5.9}$$

$$\{F[I - z^{-1}\overline{P}]^{-1}S\} \cdot C = G \tag{5.10}$$

$$\{I - z^{-1}F[I - z^{-1}\overline{P}]^{-1}[I - SH]Q\} \cdot C = D \tag{5.11}$$

$$\{H[I - z^{-1}\overline{F}]^{-1}S\} \cdot T = D \tag{5.12}$$

where these hold iff $\det[I - z^{-1}\overline{F}] = T$, and G is of order $n - 1$. So the control input (5.8) can also be written as:

$$\frac{D}{C}u(t) = \frac{G}{C}y(t) \tag{5.13}$$

the end result of which is equivalent to (3.1).

Also, by inspection it can be seen that:

$$\det[I - z^{-1}\overline{F}] = A\frac{D}{C} - B\frac{G}{C} = T \tag{5.14}$$

the end result of which is equivalent to (3.5).

To summarize the results of this section: It has been shown that the feedback controller (5.7) with $\hat{\underline{x}}(t)$ obtained from (4.6) and F obtained from (5.5), results in the same type of control algorithms. In other words, the optimal state estimate from (4.6) is effectively that used in common polynomial based pole placement controllers, such as that found in [13].

6 Innovations form

It is stated in Section 4 that an infinite number of state–space versions can be found for the system model (2.1). By far the most popular of these, see e.g. [1],[3] is the innovations form, in which

$$\begin{aligned} \underline{w}(t+1) &= P\underline{w}(t) + Qu(t) + Re(t) \\ y(t) &= H\underline{w}(t) + e(t) \end{aligned} \tag{6.1}$$

where $\underline{w}(t)$ is the state variable vector at time instant t, and is distinctly different from $\underline{x}(t)$. Matrices P, Q and H are identical to those in (4.2) and

$$R = \begin{bmatrix} 0 \\ c_n - a_n \\ \vdots \\ c_1 - a_1 \end{bmatrix} \tag{6.2}$$

Again, it is a straightforward procedure to check that the state–space description of (6.1) is a representation of the original system model (2.1), [12]. Further, the optimal state estimate [11] for this state–space model is given by

$$\underline{\hat{w}}(t+1) = P^1 \underline{\hat{w}}(t) + Qu(t) + Ry(t) \tag{6.3}$$

in which

$$P^1 = P - RH \tag{6.4}$$

The control input applied has, in this case, the form:

$$u(t) = f\underline{\hat{w}}(t) \tag{6.5}$$

where the state feedback parameters are, as before, found from (5.5).

By substitution for $\underline{\hat{w}}(t)$ from (6.3) into (6.5) it can be seen that

$$\{I - z^{-1}F[I - z^{-1}P^1]^{-1}Q\}u(t) = F[I - z^{-1}P^1]^{-1}Ry(t) \tag{6.6}$$

For a pole placement design objective, with the state feedback parameter once again found from (5.5), it follows that the disturbance/output closed loop equation is given by

$$y(t) = [1 + z^{-1}H(I - z^{-1}\overline{F})^{-1}R]e(t) \tag{6.7}$$

Analysis of equations (6.6) and (6.7) shows that the control input is of a similar form to (5.10) and that (5.11) holds once more. However, as can be seen in (6.6) the present value of output signal $y(t)$ is not itself fed directly back to form the new control input signal $u(t)$. In fact, as shown in [9], this can be seen more clearly by inspection of the D and G polynomials, where, in this case, we have that:

$$\begin{aligned} D &= 1 + d_1 z^{-1} + d_2 z^{-2} + \cdots + d_{n+1} z^{-(n+1)} \\ G &= g_1 z^{-1} + g_2 z^{-2} + \cdots + g_n z^{-n} \end{aligned} \tag{6.8}$$

The polynomial D is therefore of the same form as the general D polynomial given in (3.2) and hence is of order 1 greater than in the previous section's controller, although the g_0 term is zero, causing no direct output to input link to occur in the controller, i.e. the present output signal value is not instantly fed back.

It is in fact possible to obtain controller equivalence with common pole placement controllers [2] by feeding back the state estimate $\underline{\hat{w}}(t)$ combined with a certain amount of linear output feedback. However, this means that the optimal state $\underline{\hat{w}}(t)$

coupled with output feedback resulting overall in the feedback of a non–optimal state estimate. By employing the innovations form of state–space description to represent the model (2.1) it is thus necessary to use a non–optimal observer [12] along with the state feedback parameters in order to obtain a controller which is equivalent to the pole placement controller under consideration.

The state variable vector necessary to provide equivalence with the pole placement controllers discussed [13], is optimal in the sense of the noise free measurement state–space description and, as it is not equal to $\hat{\underline{w}}(t)$, must therefore be non–optimal in the sense of the innovations description. If the optimal, in the innovations model sense, state estimate is to be used to provide an equivalent pole placement controller, then output feedback must also be used. The result of this is that the effective state estimate employed is non–optimal.

7 Conclusions

The relationship between polynomial based and state–space based discrete–time pole placement control has been considered by restricting the problem to the univariate, unit transport delay case. It was shown that equivalence in control action could be found, despite taking different routes. In the state–space case the requirement was that an optimal state estimate from the noise free measurement model was coupled with the determined state feedback parameters in order to show equivalence with the common polynomial pole placement scheme considered.

The conclusions drawn from this work support those discovered earlier relating to optimal controllers [12], in emphasizing the necessity for using the noise free measurement state–space model, in order to obtain controller equivalence, as opposed to the innovations model.

Use of the innovations form of modelling such systems is therefore brought far more into question, and reasons for its employment appear to be due more to a popularity bandwagon rather than to practical usefulness.

References

[1] Astrom, K.J. 'Introduction to stochastic control theory', Academic Press, New York, 1970.

[2] Caines, P.E., 'Relationship between Box–Jenkins–Astrom control and Kalman Linear regulator', Proc. IEE, 119, 615–620, 1972.

[3] Goodwin, G.C. and Sin, K.S., 'Adaptive filtering, prediction and control', Prentice–Hall, 1984.

[4] Kalman, R.E., 'A new approach to linear filtering and prediction problems', Trans. ASME, Ser.D, 82, 35–45, 1960.

[5] Lam, K.P., 'Implicit and explicit self–tuning controllers', DPhil Thesis, Oxford University, OUEL Report No. 1334/80, 1980.

[6] Strejc, V.,'State–space theory of discrete linear control', John Wiley Interscience, 1981.

[7] Tsay, Y.T. and Shieh, L.S., 'State–space approach for self–tuning feedback control with pole assignment', Proc. IEE, 128, Pt. D, No. 3,93–101, 1981.

[8] Warwick, K., 'Self–tuning regulators - a state–space approach', Int.J.Control, 33, No.5, 839–858, 1981.

[9] Warwick, K., 'Self–tuning property of state–space self–tuners', Proc. IEE, 129, Pt.D, No.2, 96–100, 1982.

[10] Warwick, K., 'State–space models for self–tuning control', Proc.American Control Conference 87, Minneapolis, 1274–1279, 1987.

[11] Warwick, K., 'Optimal observers for ARMA models', Int.J.Control, 46, 1493–1503, 1987.

[12] Warwick, K., 'On the relationship between minimum variance and minimum quadratic loss controllers', Proc. Int. Conference Control 88, 587–591, 1988.

[13] Wellstead, P.E., Prager, D.L. and Zanker, P., 'Pole assignment self–tuning regulator', Proc.IEE, 126, 781–787, 1979.

[14] Wolovich, W.A., 'Linear multivariable systems', Springer–Verlag, 1974.

Robust control design using eigenstructure assignment

S.K. Spurgeon R.J. Patton

Abstract

Given the linear state–space representation of a dynamic system, a method for computing allowable eigenvector sub–spaces using Singular Value Decomposition for both real and complex eigenvalues is illustrated. A computational procedure to render assigned eigenvalues minimally sensitive to perturbations is described. The viability of using such conditioning measures for *robust* control system design is then considered using a fully non–linear simulation of a light aircraft.

1 Introduction

Many methods are available to compute accurately the feedback gains for a single–input system where at most one solution to the eigenvalue assignment problem exists. However, for the multi–input case where the feedback is underdetermined, methods do not necessarily lead to well–conditioned solutions to the problem.

A linear, time–invariant, controllable system described by the real matrix pair (A, B) and a desired self–conjugate eigenvalue set to obtain the desired damping and rise time of the system response is assumed given. The objective will be to find a linear, state–variable feedback gain where the eigenvectors corresponding to the desired eigenvalues have been assigned to obtain a *robust* solution. The freedom available to assign the eigenvectors is described and a computational method for constructing allowable eigenvector subspaces using the singular value decomposition presented. Once these spaces have been determined, two methods of eigenvector assignment are possible. If information is known about a desirable weighting of the system states for each mode, it is possible to choose a desired eigenvector specification and project this in a least–squares sense into the allowable eigenvector subspace. In this way a modal shaping method of the system response is possible. If desired weightings are unknown or unimportant a second method of eigenvector assignment is possible. This restricts the degrees of freedom available in the assignment problem in order to produce a robust solution whereby the eigenvalues are rendered insensitive to perturbations in the coefficient matrices of the system equation. This paper will describe a simple iterative procedure to produce such a robust solution. The design procedure is illustrated by considering the design of a stability augmentation system for the lateral motion model of a light aircraft. A fully non–linear simulation of the vehicle will be used to assess the performance of the control scheme; the simulation will be linearised at each simulation sample point and the eigenvalues of the corresponding locally linearised closed–loop matrix so formed investigated. This, together with a control scheme designed using purely eigenvalue placement with no eigenvector considerations, will be used to demonstrate both the effectiveness of the design approach and also the viability of using such conditioning measures to analyse closed–loop control systems designed using other techniques.

2 Characterisation of the freedom available in eigenstructure assignment

Consider the linear time–invariant system defined by the state equations:

$$\dot{\underline{x}} = A\underline{x} + B\underline{u} \tag{2.1}$$

$$\underline{y} = C\underline{x} \tag{2.2}$$

where $A \in R^{n \times n}, B \in R^{n \times m}$ and $C \in R^{r \times n}$ are the system, input and output matrices respectively. Given this system which is assumed to be controllable, apply linear state–variable feedback $\underline{u} = F\underline{x}$ where $F \in R^{m \times n}$. The following closed–loop state equation is obtained:

$$\dot{\underline{x}} = (A + BF)\underline{x} \tag{2.3}$$

It has been shown by Wonham in 1976 [18] that a full–state feedback matrix F can be found to assign an arbitrary self–conjugate set of eigenvalues if the system is controllable. In 1976, Moore [10] described the freedom available to assign eigenvectors for an arbitrary self–conjugate set of eigenvalues using state–feedback. He gives both necessary and sufficient conditions for a feedback matrix F to exist which satisfies:

$$(A + BF)\underline{v}_i = \lambda_i \underline{v}_i \qquad i = 1, \cdots, n \tag{2.4}$$

In order to present Moore's result, associate with each λ_i a matrix $S(\lambda_i) \in R^{n \times (n+m)}$ such that

$$S(\lambda_i) = [\lambda_i I - A : B] \tag{2.5}$$

and a compatibly partitioned matrix:

$$K(\lambda_i) = \begin{bmatrix} N(\lambda_i) \\ M(\lambda_i) \end{bmatrix} \tag{2.6}$$

where the columns of $K(\lambda_i)$ form a basis for the null space of $S(\lambda_i)$. The necessary and sufficient conditions to find a real matrix F satisfying (2.4) are:

1. The $\underline{v}_i \in C^n$ are linearly independent vectors
2. $\underline{v}_i = \underline{v}_j^*$ whenever $\lambda_i = \lambda_j^*$
3. $\underline{v}_i \in \text{span}(N(\lambda_i))$.

Furthermore, if such a feedback matrix F exists and rank $(B) = m$ then F is unique.

3 Computation of allowable eigenvector subspaces

From the result due to Moore it has been seen that the choice of eigenvalue λ_i determines the allowable subspace in which any chosen eigenvector must reside. This section will consider a computational method for constructing allowable eigenvector subspaces for the cases of real and complex eigenvalues.

3.1 Assignment of the Eigenvectors Corresponding to Real Eigenvalues

If it is assumed that $\lambda_i \in \mathcal{R}$, then the corresponding eigenvector $\underline{v}_i \in \mathcal{R}^n$. Equation (2.4) may be rewritten as:

$$(\lambda_i I - A)\underline{v}_i - BF\underline{v}_i = 0 \tag{3.1.1}$$

If $S(\lambda_i)$ and $K(\lambda_i)$ are as defined by equations (2.5) and (2.6) respectively, then any arbitrary vector $\underline{k}_i$ which post-multiplies $K(\lambda_i)$ will give a resulting vector which lies in the null-space of $S(\lambda_i)$. Thus:

$$[\lambda_i I - A : B]\begin{bmatrix} N(\lambda_i) \\ M(\lambda_i) \end{bmatrix}\underline{k}_i = 0 \tag{3.1.2}$$

because $K(\lambda_i)$ is a basis for the null space of $S(\lambda_i)$. Expanding equation (3.1.2) yields:

$$[\lambda_i I - A]N(\lambda_i)\underline{k}_i + BM(\lambda_i)\underline{k}_i = 0 \tag{3.1.3}$$

Comparison of equation (3.1.3) with (3.1.1) yields:

$$\underline{v}_i = N(\lambda_i)\underline{k}_i \tag{3.1.4}$$

with $\underline{v}_i \in Span[N(\lambda_i)]$ and $\underline{k}_i$ determining where in the allowable subspace $\underline{v}_i$ exists. The required null space of $S(\lambda_i)$ will be computed using the singular value decomposition which is defined as follows [5]:

If $Z \in \mathcal{R}^{m \times n}$, then there exist orthogonal matrices:

$$U = [\underline{u}_1, \cdots, \underline{u}_m] \in \mathcal{R}^{m \times m}$$

and

$$V = [\underline{v}_1, \cdots, \underline{v}_n] \in \mathcal{R}^{n \times n}$$

such that:

$$Z = U\text{diag}(\sigma_1, \cdots, \sigma_p)V^{\mathsf{T}} \tag{3.1.5}$$

where $p = \min(m, n)$ and $\sigma_1 \geq \sigma_2 \geq \cdots \geq \sigma_m \geq 0$. This is the singular value decomposition of Z. The σ_i are the singular values of Z and the vectors $\underline{u}_i$ and $\underline{v}_i$ are the ith left singular vector and the ith right singular vector respectively.

Consider now the singular value decomposition of $S(\lambda_i) \in \mathcal{R}^{n\times(n+m)}$ defined in equation (1.5):

$$\begin{aligned} S(\lambda_i) &= [\lambda_i I - A : B] \\ &= U \; {}_{n}\!\overset{\qquad n \qquad\qquad\qquad m}{\begin{bmatrix} \sigma_1 & & 0 & \cdot & \\ & \sigma_2 & & \cdot & 0 \\ & & \cdot & \cdot & \\ 0 & & \sigma_n & \cdot & \end{bmatrix}} V^{\mathsf{T}} \\ &= U[\text{diag}(\sigma_i) \; : \; 0]V^{\mathsf{T}} \end{aligned} \tag{3.1.6}$$

Post–multiplying both sides of equation (3.1.6) by the orthogonal matrix V, it is noted that the last m columns of the $S(\lambda_i)V$ product are null and so an appropriate null–space of $S(\lambda_i)$ can be found by using the last m columns of V. This method has the advantage that existing poles of the open–loop system or values near to them may still be chosen as eigenvalues of the closed–loop system. This would be important, for example, in the eigenvector assignment of a partially controllable system in which certain modes can be fixed or are invariant by feedback. Methods which utilise $(\lambda_i I - A)^{-1}$ in the subspace calculation may thus lead to numerical problems. The reader is refered to references [2,13,14] for a description of this approach.

3.2 Assignment of the Eigenvectors Corresponding to Complex Eigenvalues

If it is assumed that $\lambda_i \in \mathcal{C}$, the analysis is a little more complicated [9]. Write:

$$\lambda_i = \lambda_{re} + j\lambda_{im} \tag{3.2.1}$$

Let the corresponding eigenvector $\underline{v}_i$ be written: $\underline{v}_i = \underline{v}_{re} + j\underline{v}_{im}$ Note that there will also be a complex conjugate eigenvector $\underline{v}_{i+1}$ corresponding to the complex conjugate eigenvalue λ_{i+1}. The eigenvector (3.2.2) must satisfy the equation (1.4) and hence:

$$(A+BF)\underline{v}_{re} = \lambda_{re}\underline{v}_{re} - \lambda_{im}\underline{v}_{im} \tag{3.2.2}$$

$$(A+BF)\underline{v}_{im} = \lambda_{re}\underline{v}_{im} + \lambda_{im}\underline{v}_{re} \tag{3.2.3}$$

Equation (3.2.3) may be rewritten in matrix form as:

$$[\lambda_{re} I - A : \lambda_{im} I : B] \begin{bmatrix} \underline{v}_{re} \\ -\underline{v}_{im} \\ -F\underline{v}_{re} \end{bmatrix} \tag{3.2.4}$$

Similarly for equation (3.2.4):

$$[\lambda_{re} I - A : \lambda_{im} I : B] \begin{bmatrix} \underline{v}_{im} \\ \underline{v}_{re} \\ -F\underline{v}_{im} \end{bmatrix} \tag{3.2.5}$$

It follows that, for complex eigenvalues, $S(\lambda_i) \in \mathcal{R}^{n\times(2n+m)}$ is defined by:

$$S(\lambda_i) = [\lambda_{re}I - A : \lambda_{im}I : B] \tag{3.2.6}$$

with a compatibly partitioned $K(\lambda_i)$ whose columns constitute a set of basis vectors for the null space of $S(\lambda i)$:

$$K(\lambda_i) = \begin{bmatrix} N(\lambda_i) \\ P(\lambda_i) \\ M(\lambda_i) \end{bmatrix} \tag{3.2.7}$$

From equations (3.2.5), (3.2.6) and (3.2.8):

$$\begin{bmatrix} \underline{v}_{re} \\ -\underline{v}_{im} \\ -F\underline{v}_{re} \end{bmatrix} \in \text{span} \begin{bmatrix} N(\lambda_i) \\ P(\lambda_i) \\ M(\lambda_i) \end{bmatrix} \tag{3.2.8}$$

and:

$$\begin{bmatrix} \underline{v}_{im} \\ \underline{v}_{re} \\ -F\underline{v}_{im} \end{bmatrix} \in \text{span} \begin{bmatrix} N(\lambda_i) \\ P(\lambda_i) \\ M(\lambda_i) \end{bmatrix} \tag{3.2.9}$$

Combining the information from equations (3.2.9) and (3.2.10):

$$\begin{bmatrix} \underline{v}_{re} \\ \underline{v}_{im} \end{bmatrix} \in \text{span} \begin{bmatrix} N(\lambda_i) \\ -P(\lambda_i) \end{bmatrix} \cap \text{span} \begin{bmatrix} P(\lambda_i) \\ N(\lambda_i) \end{bmatrix} \tag{3.2.10}$$

$$\in \text{span}(\alpha^{\mathsf{T}}) \cap \text{span}(\beta^{\mathsf{T}}) \tag{3.2.11}$$

Let K_α and K_β be the matrices whose columns constitute bases for the null spaces of α and β respectively. Also let:

$$\gamma = [K_\alpha : K_\beta]^{\mathsf{T}} \tag{3.2.12}$$

and K_γ be the matrix whose columns constitute a basis for the null space of γ. K_γ is thus a basis for span $(\alpha^{\mathsf{T}}) \cap$ span (β^{T}) and a specific vector within the null space of γ is defined by postmultiplying K_γ by a vector $\underline{w}_T$. Thus:

$$\begin{bmatrix} \underline{v}_{re} \\ \underline{v}_{im} \end{bmatrix} = K_\gamma \underline{w}_T \tag{3.2.13}$$

The required null spaces are computed using the singular value decomposition as illustrated in equations (3.1.5) and (3.1.6). It should be noted that if $\lambda_1, \lambda_2 \in \mathcal{C}$ then $\lambda_1 = \lambda_2^*$ because the eigenvalues form a self-conjugate set. Furthermore, from the results of Moore [10], $\underline{v}_1 = \underline{v}_2^*$ and so the specification of one complex eigenvalue and eigenvector contains all of the information of the complex-conjugate pair. Thus for an n-state system with one complex eigenvalue pair the eigenvector matrix V is given by:

$$V = [\underline{v}_{re}, \underline{v}_{im}, \underline{v}_3, \underline{v}_4, \cdots \underline{v}_n] \tag{3.2.14}$$

Once V has been determined, the computation of a feedback matrix F is well known and described elsewhere [7].

4 Robust eigenstructure assignment

A desirable property of any closed–loop system is that the eigenvalues are rendered insensitive to perturbations in the coefficient matrices of the system equations ; such a well–conditioned solution will be termed *robust.* It has been seen that the solution to the problem of designing a state–feedback controller using eigenvalue placement is generally underdetermined with many degrees of freedom. The objective of the design process under consideration here will be to determine a robust control system by restricting the degrees of freedom of the assignment problem. Let $\underline{v}_i$ and $\underline{w}_i, i = 1, \cdots, n$ be the right and left eigenvectors of the closed–loop system matrix corresponding to an eigenvalue λ_i. Therefore:

$$(A + BF)\underline{v}_i = \lambda_i \underline{v}_i, \; \underline{w}_i^{\mathsf{T}}(A + BF) = \lambda_i \underline{w}_i^{\mathsf{T}} \tag{4.1}$$

If the closed–loop system matrix $(A+BF)$ has n linearly independent eigenvectors, then it is diagonalisable and it has been shown by Wilkinson [17] that the sensitivity of an eigenvalue λ_i to perturbations in A, B and F depends upon the magnitude of the condition number, c_i, where:

$$c_i = \frac{\|\underline{w}_i\|_2 \|\underline{v}_i\|_2}{|\underline{w}_i^{\mathsf{T}} \underline{v}_i|} \geq 1 \tag{4.2}$$

For the case of multiple eigenvalues, a particular choice of eigenvectors is assumed. A bound upon the individual eigenvalue sensitivities is given by:

$$\max c_i \leq \kappa_2(V) = \|V\|_2 \|V^{-1}\|_2 \tag{4.3}$$

where $\kappa_2(V)$ denotes the condition number of the matrix $V = [\underline{v}_1, \cdots, \underline{v}_n]$ of right eigenvectors. The feedback matrix is obtained by assigning a set of linearly independent right eigenvectors corresponding to the required eigenvalues such that the matrix of eigenvectors is as well–conditioned as possible. The robust eigenvalue assignment problem may therefore be stated as follows:

Given A, B and a set of desired eigenvalues, find a real matrix F and a non–singular matrix V satisfying:

$$(A + BF)V = V\Lambda \tag{4.4}$$

such that some measure of the conditioning, or robustness, of the eigenproblem is optimised.

Some flexibility is available in the choice of measure of the conditioning of the eigenproblem. For example, $\|\underline{c}\|_2$ where $\underline{c}^{\mathsf{T}} = [c_1, \cdots, c_n]$ is the vector of condition numbers corresponding to the matrix of eigenvectors V could be chosen. Alternatively the condition number of the matrix V defined by equation (4.3) could be used.

The degrees of freedom that are available in the choice of feedback matrix F are reflected by the degrees of freedom available in the selection of the matrix of eigenvectors. For the case of a single input system, V is uniquely determined except for scaling and the condition numbers c_i cannot be controlled. If $m = n$, then V may always be chosen to be orthogonal and hence $c_i = 1$ for all i. However, for the general multi-input system the sensitivities of the assigned eigenvalues can be controlled only to a limited extent by appropriate choice of V.

The merits of a robust eigenstructure assignment approach have been discussed and the practical implementation of the theory will now be considered. The work described is an extension of the work of Kautsky et al. [7,8,12]; their "Method O" will be augmented to enable both real and complex eigenvalue assignment using the allowable subspaces described earlier in the chapter. An algorithm results which is simple to implement and, although not guaranteed to converge, is adequate to provide considerable improvement in conditioning, usually with little effort.

The objective of "Method O" is to choose a set of right vectors such that each is maximally orthogonal to the space spanned by the remaining vectors. This is equivalent to choosing $\underline{v}_i$ and the normalised vector $\underline{\tilde{y}}_i$ orthogonal to the space $l_i =< \underline{v}_j : j \neq i >$ for all i. This is accomplished using an iterative process in which each vector $\underline{v}_i$ is replaced by a vector with maximum mis–alignment angle to the current space l_i for each $i = 1, \cdots, n$ in turn. The new vector is obtained using the QR method. The decomposition of V_i is found as follows:

$$V_i = [\underline{v}_1, \cdots, \underline{v}_{i-1}, \underline{v}_{i+1}, \cdots \underline{v}_n] \quad (4.5)$$

$$= [Q_i \underline{\tilde{y}_i}] \begin{bmatrix} R_i \\ \underline{O}^{\mathsf{T}} \end{bmatrix} \quad (4.6)$$

Thus the $\underline{y}_i$ orthogonal to l_i is found. The columns of the allowable subspace $N(\lambda_i)$ are othonormalised to give $Q(\lambda_i)$. The vector $\underline{\tilde{y}}_i$ is then projected into $Q(\lambda_i)$ to give:

$$\underline{v}_i^{\text{new}} = Q(\lambda_i) \frac{Q(\lambda_i)^{\mathsf{T}} \underline{\tilde{y}}_i}{\|Q(\lambda_i)^{\mathsf{T}} \underline{\tilde{y}}_i\|_2} \quad (4.7)$$

This vector has minimum mis–alignment angle to $\underline{\tilde{y}}_i$.

Some modification of the above is required for the case of complex–conjugate eigenvalues. This is because for a complex eigenvector the real and imaginary parts are related from equation (3.2.13) by:

$$\begin{bmatrix} \underline{v}_{re} \\ \underline{v}_{im} \end{bmatrix} = K_\gamma \underline{w}_T \quad (4.8)$$

$$= \begin{bmatrix} S_{re} \\ S_{im} \end{bmatrix} \underline{w}_T \quad (4.9)$$

where $\underline{w}_T$ is any vector of appropriate dimension and S_{re} and S_{im} denote the allowable subspaces of the real and imaginary parts of the eigenvector respectively. The vectors forming the columns of S_{re} are normalised and the corresponding columns of S_{im} are adjusted by the same factor to give:

$$\begin{bmatrix} \underline{v}_{re} \\ \underline{v}_{im} \end{bmatrix} = \begin{bmatrix} R_{re} \\ R_{im} \end{bmatrix} \underline{w}_T \quad (4.10)$$

Then if we define:

$$V = [\underline{v}_{re}, \underline{v}_{im}, \underline{v}_3, \cdots, v_n] \tag{4.11}$$

where $\underline{v}_3, \cdots, \underline{v}_n$ are real, then remove $\underline{v}_{re}$ and proceed as before to find the new real component which has minimum mis–alignment angle to the vector which is orthogonal to the set $< \underline{v}_{im}, \underline{v}_3, \cdots, \underline{v}_n >$ ie.

$$\underline{v}_{re}^{\text{new}} = R_{re} \frac{R_{re}^{\mathsf{T}} \underline{\tilde{y}}_1}{\| R_{re}^{\mathsf{T}} \underline{\tilde{y}}_1 \|_2} \tag{4.12}$$

The determination of a suitable vector $\underline{w}_T$ in equation (4.10) then defines the corresponding imaginary part of the eigenvector and so it is necessary to update $\underline{v}_{im}$ using equation (4.7) as follows:-

$$\underline{v}_{im}^{\text{new}} = R_{im} \frac{R_{re}^{\mathsf{T}} \underline{\tilde{y}}_1}{\| R_{re}^{\mathsf{T}} \underline{\tilde{y}}_1 \|_2} \tag{4.13}$$

Now proceed with the sweep and update $\underline{v}_3, \cdots, \underline{v}_n$ in turn.

Any set of linearly independent eigenvectors belonging to the required allowable subspaces can be taken to given an initial eigenvector matrix. It has been found in practice to be better to update V after the $(n-1)$th iteration as:

$$V_n = \alpha V_{n-1} + (1-\alpha) V_{n-2} \qquad 0.5 < \alpha \leq 1 \tag{4.14}$$

This is because the imaginary parts of the eigenvector are not updated using the maximum orthogonality criterion but are predetermined from the corresponding real part. Some oscillation of successive V_k results which is smoothed using the linear combination of equation (4.14). A suitable stopping criterion was found to be given by whether or not there was any improvement from $\kappa(V_{n-2})$ to $\kappa(V_{n-1})$.

5 Stability augmentation system design

The viability of using the conditioning measures which have been described here will now be considered [11,16]. The robust control system design procedure will be used to develop an aircraft stability augmentation system for a light aircraft. For the vehicle under consideration the equations of motion and the aerodynamic parameters are well defined and published elsewhere [3]. The design of a state regulator for the lateral motion is considered. The lateral and longitudinal motions are assumed decoupled for control system design and a "stick fixed" linearisation established by considering small changes about a chosen state trajectory gives the following lateral system linear model:

$$\dot{\underline{x}} = A\underline{x} + B\underline{u} \tag{5.1}$$

where

$$\begin{aligned} \underline{x}^{\mathsf{T}} &= (v, p, r, \Phi, \Psi) \\ &= (\text{sideslip velocity, roll rate, yaw rate, roll angle, yaw angle}) \\ \underline{u} &= \begin{bmatrix} \zeta_c \\ \xi_c \end{bmatrix} \begin{matrix} \text{rudder angle demand} \\ \text{aileron angle demand} \end{matrix} \end{aligned} \tag{5.2}$$

and the state A and B matrices are given by

$$A = \begin{bmatrix} y_v & 0 & y_r - U_0 & g & 0 \\ l_v & l_p & l_r & 0 & 0 \\ n_v & n_p & n_r & 0 & 0 \\ 0 & 1 & 0 & 0 & 0 \\ 0 & 0 & 1 & 0 & 0 \end{bmatrix}$$

$$B = \begin{bmatrix} y_\zeta & 0 \\ l_\zeta & l_\xi \\ n_\zeta & n_\xi \\ 0 & 0 \\ 0 & 0 \end{bmatrix} \tag{5.3}$$

The aerodynamic derivatives are denoted by y_v, n_v etc, and U_0 is the nominal forward air speed The trim flight matrices corresponding to a $33ms^{-1}$ airspeed are:

$$\begin{aligned} A &= \begin{bmatrix} -0.277 & 0 & -32.9 & 9.81 & 0 \\ -0.1033 & -8.03025 & 3.75 & 0 & 0 \\ 0.3649 & 0 & -0.639 & 0 & 0 \\ 0 & 1 & 0 & 0 & 0 \\ 0 & 0 & 1 & 0 & 0 \end{bmatrix} \\ B^{\mathsf{T}} &= \begin{bmatrix} -5.432 & 0 & -9.49 & 0 & 0 \\ 0 & -28.64 & 0 & 0 & 0 \end{bmatrix} \end{aligned} \tag{5.4}$$

The lateral motion is characterised by three modes: roll, Dutch Roll and spiral. These modes are required to be such that the spiral mode is very slow, the roll mode is fast and stable and the Dutch Roll mode has 0.7 damping ratio. A set of suitable eigenvalues for these closed–loop modes have been chosen as:

$$\begin{aligned} \lambda_1, \lambda_2 &= -1.5 \pm j1.5 && \text{Dutch Roll mode} \\ \lambda_3 &= -4 && \text{Roll mode} \\ \lambda_4 &= -0.05 && \text{Spiral mode} \\ \lambda_5 &= -1 && \text{Heading angle mode} \end{aligned} \tag{5.5}$$

eigenvalue	$-1.5 \pm j1.5$	-4	-0.05	-1
condition number when assigned using Robust technique	11.61	3.16	1.15	2.48
condition mumber when assigned with no Robust consideration	50.44	13.22	7.95	62.75

Table 1. Individual Eigenvalue Sensitivities

Using the robust assignment technique, a full-state feedback matrix is generated with corresponding eigenvector matrix given by:

$$V_{ROB} = \begin{bmatrix} 0.9936 & 0 & -0.1563 & 0.0559 & 0.0631 \\ 0.0640 & -0.0641 & 0.9546 & 0.0080 & -0.6764 \\ 0.0188 & -0.0441 & -0.0830 & -0.0492 & 0.2014 \\ -0.0427 & 0 & -0.2387 & -0.1609 & 0.6764 \\ -0.0209 & 0.0085 & 0.0208 & 0.9841 & -0.2014 \end{bmatrix} \quad (5.6)$$

where:

$$\kappa_2(V_{ROB}) = 25.25 \quad (5.7)$$

The individual eigenvalue sensitivities are shown in the second row of Table 3.1.

A second full-state feedback matrix is generated with no consideration of the robustness of the eigenstructure. In other words, the gains are determined using pole placement. The corresponding eigenvector matrix V_{NONROB} is given by:

$$V_{NONROB} = \begin{bmatrix} 0.0869 & 2.6890 & -0.6376 & 0.4718 & -0.0155 \\ 0.8923 & 0 & 0.7382 & 0.0062 & -0.6779 \\ -0.012 & -0.0926 & -0.1163 & -0.0436 & 0.2006 \\ -0.2974 & -0.2974 & -0.1846 & -0.1237 & 0.6780 \\ -0.0038 & 0.0579 & 0.0291 & 0.8719 & -0.2007 \end{bmatrix} \quad (5.8)$$

where

$$\kappa_2(V_{NONROB}) = 240.53 \quad (5.9)$$

and the individual eigenvalue sensitivities are given in the final row of Table 1. The improvement gained by a reduction of an order of magnitude in the condition number κ_2 is being considered. The desired modal shaping for the lateral motion of an aircraft configuration is well known [1,6,15]. For example, to conform to handing quality criteria, it is necessary to decouple the rolling motions from the yawing motions; if the aircraft is banked a side force is created due to the tilting of the lift vector which will create sideslip or yaw. The converse is also true. Dutch Roll is a phenomenon associated with the (v, r) subsystem of the motion and so no coupling of Dutch roll into either roll angle or roll rate is desirable. The roll mode should appear dominantly on roll rate and not on yaw rate or sideslip velocity to ensure acceptable behaviour in banking manoeuvres. The spiral mode should not appear in sideslip velocity in order to avoid sideslip in steady turns. This gives a desired modal structure to the eigenvector matrix as shown below:

$$V_{MODAL} = \begin{bmatrix} 1 & X & 0 & 0 & X \\ 0 & 0 & 1 & X & X \\ X & 1 & 0 & X & X \\ 0 & 0 & X & 1 & X \\ X & X & X & X & X \end{bmatrix} \tag{5.10}$$

where X denotes that the magnitude of the element is unimportant. Equation (5.10) should be compared with the robust and non–robust eigenvector matrices given in equations (5.6) and (5.8) respectively. It can be seen that desired modal shaping results from the robust eigenvalue assignment technique as may be expected. However, for the eigenvector matrix resulting from purely pole–placement and seen to be lacking in robustness from Table 1, the modal weightings are undesirable.

A non–linear simulation incorporating the full force and moment lateral and longitudinal dynamics together with cross–coupling effects and wind turbulence in the form of a Dryden Spectrum [4] will now be to assess the performance of the robustly designed controller. This will be linearised at each simulation sample point and the eigenstructure of the corresponding locally linearised closed–loop matrix so formed investigated. In this way the eigenvalue variations can be displayed graphically in order to assess quantitatively the performance of the control scheme in the presence of uncertainty. Figure 1 shows the variations of the real parts of the eigenvalues during a simulation for both control schemes developed here. The control scheme designed using the robust assignment algorithm is seen to produce an almost invariant response. However, the control scheme derived using eigenvalue placement alone is seen to cause the eigenvalues to change significantly during the simulation resulting in an instability in the spiral mode; the aircraft was in fact found to stall at approximately $t = 4.5$ seconds using this second approach.

6 Conclusion

A method of eigenstructure assignment which restricts the degrees of freedom available in the assignment problem in order to produce a well–conditioned solution whereby the chosen eigenvalue set is rendered insensitive to perturbations in the coefficient matrices of the system equations has been discussed. The power of this robust eigenstructure assignment approach has been demonstrated by displaying the closed–loop eigenvalue variations during a fully non–linear simulation with stochastic disturbance. A control scheme designed using purely eigenvalue placement with no robust eigenstructure considerations has been shown to render the same system unstable. The conditioning measures used are found to be worthy of consideration when analysing closed–loop control systems designed using other techniques.

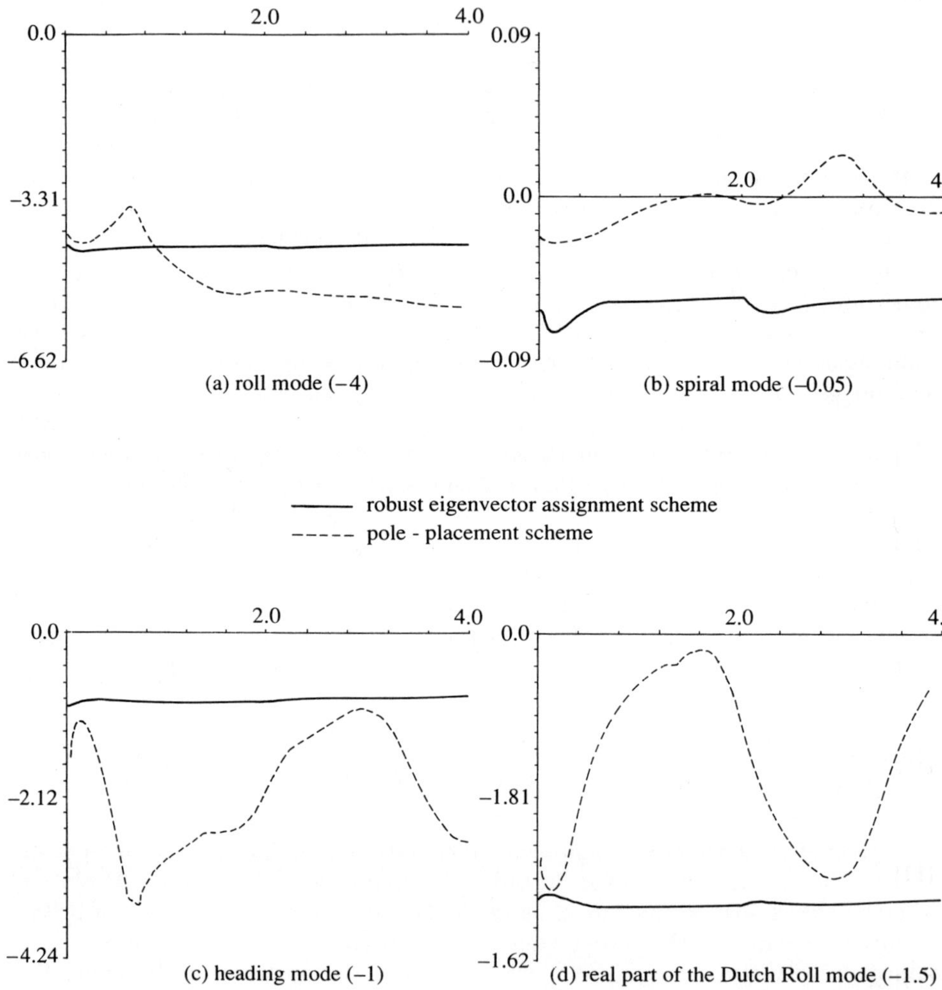
0.0
2.0
4.0
-3.31
-6.62
(a) roll mode (–4)
0.09
0.0
2.0
4.0
–0.09
(b) spiral mode (–0.05)
robust eigenvector assignment scheme
pole - placement scheme
0.0
2.0
4.0
-2.12
-4.24
(c) heading mode (–1)
0.0
2.0
4.0
-1.81
-1.62
(d) real part of the Dutch Roll mode (–1.5)

Figure 1.

References

[1] Alag, G.S. and Duke, E.L., Development of Control Laws for a Flight Test Maneuver Autopilot, J. Guidance, Control and Dynamics, *9*, 4, 441–445, July–August 1986.

[2] Andry, A.N., Shapiro, E.Y., and Chung, J.C., Eigenstructure Assignment for Linear Systems, IEEE Trans. on Aerospace and Electronic Systems, AES–19, 5, 711–729, September 1983.

[3] Aslin, P.P., Aircraft Simulation and Robust Flight Control System Design, D.Phil. Thesis, University of York, 1985.

[4] Gault, J.D., and Gunter, D.E., Atmospheric Turbulence Considerations for Future Aircraft Designed to Operate at Low Altitudes, J. Aircraft, 5, (2), 1968.

[5] Golub, G., and Van Loan, C., Matrix Computations, North Oxford Academic, 1983.

[6] Harvey, C.A., and Pope, R.E., Design Techniques for Multivariable Flight Control Systems, in Control and Dynamic Systems, 18, Academic Press, 1982.

[7] Kautsky, J., and Nichols, N.K., and Van Dooren, P., Robust Pole Assignment in Linear State Feedback, Int. J. Control, *41*, 5, 1129–1155, 1985.

[8] Kautsky, J., Nichols, N.K., Robust Eigenstructure Assignment in State Feedback Controls, University of Reading, Dept. of Mathematics, Numerical Analysis Report NA/2/83.

[9] Mielke, R.R., Tung, L.J., and Carraway, P.I., Design of Multivariable Feedback Control Systems via Spectral Assignment Using Reduced–Order Models and Reduced–Order Observers, NASA Contractor Report 3889.

[10] Moore, B.C., On the Flexibility Offered by State Feedback in Multivariable Systems Beyond Closed–loop Eigenvalue Assignment, IEEE Trans. Automatic Control, AC–21, 689–692, 1976.

[11] Mudge, S.K., and Patton, R.J., An Analysis of the Technique of Robust Eigenstructure Assignment with Application to Aircraft Control, IEE Proceedings Part D, Control Theory and Applications, 135, 4, 275-281, 1988

[12] Nichols, N.K., Robustness in Partial Pole Placement, IEEE Trans. Automatic Control, AC–23, 8, 728–732, August 1987.

[13] Sobel, K.M., and Shapiro, E.Y., A Design Methodology for Pitchpointing Flight Control System, J. Guidance, Control and Dynamics, *8*, 2, 181–187, March–April 1985.

[14] Sobel, K.M., and Shapiro, E.Y., Eigenstructure Assignment: A Tutorial–Part 1 Theory, Proceedings of the 1985 American Control Conference, Boston, MA, 456–460, June 1985.

[15] Sobel, K.M., and Shapiro, E.Y., Eigenstructure Assignment: A Tutorial–Part 2 Applications, Proceedings of the 1985 American Control Conference, Boston, MA, 461–467, June 1985.

[16] Sobel, K.M., and Shaliro, E.Y., Application of Eigenstructure Assignment to Flight Control Design: Some Extensions. Journal Guidance, Control and Dynamica, *10*, 1, 73–81, Jan–Feb 1987.

[17] Wilkinson, J.H., The Algebraic Eigenvalue Problem, Oxford University Press, 1965.

[18] Wonham, W.M., On Pole Assignment in Multi–input, Controllable Linear Systems, IEEE Trans. Automatic Control, AC–12, 660–665, 1967.

A geometric approach on output feedback pole assignment

W.C.D. Ho

Abstract

In this paper the concept of controllability subspaces for a pair (A,B) or observability subspaces for (A,C) can be described in terms of the Kernel of the matrices $(A-\lambda I,-B)$ or $(A'-\lambda I,-C')$ or $(A'-\lambda I,-C')$ in the form of a particular polynomial matrix. Although Warren and Eckberg [9] have studied the dimensions of controllability spaces, there are no numerical examples to support the use of this theory in practice. This paper will extend Warren and Eckberg's approach and will apply to output feedback pole assignment by satifying some necessary and sufficient conditions developed by Fletcher et al [2]. A complementary set of left and right eigenvalues can be established by a simple substitution of some eigenvalues into the polynomial matrix. This makes it convenient to assign all the poles to any location without changing the controllability and observability subspaces. The orthogonal properties of both subspaces have important implications for robust eigenstructure assigned by output feedback.

1 Introduction

Consider a time–invariant linear multivariable control system with state and output equations.

$$\begin{aligned} \dot{x} &= Ax + Bu \\ y &= Cx \end{aligned} \tag{1.1}$$

where $x(t), u(t), y(t)$ are n, m, p dimensional vectors; A, B, C are real constant matrices of appropriate dimensions with B, C of full rank. (A, B) is controllable and (A, C) is observable. The definition of controllability and observability are given in the next section.

Output feedback is applied to the system in accordance with the control low,

$$u(t) = -Ky(t) \tag{1.2}$$

It is important to find an appropriate choice of real matrix K such that the matrix $A - BKC$ has given self–conjugate distinct eigenvalues.

Recently Fletcher, Kautsky, Kolka and Nichols [2] have reported some necessary and sufficient conditions for this problem by employing the concept of inner inverse of a matrix and complimentary sets of left and right eigenvectors,

$S_o = < s_1, s_2, \cdots, s_q >$ and $T_o = < t_{q+1}, \cdots, t_n >$. Their approach does not impose the restriction $n < m + p$ [7], but S_o, T_o have to satisfy the following conditions.

$$\text{rank } S_0 = q, \text{ rank } T_0 = n - q \tag{1.3}$$
$$AS_0 - S_0\Lambda_q = BV_0 \text{ for some } V_0 \tag{1.4}$$
$$T_0'A - \Lambda_{n-q}T_0' = U_0C \text{ for some } U_0 \tag{1.5}$$
$$\text{if } CS_0z = 0 \text{ then } BV_0z = 0 \tag{1.6}$$
$$\text{if } y'T_0B = 0 \text{ then } y'U_0'C = 0 \tag{1.7}$$
$$T_0'S_0 = 0 \tag{1.8}$$

In (1.4) Λ_q is a $q \times q$ diagonal matrix with $\lambda_1, \lambda_2, \cdots, \lambda_q$ along the diagonal. In (1.5) Λ_{n-q} is a $(n-q) \times (n-q)$ diagonal matrix with $\lambda_{q+1}, \cdots, \lambda_n$ along the diagonal. Fletcher at al[2] have shown that if (1.4) and (1.6) hold then there are matrices K_s which satisfy

$$(A - BK_sC)S_o = S_o\Lambda_q \tag{1.9}$$

Similarly (1.5) and (1.7) ensure there are solutions K_t of

$$T_o'(A - BK_tC) = \Lambda_{n-q}T_0' \tag{1.10}$$

The general solution of K_s and K_t are given as

$$\begin{aligned} K_s &= V_o(CS_o)^- + (T_o'B)^-U_o' - (T_o'B)^-U_o'(CS_o)(CS_o)^- \\ &\quad +\{I - (T_o'B)^- - (T_o'B)\}Y\{I - (CS_o)(CA_o)^-\} && (1.11) \\ K_t &= V_o(CS_o)^- + (T_o'B)^-U_o' - (T_o'B)^-(T_o'B)V_o(CS_o)^- \\ &\quad +\{I - (T_o'B)^-(T_o'B)\}Z\{I - (CS_o)(CS_o)^-\} && (1.12) \end{aligned}$$

The matrices Y, Z in [11] and [12] are arbitrary.

Fletcher et al [2] have shown that if (1.8) holds then $K_s = K_t$ and pole assignment is achieved.

Some approximate results have been obtained by setting $T_o'S_o = E$ and reducing $||E||_F$ as much as possible [3]. The effect on pole assignment of these equations being "almost satisfied" has also been described [4].

The purpose of this paper is to relate the conditions (1.3) - (1.8) to controllability and observability subspaces. S_o and T_o can be chosen by a simple substitution of the eigenvalues in a particular form of polynomial matrix. In this paper we define bold letter to be subspaces, for example $\mathbf{B}$ to be $I_m(B)$ (i.e. Image of B). In section 2 we shall describe some definitions and the main results of this work. In section 3 some properties of the results and an example will be given. In section 4 the orthogonal condition of the controllability and observability subspaces are discussed.

2 Main results

In this section some definitions from Wonham [11] will be described, and also some results of this work will also be outlined. Proofs of theorems and lemmas are omitted in this short paper and can be found in [5].

Definition [11] 1 *If* $A : \mathbf{X} \longrightarrow \mathbf{X}, \mathbf{B} \subseteq \mathbf{X}$ *and* $dim(X) = n$ *then*

$$A|B = B + AB + \cdots + A^{n-1}B \tag{2.1}$$

which means $A|B$ *is the space reachable under the action of* A *from inputs to* B*. If* $\{A|B\} = \mathbf{X}$ *then (A,B) is a controllable pair.*

Theorem [11] 1 *The following are equivalent for a subspace* $\mathbf{Y}$ *of* $\mathbf{X}$*:*

$$(i) A\mathbf{Y} \subseteq \mathbf{Y} + \mathbf{B} \qquad (ii)(A + BF)\mathbf{Y} \subseteq \mathbf{Y} \text{ for some } F. \tag{2.2}$$

The subspaces Y *satisfying (i) and (ii) in theorem 1 are called* (A, B)*–invariant.*

Definition [11] 2 *A subspace* $\mathbf{R}$ *is controllability subspace (c.s) of* (A, B) *if*

(i) $\mathbf{R}$ *is* (A, B)*-invariant and*

(ii) $\mathbf{R} = \{A + BF|\mathbf{B} \cap \mathbf{R}\}$ *for some* F*.*

It is clear that $A\mathbf{R} \subseteq \mathbf{R} + \mathbf{B}$ (i.e (A, B)–invariant) such that

$$\begin{aligned} \mathbf{R} &= \mathbf{B} \cap \mathbf{R} + (A + BF)\mathbf{R} \cap \mathbf{B} + (A + BF)^2 R \cap B \cdots + \\ &\quad (A + BF)^{r-1}\mathbf{R} \cap \mathbf{B} \end{aligned} \tag{2.3}$$

$$\dim(\mathbf{R}) = r \tag{2.4}$$

If we take, for example

$$\mathbf{R} = \text{span} < s_1, s_2, \cdots, s_q > \tag{2.5}$$

$$(A - BK_sC)\mathbf{R} \subseteq \mathbf{R} \tag{2.6}$$

so condition (i) of definition 2 is satisfied. It is possible by careful choice of $s_1, s_2, \cdots, s_q$ to have (ii) of definition 2 not satisfied but in all practical cases this $\mathbf{R}$ will be a controllability subspace. Precise conditions to ensure this can be found in Fletcher [1]. $\mathbf{R}$ is assumed to be a controllability subspace.

$$\text{Write } A_s = A - BK_sC \tag{2.7}$$

A_s is the closed loop matrix. We now investigate how to improve the conditioning of the closed loop eigenvalues by first setting up the following theorems.

Theorem 2 *Given* $A_s = (A - BK_sC)$ *as defined in (2.7).* $\mathbf{R}$ *is a controllability subspace defined in definition 2. For any value* λ*, we have*

$$\mathbf{S}(\lambda) = \{s \mid (A - \lambda I)s \in \mathbf{B}\} \tag{2.8}$$

such that

$$\mathbf{S}(\lambda) \cap \mathbf{R} = \{s \mid (A_s - \lambda I)s \in \mathbf{B} \cap \mathbf{R}\} \tag{2.9}$$

Theorem 3 *Given $A_s\mathbf{R} \subseteq \mathbf{R}$ satisfies definition 2(ii). A subspace $\mathbf{R} \in R^n$ of dimension r is a controllability subspace if and only if there exist $s(\lambda) \in R^n[\lambda]$ and $u(\lambda) \in R^m[\lambda]$ such that*

1. *deg $u(\lambda) = k$ and deg $s(\lambda) = k - 1$ for $k \geq r$*
2. *$(A_s - \lambda I)s(\lambda) = B_* u(\lambda)$ for some G, $BG = B_*$, $Im(BG) = \mathbf{B} \cap \mathbf{R}$*
3. *If $s(\lambda) = \sum_1^k \lambda^{i-1} x_{i-1}$ then $\mathbf{R} = \{x_{i-1}, i \in \underline{k}\}$*

The main differences between theorem 3 and Warren and Eckberg's approach is that a smaller subspace $\mathbf{R} \cap \mathbf{S}(\lambda)$ has been established rather than a general subspace $\mathbf{R}$. Subspace $\mathbf{R} \cap \mathbf{S}(\lambda)$ gives a more precise result in numerical work [5] and the significant advantage of $\mathbf{R} \cap \mathbf{S}(\lambda)$ will be shown in sections 3 and 4. x_{r-1} is a cyclic generator for $\mathbf{R}$ with respect to the matrix $(A_s + B_* F_*)$ for some F_*. $\mathbf{R}$ is also a characterization of controllability subspaces in terms of elements in $ker(A_s - \lambda I; -B_*)$ by the method outlined by Warren and Eckberg [9]. Now we shall extend our study to polynomial matrices for observability subspaces. From Wonham [11], let $\dim(X) = n$. The pair $(C, A)$is observable if

$$\sum_{i=1}^{n} \ker(CA^{i-1}) = 0 \tag{2.10}$$

It is said that the system (1.1) is observable if and only if (C, A) is observable. The following lemma is from Wonham [11].

Lemma 1 *Let $C : \mathbf{X} \longrightarrow \mathbf{Y}$and $A : \mathbf{X} \longrightarrow \mathbf{X}$ be maps with duals $C' : \mathbf{Y}' \longrightarrow \mathbf{X}'$ and $A' : \mathbf{X}' \longrightarrow \mathbf{X}$. Then (C, A) is observable if and only if (C', A') is controllable.*

Definition 3 *A subspace $\mathbf{Q}$ is an observability subspace (o.s) of (A', C') if*

(i) *$\mathbf{Q}$ is (A', C')-invariant*

(ii) *$\mathbf{Q} = \{(A + JC)' | Im(C') \cap \mathbf{Q}\}$ for some J.*

If we take, for example

$$\begin{aligned} \mathbf{Q} = span < t_{q+1}, t_{q+2}, \cdots, t_n > \\ (A - BK_t C)'\mathbf{Q} \subseteq \mathbf{Q} \end{aligned} \tag{2.11}$$

so condition (i) of definition 3 is satisfied. It is possible by careful choice $t_{q+1}, t_{q+2}, \cdots, t_n$ to have (ii) of definition 3 not satisfied but in all practical cases this $\mathbf{Q}$ will be the observability subspace. More detail can be found in Fletcher [1]. $\mathbf{Q}$ is assumed to be an observability subspace. We write $A_t = A - BK_t C$.

The following two theorems are similar to theorems 2 and 3.

Theorem 4 *Given $A_t = A - BK_t KC$ as defined in (14). $\mathbf{Q}$ is the observability subspace defined in definition 5.3. For a given λ, we have*

$$\mathbf{T}(\lambda) = \{t \mid (A' - \lambda I)t \in Im(C')\} \tag{2.12}$$

such that

$$\mathbf{T}(\lambda) \cap \mathbf{Q} = \{t \mid (A'_t - \lambda I)t \in Im(C') \cap \mathbf{Q}\} \tag{2.13}$$

Theorem 5 *Given $A'_t\mathbf{Q} \subseteq \mathbf{Q}$ satisfies definition 3(ii).*

A subspace $Q \in R^n$ of dimension r is an observable subspace if and only if there exist $t(\lambda) \in R^n[\lambda]$ and $v(\lambda) \in R^p[\lambda]$ that

1. *$deg\ u(\lambda) = k$ and $deg\ t(\lambda) = k - 1$ for some $k \geq r$*
2. *$(A'_t - \lambda I)t(\lambda) = C'_* v(\lambda)$ for some H, $C'H = C'_*$*
3. *If $t(\lambda) = \sum_1^k \lambda^{i-1} y_{i-1}$ then $\mathbf{Q} = span\{y_{i-1}, i \in \underline{k}\}$*

3 Some properties of $S(\lambda) \cap R$ and $T(\lambda) \cap Q$

Let us describe the subspaces $\mathbf{S}(\lambda)$ as defined in theorems 2 and 5.

$$\mathbf{S}(\lambda) = [s(\lambda) : (A - \lambda I)s(\lambda) \in \text{Im}(B)] \text{ of dimension } m \tag{3.1}$$

$$\mathbf{T}(\lambda) = [t(\lambda) : (A' - \lambda I)t(\lambda) \in \text{Im}(C')] \text{ of dimension } p \tag{3.2}$$

where λ is a variable. Let $\mathbf{R}$, $\mathbf{Q}$ be the controllability and observability subspaces mentioned in definition 2 and 3. From theorem 2 and 5, $\mathbf{R} \cap \mathbf{S}(\lambda)$ and $\mathbf{Q} \cap \mathbf{T}(\lambda)$ can be expressed as polynomial vector spaces as we now describe: Let $\dim(\mathbf{B} \cap \mathbf{R}) = m_o$.

$$\mathbf{S}(\lambda) \cap \mathbf{R} = < W_1(\lambda), \cdots, W_{m_o}(\lambda) > \tag{3.3}$$

$$W_i(\lambda) = \sum_{j=0}^{v_i - 1} \lambda^j w_{ij} \qquad i = 1, 2, \cdots, m_o \tag{3.4}$$

w_{ij} is a constant vector for all $i \in \underline{m_o}$, $j \in \underline{v_i - 1}$

$$\sum_{i-1}^{m_o} v_i = dim(\mathbf{R}) \tag{3.5}$$

Let $dim(\mathbf{C}' \cap \mathbf{Q}) = p_o$. Then

$$\mathbf{T}(\lambda) \cap \mathbf{Q} = < Z_1(\lambda), \cdots, Z_{p_o}(\lambda) > \tag{3.6}$$

$$Z_i(\lambda) = \sum_{j=0}^{v_i - 1} \lambda^j z_{ij} \tag{3.7}$$

$$\sum_{i=1}^{p_o} v_i = \dim(\mathbf{Q}) \tag{3.8}$$

z_{ij} is a constant vector for all $j \in \underline{v_i - 1}$, $i \in p_o$. From [11] theorem 5.2, the eigenvalues $\lambda_1, \cdots, \lambda_{m_o}$ can be chosen with linear independent eigenvectors in $\mathbf{R}$. Here this idea can be extended to $\mathbf{S}(\lambda) \cap \mathbf{R}$.

For given n distinct eigenvalues $\lambda_i \in C^n, i \in \underline{m}_o$, a vector $s(\lambda)$ can be chosen in subspace $\mathbf{S}(\lambda)\cap\mathbf{R}$ such that $s(\lambda)$ is a linear combination of $W_1(\lambda), W_2(\lambda), \cdots, W_{m_o}(\lambda)$ for all $i \in \underline{n}$. $s(\lambda_1), s(\lambda_2), \cdots, s(\lambda_{m_o})$ are linearly independent and also w_{ij} are linearly independent by multiplying a Van der Monde matrix.

Now the properties of $W_i(\lambda)$ (for all $i \in \underline{m}_o$) related to $\mathrm{Im}(B) \cap \mathbf{R}$ can be deduced.

$$w_{i,v_i-1}, (A_s w_{i,v_i-1} - w_{i,v_{i-2}}), \cdots, A_s w_{i,0} \ \mathrm{Im}(B) \cap \mathbf{R} \tag{3.9}$$

for all $i \in \underline{m_o}$. Similarly we apply the same procedure on $\mathbf{T}(\lambda) \cap \mathbf{Q}$ and obtain

$$z_{i,v_i-1}, (A_t z_{i,v_i-1} - z_{i,v_i-2}), \cdots, A_t z_{i,0} \ \mathrm{Im}(C') \cap \mathbf{Q} \tag{3.10}$$

for all $i \in \underline{p}_o$. These properties are well–known for $\mathbf{S}(\lambda)$, $\mathbf{T}(\lambda)$ [7] that is when $\mathbf{R}$ is the entire state space but are new for $\mathbf{S}(\lambda) \cap \mathbf{R}$, $\mathbf{T}(\lambda) \cap \mathbf{Q}$. The following examples demonstrate the construction of $\mathbf{S}(\lambda)$ shown in theorem 3 with the properties of $W_i(\lambda)$.

Example 1 *Suppose $dim(\mathbf{R} \cap \mathbf{B}) = 2$ and $dim(R) = 4$*
$\mathbf{S}(\lambda) \cap \mathbf{R}$ can be easily expressed as $< W_1(\lambda), W_2(\lambda) >$ where

$$W_1(\lambda) \quad = \quad w_{1,0} + \lambda w_{1,1} \tag{3.11}$$
$$W_2(\lambda) \quad = \quad w_{2,0} + \lambda m_{2,1} \tag{3.12}$$

Thus $w_{1,0}$, $w_{1,1}$, $w_{2,0}$ and $w_{2,1}$ are linearly independent vectors $\in \mathbf{R}$. We can choose them as a set of constant vector such that

$$s(\lambda) = w_{1,0} + \lambda w_{1,1} + \lambda^2 w_{2,0} + \lambda^3 w_{2,1} \in \mathbf{S}(\lambda) \cap \mathbf{R} \tag{3.13}$$

Hence all other right eigenvectors $s(\lambda_j)$ can be constructed just replacing λ by λ_j in (3.13).

3.1 Advantages of expressing subspaces in vector space format

Expressing $\mathbf{S}(\lambda) \cap \mathbf{R}$ and $\mathbf{T}(\lambda) \cap \mathbf{Q}$ as (18)–(19) and (21)–(22) is advantageous in choosing the vectors in S_o and T_o to keep $T_o'S_o = 0$. Moreover matrices CS_o and $T_o'B$ can also be kept as full rank which satisfy the sufficient conditions of the perturbation theory in [6]. Now we show two lemmas which describe this:

Lemma 2 *Given $T_o'S_o = 0$ (i.e. $< t_{q+1}, t_{q+2}, \cdots, t_n >^{\perp} = < s_1, s_2, \cdots, s_q >$). CS_o does not have full row rank if and only if there exists a non–zero vector in $< t_{q+1}, t_{q+2}, \cdots, t_n > \in Im(C')$.*

Lemma 3 *Given $T_o'S_o = 0$ (i.e. $< t_{q+1}, t_{q+2}, \cdots, t_n >^{\perp} = < s_1, s_2, \cdots, s_q >$). $T_o'B$ does not have full column rank if and only if there exists a vector in $< s_1, s_2, \cdots, s_q > \in Im(B)$.*

These two limmas can be easily proved [5] and the next corollary follows from the above results.

Corollary 1 *Given* $T_o'S_o = 0$

(i) If any vector in T_o *lies in* $Im(C')$*, then* CS_o *does not have full row rank.*

(ii) If any vector in S_o *lies in* $Im(B)$*, then* $T_o'B$ *does not have full column rank.*

Corollary 2 *Given* $T_o'S_o = 0$

(i) If any vector in S_o *is chosen as* w_{i,v_i-1} *which is the vector with the largest leading power of* λ *(for every* $i \in \underline{m}_o$*), then* $T_o'B$ *does not have full column rank.*

(ii) If any vector in T_o *is chosen as* z_{i,v_i-1} *which is the vector with the largest leading power of* λ *(for every* $i \in \underline{p}_o$*), then* CS_o *does not have full row rank.*

Proof 1 From (3.9) and (3.10) $w_{i,v_i-1} \in \mathrm{Im}(B) \cap R$ for all $i \in \underline{m}_o$ and $z_{i,v_i-1} \in \mathrm{Im}(C') \cap \mathbf{Q}$ for all $i \in \underline{p}_o$.

And from the above two lemmas, (i) and (ii) are proved.

It is worth mentioning that the rank of matrices $CS_o(p \times q)$ and $T_o'B(n - q \times m)$ are dependent on the value q.

In $m + q > n$ case [5], q is bounded by $n - m \leq q \leq p$, in practice CS_o is required to be full column rank of q and $T_o'B$ is required to be full row rank $n - q$ [5]. Thus, the situations in Corollaries 1 and 2 have no serious effect on the rank of CS_o and $T_o'B$ except when they are square matrices (i.e. $p = q, n - q = m$). In $m + p \leq n$ case, the rank of CS_o or $T_o'B$ would be seriously affected. Precautions should be taken to ensure that the initial chosen vectors in S_o do not lie in $Im(B)$ and vectors in T_o do not lie in $\mathrm{Im}(C')$.

From these lemmas and corollaries, some vectors are not suitable to be the initial chosen vectors for the pole assignment algorithm in [4] because it may cause the CS_o and $T_o'B$ to be close to rank deficient situation. This also explains the conditions (17v) in [1] which have been rather mysterious.

4 Orthogonal condition of $S(\lambda) \cap R$ and $T(\lambda) \cap Q$

If $s_i \in \mathbf{S}(\lambda_i) \cap \mathbf{R}$, $i = 1, 2, \cdots, q$ and $t_j \in \mathbf{T}(\lambda_j) \cap \mathbf{Q}$, $j = q + 1, \cdots, n$ it has been noted in Fletcher [1] that $s_1, s_2, \cdots, s_q; t_{q+1}, t_{q+2}, \cdots, t_n$ can be chosen so as to be linearly independent.

If $\mathbf{R}^\perp \subseteq \mathbf{Q}$, it implies that all vectors in $S_o = [s_1, s_2, \cdots, s_q]$ are orthogonal to those vectors in $T_o = [t_{q+1}, t_{q+2}, \cdots, t_n]$, such that

$$T_o'S_o = 0 \tag{4.1}$$

Hence, T_o, S_o satisfy all the conditions for pole assignment mentioned in Fletcher et al (1985). From theorems 3 and 5 two sets of constant vectors $\{x_0, x_1, \cdots, x_{r-1}\}$,

$\{y_0, y_1, \cdots, y_{r-1}\}$ span $\mathbf{R}$ and $\mathbf{Q}$ respectively. These two sets of vectors are orthogonal to each other because $\mathbf{R}^\perp \subseteq \mathbf{Q}$. Hence by substitution of different eigenvalues into λ in theorems 3 and 5, S_o and T_o can be obtained and poles assignment can be achieved (Fletcher et al 1985). Further, some eigenvalues can be shifted to other regions without losing the orthogonal properties of $\mathbf{R}$ and $\mathbf{Q}$.

If $\mathbf{R}$ is almost a complement of $\mathbf{Q}$, then

$$T_o' S_o = E \tag{4.2}$$

where norm of E is small for whatever eigenvalue $\lambda_1, \lambda_2, \cdots, \lambda_n$.

Some approximate solution has been produced by Fletcher and Ho [4] to reduce E to zero in order to achieve the pole assignment by output feedback. The approach of establishing controllability and observability subspaces mentioned in this paper gives a deeper understanding of reducing the matrix E.

5 Conclusions

The concept of controllability subspaces and observability subspaces, $\mathbf{S}(\lambda) \cap \mathbf{R}$ and $\mathbf{T}(\lambda) \cap \mathbf{Q}$ are described in theorems 2 and 4 these subspaces can be expressed by two sets of constant vectors S_o and T_o. If $\mathbf{R}^\perp \subset \mathbf{Q}$, it implies all vectors in S_o are orthogonal to those in T_o and pole placement by output feedback can be achieved by satisfying the condition in (1.8).

From theorems 3 and 5, it is important to note that the constant vectors in S_o and T_o form a cyclic generator for $\mathbf{Q}$ and $\mathbf{R}$. Hence any right eigenvector $s(\lambda_j)$ of $A - BK_sC$ in $\mathbf{Q}$ can be constructed just replacing λ by the assigned eigenvalues λ_j in theorem 3(iii); also any left eigenvector in $\mathbf{R}$ can be calculated in the same way by applying theorem 5(iii). Both left and right eigenvectors satisfying conditions (3) to (8) can then be worked out without further computation.

This geometric approach will further explore the output feedback problem, with the possibility of converting the problem into robust eigenstructure assignment by state feedback in the subspace $\mathbf{Q}$ and $\mathbf{R}$.

References

[1] Fletcher L.R., An intermediate algorithm for pole assignment by output feedback . Int. J. Control, *31*, 1121–1136,1980.

[2] Fletcher, L.R., Kautsky, J., Kolka, G.K.G. and Nichols, N.K. Some necessary and sufficient conditions for eigenstructure assignment by output feedback, Int. J. Control, *42*, 1457–1468, 1985.

[3] Fletcher, L.R. and Ho, W.C. On approximation eigenstructure assignment by output feedback, IMA. Symp., University of Oxford, 3–5 September, 1986a.

[4] Fletcher, L.R. and Ho, W.C., Approximation results in output feedback pole assignment, 25th IEEE Conference on Decision and Control, 1986b.

[5] Ho, W.C., Computational analysis of pole assignment by output feedback, Ph.D thesis, University of Salford, 1985.

[6] Ho, W.C. and Fletcher, L.R., Perturbation theory on output feedback pole assignment, Int. J. Control, 1988 (to appear).

[7] Kimura, H., Pole assignment by gain output feedback, IEEE Trans. Auto Control, AC22, 509–516, 1975.

[8] Moore, B.C. and Laub, A.J. Computation of supremal (A,B) invariant and controllability subspaces', IEEE Trans. Auto. Control, *235*, 783–792, 1978.

[9] Warren, M.E. and Eckberg, A.E., Jr., On the dimensions of controllability subspaces: 'A Characterization via polynomial matrices and Lounecker invariants', Siam J. Control, *13*, 2, 434–445, 1975.

[10] Wonham, W.M., On pole assignment in multi input controllable linear system, IEEE Trans Auto Control, *12*, 16, 660–665 1967.

[11] Wonham, W.M., *Linear Multivariable Control Theory*: A Geometric Approach, New York, Springer–Verlag 1979.

Invertibility of linear and nonlinear control systems: a survey

D.J. Bell

Abstract

This survey paper briefly describes the so-called left-inverse problem in which it is required to find the inputs to a system when only the outputs are available. Much of the literature for this problem on both linear and nonlinear systems is reviewed and some indication of the methods used to obtain necessary and sufficient conditions for the existence of an inverse and ways of construction when it does exist is given. Linear algebra forms the basis for the analysis of linear systems whilst the techniques of differential geometry are used for nonlinear systems. A recent approach to nonlinear systems using differential algebra is also mentioned. Papers describing right inverses are also listed in the Introduction.

1 Introduction

When it exists, the inverse of a dynamical system can arise from two different but related problems. First, there is the question: when is it possible to determine the input to the system when only the outputs are observed? When such an input exists the system is said to be *input observable* and gives rise to a left *inverse system.* The second problem is that of finding an input which produces an output such that the latter follows any prescribed continuous path (the tracking problem). If such an input exists for the tracking problem then the system is said to be *output controllable* and leads to a *right inverse system.* For the last twenty five years, work has progressed on both types of problem.

Naturally, inverses to linear systems were investigated first and initially restricted to single-input and single-output (SISO). One of the earliest books to mention inverse systems was [47]. An early paper [4] investigated the existence of an inverse to a SISO linear system using a frequency domain representation. A companion paper [5] gave a necessary and sufficient condition for output controllability for a multivariable linear system, with m inputs and p outputs in terms of a state space representation. Simplified versions of the condition in [5] were given for the special case $m = p$ [9] and for when $m \neq p$ [33], the duality of left-and right-invertibility being stressed in the latter reference. Incidentally, an inverse system will never exist if $m > p$ so that the primal system must always have at least as many outputs as there are inputs. This is true for both linear systems [33] and nonlinear systems [24].

An alternative approach to that given in [33] but for $m = p$ was presented in [35]. The inversion algorithm in [4] was generalized to a time–variable system with one output in [34] and inversion of multivariable time–varying systems discussed in [35]. Reproducibility, meaning the tracking problem, was also considered for time–varying systems in [1] where necessary and sufficient conditions are also given for nonlinear input/output systems. A further modification to the results of [33] was given in [44]. In [43] conditions on right and left invertibility are given as a rank condition which differs in form from that given in [5] and [33]. Existence conditions, construction and applications of inverse systems can be found in [3], [10], [21], [22], [28] and[32]. A further paper [45] on output controllability builds upon the concepts first introduced in [5]. A good review of the state–of–the–art for left and right inverses of linear, time–invariant dynamical systems is given in [46].

Although nonlinear systems were discussed briefly in [5] and again in [29], the first major attempts at analysing inverses of nonlinear systems are described in [15], [16] and [17]. In [16] attention was restricted to linear analytic systems with single input but a companion paper [17] investigated the multi–input case using an approach based on [35]. The inversion of nonlinear analytic systems with single input was discussed in [42] and for multivariable systems in [11]. Further contributions to right–invertibility for multi–input/multi–output linear analytic systems have been given in [18], [36], [39], [40] and [41]. A modification to the algorithm of [17] is given in [37] and [18] in particular is based upon the class of systems related to those in [30]. All of the above mentioned papers on the invertibility of nonlinear systems use the concepts of differential geometry. Other papers using a geometric approach are [23], [24] and [25].

The decoupling problem is targeted in both [7], [8] and [38], the results of [7] being summarized in [20]. The relevance of the notion of zeros at infinity to the nonlinear decoupling problem is described in [26]. This notion is also discussed at some length in [25] where a good description is given of the linear situation. Recently, a very general analytic nonlinear system has been considered [6] giving a general inversion algorithm and also the construction of a right–inverse system. Right–invertibility is also discussed in [27] and [31]. At least two books [19], [2] are now available which review the geometric approach to nonlinear systems theory.

Finally, in the last few years an alternative approach has been suggested to the differential geometric methods used over the last twentyfive years. This is the application of another branch of pure mathematics known as differential algebra and the idea was first floated in a paper by Fliess [13]. Roughly speaking, differential algebra is to differential equations what commutative algebra and Galois theory is to algebraic equations. Right–and left–invertibility in this new framework is discussed in [14]. The concept of zeros at infinity for nonlinear systems are discussed in a differential algebraic framework in [12].

The present survey is by no means exhaustive and further references may be found in the above mentioned papers, including those for the inversion of linear and nonlinear discrete–time systems [11], [33].

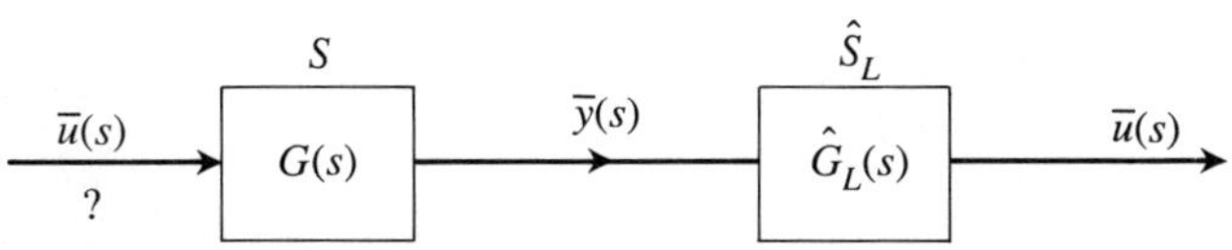

Figure 1. Input function observability

2 Linear systems[46]

A multivariable, time–invariant, linear control system S is considered to have the form

$$S: \dot{x} = Ax + Bu, x(0) = 0 \quad (2.1)$$
$$y = Cx + Du \quad (2.2)$$

where $x(t) \in \Re^n$, $u(t) \in \Re^m$, $y(t) \in \Re^p$, $A \in \Re^{n\times n}$, $B \in \Re^{n\times m}$, $C \in \Re^{p\times n}$. It has a $p \times m$ rational transfer function $G(s)$ given by

$$G(s) = C(sI - A)^{-1}B + D \quad (2.3)$$

so that the input/output map is defined by

$$\overline{y}(s) = G(s)\overline{u}(s). \quad (2.4)$$

A matter of considerable importance in automatic control theory and practice is to observe the output $y(\cdot)$ from system S and to deduce the input producing that output. If this can be done then the input is said to be ***functionally reproducible*** and the system is called ***input function observable.*** What is required is a linear system $\hat{S}_L$ with transfer function $\hat{G}_L(s)$ such that when the observed output from S is fed as input into $\hat{S}_L$ the output of $\hat{S}_L$ is $u(\cdot)$, the required input of S (Fig. 1).

From Figure 1,

$$\overline{u}(s) = \hat{G}_L(s)\overline{y}(s) \quad (2.5)$$

and (2.4), (2.5) together yield

$$\hat{G}_L(s)G(s) = I_m. \quad (2.6)$$

The $m \times p$ transfer matrix $\hat{G}_L(s)$ is called a *left inverse* of $G(s)$ and the system S is said to have a left inverse. System $\hat{S}_L$ is called a *left inverse system* to S.

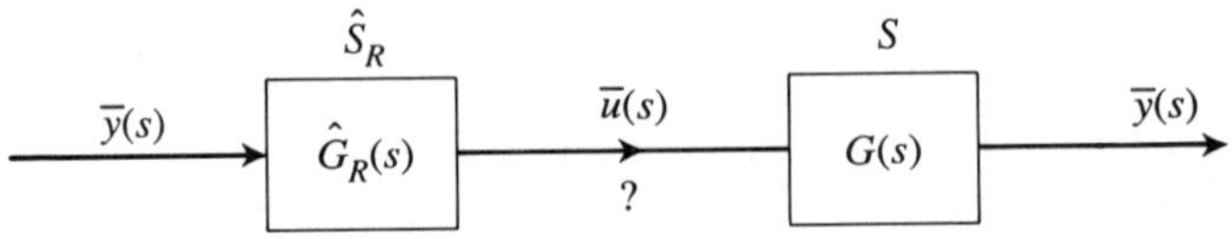

Figure 2. Output function controllability

A slightly different but related problem is to find a control $u(\cdot)$ such that system S will yield a specified output $y(\cdot)$. If there exists such a control then the output is said to be ***functionally reproducible*** and S is called ***output function controllable.*** Here a linear system $\hat{S}_R$ with transfer function $\hat{G}_R(s)$ is required such that when the specified output is fed as input into $\hat{S}_R$ the output of $\hat{S}_R$ is the required input to S (Fig. 2).

From Figure 2,

$$\overline{u}(s) = \hat{G}_R(S)\overline{y}(s) \tag{2.7}$$

and (2.4), (2.7) together yield

$$G(s)\hat{G}_R(s) = I_p. \tag{2.8}$$

The $m \times p$ transfer matrix $\hat{G}_R(s)$ is called a ***right inverse*** of $G(s)$ and the system S is said to have a right inverse. System $\hat{S}_R$ is called a ***right inverse system*** to S.

The major part of this paper will be concerned only with left inverse systems for which the basic problem is solving (2.1), (2.2) for control u in terms of output y and time derivatives of y.

2.1 Single input/single output systems [4]

For this case, $m = 1 = p$ and (2.1), (2.2) are written as

$$S' : \dot{x} = Ax + bu \tag{2.9}$$

$$y = cx + du \tag{2.10}$$

where $u(t) \in \Re$, $y(t) \in \Re$, $b \in \Re^n$, $c^t \in \Re^n$ and $d \in \Re$. The transfer function (2.3) is then given by

$$g(s) = c(sI - A)^{-1}b + d \tag{2.11}$$

and (2.4) becomes

$$\overline{y}(s) = g(s)\overline{u}(s). \tag{2.12}$$

For simplicity, assume that $d = 0$. Then differentiating y w.r.t. time yields

$$\begin{aligned} \dot{y} &= c\dot{x} \\ &= c(Ax + bu), \\ \text{i.e. } \dot{y} &= cAx + cbu. \end{aligned} \tag{2.13}$$

If $cb \neq 0$ then

$$u = (\dot{y} - cAx)/(cb) \tag{2.14}$$

and, provided $\dot{y}$ is measurable, the problem is solved. However, if $cb = 0$ then differentiate in (2.13) to give

$$\begin{aligned} \ddot{y} &= cA\dot{x} \\ &= cA(Ax + bu), \\ \text{i.e. } \ddot{y} &= cA^2x + cAbu. \end{aligned} \tag{2.15}$$

Again, if $cAb \neq 0$ then the problem is solved with

$$u = \left(\ddot{y} - cA^2x\right)/(cAb). \tag{2.16}$$

On the other hand, if $cAb = 0$ then repeat the procedure by differentiating (2.15) and so on. Two possibilities arise; either (i) $cb, cAb, cA^2b, \cdots, cA^{n-1}b$ are all zero in which case u has no effect on y and the transfer function $g(s)$ is zero so that the system S' is *not* invertible, or (ii) $\exists\alpha < n$ such that $cb, cAb, cA^2b, \cdots, cA^{\alpha-2}b$ are all zero but $cA^{\alpha-1}b \neq 0$ in which case

$$u = \left(y^{(\alpha)} - cA^{\alpha}x\right) / \left(cA^{\alpha-1}b\right) \tag{2.17}$$

and system S' is invertible. The parameter α is called the *relative order* of S', a parameter akin to the order of a singular control.

Substituting (2.17) into (2.9),

$$\begin{aligned} \dot{x} &= Ax + b\left(y^{(\alpha)} - cA^{\alpha}x\right) / \left(cA^{\alpha-1}b\right) \\ \text{i.e. } \dot{x} &= \left(A - \frac{1}{cA^{\alpha-1}b}bcA^{\alpha}\right)x + \frac{1}{cA^{\alpha-1}b}by^{(\alpha)} \end{aligned} \tag{2.18}$$

and by writing $cA^{\alpha-1}b = 1/q \in \Re$, (2.18) and (2.17) can be written as

$$\dot{x} = (A - qbcA^{\alpha})\,x + qby^{(\alpha)} \tag{2.19}$$

$$u = -qcA^{\alpha}x + qy^{(\alpha)}. \tag{2.20}$$

Equation (2.19), (2.20) define the inverse system to S', in which the input is the time derivative $y^{(\alpha)}$ of the output y.

Taking Laplace transforms of this inverse system yields

$$\overline{u}(s) = s^{\alpha}\left[-q^2cA^{\alpha}\left(sI - A + qbcA^{\alpha}\right)^{-1}b + q\right]\overline{y}(s) \tag{2.21}$$

$$= s^{\alpha}\hat{g}(s)\overline{y}(s), \text{ say.} \tag{2.22}$$

(2.12) and (2.22) together yield

$$\hat{g}(s)g(s) = \frac{1}{s^{\alpha}} \cdot 1 \tag{2.23}$$

Invertibility does not depend upon the initial state.
Example 2.1

$$\begin{aligned}\dot{x}_1 &= -3x_1 + 2u\\ \dot{x}_2 &= -x_2 - u\\ y &= x_1 + x_2\end{aligned}$$

$$A = \begin{bmatrix} -3 & 0 \\ 0 & -1 \end{bmatrix},\ b = \begin{bmatrix} 2 \\ -1 \end{bmatrix}, c = (1,1), d = 0.$$

$cb = 1 \neq 0$ so $\alpha = 1$ and $q = 1$.

2.2 Multi–input/multi–output systems[4],[33],[35]

Equations (2.6) and (2.8) define left and right inverses $\hat{G}_L(s)$ and $\hat{G}_R(s)$ of the transfer matrix $G(s)$ corresponding to system S. A difficulty with these definitions is that in general neither $\hat{G}_L(s)$ nor $\hat{G}_R(s)$ will be a *proper* transfer matrix and so will not give rise to linear inverse systems. Guided by (2.23), the left inverse $\hat{G}_L(s)$ is modified so that system S is said to have an α–integral left inverse system $\hat{S}_L$ if

$$\hat{G}_L(s)G(s) = \frac{1}{s^{\alpha}} I_m \tag{2.24}$$

where $\hat{G}_L(s)$ is now a proper transfer matrix and leads to a linear left inverse system $\hat{S}_L$.

Similarly, system S is said to have a β–integral right inverse system $\hat{S}_R(s)$ if

$$G(s)\hat{G}_R(s) = \frac{1}{s^{\beta}} I\hat{p} \tag{2.25}$$

where again $\hat{G}_R(s)$ is a proper transfer matrix giving a linear right inverse system $\hat{S}_R$.

Necessary and sufficient conditions for the existence of an α–integral left inverse for system S can be obtained by first noting that the solution of (2.1) is given by

$$x(t) = e^{At} \int_0^t e^{-A\tau} Bu(\tau) d\tau$$

so that the output in (2.2) can be written as

$$y(t) = Ce^{At} \int_0^t e^{-A\tau} Bu(\tau) d\tau + Du$$

$$= C\left(I + At + \frac{A^2t^2}{2!} + \cdots\right)\left(\int_0^t Bu(\tau)d\tau - AB\int_0^t su(s)ds + \right.$$

$$\left.\frac{A^2B}{2!}\int_0^t s^2u(s)ds - \cdots\right) + Du \tag{2.26}$$

i.e. $$y(t) = Du + CB\int_0^t u(\tau)d\tau + CAB\int_0^t (t-\tau)u(\tau)d\tau + \cdots$$

Now proceed as in Section 2.1 by differentiating (2.26) w.r.t. time to give

$$\dot{y}(t) = D\dot{u}(t) + CBu(t) + CAe^{At}\int_0^t e^{-A\tau}Bu(\tau)d\tau$$

$$= D\dot{u}(t) + CBu(t) + CA\left(I + At + \frac{A^2t^2}{2!} + \cdots\right)\cdot$$

$$\left(\int_0^t Bu(\tau)d\tau - AB\int_0^t su(s)ds + \frac{A^2B}{2!}\int_0^t s^2u(\tau)d\tau \cdots\right) \tag{2.27}$$

i.e. $$\dot{y}(t) = D\dot{u}(t) + CBu(t) + CAB\int_0^t u(s)ds + CA^2B\int_0^t (t-s)u(s)ds + \cdots$$

Similarly (2.27) can be differentiated to give

$$\ddot{y}(t) = D\ddot{u}(t) + CB\dot{u}(t) + CABu(t) + CA^2B\int_0^t u(s)ds$$

$$+ CA^3B\int_0^t (t-s)u(s)ds + \cdots \tag{2.28}$$

and so on. Equations (2.26), (2.27) (2.28) and the rest may then be written in block matrix form as

$$\begin{bmatrix} y \\ \dot{y} \\ \ddot{y} \\ \vdots \end{bmatrix} = \begin{bmatrix} \cdots & CAB & CB & D & \cdots & & \\ \cdots & CA^2B & CAB & CB & D & \cdots & \\ \cdots & CA^3B & CA^2B & CAB & CB & D & \cdots \\ & & & \vdots & & & \end{bmatrix} \begin{bmatrix} \int_0^t (t-\tau)u(\tau)d\tau \\ \int_0^t u(\tau)d\tau \\ u(t) \\ \dot{u}(t) \\ \ddot{u}(t) \\ \vdots \end{bmatrix} \tag{2.29}$$

By defining $M_o \triangleq D$ and for $i = 1, 2, \cdots$

$$M_i \triangleq \begin{bmatrix} D & 0 & 0 & & \cdots & 0 \\ CB & D & 0 & & & \\ CAB & CB & D & & & \\ CA^2B & CAB & CB & D & & \vdots \\ & \vdots & & & \ddots & \\ & \vdots & & & & \ddots \\ CA^{i-1}B & CA^{i-2}B & & & \cdots & CB & D \end{bmatrix} \tag{2.30}$$

System S is invertible iff $\exists\ k \in N$ s.t.

$$\text{rank}(M_k) - \text{rank}(M_{k-1}) = m. \tag{2.31}$$

If such a k exists, then $k = \alpha$, the relative order of the system S.
Example 2.2

$$\begin{aligned} \dot{x}_1 &= x_2 & y_1 &= x_1 \\ \dot{x}_2 &= x_3 + u_1 & y_2 &= x_2 \\ \dot{x}_3 &= u_2 & y_3 &= x_3 \end{aligned}$$

$$B = \begin{bmatrix} 0 & 0 \\ 1 & 0 \\ 0 & 1 \end{bmatrix}, \ C = I_3,\ D = O_{3\times 2},\ n = 3,\ m = 2,\ p = 3.$$

The system is invertible because $\dot{y}_2 = \dot{x}_2 = x_3 + u_1 \Rightarrow u_1 = \dot{y}_2 - y_3$ and $\dot{y}_3 = \dot{x}_3 = u_2 \Rightarrow u_2 = \dot{y}_3$. This is confirmed by the above condition as follows: $M_o = D = O_{3\times 2}$ and $\text{rank}(M_o) = 0$.

$$M_1 = \begin{bmatrix} D & O_{3\times 2} \\ CB & D \end{bmatrix} = \begin{bmatrix} O_{4\times 2} & O_{4\times 2} \\ I_2 & O_{2\times 2} \end{bmatrix}$$

$$\text{rank}(M_1) = 2$$

so that $\text{rank}(M_1) - rank(M_o) = 2$.

An alternative approach for the case $m = p$ is due to Silverman [35]. Suppose $\text{rank}(D) = q_o < p$ (i.e. D is a singular matrix). Then there exists a nonsingular $p \times p$ matrix $S_o \in \Re^{p\times p}$ such that

$$D_o \triangleq S_0 D = \begin{bmatrix} \overline{D}_0 \\ O \end{bmatrix} \in \Re^{p\times p}$$

where $\overline{D}_o \in \Re^{q_o\times p}$ and $\text{rank}(\overline{D}_o) = q_o$. From (2.2) it follows that

$$S_o y = S_o C x + S_o D u$$

or

$$y_o(t) = C_o x(t) + D_o u(t) \tag{2.32}$$

where $y_o = S_o y \in \Re^p$ and $C_o = S_o C \in \Re^{p\times n}$.

Write

$$C_0 = \begin{bmatrix} \overline{C}_0 \\ \tilde{C}_0 \end{bmatrix} \text{ and } y_0 = \begin{bmatrix} \overline{y}_0 \\ \tilde{y}_0 \end{bmatrix}$$

where $\overline{C}_0 \in \Re^{q_0\times n}$, $\tilde{C}_0 \in \Re^{(p-q_0)\times n}$, $\overline{y}_0 \in \Re^{q_0}$, $\tilde{y}_0 \in \Re^{(p-q_0)}$. Then

$$\begin{bmatrix} \overline{y}_0 \\ \tilde{y}_0 \end{bmatrix} = \begin{bmatrix} \overline{C}_0 \\ \tilde{C}_0 \end{bmatrix} x + \begin{bmatrix} \overline{D}_0 \\ 0 \end{bmatrix} u \tag{2.33}$$

Now pre-multiply (2.35) by the differential operator

$$M_0 = \begin{bmatrix} I_{q_0} & 0 \\ 0 & I_{p-q_0}(d/dt) \end{bmatrix} \tag{2.34}$$

to give

$$\begin{bmatrix} \overline{y}_0 \\ \tilde{y}_0 \end{bmatrix} = \begin{bmatrix} \overline{C}_0 \\ \tilde{C}_0 A \end{bmatrix} x + \begin{bmatrix} \overline{D}_0 \\ \tilde{C}_0 B \end{bmatrix} u. \tag{2.35}$$

Let

$$\overline{\overline{y}} \triangleq \begin{bmatrix} \overline{y}_0 \\ \tilde{y}_0 \end{bmatrix}, \ \overline{\overline{C}} = \begin{bmatrix} \overline{C}_0 \\ \tilde{C}_0 A \end{bmatrix}, \ \overline{\overline{D}} = \begin{bmatrix} \overline{D}_0 \\ \tilde{C}_0 B \end{bmatrix}.$$

Then

$$\overline{\overline{y}} = \overline{\overline{C}}x + \overline{\overline{D}}u. \tag{2.36}$$

Let $q_1 = \text{rank}(\overline{\overline{D}})$. If $q_1 = p$ then

$$u(t) = \overline{\overline{D}}^{-1}\overline{\overline{y}}(t) - \overline{\overline{D}}^{-1}\overline{\overline{C}}x(t) \tag{2.37}$$

and (2.1) becomes

$$\dot{x}(t) = \left(A - B\overline{\overline{D}}^{-1}\overline{\overline{C}}\right)x(t) + B\overline{\overline{D}}^{-1}\overline{\overline{y}}(t). \tag{2.38}$$

The inverse system to (2.1), (2.2) is thus given by (2.38) and (2.37) with $\overline{\overline{y}}(t)$ as input and $u(t)$ as output.

If, on the other hand, $q_1 < p$ then the above procedure can be repeated on (2.36) until an equation such as (2.36) can be solved for u. Full details can be found in Silverman [35].

3 Nonlinear systems–geometric approach[19]

Not surprisingly linear algebra is the key to unlock the mysteries of invertibility of linear control systems. To analyse that of nonlinear systems the methods of differential geometry have been used. Major results now known are for the so-called linear analytic system

$$\dot{x}(t) = f(x(t)) + \sum_{j=1}^{m} u_j(t)g_j(x(t)), x(0) = x_0 \tag{3.1}$$

$$y(t) = h(x(t)) + \sum_{j=1}^{m} u_j(t)k_j(x(t)) \tag{3.2}$$

in which the state $x(t)$ belongs to a real, connected, analytic manifold M, the vector fields $f, g_1, g_2, \cdots, g_m : M \to T(M)$ on M are real, analytic, and belong to an $\Re$–*vector* space $V(M)$, the maps $h, k_1, k_2, \cdots, k_m : M \longrightarrow \Re^p$ are real, analytic, and u_j belongs to the class Ω of real analytic functions from $[0, \infty)$ into $\Re$.

3.1 Single–input/single–output systems[16]

For simplicity assume that the output does not depend explicitly on the input . Then

$$\dot{x}(t) = f(x(t)) + u(t)g(x(t)), \quad x(0) = x_0 \in M \tag{3.3}$$

$$y(t) = h(x(t)), \qquad y(t) \in \Re, u(t) \in \Re. \tag{3.4}$$

Let $x(t, u, x_0)$ denote the solution of (3.3) and $y(t, u, x_0)$ the corresponding output. The nonlinear system (3.3),(3.4) is said to be *invertible at* x_0 if, whenever $u'(\cdot)$, $u''(\cdot) \in \Omega$ are distinct controls, $y(t, u', x_0) \neq y(t, u'', x_0)$, i.e. $u'(\cdot) \neq u''(\cdot) \Longrightarrow y(\cdot, u', x_0) \neq y(\cdot, u'', x_0)$ (injective mapping). The same nonlinear system is *strongly invertible at* x_0 if there exists an open neighbourhood V of x_0 such that for all $x \in V$, the system is invertible at x. The system is *strongly invertible* if there exists an open and dense submanifold M_o of M such that for all $x_0 \in M_o$ the system is strongly invertible at x_0. The output map $h : M \to \Re$ belongs to a ring $C^\omega(M)$ of real analytic functions on M. The vector field $[f, g]$ is the Lie bracket of f and g;

$$ad^i(f)g = [f, ad^{i-1}(f)g]$$

and

$$ad^0(f)g = g. \; ad^i(f)g \in V(M) \; \forall \, i \geq 0, \;\; h \in C^\omega(M)$$

so that

$$(ad^i(f)g)(h) \in C^\omega(M) \text{ for } i \geq 0.$$

$O \in C^\omega(M)$ denotes the zero function in $C^\omega(M)$.

The *relative order* α of the nonlinear system (3.3), (3.4), is the least positive integer such that $(ad^{\alpha-1}(f)g)(h) \neq 0$. A necessary and sufficient condition for strong invertibility is that α is finite.

Consider now the system

$$\dot{z} = \hat{f}(z) + v\hat{g}(z), \;\; z(0) = z_0 \in N \tag{3.5}$$

$$w = \hat{h}(z) + v\hat{k}(z) \tag{3.6}$$

where N is a real, analytic manifold, $\hat{f}$, $\hat{g} \in V(N)$, $\hat{h}$, $\hat{k} \in C^\omega(N)$ and $v \in \Omega$. The system (3.5), (3.6) is called a left–inverse for the system (3.3), (3.4) if

$$w(\cdot, y^{(\alpha)}, z_0) = u(\cdot),$$

i.e. if the input v to the inverse system is chosen to be the αth derivative of the output $y(\cdot, u, x_0)$ of the original system, then the left–inverse system produces $u(\cdot)$ for its output.

The next example illustrates the fact that invertibility of nonlinear systems depends on the initial state.

Example 3.1

$$\begin{aligned} \dot{x}_1 &= x_1, & x_1(0) &= 1 \\ \dot{x}_2 &= u, & x_2(0) &= 0 \\ y &= x_1^2 + x_2^2. \end{aligned}$$

The solutions for this system are easily found to be

$$\begin{aligned} x_1(t) &= e^t, \\ x_2(t) &= \int_0^t u(s)ds \\ y(t) &= e^{2t} + \left(\int_0^t u(s)ds\right)^2 . \end{aligned}$$

Choose two controls u' and u'' and write

$$v_1 = \int_0^t u'(s)ds$$

and

$$v_2 = \int_0^t u''(s)ds.$$

Then $y' - y'' = v_1^2 - v_2^2$ and so $v_1 = -v_2 \Rightarrow y' = y''$. It follows that the system is not invertible. However, if the initial states are changed to $x_1(0) = 0$ and $x_2(0) = 1$, then

$$\begin{aligned} y' - y'' &= (v_1 + 1)^2 - (v_2 + 1)^2 \\ &= (v_1 - v_2)(v_1 + v_2 + 2) \end{aligned}$$

and $y' = y''$ only if $u' = u''$ so that the system with these new initial states is invertible.

If (3.3), (3.4) is invertible then its relative order α is finite, $(ad^{\alpha-1}(f)g)(h)(x) \neq 0$ and $M_\alpha = \{x \in M : (ad^{\alpha-1}(f)g)(h)(x) \neq 0\}$ is an open dense submanifold of M, called the *inverse submanifold* for the system. This submanifold M is the state space for the inverse system (3.5), (3.6).

Suppose that (3.3), (3.4) is strongly invertible with relative order α, and inverse submanifold M_α. If $x_0 \in M$ then (3.5), (3.6) is a left–inverse for (3.3), (3.4) with $z \in M_\alpha$, $v \in \Omega$,

$$\begin{aligned} \hat{k}(z) &= 1/(gf^{\alpha-1}h)(z), & \hat{h}(z) &= -\hat{k}(z)(f^\alpha h)(z), \\ \hat{f}(z) &= f(z) + \hat{h}(z)g(z), & \hat{g}(z) &= \hat{k}(z)g(z) \end{aligned}$$

Here it should be noted that

$$(ad^i(f)g) = (-1)^i \sum_{o \le j \le i} (-1)^j \binom{i}{j} f^j g f^{i-j} h$$

so that, for example,

$$(ad^2(f)g)h = (-1)^2(gf^2)h = (gf^2)h.$$

3.2 Multivariable systems[17]

The generalization of the results quoted in Section 3.1 to the case of one input but several outputs is straightforward [16]. For a multi–input/multi–output system sufficient conditions for the invertibility of system (3.1), (3.2) with $k_j(x(t)) = 0$, $j = 1, 2, \cdots, m$, have been obtained in [17] by generalizing the procedure used for linear systems [35]. As described briefly in Section 2.2 this procedure generates a sequence of linear systems, terminating when an output equation is obtained which can be solved explicitly for u. Similarly, in the nonlinear case, a sequence of nonlinear systems is generated together with a sequence of invertibility indices

$$r_1 \leq r_2 \leq \cdots \leq m. \tag{3.7}$$

The relative order α for system (3.1), (3.2) with $k_j(x(t)) = 0$, $j = 1, 2, \cdots, m$, is the least positive integer i such that $r_i = m$. The invertibility of the generated systems infers the invertibility of system (3.1), (3.2) through a sequence of matrices

$$F_o(x), F_1(x), F_2(x), \cdots \tag{3.8}$$

For all $k \geq \alpha$, the kth system is invertible at x_0 for all $x_0 \in M_k$ provided the relative order α is finite. Furthermore, system (3.1), (3.2) is strongly invertible and is strongly invertible at x_0 for all $x_0 \in M_\alpha$ if $\alpha = 1$ or $\alpha > 1$ and for $i = 1, 2, \cdots, m$, $0 \leq k \leq \alpha - 2$, $0 \leq j \leq \alpha - 2 - k$,

$$g_i f^j F_k(\cdot) = 0 \text{ on } M.$$

Three examples are given in [17].

4 Nonlinear systems–algebraic approach[13],[14]

In the last few years a new approach has been developed for the analysis of several nonlinear control problems, including the inverse problem. This analysis is based upon a branch of pure mathematics known as differential algebra which plays the same role with respect to differential equations as commutative algebra does with respect to algebraic geometry.

If K is a subfield of L so that $K \subset L$, then L is called an *extension of* K, denoted by $L : K$. In particular, if $L : K$ is an extension and $\{y_1, y_2, \cdots, y_n\}$ is a subset of L, then the subfield of L generated by $K \cup \{y_1, y_2, \cdots, y_n\}$ is written $K(y_1, y_2, \cdots, y_n)$. As an example, the subfield of $\Re$ consisting of all elements $p + q\sqrt{2}$ where $p, q \in \mathcal{Q}$, is the field extension $\mathcal{Q}(\sqrt{2})$.

A simple extension is an extension $L : K$ such that $L = K(\alpha)$ for some $\alpha \in L$. Suppose $K(\alpha) : K$ is a simple extension. If there does not exist a non–zero polynomial X over K such that $X(\alpha) = 0$ then α is a *transcendental element* over K. An extension $L : K$ is finitely generated if $L = K(y_1, y_2, \cdots, y_n)$ where n is finite. If $L : K$ is finitely generated then there exists an intermediate field M such that $M = K(y_1, y_2, \cdots, y_r)$ where the y_i are independent transcendental elements over K. The integer r is the *transcendence degree* of $L : K$ written $\partial(L : K)$. The differential field $\Re < t >$ is the field $\Re(t)$ of rational functions of t with real coefficients on which is superimposed a derivation $\Re(t) \longrightarrow \Re(t)$ given by $a \mapsto da/dt = \dot{a}$ such that, $\forall a, b \in \Re(t)$,

$$\begin{aligned} \frac{d}{dt}(a+b) &= \dot{a} + \dot{b} \\ \frac{d}{dt}(ab) &= \dot{a}b + a\dot{b}. \end{aligned}$$

4.1 Differential output rank

Consider the very general input/output system

$$\begin{aligned} \dot{x}(t) &= f(x(t)), u(t), \cdots, u^{(s)}(t)) \in X \subset \Re^n \\ y(t) &= h(x(t)), u(t), \cdots, u^{(s)}(t)) \in \Re^p \end{aligned}$$

Write $\mathcal{Q} < y >$ to be the smallest differential field containing $\mathcal{Q}, y_1, y_2, \cdots, y_p$ An example of an element in $\mathcal{Q} < y >$ is

$$\frac{y_1\dot{y}_2 - \ddot{y}_1 y_2}{3\ddot{y}_2^3}.$$

The differential transcendence degree of the differential field $\mathcal{Q} < y >$ over $\mathcal{Q}$ is called the ***differential output rank***. For a constant linear system the differential output rank is equal to the rank of the transfer matrix. As with linear systems a system is said to be differentially left invertible iff its differential output rank is equal to m.

Given differential fields $\mathcal{Q} \subset \mathcal{Q} < y > \subset \mathcal{Q} < u, y >$, it is known that

$$(\mathcal{Q}, < u, y >: \mathcal{Q}) = (\mathcal{Q} < u, y >: \mathcal{Q} < y >) + (\mathcal{Q} < y >: \mathcal{Q}). \tag{4.1}$$

From this it is possible to show that if a system is left invertible then it is possible to recover the control from the output by a finite set of equations.

References

[1] Albrecht, F., Gasse, K.A. and Wax, N. Reproducibility of Linear and Nonlinear Input–Output Systems, Journal of Mathematical Analysis and Application, 79, 178–202, 1981.

[2] Banks, S.P. "Mathematical Theories of Nonlinear Systems", Prentice Hall, London.

[3] Behn R.D. and Ho, Y.C. On a Class of Linear Stochastic Differential Games, IEEE Transactions on Automatic Control, 13, 227–240, 1968.

[4] Brockett, R.W. Poles, Zeros and Feedback: State Space Interpretation, IEEE Transactions on Automatic Control, 10, 129–135, 1965.

[5] Brockett, R.W. and Mesarovic, M.D. The Reproducibility of Multivariable Systems, Journal of Mathematical Analysis and Applications, 11, 548–563, 1965.

[6] Li, C.W. and Feng, Y.K. Functional Reproducibility of General Multivariable Analytic Nonlinear Systems, International Journal of Control, 45, 255–268, 1987.

[7] Descusse, J. and Moog, C.H. Decoupling with Dynamic Compensation for Strong Invertible Affine Non–Linear Systems, International Journal of Control, 42, 1387–1398, 1985.

[8] Descusse, J. and Moog, C.H. Dynamic Decoupling for Right–Invertible Nonlinear Systems, Systems and Control Letters, 8, 345–350, 1987.

[9] Dorato, P. On the Inverse of Linear Dynamical Systems, IEEE Transactions on Systems Science and Cybernetics, 5, 43–48, 1969.

[10] Falb, P.L. and Wolovich, W.A. Decoupling in the Design and Synthesis of Multivariable Control Systems, IEEE Transactions on Automatic Control, 12, 651–659, 1967.

[11] Fliess, M. On the Inversion of Nonlinear Multivariable Systems, in "Mathematical Theory of Networks and Systems", Lecture Notes in Control and Information Sciences, No. 58, Spinger–Verlag, Berlin, 1984.

[12] Fliess, M. A New Approach to the Structure at Infinity of Nonlinear Systems, Systems and Control Letters, 7, 419–421, 1986.

[13] Fliess, M. Some Remarks on Nonlinear Invertibility and Dynamic State–Feedback, in "Theory and Applications of Nonlinear Control Systems" (Eds. C.I.Byrnes and A.Lindquist), North Holland, Amsterdam, 1986.

[14] Fliess, M. A Note on the Invertibility of Nonlinear Input–Output Differential Systems, Systems and Control Letters, 8, 147–151, 1986.

[15] Hirschorn, R.M. Invertibility of Control Systems on Lie Groups, SIAM Journal on Control and Optimization, 15, 1034–1049, 1977.

[16] Hirschorn, R.M. Invertibility of Nonlinear Control Systems, SIAM Journal of Control and Optimization, 17, 289–297, 1979.

[17] Hirschorn, R.M. Invertibility of Multivariable Nonlinear Control Systems, IEEE Transactions on Automatic Control, 24, 855–865, 1979.

[18] Hirschorn, R.M. Output Tracking in Multivariable Nonlinear Systems, IEEE Transactions on Automatic Control, 26, 593–595, 1981.

[19] Isidori, A."Nonlinear Control Systems: An Introduction", Springer–Verlag, Berlin,1985.

[20] Isidori, A. Control of Nonlinear Systems via Dynamic State–Feedback, in "Algebraic and Geometric Methods in Nonlinear Control Theory", (Eds. M.Fliess and M.Hazewinkel),Reidel, Dordrecht, 1986.

[21] Massey, J.L. and Swain, M.K. Inverses of Linear Sequential Circuits, IEEE Transactions on Computers, 17, 330–337, 1968.

[22] Mehra, R.K. Inversion of Multivariable Linear Dynamic Systems using Optimum Smoothing, IEEE Transactions on Automatic Control, 15, 252, 1970.

[23] Nijmeijer, H. Invertibility of Affine Nonlinear Control Systems: A Geometric Approach, Systems and Control Letters, 2, 122–129, 1982.

[24] Nijmeijer, H. and Schumacher, J.M. Les Systemes Non Linéaires à plus d'Entrées que de Sorties ne sont pas Inversibles, C.R. Acad. Sci. Paris Ser I, 299, 791–794, 1984.

[25] Nijmeijer, H. and Schumacher, J.M. On the Inherent Integration Structure of Nonlinear Systems, IMA Journal of Mathematical Control and Information, 2, 87–107, 1985.

[26] Nijmeijer, H. and Schumacher, J.M. Zeros at Infinity for Affine Nonlinear Control Systems, IEEE Transactions on Automatic Control, 30, 566–573, 1985.

[27] Nijmeijer, H. Right–Invertibility for a Class of Nonlinear Control Systems: A Geometric Approach, Systems and Control Letters, 7, 125–132, 1986.

[28] Orner, P.A. Construction of Inverse Systems, IEEE Transactions on Automatic Control, 17, 151–153, 1972.

[29] Porter, W.A. Diagonalization and Inverses for Nonlinear Systems, International Journal of Control, 11, 67–76, 1970.

[30] Rebhuhn, D. Invertibility of C^∞ Multivariable Input–Output Systems, IEEE Transactions on Automatic Control, 25, 207–212, 1980.

[31] Respondek, W. and Nijmeijer, H. On Local Right–Invertibility of Nonlinear Control Systems, The International Journal of Control–Theory and Advanced Technology, 4, 325–348, 1988.

[32] Sain, M.K. Functional Reproducibility and the Existence of Classical Sensitivity Matrices, IEEE Transactions on Automatic Control, 12, 458, 1967.

[33] Sain, M.K. and Massey, J.L. Invertibility of Linear Time–Invariant Dynamical systems, IEEE Transactions on Automatic Control, 14, 141–149, 1969.

[34] Silverman, L.M. Properties and Application of Inverse systems, IEEE Transactions on Automatic Control, 13, 436–437, 1968.

[35] Silverman, L.M. Inversion of Multivariable Linear Systems, IEEE Transactions on Automatic Control, 14, 270–276, 1969.

[36] Singh, S.N. Decoupling of Invertible Nonlinear Systems with State Feedback and Precompensation, IEEE Transactions on Automatic Control, 25, 1237–1239, 1980.

[37] Singh, S.N. A Modified Algorithm for Invertibility in Nonlinear Systems, IEEE Transactions on Automatic Control, 26, 595–598, 1981.

[38] Singh, S.N. Generalized Decoupled–Control Synthesis for Invertible Nonlinear Systems, IEE Proceedings, Part D, 128, 157–161, 1981.

[39] Singh, S.N. Reproducibility in Nonlinear Systems using Dynamic Compensation and Output Feedback, IEE Transactions on Automatic Control, 27, 955–958, 1982.

[40] Singh, S.N. Generalized Functional Reproducibility Condition for Nonlinear Systems, IEEE Transactions on Automatic Control, 27, 958–960, 1982.

[41] Singh, S.N. Invertibility of Observable Multivariable Nonlinear Systems, IEEE Transactions on Automatic Control, 27, 487–489, 1982.

[42] Tsinias, J. and Kalouptsidis, N. Invertibility of Nonlinear Analytic Single–Input Systems, IEEE Transactions on Automatic Control, 28, 931–933, 1983.

[43] Wang, S.H. and Davison, E.J. A New Invertibility Criterion for Linear Multivariable Systems, IEEE Transactions on Automatic Control, 18, 538–539, 1973.

[44] Willsky, A.S. On the Invertibility of Linear Systems, IEEE Transactions on Automatic Control, 19, 272–274, 1974.

[45] Wohltmann, H.W. Function Space Dependent Criteria for Target Path Controllability of Dynamical Systems, International Journal of Control, 41, 709–715, 1985.

[46] Wolovich, W.A. "Linear Multivariable Systems", Springer–Verlag, New York, 1974.

[47] Zadeh, L.A. and Desoer, C.A. "Linear System Theory", McGraw–Hill, New York, 1963.

On frequency response for nonlinear systems

S.P. Banks B. Chanane

Abstract

In this paper we define a 'generalized frequency response' for a nonlinear input–output map S_{x_0} as the mapping $\mathcal{I}o\mathcal{F}o\mathcal{F}^{-1}o\mathcal{I}^{-1}$ from ℓ^2 to ℓ^2 where $\mathcal{F}$ is the Fourier transform and $\mathcal{I}$ the usual isomorphism from $L^2[-\infty,\infty]$ to ℓ. Realization results relative to linear and bilinear systems are presented. Also sufficient conditions for $\mathcal{L}^2$–stability of bilinear systems are derived.

1 Introduction

In this paper we define a 'Generalized Frequency Response' (subsequently called 'GFR') for a nonlinear system. Previous attempts to define some kind of 'frequency domain' theory for nonlinear systems (e.g., [5]) were based on the Volterra series expansion of the input–output map of the system and then associating a sequence of transfer functions $H_k(s_1,\cdots,s_k)$, $k \geq 1$, each one being the multidimensional Laplace transform of the corresponding Volterra series kernel. Setting $s_j = iw_j$, ($i=\sqrt{-1}$) gives the 'frequency response' of the kth kernel $H_k(iw_1,\cdots,iw_k)$. However, this approach has two immediate difficulties. Firstly, we obtain frequency spaces of increasing dimension and secondly, it is not at all clear how the functions $H_k(iw_1,\cdots,iw_k)$ relate to the responses of the system to standard inputs (e.g., complex exponentials).

In a recent paper [3], we introduced the notion of a 'GFR' for a nonlinear input–output map S_{x_0} by using the natural isomorphism $\mathcal{I}$ between $\mathcal{L}^2_m[0,\infty]$ and ℓ^2 in the time domain and then defining the 'GFR' as the map $\mathcal{I}oS_{x_0}o\mathcal{I}^{-1}$ from ℓ^2 to ℓ^2. Here, we shall make use of the isomorphism $\mathcal{I}$ between $L^2[-\infty,\infty]$ and ℓ^2 in the frequency domain (i.e., we operate on the Fourier transform of the input and output) and then defining the 'GFR' as the induced map from ℓ^2 to ℓ^2 defined by $\mathcal{I}o\mathcal{F}oS_{x_0}\mathcal{F}^{-1}o\mathcal{I}^{-1}$. Using the invariance property of the scalar product under the Fourier transform, we can see that the approach introduced here and the one used in [3] are equivalent.

The paper is organised as follows: In section 2, we shall consider sufficient conditions for a bilinear system to be $\mathcal{L}^2$–stable. In section 3, we define a 'GFR' for a nonlinear input–output map. We shall illustrate the 'GFR' for linear and bilinear systems. Finally, a realization theory will be presented in section 4.

We shall make use in the sequel of the following notation: The Fourier transform F of a square integrable function f is defined by

$$F(i\omega) = \frac{1}{\sqrt{2\pi}} \int_0^\infty f(t)e^{-i\omega t}dt \tag{1.1}$$

If x is an n–vector and A is an $n \times n$–matrix, we take as compatible norms

$$||x|| = max_{i=1,n}|x_i| \text{ and } ||A|| = max_{i=1,n} \sum_{k=1,n} |a_{i,k}| \tag{1.2}$$

$||.||_1$ and $||.||_{2\omega}$ are the norms associated with the standard L^1 and L^2_ω spaces, the space of absolutely integrable functions and the space of square integrable functions (with weight w) respectively, Whereas ℓ^2 designates the space of square summable sequences.

2 $\mathcal{L}^2$–stability of bilinear systems:

In this section we shall present sufficient conditions for the $\mathcal{L}^2$–stability of bilinear systems. By $\mathcal{L}^2$–stability we understand that:

$$\forall u \in \Omega \subset L^2_\omega[0,\infty] \to y \in L^2_w[0,\infty]$$

where u and y are respectively the input and the output of the system and Ω is defined by the sufficient conditions. The weight function w is such that $w(t)e^{-2\alpha t} \in L^1[0,\infty]$ for some $\alpha \geq 0$, $w(t) \geq 0$, for $t \geq 0$.

Consider for simplicity the single–input single–output system described by

$$\begin{aligned} \dot{x} &= Ax + uNx + Bu, \; x(0) = 0 \\ y &= Cx \end{aligned} \tag{2.1}$$

where A, N, B, C are constant matrices of suitable dimensions.

As is well known [1], the input–output map is given by:

$$\begin{aligned} y(t) &= Ce^{At}x_0 \\ &+ \sum_{j\geq 1} \int_0^t \int_0^{\sigma_j} \cdots \int_0^{\sigma_2} v_j(t,\sigma_1,\cdots,\sigma_j)u(\sigma_1)\cdots u(\sigma_j)d\sigma_1\cdots\sigma_j \\ &+ \sum_{j\geq 1} \int_0^t \int_0^{\sigma_j} \cdots \int_0^{\sigma_2} w_j(t,\sigma_1,\cdots,\sigma_j)u(\sigma_1)\cdots u(\sigma_j)d\sigma_1\cdots\sigma_j \end{aligned} \tag{2.2}$$

where

$$\begin{aligned} v_1(t,\sigma_1) &= Ce^{A(t-\sigma_1)}B \\ v_j(t,\sigma_1,\cdots,\sigma_j) &= Ce^{A(t-\sigma_j)}Ne^{A(\sigma_j-\sigma_{j-1})}N\cdots Ne^{A(\sigma_2-\sigma_1)}B, j > 1 \end{aligned}$$

and

$$\begin{aligned} w_1(t,\sigma_1) &= Ce^{A(t-\sigma_1)}Ne^{A\sigma_1}x_0 \\ w_j(t,\sigma_1,\cdots,\sigma_j) &= Ce^{A(t-\sigma_j)}Ne^{A(\sigma_j-\sigma_{j-1})}N\cdots Ne^{A(\sigma_2-\sigma_1)}Ne^{A\sigma_1}x_0, j > 1 \end{aligned}$$

Theorem 1 *A sufficient condition for the system (2.1) to be $\mathcal{L}^2$-stable is that the following holds:*

1. *The eigenvalues of A have real parts less than $-\alpha$*
2. *u, $e^{\alpha t}u \in L^1[0,\infty]$.*

Proof

It follows from (i) as a standard result in the theory of differential equations that there exists a positive constant K such that

$$\|e^{At}\| \leq Ke^{-\alpha t}, \ t \geq 0.$$

Equation (2.1) yields

$$x(t) = e^{At}x_0 + \int_0^t e^{A(t-\tau)}Bu(\tau)d\tau + \int_0^t e^{A(t-\tau)}Nx(\tau)u(\tau)d\tau \tag{2.3}$$

from which we obtain,

$$\begin{aligned} \|x(t)\| &\leq Ke^{-\alpha t}\|x_0\| + K\|B\| \int_0^t e^{-\alpha(t-\tau)}|u(\tau)|d\tau \\ &+ K\|N\| \int_0^t e^{-\alpha(t-\tau)}|u(\tau)|\|x(\tau)\|d\tau \\ &\leq Ke^{-\alpha t}\|x_0\| + K\|B\|e^{-\alpha t} \int_0^\infty e^{\alpha\tau}|u(\tau)|d\tau \\ &+ K\|N\|e^{-\alpha t} \int_0^t e^{\alpha\tau}|u(\tau)|\|x(\tau)\|d\tau \end{aligned} \tag{2.4}$$

Therefore,

$$\begin{aligned} e^{\alpha t}\|x(t)\| &\leq K\left[\|x_0\| + \|B\| \int_0^\infty e^{\alpha\tau}|u(\tau)|d\tau\right] \\ &+ K\|N\| \int_0^t |u(\tau)| \left[e^{\alpha\tau}\|x(\tau)\|\right] d\tau \end{aligned} \tag{2.5}$$

Thus, using Gronwall lemma, we obtain,

$$e^{\alpha t}\|x(t)\| \leq K\left[\|x_0\| + \|B\| \int_0^\infty e^{\alpha\tau}|u(\tau)|d\tau\right] e^{K\|N\| \int_0^t |u(\tau)|d\tau} \tag{2.6}$$

Hence,

$$\begin{aligned} \|y\|_{2w}^2 &= \int_0^\infty w(t)y^2(t)dt \\ &\leq \int_0^\infty w(t)\|C\|^2\|x(t)\|^2dt \\ &\leq K^2\|C\|^2 \left[\|x_0\| + \|B\| \int_0^\infty e^{\alpha\tau}|u(\tau)|d\tau\right]^2 e^{2K\|N\| \int_0^\infty |u(\tau)|d\tau} \int_0^\infty e^{-2\alpha t}w(t)dt \\ &< \infty \end{aligned} \tag{2.7}$$

3 Generalized frequency response of nonlinear input–output maps:

Consider a system S given in terms of an input–output map

$$S : R^n \times L^2_w[0,\infty] \longrightarrow L^2_w[0,\infty] \tag{3.1}$$

defined by

$$y(t) = S(x_0, u(\cdot))(t) \tag{3.2}$$

where u and y are respectively the input and the output of the system and x_0 is the initial state in some given state–space realization. For each fixed initial state x_0, we have a map

$$S_{x_0} = S(x_0, \cdot) : L^2_w \longrightarrow L^2_w[0,\infty] \tag{3.3}$$

For simplicity, we have assumed scalar input and scalar output.

In a recent paper [3], we introduced the notion of a 'Generalized Frequency Response' by using the natural isomporhism $\mathcal{I}$ between $L^2_w[0,\infty]$ and ℓ^2 in the time–domain, and then defining the 'GFR' as the induced map from ℓ^2 to ℓ^2 such that the diagram (A)

$$\begin{array}{ccc} L^2_w[0,\infty] & \xrightarrow{S_{x_0}} & L^2_w[0,\infty] \\ \mathcal{I}\downarrow & & \downarrow\mathcal{I} \\ \ell^2 & \xrightarrow{s_{x_0}} & \ell^2 \end{array}$$

commutes.

Alternatively, we can use the natural isomorphism between $L^2[-\infty,\infty]$ and ℓ^2 in the frequency domain (i.e., we operate on the Fourier transforms of the input and output) and then defining the 'GFR' as the induced map from ℓ^2 to ℓ^2 such that the diagram (B)

$$\begin{array}{ccc} L^2_w[0,\infty] & \xrightarrow{S_{x_0}} & L^2_w[0,\infty] \\ \mathcal{F}\downarrow & & \downarrow\mathcal{F} \\ L^2_w[-\infty,\infty] & \xrightarrow{\hat{S}_{x_0}} & L^2_w[-\infty,\infty] \\ \mathcal{I}\downarrow & & \downarrow\mathcal{I} \\ \ell^2 & \xrightarrow{s_{x_0}} & \ell^2 \end{array}$$

commutes.

Let $\{e_j\}_{j\geq 0}$ be a basis of $L^2[0,\infty]$ and $E_j = \mathcal{F}\{e_j\}, j \geq 0$. Recalling the fact that the scalar product is invariant under the Fourier transform, we deduce that $\{E_j\}_{j\geq 0}$ is a basis of $L^2[-\infty,\infty]$. Let $\mathcal{I}$ denotes the usual isomorphism

$$\mathcal{I} : L^2[-\infty,\infty] \rightarrow \ell^2 \tag{3.4}$$

given by

$$\mathcal{I}(F) = \{F_j\}_{j\geq 0} \tag{3.5}$$

where

$$F \in L^2[-\infty, \infty], \ F = \sum_{j \geq 0} F_j E_j \tag{3.6}$$

explicitly s_{x_0} is given by

$$s_{x_0} = \mathcal{I} o \mathcal{F} o S_{x_0} o \mathcal{F}^{-1} o \mathcal{I}^{-1} \tag{3.7}$$

that is

$$s_{x_0}(\{U_k\}_{k \geq 0}) = \{Y_j\}_{j \geq 0} \tag{3.8}$$

where

$$U_k =< \mathcal{F}(u), E_k >, Y_j =< \mathcal{F}(S_{x_0}(u)), E_j > \tag{3.9}$$

and $< \cdot, \cdot >$ denotes the scalar product in $L^2[-\infty, \infty]$. We have

$$U_k =< u, e_k > \text{ and } Y_j =< S_{x_0}(u), e_j >$$

Let assume the systems at hand are $\mathcal{L}^2$–stable. We shall illustrate the expression(3.7) for the linear and bilinear input–output maps.

Example 1: Linear systems

Consider the linear system

$$\begin{aligned} \dot{x} &= Ax + Bu, \ x(0) = x_0 \\ y &= Cx \end{aligned} \tag{3.10}$$

then the input–output map is given by

$$y(t) = Ce^{At}x_0 + \int_0^t Ce^{A(t-\tau)}Bu(\tau)d\tau \tag{3.11}$$

In this case

$$S(x_0, u(.))(t) = g_0(t) + (g \star u)(t) \tag{3.12}$$

where

$$g_0(t) = Ce^{At}x_0, \ g(t) = Ce^{At}B$$

and $\star$ denotes the convolution operator.

Taking the Fourier transform of both sides of (3.11) we obtain

$$Y(i\omega) = G_0(i\omega) + G(i\omega)U(i\omega) \tag{3.13}$$

Let $x_0 = 0$ and introduce a basis $E_j, j \geq 0$ of $L^2[-\infty, \infty]$. We obtain,

$$Y_l = \sum_{j \geq 0} G_{lj} U_j, l \geq 0 \tag{3.14}$$

where

$$Y(i\omega) = \textstyle\sum_{l \geq 0} Y_l E_l(i\omega), \ U(i\omega) = \sum_{j \geq 0} U_j E_j(i\omega)$$
$$G_{lj} =< G(i\omega)E_j(i\omega), E_l(i\omega) >$$

We therefore see that the matrix representation of the linear operator $s : \ell^2 \longrightarrow \ell^2$ for the linear system above, with respect to the basis $(1,0,0,\cdots),(0,1,0,\cdots),\cdots$ of ℓ^2 is $(G_{lj})_{l,j\geq 0}$.

Example 2: Bilinear systems Consider the bilinear system (2.1). Let $N = N_1N_2$ where the dimensions of N, N_1, N_2 are respectively $n \times n, n \times m, m \times n$, and

$$\begin{array}{ll} g_0(t) = Ce^{At}B, & g_1(t) = Ce^{At}N_1 \\ g_2(t) = N_2e^{At}N_1, & g_3(t) = N_2e^{At}B \\ g(t) = Ce^{At}x_0, & g_4(t) = N_2e^{At}x_0 \end{array} \tag{3.15}$$

Equation (2.2) can therefore be rewritten as

$$\begin{aligned} y(t) &= g(t) + (g_0 \star u)(t) \\ &+ \sum_{j\geq 2}(g_1 \star [u(g_2 \star [u(\cdots u(g_2 \star [u(g_2 \star [u(g_3 \star u)])]) \cdots)])])(t) + (g \star u)(t) \\ &+ \sum_{j\geq 2}(g_1 \star [u(g_2 \star [u(\cdots u(g_2 \star [u(g_2 \star [u(g_4 \star u)])]) \cdots)])])(t) \end{aligned} \tag{3.16}$$

where g_2 appears $j-2$ times in the first summation and $(j-1)$ times in the second. Taking the Fourier transform of both sides of (3.16) and using its properties we obtain

$$\begin{aligned} Y(i\omega) &= G(i\omega) + G_0(i\omega)U(i\omega) + \sum_{j\geq 2}\frac{1}{\sqrt{2\pi}^{j-1}}(G_1[U \star (G_2[U \star (\cdots \\ &\quad \cdots U \star (G_2[U \star (G_2[U \star (G_3U)])]) \cdots)])])(i\omega) + G(i\omega)U(i\omega) \\ &+ \sum_{j\geq 2}\frac{1}{\sqrt{2\pi}^{j}}(G_1[U \star (G_2[U \star (\cdots \\ &\quad \cdots (G_2[U \star (G_2[U \star (G_2(G_4 \star U))])]) \cdots)])])(i\omega) \end{aligned} \tag{3.17}$$

Let $x_0 = 0$ and introduce a basis $E_j j \geq 0$ of $L^2[-\infty,\infty]$. If $U(i\omega) = \sum_{j\geq 0} U_jE_j(i\omega)$ then the previous equality becomes

$$\begin{aligned} Y(i\omega) &= \sum_{k_1\geq 0} U_{k_1}G_0(i\omega)E_{k_1}(i\omega) \\ &+ \sum_{j\geq 2}\sum_{k_1\geq 0}\cdots\sum_{k_j\geq 0} U_{k_1}\cdots U_{k_j}\frac{1}{\sqrt{2\pi}^{j-1}}(G_1[E_{k_j} \star (\cdots \\ &\quad \cdots E_{k_4} \star (G_2[E_{k_3} \star (G_3[E_{k_2} \star (G_3E_{k_1})])]) \cdots)])])(i\omega) \end{aligned} \tag{3.18}$$

Therefore,

$$Y_l = \sum_{j\geq 1}\sum_{k_1\geq 0}\cdots\sum_{k_j\geq 0} V_{jk_1\cdots k_jl}U_{k_1}\cdots U_{k_j} \tag{3.19}$$

where

$$V_{1k_1l} =< G_0(i\omega)E_{k_1}(i\omega), E_l(i\omega) > \tag{3.20}$$

and

$$V_{jk_1\cdots k_jl} = \frac{1}{\sqrt{2\pi}^{j-1}} < (G_1[E_{k_{j-1}} \star (G_2[E_{k_{j-1}} \star (\cdots$$
$$\cdots E_{k_4} \star (G_2[E_{k_3} \star G_2[E_{k_2} \star (G_3 E_{k_1}])]) \cdots$$
$$)])])(i\omega),\ E_l(i\omega) >,\ j > 1 \tag{3.21}$$

Hence, the diagram (B) induces the map $s_{x_0} : \ell^2 \longrightarrow \ell^2$ given by

$$s_{x_0}((U_0, U_1, \cdots))_l = \sum_{j\geq 1} \sum_{k_1\geq 0} \cdots \sum_{k_j\geq 0} V_{jk_1} \cdots {}_{k_j\ell} U_{k_1} \cdots U_{k_j}$$

4 Realization theory:

In this section we shall consider the problem of the realizability and the state-space realization of an analytic map $s : \ell^2 \longrightarrow \ell^2$ which defines a 'Generalized Frequency Response'. We shall present conditions under which such s is realized by a linear or a bilinear system.

4.1 Linear systems:

Theorem 2:

A necessary and sufficient condition for a sequence of numbers $\{G_{lj}\}_{l,j\geq 0}$ to be the 'Generalized Response' of a linear system with zero initial condition (with respect to a given basis $E_k k \geq 0$ of $L^2[-\infty, \infty]$) is that there exists a strictly proper rational function $G(i\omega)$ such that

$$\sum_{l\geq 0} G_{lj} E_l(i\omega) = G(i\omega) E_j(i\omega) \tag{4.1}$$

for all $j \geq 0$. $G(i\omega)$ is then the Fourier transform of the impulse response of the linear system.

Proof:

Immediate and shall be omitted.

4.2 Bilinear systems:

Let e_k, $k \geq 0$ be the Laguerre functions defined by

$$e_k(t) = e^{-\frac{t}{2}} \sum_{m=0} \frac{(-1)^m}{m!} \binom{k}{m} t^m$$

They constitute a complete orthonormal basis for $L^2[0, \infty]$.

Consider E_k the Fourier transform of e_k, we have

$$E_k(i\omega) = \frac{1}{\sqrt{2\pi}} \frac{1}{\frac{1}{2} + i\omega} \left[\frac{-\frac{1}{2} + i\omega}{\frac{1}{2} + i\omega}\right]^k$$

$E_k k \geq 0$ is therefore a complete basis for $L^2[-\infty, \infty]$.

Remark 1:

The coefficients in the expansion of $F(i\omega)$ with respect to the basis $E_k k \geq 0$ are the coefficients in the Taylor expansion of $\overline{F}$

$$\overline{F}(z) = \sqrt{2\pi}\frac{1}{1-z}F\left(\frac{1}{2}\frac{1+z}{1-z}\right)$$

at $z = 0$, So $F_k = \frac{\sqrt{2\pi}}{k!}\left[\frac{1}{1-z}F\left(\frac{1}{2}\frac{1+z}{1-z}\right)\right]^{(k)}_{z=0}$ where $F(i\omega) = \sum_{k\geq 0} F_k E_k(i\omega)$.

Theorem 3:

A necessary and sufficient condition for a sequence of numbers $\{V_{jk_1}\cdots_{k_jl}\}_{j\geq 1, k_1,\cdots,k_j,l\geq 0}$ to be the 'Generalized Frequency Response' of a bilinear system with zero initial condition (with respect to the given basis $E_k k \geq 0$ of $L^2[-\infty, \infty]$) is that there exist four matrices $G_0(i\omega), G_1(i\omega), G_2(i\omega), G_3(i\omega)$ with dimensions respectively 1×1, $1 \times m, m \times m, m \times 1$ of strictly proper rational functions such that

1.
$$\sum_{l\geq 0} V_{1k_1l}E_l(i\omega) = G_0(i\omega)E_{k_1}(i\omega) \tag{4.2}$$

2.
$$\begin{aligned}\textstyle\sum_{l\geq 0} V_{jk_1\cdots k_jl}E_l(i\omega) = \frac{1}{k_j!\cdots k_2!}D_j^{k_j}\cdots D_2^{k_2}\\ [G_1(i\omega)G_2(i\omega - z_j)G_2(i\omega - z_j - z_{j-1})\cdots\\ \cdots G_2(i\omega - z_j - \cdots - z_3)G_3(i\omega - z_j - \cdots - z_2)\\ E_{k_1}(i\omega - z_j - \cdots - z_2)(-\tfrac{1}{2} + z_2)^{k_2}\cdots(-\tfrac{1}{2} + z_j)^{k_j}\big]_{z_2=\cdots=z_j=-\frac{1}{2}}\\ k_1, \cdots, k_j \geq 0, j > 1\end{aligned} \tag{4.3}$$

where $D_j^{k_j} = \partial^{k_j}/\partial z_j^{k_j}$.

Proof:

1. (4.2) can readily be obtained from (3.20) by premultiplying by $E_l(i\omega)$ and summing over $l \geq 0$. Whereas (3.15) yields $G_0(i\omega) = C(i\omega - A)^{-1}B$.

2. From (3.21) we obtain

$$\sum_{l\geq 0} V_{jk_1}\cdots_{k_jl}E_l(i\omega) = \frac{1}{\sqrt{(2\pi)^{j-1}}}(G_1[E_{k_j} \star (G_2[E_{k_{j-1}} \star (\cdots$$
$$\cdots E_{k_4} \star (G_2[E_{k_3} \star (G_2[E_{k_2} \star (G_3E_{k_1})])]) \cdots)])])(i\omega)$$

Starting from the inner square bracket and proceeding outward, we obtain, after a change of variable $i\Omega = z$ and the use of residue theory,

- $$(E_{k_2} \star (G_3 E_{k_1}))(i\omega) = \frac{\sqrt{2\pi}}{k_2!} D_2^{k_2}[G_3(i\omega - z_2)E_{k_1}(i\omega - z_2) \left(-\frac{1}{2} + z_2\right) k_2]_{z_2=-\frac{1}{2}} \tag{4.5}$$

- $$\begin{aligned}(E_{k_3} \star [G_2[E_{k_2} \star (G_3 E_{k_1})]])(i\omega) &= \textstyle\int_{-\infty}^{\infty} E_{k_3}(i\Omega) G_2(i\omega - i\Omega) \frac{\sqrt{2\pi}}{k_2!} \\ &\quad D_2^{k_2}[G_3(i\omega - i\Omega - z_2)E_{k_1}(i\omega - i\Omega - z_2)(-\tfrac{1}{2} + z_2)^{k_2}]_{z_2=-\frac{1}{2}} d\Omega \\ &= \tfrac{\sqrt{2\pi}}{k_2!}(-i) \textstyle\int_{-i\infty}^{i\infty} E_{k_3}(z_3) G_2(i\omega - z_3) \\ &\quad D_2^{k_2}[G_3(i\omega - z_3 - z_2)E_{k_1}(i\omega - z_3 - z_2)(-\tfrac{1}{2} + z_2)^{k_2}]_{z_2} = -\tfrac{1}{2} dz_3 \\ &= \tfrac{\sqrt{2\pi}^2}{k_3! k_2!} D_3^{k_3}[G_2(i\omega - z_3) \\ &\quad D_2^{k_2}[G_3(i\omega - z_3 - z_2)E_{k_1}(i\omega - z_3 - z_2)(-\tfrac{1}{2} + z_2)^{k_2}]_{z_2=-\frac{1}{2}} \\ &\quad (-\tfrac{1}{2} + z_3)^{k_3}]_{z_3=-\frac{1}{2}}\end{aligned} \tag{4.6}$$

- $$\begin{aligned}&(E_{k_j} \star (G_2[E_{k_{j-1}} \star (\cdots E_{k_3} \star (G_2[E_{k_2} \star (G_3 E_{k_1})] \cdots)]))(i\omega) = \\ &\tfrac{\sqrt{(2\pi)^{j-1}}}{k_j! \cdots k_2!} D_j^{k_j}[G_2(i\omega - z_j) D_{j-1}^{k_{j-1}}[G_2(i\omega - z_j - z_{j-1}) \cdots \\ &\cdots D_3^{k_3}[G_2(i\omega - z_j - \cdots - z_3) D_2^{k_2}[G_3(i\omega - z_j - \cdots - z_2) \\ &E_{k_1}(i\omega - z_j - \cdots - z_2)(-\tfrac{1}{2} + z_2)^{k_2}](-\tfrac{1}{2} + z_3)^{k_3}] \cdots \\ &\cdots (-\tfrac{1}{2} + z_{j-1})^{k_{j-1}}(-\tfrac{1}{2} + z_j)^{k_j}]_{z_1 = \cdots = z_j = -\frac{1}{2}}\end{aligned}$$

Hence, (4.3). From (3.15) we get

$$\begin{aligned}G_0(i\omega) = C(i\omega I - A)^{-1}B, \; G_1(i\omega) = C(i\omega I - A)^{-1}N_1 \\ G_2(i\omega) = N_2(i\omega I - A)^{-1}N_1, \; G_3(i\omega) = N_2(i\omega I - A)^{-1}B\end{aligned} \tag{4.8}$$

5 Conclusion

In this paper we have presented the idea of a 'Generalized Frequency Response' for a nonlinear system. The theory has been illustrated in the linear and bilinear cases. Realization results were obtained regarding these two cases. Sufficient conditions for the $\mathcal{L}^2$–stability of bilinear systems were derived. For simplicity, we considered single–input single–output systems.

References

[1] P.D'Alessandro, A. Isidori and Ruberti, 'Realization and Structure Theory of Bilinear Systems' S.I.A.M.J. Cont. & Opt. 12,pp517–535, 1974.

[2] S.P.Banks, Mathematical Theories Of Nonlinear Systems, Prentice–Hall, 1988.

[3] S.P.Banks and B. Chanane, 'A generalized frequency response for nonlinear systems', I.M.A.J. Of Math. Cont. & Inf., 5, 147–166, 1988.

[4] C.M. Lesiak and A. J. Krener, 'The existence and uniqueness of Volterra series for nonlinear systems ', IEEE Trans. Auto. Cont., AC–23 pp 1090–1095, 1978.

[5] G.E. Mitzel, S.J. Clancy and W.J. Rugh, 'On Transfer Function Representations For Homogeneous Nonlinear Systems', I.E.E.E. Trans. Aut. Cont. AC–24,pp. 242–249, 1979.

[6] W.J.Rugh, Nonlinear System Theory: the Volterra/Wiener Approach, Johns Hopkins Univ. Baltimore, MD, 1981.

Composite control of nonlinear singularly perturbed systems: a geometric approach

P. M. Sharkey J. O'Reilly

Abstract

Using a geometric approach, a composite control, the sum of a slow control and a fast control, is derived for a general class of nonlinear singularly perturbed systems. A new and simpler method of composite control design is proposed whereby the fast control is completely designed at the outset. The slow control is then free to be chosen such that the slow integral manifold of the original system approximates a desired design manifold to within any specified order of ε accuracy.

1 Introduction

The state feedback control of nonlinear systems remains a difficult problem area compared with its linear counterpart. It is also a fact that systems, for the most part, contain both slow and fast dynamical phenomena. So the problem of feedback control for nonlinear systems appears to be compounded by ill–conditioning due to be interaction of dynamic variables in separate time–scales.

What might, at first sight, seem to be a grave disadvantage, namely the co–presence of slow and fast system dynamics, can in fact be turned to great advantage. The key to this reversal of fortune is the singular perturbation approach. The approach is an analytic one which directly exploits the separation of system time–scales, made explicit by the small singular perturbation parameter ε, so as to derive simpler well conditioned controls [6]. Indeed, progress in this area has been enhanced by two major developments.

The first development is the introduction of the notion of composite control as the sum of a slow control and a fast control, each respectively designed on the basis of reduced–order nonlinear slow and fast system models in separate time scales [1], [8]. Whereas state feedback control designed for the original high–order system without the use of singular perturbation methods might well be ill–conditioned, the alternative composite control is well–conditioned and much simpler to implement on the original system.

The second significant recent development in the use of singular perturbation methods for the control of nonlinear systems is the integral manifold approach [5], [6]. In the integral manifold approach, the dynamic behaviour of nonlinear singularly perturbed system is captured geometrically by the rapid approach of the fast system states to the attractive slow manifold. The fundamental property of the integral manifold, characterised by the so–called integral manifold condition, is that once the fast system states hit the integral manifold, they remain on the manifold thereafter. For general nonlinear singularly perturbed systems under composite state feedback control, the fast control, which steers the fast system states to the slow integral manifold, is designed to be inactive on the manifold. By equating coefficients of like powers of the small parameter ε in the integral manifold condition, it is, in principle, possible to design a slow control that will steer the system along a desired manifold to within any order of ε accuracy of the slow integral manifold.

For a particular class of nonlinear singularly perturbed systems known as fast actuator–type systems Khorasani and Kokotovic [4], Sharkey and O'Reilly [9], have shown, through equating coefficients of ε in the integral manifold condition, that an ε–correction of the slow control compensates for the first–order deviation of the slow integral manifold from the desired slow manifold or design manifold.

In this paper, we treat general nonlinear singularly perturbed systems where the control input excites both slow and fast system states in a nonlinear fashion. Unlike earlier work [5], we do not successively design for a slow control, then a fast control, then for a corrected slow and a corrected fast control, and so on. Rather we adopt a markedly different method by designing completely the fast control from the outset: then the slow control can be designed to any order of ε accuracy without readjustment of the fast control. By initially evaluating a general 'control' in terms of both original states, x and z, and the perturbation parameter ε, we ensure that the fast transient dynamics of the system, if stabilizable, are stable, that the slow integral manifold exists and, moreover, that the system states z are attracted to that manifold. In addition, the slow component of the chosen initial 'fast' control can have a beneficial effect on the shaping of the design manifold, parameterized by the slow control, prior to slow control design. The slow control is then free to be chosen such that the actual integral manifold of the original singularly perturbed system approximates a desired design manifold to within any specified order of ε accuracy.

2 Two–time–scale properties and the integral manifold

Consider the nonlinear system model

$$\dot{x} = f(x, z, u, \varepsilon), \quad x(t_o) = x_o, \tag{2.1a}$$

$$\varepsilon\dot{z} = g(x, z, u, \varepsilon), \quad z(t_o) = z_o, \tag{2.1b}$$

where the control u will assume the general state feedback form

$$u = u_s(x, \varepsilon) + u_f(x, z, \varepsilon) \tag{2.1c}$$

with $x \in B_x \subset R^n$, $z \in B_z \subset R^m$, $u \in R^p$ and where the small perturbation parameter $\varepsilon > 0$ represents the speed ratio of the slow versus the fast dynamical phenomena of the system. The functions f, g, u_s and u_f are sufficiently many times continuously differentiable functions of their arguments and are bounded functions as ε tend to zero; $B_x \times B_z$ represents a closed region enclosing the equilibrium point of interest $(x^T, z^T) = (\underline{x}^T, \underline{z}^T)$. Dividing both sides of (2.1b) by ε we have

$$\dot{z} = \frac{1}{\varepsilon} g(x, z, u, \varepsilon), \tag{2.2}$$

where it is observed that, when ε is small, the system state, $z(t)$, contains fast dynamics. In particular, by setting ε equal to zero, the fast transient dynamics of $z(t)$ become instantaneous for $g \neq 0$. Assuming that the fast transient does not escape to infinity but that $z(t)$ converges to some quasi-steady state, we are left with an nth order system model containing the slow dynamics

$$\dot{x} = f(x, z, u_s(x,0) + u_f(x,z,0), 0), \quad x(t_0) = x_0, \tag{2.3a}$$
$$0 = g(x, z, u_s(x,0) + u_f(x,z,0), 0), \tag{2.3b}$$

or, equivalently, the *reduced-order slow dynamical model*

$$\dot{x} = f(x, h(x, u_s(x,0), 0), u_s(x,0) + u_f(x, h(x, u_s(x,0), 0), 0), 0), \tag{2.4}$$

where the vector

$$z = h(x, u_s(x,0), 0) \tag{2.5}$$

is a solution of the now algebraic vector equation (2.3b). This collapse of the dynamical system order from $n + m$ in the full system model (2.1) to n in the reduced order system model (2.4), (2.5) is termed a *singular perturbation*, and the system (2.1) is known, correspondingly, as a *singularly perturbed system* [6].

In reality, the positive parameter ε is small but non–zero, so that the dynamical behaviour of the singularly perturbed system (2.1) is characterized by the decay of the fast transient in $z(t)$ followed by a slow state evolution in $x(t)$. This intrinstic *two–time–scale–behaviour* of the singularly perturbed system (2.1) is further exhibited by the definition of a fast time scale

$$\tau = \frac{(t - t_0)}{\varepsilon}, \quad \frac{d}{d\tau} = \varepsilon \frac{d}{dt} \tag{2.6}$$

where $\tau = 0$ when $t = t_0$, so that, in the *fast time–scale*, the original system model (2.1) assumes the form

$$\tfrac{dx}{d\tau} = \varepsilon f(x, z, u, \varepsilon), \quad x(0) = x_0, \tag{2.7a}$$
$$\tfrac{dz}{d\tau} = g(x, z, u, \varepsilon), \quad z(0) = z_0. \tag{2.7b}$$

It is observed that, while the dynamics of $z(\tau)$ evolve normally, according to (2.7b), the dynamics of the slow state $x(\tau)$ are almost stationary, for small ε, according to (2.7a). By setting ε equal to zero, as before, the slow state vector $x \in B_x$ and the slow control $u_s \in R^p$ become fixed parameters according to the boundary–layer dynamical model ($\varepsilon = 0$)

$$\frac{dx}{d\tau} = 0, \qquad x(0) = x_0, \tag{2.8a}$$

$$\frac{dz}{d\tau} = g(x, z, u_s(x,0) + u_f(x,z,0), 0), \quad z(0) = z_0, \tag{2.8b}$$

with $x(\tau) = x_0$ and $u_s(x(\tau), 0) = u_s(x_0, 0)$, for $\tau \geq 0$ according to (2.8a). An important result of transforming the original singularly perturbed system (2.1) into the fast time–scale, as (2.7), is that we have readily converted a singular perturbation problem in the slow time scale (with time derivatives multiplying ε on the left–hand–side) into a regular perturbation problem (2.7) (with ε entering into the right–hand–side as a regular perturbation only) in the fast time–scale, τ. In other words, the original system (2.1) whose solution $z(t)$ is discontinuous at $\varepsilon = 0$ has been transformed into an equivalent system (2.7) whose solution, $z(\tau)$, is continuous in the neighbourhood of $\varepsilon = 0$.

This *regularization* of the singular perturbation problem is central to earlier studies of singularly perturbed differential equations [2], [3], [11], [12]. The importance of this regularization lies in the fact that regular perturbation problems are rather easier to solve than singular perturbation problems.

We now turn our attention to a most useful geometric characterization of the two–time–scale dynamic behaviour of the nonlinear singularly perturbed system (2.1) : the *integral manifold.* We have already seen that, at $\varepsilon = 0$, the graph $z = h(x, u_s(x,0), 0)$ defines a slow manifold which is an equilibrium manifold of the infinitely fast (assumed stable) states in (2.2). Thus, for small $\varepsilon > 0$, we could propose that a graph of the form

$$z = h(x, u_s(x,\varepsilon), \varepsilon), \tag{2.9}$$

which is a regular perturbation of (2.5), be a manifold of the original system (2.1) or (2.7). Differentiating both sides of (2.9) with respect to time t we have

$$\dot{z} = \left(\frac{\partial h}{\partial x} + \frac{\partial h}{\partial u_s}\frac{\partial u_s}{\partial x}\right)\dot{x} \tag{2.10}$$

Now, multiplying both sides of (2.10) by ε and substituting for $\dot{x}$, $\dot{z}$ and z from equations (2.1a), (2.1b) and (2.9) respectively, we arrive at an ε–dependent partial differential equation

$$\begin{aligned} &g(x, h(x, u_s, \varepsilon), u_s + u_f(x, h(x, u_s, \varepsilon), \varepsilon), \varepsilon) \\ &= \varepsilon \left(\tfrac{\partial h}{\partial x} + \tfrac{\partial h}{\partial u_s}\tfrac{\partial u_s}{\partial x}\right) f(x, h(x, u_s, \varepsilon), u_s + u_f(x, h(x, u_s, \varepsilon), \varepsilon), \varepsilon) \end{aligned} \tag{2.11}$$

known as the *integral manifold condition*, where

$$u_s \triangleq u_s(x, \varepsilon) \tag{2.12}$$

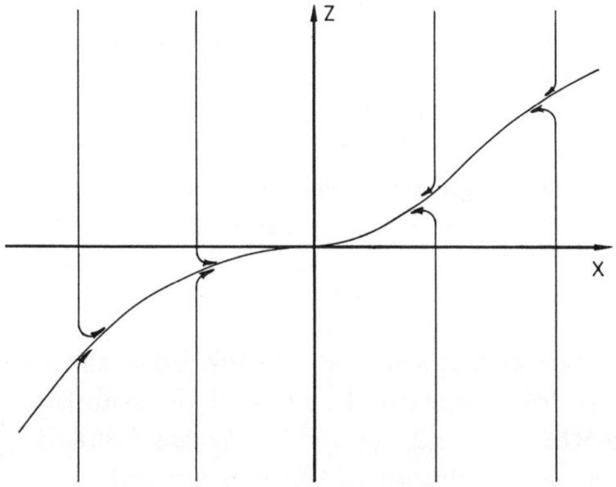

Figure 1. Slow ($z = h(x, u_s(x,\varepsilon),\varepsilon)$) and fast ($x \simeq \bar{x}$) manifolds for the singularly perturbed system (2.1).

Definition 2.1: A solution (2.9) of equation (2.11), assuming it exists, is known as an integral manifold of the singularly perturbed system (2.1) in that, if (2.9) holds for some time $t = t^* \geq t_0$, it holds for all $t \geq t^*$.

The time interval $t \in [t_0, t^*]$, known as the boundary–layer, is the time it takes for the system state $z(t)$, assumed stable, to hit the integral manifold, its quasi–steady–state. The exact reduced slow system model is obtained by substituting the integral manifold $h(x, u_s(x,\varepsilon),\varepsilon)$ for z in (2.1) to give

$$\dot{x} = f(x, h(x, u_s(x,\varepsilon),\varepsilon) + u_s(x,\varepsilon) + u_f(x, h(x, u_s(x,\varepsilon),\varepsilon),\varepsilon),\varepsilon), \tag{2.13}$$

for time $t \geq t^*$.

Geometrically we can think of the fast state, $z(t)$, of the fast system model (2.7) (confined to a fast manifold parameterized by the 'frozen' state vector $x(\tau)$ and slow control $u_s(x(\tau),\varepsilon)$), as being rapidly attracted to (or repelled from) the equilibrium slow manifold (2.9). Fig. 2.1 gives a graphical interpretation of the system model (2.1) with z stable. As we shall only be interested in controlled fast motions which are asymptotically stable, the rapid 'descent' of $z(\tau)$ to hit the equilibrium integral manifold (2.9) at time $t = t^*$ is followed by a 'slide', under control action $u_s(x,\varepsilon)$ according to (2.13), along this integral manifold, as illustrated in Fig. 2.1.

3 Existence of the Integral Manifold for Composite Control

In Section 2 we reviewed the two–time–scale dynamic characteristics instrinic to general nonlinear singularly singularly perturbed systems and intimate as to what, ideally, our feedback control strategy should be. Namely, that we should take advantage of the two–time–scale decomposition of the system (2.1) into separate lower–order slow and fast system models, (2.13) and (2.7b), respectively, where, in

(2.7b) the vectors x and u_s are treated as slowly varying parameters. The control strategy involves the separate designs of a fast control, u_f, for the fast system (2.7b), followed by a slow control u_s for the slow system (2.13), culminating in their addition to form a well–conditioned composite control, $u = u_s + u_f$, for direct application to the original singularly perturbed system (2.1). In this paper, we consider the nonlinear composite state feedback of the general form

$$u = u_s(x, \varepsilon) + u_f(x, z, \varepsilon). \tag{3.1}$$

Such a design strategy does, however, require both the existence of a solution (2.9) to the integral manifold condition (2.11) and the stability of the fast transients of the system state vector, z. Both requirements are fulfilled by the following two fundamental conditions on the design of the fast control

$$u_f = N(x, z, \varepsilon). \tag{3.2}$$

Condition 3.1: The algebraic vector equation

$$g(x, z, u_s(x, 0) + N(x, z, 0), 0) = 0 \tag{3.3a}$$

has the real isolated solution

$$z = h_0(x, u_s(x, 0)) \tag{3.3b}$$

for all $(x^T, z^T) \in B_x \times B_z$, $u_s \in R^p$, that is,

$$g(x, h_0(x, u_s(x, 0)), u_s(x, 0) + N(x, h_0(x, u_s(x, 0)), 0), 0) = 0. \tag{3.3c}$$

Condition 3.2: The eigenvalues $\lambda_i(x, u_s(x, 0))$, $i = 1, 2, \cdots, m$, of the $m \times m$ Jacobian function matrix

$$B(x, u_s(x, 0)) = \left(\frac{\partial g}{\partial z}(x, z, u, \varepsilon) + \frac{\partial g}{\partial u_f}(x, z, u, \varepsilon) \frac{\partial N}{\partial z}(x, z, \varepsilon) \right) \Bigg|_{\substack{\varepsilon = 0 \\ z = h_0(x, u_s(x, 0))}} \tag{3.4a}$$

satisfy for all $x \in B_x$, the inequality

$$Re\{\lambda_i(x, u_s(x, 0))\} \leq -\beta, \quad \beta > 0. \tag{3.4b}$$

Condition 3.1 ensures that the reduced system model

$$\dot{x} = f(x, h_0(x, u_s(x, 0)), u_s(x, 0) + N(x, h_0(x, u_s(x, 0)), 0), 0) \tag{3.5}$$

is well defined. When Condition 3.1 is satisfied the original system (2.1) under the composite nonlinear state feedback (3.1), (3.2) is said to be in *standard form* [6]. Condition 3.2 imposes a strong stability requirement on the fast system (2.8b) at $\varepsilon = 0$ such that the equilibrium point of interest (2.9) is an asymptotically stable equilibrium point of the original fast system (2.7b), for all $\varepsilon \in [0, \varepsilon_1]$, with

$\varepsilon_1 > 0$. Most initial value (Tikhonov–type) state approximations and stability theorems in the singular perturbations literature rely on assumptions closely related to Conditions 3.1 and 3.2. Here, moreover, in the following theorem, essentially due to Zadiraka [12], both conditions provide for the existence of a unique solution to the integral manifold condition (2.11), rewritten with the composite state feedback control (3.1), (3.2) as

$$\begin{aligned} &g(x, h(x, u_s, \varepsilon), u_s + N(x, h(x, u_s, \varepsilon), \varepsilon), \varepsilon) \\ &= \varepsilon \left(\tfrac{\partial h}{\partial x}(x, u_s, \varepsilon) + \tfrac{\partial h}{\partial u_s}(x, u_s, \varepsilon)\tfrac{\partial u_s}{\partial x}(x, \varepsilon)\right) f(x, h(x, u_s, \varepsilon), u_s \\ &\qquad + N(x, h(x, u_s, \varepsilon), \varepsilon), \varepsilon). \end{aligned} \tag{3.6}$$

where

$$u_s \triangleq u_s(x, \varepsilon).$$

Theorem, 3.1: Under Conditions 3.1 and 3.2 there exists an $\varepsilon_2 > 0$ such that the nonlinear singularly perturbed system (2.1) under the composite state feedback control (3.1), (3.2) has the unique integral manifold

$$z = h(x, u_s(x, \varepsilon), \varepsilon) \tag{3.7}$$

for all $(x^T, z^T) \in B_x \times B_z$ and all $\varepsilon \in [0, \varepsilon_2]$

Moreover, if the functions f, g, u_s and N are sufficiently many times continuously differentiable functions of their arguments, then the integral manifold (3.7) may be represented by an asymptotic expansion about $\varepsilon = 0$

$$h(x, u_s, \varepsilon) = h_0(x, u_s) + \varepsilon h_{01}(x, u_s) + \varepsilon^2 h_{02}(x, u_s) + \cdots + \varepsilon^k h_{0k}(x, u_s) + \cdots, \tag{3.8}$$

where

$$h_0(x, u_s(x, 0)) \triangleq h(x, u_s(x, 0), 0). \tag{3.9}$$

A statement of Theorem 3.1 for systems without control is given by Sobolev [10], while its proof, again for unforced systems is to be found in Zadiraka, [12]. This proof is lengthy and makes extensive use of contraction mapping arguments. However, its present extension to the system (2.1) under the composite control (3.1), (3.2) is straightforward since the fast control u_f is known and u_s in (3.6) and (3.7) is a continuous and bounded function of x and ε.

Remark 3.1: If conditions 3.1 and 3.2 are satisfied for all $x \in R^n$ and all $z \in R^m$ then results of Theorem 3.1 hold globally.

Remark 3.2: While the fast control plays an important role in the definition of the integral manifold, the fast transient dynamics of the control, $u_f(x, z, \varepsilon)$, introduced through the system state, z, are inactive on the integral manifold for time $t \geq t^*$.

4 Control on the Design Manifold

Theorem 3.1 shows that, not only does the integral manifold, (3.7) of the original singularly perturbed system, (2.1), exist, but that, in principle, it can be approximated to any order of ε accuracy. Usually $O(\varepsilon^2)$ or even $O(\varepsilon)$ accuracy will suffice for most applications; but, in general, we shall refer to the $O(\varepsilon^{k+1})$ approximation of the integral manifold $h(s, u_s(x,\varepsilon), \varepsilon)$ as that provided by the *design manifold* $h_k(x, u_s(x,\varepsilon), \varepsilon)$ according to

$$h_k(x, u_s(x,\varepsilon), \varepsilon) = h(x, u_s(x,\varepsilon), \varepsilon) + O(\varepsilon^{k+1}), \tag{4.1}$$

for any $k = 0, 1, 2, \cdots$

The basic design procedure is as follows: first, steer the system state, z, to the design manifold (4.1) in accordance with Condition 3.1 and 3.2 by way of a fast control design $u_f = N(x, z, \varepsilon)$; then, control the slow motions of the system (2.1) by the design of a slow control $u_s = M(x, \varepsilon)$ for the *kth order slow design model*

$$\dot{x} = f(x, h_k(x, u_s(x,\varepsilon), \varepsilon), u_s(x,\varepsilon) + N(x, h_k(x, u_s(x,\varepsilon), \varepsilon), \varepsilon), \varepsilon). \tag{4.2}$$

Finally, add the slow and fast controls to form the *composite state feedback control*

$$u = M(x, \varepsilon) + N(x, z, \varepsilon) \tag{4.3}$$

for application to the original system (2.1).

Thus, after the design of the fast control $u_f = N(x, z, \varepsilon)$, we need to determine the design manifold $h_k(x, u_s, \varepsilon)$, parameterized by x, u_s and ε. A step by step procedure for this, based on the integral manifold condition (3.6) (and equivalent to the matched asymptotic expansions of Fenichel [2]) is as follows:

Step 4.0: With the fast control $u_f = N(x, z, \varepsilon)$ explicitly known, solve the equation

$$g(x, \phi, u_s + N(x, \phi, 0), 0) = 0 \tag{4.4a}$$

for the zeroth–order manifold

$$M_0^{u_s} : \phi = h_0(x, u_s). \tag{4.4b}$$

where $O(\varepsilon)$ terms in (4.4a) are neglected.

Step 4.1: Using $h_0(x, u_s)$ and its partial derivatives with respect to x and u_s, obtained form Step 4.0, solve the equation

$$g(x, \phi, u_s + N(x, \phi, \varepsilon), \varepsilon) = \varepsilon \left(\frac{\partial h_0}{\partial x} + \frac{\partial h_0}{\partial u_s} \frac{\partial u_s}{\partial x} \right) f(x, h_0, u_s + N(x, h_0, 0), 0) \tag{4.5a}$$

for the first–order manifold

$$M_1^{u_s} : \phi = h_1(x, u_s, \varepsilon) \tag{4.5b}$$

where $O(\varepsilon^i)$, $i \geq 2$, terms in (4.5a) are neglected.

Continuing on in this manner $\cdots$

Step 4.k: Using $h_{k-1}(x, u_s, \varepsilon)$ and its partial derivatives with respect to x and u_s, obtained from Step $4(k-1)$, solve the equation

$$g(x, \phi, u_s + N(x, \phi, \varepsilon), \varepsilon)$$
$$= \varepsilon \left(\tfrac{\partial h_{k-1}}{\partial x} + \tfrac{\partial h_{k-1}}{\partial u_s} \tfrac{\partial u_s}{\partial x} \right) f(x, h_{k-1}, u_s + N(x, h_{k-1}, \varepsilon), \varepsilon) \tag{4.6a}$$

to give the kth order manifold

$$M_k^{u_s} : \phi = h_k(x, u_s, \varepsilon), \tag{4.6b}$$

where $O(\varepsilon^i), i \geq k+1$, terms in (4.6a) are neglected.

It is observed that the design procedure can be terminated at any given Step $4.k, k = 0, 1, 2, \cdots$ Each step is solved complete rather than solving for successive coefficients of a power series expansion. A great advantage of this design procedure is that the slow control, u_s need not be specified or approximated at any stage, as the design manifold is parameterized by u_s as well as by x and ε.

Having determined the design manifold (4.6b), we design the slow control $u_s = M(x, \varepsilon)$ for the kth order slow design model (4.2) and implement the composite control (4.3) on the original system (2.1).

5 Conclusion

Using a geometric approach, we have derived a composite control, the sum of a slow control and a fast control, for a general class of nonlinear singularly perturbed systems. A new and simpler method of composite control design is proposed whereby the fast control is completely designed at the outset: subsequently, the slow control can be designed to any order of ε accuracy without readjustment of the fast control. An important advantage of the new method over earlier methods is that the designer can immediately ascertain as to whether the fast system dynamics are stable (stabilizable by fast control) or not and so determine as to whether the development of a slow control is possible or not. In other words, the initial 'fast' control, designed under Conditions 3.1 and 3.2, provides for the existence of a slow integral manifold. Moreover, the slow component of the 'fast' control, can have a beneficial effect on the shaping of the integral manifold prior to the design of the slow control. The slow control is then free to be chosen such that the slow integral manifold of the original singularly perturbed system approximates a desired design manifold to within any specified order of ε accuracy.

6 Acknowledgements

This work was funded by the U.K. Science and Engineering Research Council under Grant Number GR/D/33120.

References

[1] Chow, J.H., and Kokotovic, P.V., Two–Time–Scale Feedback Design of a Class of Nonlinear Systems, IEEE Trans. Autom. Control, 23, 438–443, 1978.

[2] Fenichel, N., Geometric Singular Perturbation Theory for Ordinary Differential Equations, J. Diff. Equations, 31, 53, 1979.

[3] Hoppensteadt, F., Properties of Solutions of Ordinary Differential Equations with Small Parameters, Comm. Pure Appl. Math. 34, 807–840, 1971.

[4] Khorasani, K., and Kokotovic, P.V., A Corrective Feedback Design for Nonlinear Systems with Fast Actuators, IEEE Trans. Autom. Control, 31, 67–69, 1986.

[5] Kokotovic, P.V., Recent Trends in Feedback Design : An Overview, Automatica, 21, 3, 225–236, 1985.

[6] Kokotovic, P.V., Khalil, K.H., and O'Reilly, J., 'Singular Perturbation Methods in Control: Analysis and Design', Academic Press, New York, 1986.

[7] Marino, R., and Kokotovic, P.V., A Geometric Approach to Composite Control of Two–Time–Scale Systems, Proc. 25th Conf. Decision and Control, Athens, TP11-5, 1397–1399, 1986.

[8] Saberi, A., and Khalil, H.K., Stabilization and Regularization of Nonlinear Singularly Perturbed Systems–Composite Control, IEEE Trans. Autom. Control, 30, 739–746, 1984.

[9] Sharkey, P.M., and O'Reilly, J., Exact Design Manifold Control of a Class of Nonlinear Singularly Perturbed Systems, IEEE Trans. Autom. Control, 32, 933–935, 1987.

[10] Sobolev, V.A., Integral Manifolds and Decomposition of Singularly Perturbed Systems, Syst. Control Lett. 5,3, 169–179, 1984.

[11] Vasil'eva, A.B., Asymptotic Behaviour of Solutions to Certain Problems involving Nonlinear Differential Equations Containing A Small Parameter Multiplying the Highest Derivatives, Russian Math. Surveys, 18, 13–81, 1963.

[12] Zadiraka, K.V., On a Nonlocal Integral Manifold of a Nonregularly Perturbed Differential System, Amer. Math. Soc. Transl., 2, 89, 29–49, 1970.

Asymptotic stabilization of a class of nonlinearly coupled uncertain dynamical systems

D.P. Goodall E.P. Ryan

Abstract

Discontinuous feedback controls, which guarantee global asymptotic stability of the zero state of a class of nonlinearly coupled uncertain dynamical systems, are presented. A formulation based on differential inclusions is adopted wherein system uncertainty is modelled by set–valued maps.

1 Introduction

The prototype control system to be considered has the structure of two bilinearly coupled subsystems and has a non–asymptotically–stabilizable linearization. This system is assumed to be subject to uncertainty, modelled by additional nonlinearities in the coupling terms and by augmenting the nominal differential equations by a set–valued map, thereby giving rise to a controlled differential inclusion (Aubin and Cellina [2]). Within this framework, the problem of stabilization is considered and a class of discontinuous feedback controls is developed which renders the system zero state globally attractive.

The approach taken here is to break down the problem into two separate stages. In the first stage a nonlinear manifold M is constructed with the property that, if it is rendered invariant by suitable feedback, then all solutions in M tend to the state origin. The results of Slemrod [7] on bilinear stabilization are invoked at this stage. In the second stage a feedback control is designed which renders (i) M both invariant and globally finite–time attractive, (ii) the zero state of the differential inclusion system globally attractive. In designing the feedback control law, the procedure follows closely that used by Goodall and Ryan [5].

Stability of systems, similar in nature to the prototype, have also been investigated by Aeyels [1], Brockett [3], and Carr [4]. In the context of their work, the manifold M can be interpreted as a global *centre manifold.*

2 The system

The prototype control system (with control input $u(t)$) consists of two bilinearly coupled subsystems:

$$\dot{x}(t) = A_0 x(t) + \sum_{i=1}^{q} y_i(t) B_i x(t), \quad x(t) \in \Re^p, \quad y(t) \in \Re^q, \tag{2.1a}$$

$$\dot{y}(t) \quad = \quad A_1 y(t) + u(t), \ u(t) \in \Re^q, \tag{2.1b}$$

where $y(t) = [y_1(t), \cdots, y_q(t)]^\top$, A_0, $B_i \in \Re^{p \times p}$ and $A_1 \in \Re^{q \times q}$ are known matrices defining the nominal bilinearly coupled system. Attention is restricted to the subclass of systems (2.1a)–(2.1b) for which the following hypothesis holds:

A1: There exists a real symmetric matrix $K > 0$ such that

$$KA_0 + A_0^\top K = 0.$$

This hypothesis implies that A_0 is stable, but not asymptotically stable. Therefore the nominal system has a stabilizable, but not asymptotically stabilizable, linearization.

Consider now the above system subject to uncertainty. It is supposed that the uncertainty can be modelled by a nonlinear perturbation to equation (2.1). Specifically, the controlled uncertain dynamical system is of the form:

$$\dot{x}(t) \quad = \quad A_0 x(t) + \sum_{i=1}^{q} (y_i(t) + f_i(x(t), y(t))) B_i x(t), \tag{2.2a}$$

$$x(t) \quad \in \quad \Re^p, \ y(t) \in \Re^q,$$

$$\dot{y}(t) \quad \in \quad A_1 y(t) + u(t) + \mathcal{G}(x(t), y(t), u(t)), \ u(t) \in \Re^q, \tag{2.2b}$$

where $f_i : \Re^p \times \Re^q \to \Re$ are unknown, real–valued functions and $\mathcal{G} : \Re^p \times \Re^q \times \Re^q \rightrightarrows \Re^q$ is a known multifunction (set–valued map), encompassing all possible perturbations to the nominal differential equations, modelling the uncertainty in the system.

Remark: The following notation will be adopted. For $\nu \in \Re^n$ and $S \subset \Re^n$, $\nu + S$ is defined as $\{\nu + s : s \in S\} \subset \Re^n$. For $S_1, S_2 \subset \Re^n$, $S_1 + S_2$ denotes the set $\{s_1 + s_2 : s_1 \in S_1; s_2 \in S_2\}$.

3 Problem statement

A class $\mathcal{U}$ of admissible feedbacks, termed *generalized feedbacks*, is first defined in the following manner:

Definition 1: A multifunction $\mathcal{F} : \Re^p \times \Re^q \rightrightarrows \Re^q$ is a *generalized feedback* if:

1. $\mathcal{F}$ is upper semicontinuous (see Aubin and Cellina [2]) with nonempty, convex and compact values;

2. $\mathcal{F}$ is singleton–valued except on a set $\sum_{\mathcal{F}}$ of Lebesgue measure zero in $\Re^p \times \Re^q$.

Remark: In the present context, the set $\sum_{\mathcal{F}}$ may be interpreted as a switching surface of control discontinuities.

The objective may now be loosely stated as that of determining a generalized feedback law $u(t) \in \mathcal{F}(x(t), y(t))$ which renders the zero state of the feedback controlled differential inclusion system:

$$\dot{x}(t) = A_0 x(t) + \sum_{i=1}^{q} (y_i(t) + f_i(x(t), y(t))) B_i x(t) \tag{3.1a}$$

$$\dot{y}(t) \in A_1 y(t) + \mathcal{F}(x(t), y(t)) + \mathcal{G}(x(t), y(t), \mathcal{F}(x(t), y(t))), \tag{3.1b}$$

where $\mathcal{G}(x, y, \mathcal{F}(x, y)) := \bigcup_{u \in \mathcal{F}(x,y)} \mathcal{G}(x, y, u)$, *globally asymptotically stable* in the sense of the following definition.

Definition 2: The feedback controlled differential inclusion system (3.1) is *globally asymptotically stable* if the following holds:

1. *Existence and indefinite continuation of solutions:* for each

$$\begin{bmatrix} x^0 \\ y^0 \end{bmatrix} \in \Re^n, \text{ where } n = p + q,$$

there exists a local solution

$$\begin{bmatrix} x \\ y \end{bmatrix} : [0, \tau) \rightarrow \Re^n$$

(i.e. an absolutely continuous function satisfying (3.1) a.e.) with

$$\begin{bmatrix} x \\ y \end{bmatrix}(0) = \begin{bmatrix} x^0 \\ y^0 \end{bmatrix}$$

and every such solution can be extended into a solution on $I := [0, \infty)$;

2. *Boundedness of solutions:* for each $\delta > 0$, $\exists d(\delta) > 0$ such that

$$\begin{bmatrix} x \\ y \end{bmatrix}(t) \in d(\delta)\mathcal{B}_n$$

where $\mathcal{B}_n$ is the open unit ball centred at the origin in $\Re^n$, for all t on every solution

$$\begin{bmatrix} x \\ y \end{bmatrix} : I \rightarrow \Re^n$$

with

$$\begin{bmatrix} x \\ y \end{bmatrix}(0) \in \delta\mathcal{B}_n;$$

3. *Stability of the origin in* $\Re^n$: for each $\delta > 0$, $\exists D(\delta) > 0$ such that

$$\begin{bmatrix} x \\ y \end{bmatrix}(t) \in \delta\mathcal{B}_n$$

for all t on every solution

$$\begin{bmatrix} x \\ y \end{bmatrix} : I \to \Re^n$$

with

$$\begin{bmatrix} x \\ y \end{bmatrix}(0) \in D(\delta)\mathcal{B}_n;$$

4. *Global attractivity of the origin in* $\Re^n$: for each $\delta > 0$ and $\varepsilon > 0$, $\exists\, T(\delta, \varepsilon) \geq 0$ such that

$$\begin{bmatrix} x \\ y \end{bmatrix}(t) \in \varepsilon\mathcal{B}_n \ \forall\, t \geq T(\delta, \varepsilon)$$

on every solution

$$\begin{bmatrix} x \\ y \end{bmatrix} : I \to \Re^n$$

with

$$\begin{bmatrix} x \\ y \end{bmatrix}(0) \in \delta\mathcal{B}_n.$$

4 Outline of method

For this problem, the method for determining a generalized feedback control law is broken down into two stages.

4.1 Stage I

In the first stage, subsystem (3.1a) is regarded as an isolated system with input y and a smooth feedback function $h : x \mapsto y = h(x), \Re^p \to \Re^q$, is sought to stabilize this system. This is achieved, under the following hypotheses, by an approach akin to that of Slemrod [7].

A2:

1. For each i, f_i is continuous and $f_i(x, 0) = 0\ \forall x$;

2. There exist known real constants $\beta \in [0,1)$, $\delta > 0$ and known function ϕ : $\Re \rightarrow \Re$ satisfying
   (i) $\phi \in C^1(\Re)$ and (ii) $\delta v^2 \leq v\phi(v)$, $\forall v \in \Re$,
   such that, for all i,

$$|f_i(x,y)| \leq |\phi(\langle x, KB_i x\rangle)| + \beta |y_i|;$$

3. There exists a nonempty set $\Omega \subset \Re^p \setminus \{0\}$ such that
   (i) for each $x \in \Omega$ there exists a positive integer k such that $\text{span}\{A_0 x, ad^0(A_0,B_i)x, ad^1(A_0,B_i)x, \cdots, ad^k(A_0,B_i)x; i = 1,\cdots,q\} = \Re^p$ and (ii) $\{0\}$ is the only subset of $\Omega^c \cap \Gamma$ which is invariant under $exp(A_0 t)$, where Ω^c denotes the complement of Ω and

$$\Gamma := \{x \in \Re^p : \langle x, KB_i x\rangle = 0, i = 1,\cdots,q\}.$$

Remark: Conditions (1) and (2) are strong structural conditions on the uncertainties f_i; condition (3) is essentially that of Ryan and Buckingham [6], a weaker version of that of Slemrod [7]. The operators $ad^j(A_0,B_i)$ are defined recursively as follows:

$$ad^0(A_0,B_i) := B_i; ad^1(A_0,B_i) \equiv ad(A_0,B_i) := A_0 B_i - B_i A_0,$$
$$ad^j(A_0,B_i) := ad(A_0, ad^{j-1}(A_0,B_i)),\ j = 2,3,\cdots.$$

4.2 Stage II

For the overall system, the relation $y = h(x)$ (with h as in §3.1) defines a smooth manifold M in the state space $\Re^p \times \Re^q$. The second stage of the method is to determine a generalized feedback law $(x,y) \mapsto \mathcal{F}(x,y)$ which renders the state origin globally asymptotically stable. More specifically, $\mathcal{F}$ is designed so as to render the manifold M globally finite–time attractive and invariant. Thus, all solutions of the feedback controlled system ultimately attain the manifold M; motion thereafter is constrained to M and hence (by the analysis of stage I) tends asymptotically to the origin. This feedback design is achieved, under the following additional hypothesis (a structural condition on $\mathcal{G}$), by a procedure that follows closely that used by Goodall and Ryan [5].

A3: There exists a multifunction $\mathcal{G}_s : \Re^p \times \Re^q \rightrightarrows \Re^q$, which is upper semicontinuous with convex and compact values, and a real constant $\kappa \in [0,1)$ such that $\mathcal{G}(x,y,u) = \mathcal{G}_s(x,y) + \kappa\|u\|\overline{\mathcal{B}}_q$, where $\mathcal{B}_q$ denotes the open unit ball in $\Re^q$ (with closure $\overline{\mathcal{B}}_q$).

5 Determination of the manifold M

In this section, a feedback function $x \longmapsto y = h(x) \in \Re^q$ is constructed which renders the zero state of system (3.1a) globally asymptotically stable. The approach is akin to that used by Slemrod [7] which uses Lyapunov theory and the *invariance principle* of LaSalle. This feedback control law is then used to define the manifold M.

Lemma 1: Under the conditions stated in the hypotheses A1 and A2, the feedback function $x \longmapsto y = [y_1, y_2, \cdots, y_q]^T$, where

$$y_i := -\alpha(1-\beta)^{-1}\phi(\langle x, KB_i x\rangle)$$

and $\alpha > 1$, renders the zero state of system (3.1a) globally asymptotically stable.

Proof: The hypothesis $A2(1)$ ensures that, for each $x^0 \in \Re^p$, there exists at least one (local) solution $x(\cdot) : [0.\tau) \longrightarrow \Re^p$ with $x(0) = x^0$. The behaviour of the function $V_1 : R^p \longrightarrow \Re_0^+$ along solutions is examined, where V_1 is defined by $V_1(x) := 1/2\langle x, Kx\rangle$. Under the conditions stated in hypothesis A1 and A2(2), a straightforward calculation shows that, along every local solution of (3.1a), the following holds for almost all $t \in [0, \tau)$:

$$\dot{V}_1(x(t)) \leq -(\alpha-1)\sum_{i=1}^{q} |\phi(\langle x(t), KB_i x(t)\rangle)||\langle x(t), KB_i x(t)\rangle| \leq 0,$$

which implies that all solutions can be continued indefinitely. Also, properties of boundedness of solutions and stability clearly hold. Invoking the invariance principle of LaSalle, it may be concluded that all solutions of (3.1a) approach (as $t \longrightarrow \infty$) a set Λ, which is the largest invariant subset of Γ (defined in A2(3)). For system (3.1a), with $x^0 \in \Lambda$, the state $x(t)$ satisfies: $\langle x(t), KB_i x(t)\rangle = 0$ for all i and $\dot{x}(t) = A_0 x(t)$ for all $t \in \Re$. Slemrod [7] has shown that this characterizes the set Λ in the form

$$\Lambda := \{x \in \Re^p : \langle x, KA_0 x\rangle = 0, \langle x, K ad^j(A_0, B_i)x\rangle = 0; j \in Z_0^+, i = 1, \cdots q\}.$$

As a consequence of A2(3), following an argument applied by Ryan and Buckingham [6], one may deduce that $\Lambda = \{0\}$ and hence global asymptotic stability of the origin may be concluded. □

Introducing the function $x \longmapsto h(x) = [h_1(x), h_2(x), \cdots, h_q(x)]^{\mathsf{T}}$, where

$$h_i(x) := -\alpha(1-\beta)^{-1}\phi(\langle x, KB_i x\rangle)$$

and $\alpha > 1$, the manifold $M \subset \Re^p \times \Re^q$ is defined as $M := \{(x, y) \in \Re^p \times \Re^q : y = h(x)\}$.

6 Class of generalized feedback controls U

It may be assumed (w.l.o.g.) that the eigenvalues of A_1 lie in the open left half of the complex plane. Let $P > 0$ be the unique symmetric solution of the Lyapunov equation

$$PA_1 + A_1^{\mathsf{T}}P + Q = 0 \tag{6.1}$$

where $Q > 0$ is a design parameter. For any compact set $W \neq \emptyset$, define $\xi(W) := \max\{\|w\| : w \in W\}$, with $\xi(\emptyset) := 0$. Also, the multifunctions $\mathcal{P} : \Re^p \times \Re^q \rightrightarrows \Re^q$

and D: $\Re^q \rightrightarrows \Re^q$ are defined by

$$(x,y) \longmapsto \mathcal{P}(x,y) := \{\nu \in \Re^q : \langle \nu, Py\rangle \geq \langle \nu, Ph(x)\rangle\}$$
$$u \longmapsto \mathcal{D}(u) := \begin{cases} \|u\|^{-1}u, & u \neq 0 \\ \overline{\mathcal{B}}_q, & u = 0 \end{cases}$$

The proposed feedback, for stabilization of the system (3.1), is taken to be the multifunction

$$(x,y) \longmapsto \mathcal{F}(x,y) := -A_1 h(x) + Dh(x)\{A_0 x + \sum_{i=1}^{q} y_i B_i x\} -\rho(x,y)\mathcal{D}(P(y-h(x))), \tag{6.2}$$

where $Dh(x) \in \Re^{q\times p}$ is the Frèchet derivative of $h : \Re^p \to \Re^q$ at x, ρ is any continuous functional satisfying

$$\rho(x,y) \geq \rho_0(x,y) := (1-\kappa)^{-1}\{\kappa\| - A_1 h(x) + Dh(x)\{A_0 x + \sum_{i=1}^{q} y_i B_i x\}\| + \sum_{i=1}^{q}(\beta|y_i| + \alpha^{-1}(1-\beta)|h_i(x)|)\|Dh(x)B_i x\| + \xi(\mathcal{G}_\xi(x,y)\bigcap\mathcal{P}(x,y)) + \gamma\}, \tag{6.3}$$

and $\gamma > 0$ is a design parameter. Note that the continuity of ρ and the upper semicontinuity of $\mathcal{D}$ ensure that $\mathcal{F}$ is upper semicontinuous, and clearly takes convex and compact values; also, $\mathcal{F}$ is singleton–valued except on a set $\Sigma_{\mathcal{F}}$. Hence $\mathcal{F}$ qualifies as a generalized feedback.

Remarks:

(i) If A_1 is not asymptotically stable append the generalized feedback $\mathcal{F}$ with a term $(A^* - A_1)(y - h(x))$, where A^* is an asymptotically stable matrix, and replace A_1 in (6.1) by A^*.

(ii) The intersection $\mathcal{G}_\xi(x,y)\bigcap\mathcal{P}(x,y)$ is adopted in (6.3) in order to economize on the gain ρ by exploiting the possible occurrence of "stability enhancing" uncertainties.

7 Asymptotic stability of the feedback controlled differential inclusion system

Defining $e(t) := y(t) - (h_0 x)(t)$, the function $t \longmapsto e(t) \in \Re^q$ satisfies the differential inclusion

$$\dot{e}(t) \in A_1 e(t) + \mathcal{H}(x(t), e(t), \mathcal{F}(x(t), e(t) + h(x(t)))) \tag{7.1}$$

where

$$\begin{aligned}\mathcal{H}(x,e,\mathcal{F}(x,e+h(x))) &:= \mathcal{G}(x,e+h(x),\mathcal{F}(x,e+h(x))) - \rho(x,e+h(x))\mathcal{D}(Pe) \\ &- Dh(x)\sum_{i=1}^{q} f_i(x,e+h(x))B_i x\end{aligned}$$

and $\mathcal{F}$ is given by (6.2). It can be readily verified that the multifuntion $\mathcal{G}$ is upper semicontinuous with convex and compact values. Consider the system

$$\begin{bmatrix} \dot{x}(t) \\ \dot{e}(t) \end{bmatrix} \in \tilde{\mathcal{G}}(x(t),e(t)) \tag{7.2a}$$

with initial condition

$$\begin{bmatrix} x(0) \\ e(0) \end{bmatrix} = \begin{bmatrix} x^0 \\ e^0 \end{bmatrix} \tag{7.2b}$$

where the multifunction $\tilde{\mathcal{G}} : \Re^p \times \Re^q \rightrightarrows \Re^n$, $n = p + q$, is defined by

$$\tilde{\mathcal{G}}(x,e) := \tag{7.2c}$$
$$\left\{ \begin{bmatrix} A_0 x + \sum_{i=1}^{q}(e_i + h_i(x) + f_i(x,e+h(x)))B_i x \\ A_1 e + \eta \end{bmatrix} : \eta \in \mathcal{H}(x,e,\mathcal{F}(x,e+h(x))) \right\}$$

Clearly the multifunction $(x,e) \longmapsto \tilde{\mathcal{G}}(x,e)$ is upper semicontinuous with convex and compact values. Hence, for each pair $\begin{bmatrix} x^0 \\ e^0 \end{bmatrix}$, the initial value problem (7.2) admits a local solution $\begin{bmatrix} x \\ e \end{bmatrix} : [0,\tau) \to \Re^n$ (see Aubin and Cellina [2]).

Define $V_2 : \Re^n \to \Re_0^+$, a Lyapunov function candidate, as

$$V_2\left(\begin{bmatrix} x \\ e \end{bmatrix}\right) := \frac{1}{2}\langle \begin{bmatrix} x \\ e \end{bmatrix}, L\begin{bmatrix} x \\ e \end{bmatrix}\rangle, \qquad L = \begin{bmatrix} K & 0 \\ 0 & \zeta P \end{bmatrix},$$

where ζ is a positive constant to be specified.

Along each local solution $\begin{bmatrix} x \\ e \end{bmatrix} : [0,\tau) \to \Re^n$,

$$\dot{V}_2\left(\begin{bmatrix} x(t) \\ e(t) \end{bmatrix}\right) \in \mathcal{L}(x(t),e(t)) = \mathcal{L}_1(x(t),e(t)) + \mathcal{L}_2(x(t),e(t)),$$

where

$$\mathcal{L}_1(x,e) := \{\langle A_0 x + \sum_{i=1}^{q}(e_i + h_i(x) + f_i(x,e+h(x)))B_i x, Kx\rangle\} \subset \Re$$

and

$$\mathcal{L}_2(x,e)^1 := \zeta\langle\langle A_1 e + \mathcal{H}(x,e,\mathcal{F}(x,e+h(x))), Pe\rangle\rangle \subset \Re.$$

[1] For $x \in X$, and $S \subset X$, $\langle\langle x,S\rangle\rangle := \{\langle x,s\rangle : s \in S\} \subset \Re$.

Thus it is required to show that $\mathcal{L}(x,e)\bigcup(0,\infty)=\emptyset$, for all $(x,e)\in\Re^p\times\Re^q$. It is easily shown that

$$\max\mathcal{L}_1(x,e)\leq\sum_{i=1}^{q}\{-\delta(\alpha-1)|<x,KB_ix>|^2+(1+\beta)|e_i||<x,KB_ix>|\}.$$

Using (6.1) and the hypotheses A2–A3,

$$\begin{aligned}\max\mathcal{L}_2(x,e) &\leq \zeta[-1/2\langle e,Qe\rangle+\kappa\xi(\mathcal{F}(x,e+h(x)))\|Pe\|\\ &+ \langle\langle\mathcal{G}_s(x,e+h(x))-\rho(x,e+h(x))\mathcal{D}(Pe)\\ &\quad -Dh(x)\sum_{i=1}^{q}f_i(x,e+h(x))B_ix,Pe\rangle\rangle],\end{aligned}$$

whence, in view of (6.2) and (6.3),

$$\begin{aligned}\max\mathcal{L}_2(x,e) &\leq -1/2\zeta(\langle e,Qe\rangle+2\gamma\|Pe\|) \\ &\leq -1/2\zeta\sigma_{min}(Q)\sum_{i=1}^{q}|e_i|^2,\end{aligned}\tag{7.3}$$

where σ denotes spectrum.
Hence

$$\max\mathcal{L}(x,e)\leq-1/2\sum_{i=1}^{q}\langle\begin{bmatrix}|<x,KB_ix>|\\|e_i|\end{bmatrix},E_\zeta\begin{bmatrix}|<x,KB_ix>|\\|e_i|\end{bmatrix}\rangle$$

where $E_\zeta:=\begin{bmatrix}2\delta(\alpha-1) & -(1+\beta)\\ -(1+\beta) & \zeta\sigma_{min}(Q)\end{bmatrix}$.
Chosing $\zeta=(1+\beta)^2[2\delta(\alpha-1)\sigma_{min}(Q)]^{-1}$ ensures that, along all local solutions of (7.2), $\dot{V}_2\left(\begin{bmatrix}x(t)\\e(t)\end{bmatrix}\right)\leq 0$, a.e.

Therefore every local solution $\begin{bmatrix}x\\e\end{bmatrix}:[0,\tau)\to\Re^n$ can be continued indefinitely and the property of boundedness of solutions readily follows.
Remark: Taking $\zeta>(1+\beta)^2[2\delta(\alpha-1)\sigma_{min}(Q)]^{-1}$, $\dot{V}_2<0$ along all nontrivial solutions and hence the state origin is globally attractive. This implies that the manifold M is attractive and the zero of the system governed by (3.1), with control given by (6.2), is globally asymptotically stable.

It remains to show that manifold M is finite–time attractive and invariant. Consider the behaviour of the function $V_3:\Re^q\to\Re_0^+$,

$$e\mapsto V_3(e):=1/2\langle e,Pe\rangle$$

along solutions of (7.1).

Lemma 2: For each $(x^0, y^0) \in \Re^p \times \Re^q$, the manifold M is attained in finite time t_f, satisfying $t_f < \gamma^{-1}[2||P^{-1}||V_3(e^0)]^{1/2}$, and $(x(t), y(t)) \in M$ for all $t > t_f$.
Proof: From (7.3) it follows that, for $Pe \neq 0$,

$$\begin{aligned} \dot{V}_3(t) &< -\gamma||Pe(t)|| \\ &< -\gamma\sqrt{2}||P^{-1}||^{-1/2}\{V_3(e(t))\}^{1/2}, \end{aligned}$$

a.e. along all solutions of (7.1).
Integration shows that the time, t_f, taken to attain M satisfies

$$t_f < \gamma^{-1}\{2||P^{-1}||V_3(e^0)\}^{1/2}$$

(which is inversely proportional to the design parameter γ). Moreover, $V_3(e(t)) = 0$ for all $t > t_f$ and hence $x(t) \in M$ for all $t > t_f$. □
Remark: The manifold M is in fact smooth and qualifies as a *centre manifold* for system (3.1) (see Carr [4]).

Hence as a consequence of the above analysis, the following theorem can be concluded.
Theorem: The generalized feedback $\mathcal{F} \in \mathcal{U}$, given by (6.2), renders the zero state of the differential inclusion system (3.1) globally asymptotically stable.

References

[1] Aeyels, D., Stabilization of a class of nonlinear systems by a smooth feedback control, *Systems and Control Letters*, 5, 289–294, 1985.

[2] Aubin, J–P., and Cellina, A., "*Differential Inclusions*", Springer Verlag, New York, 1984.

[3] Brockett, R.W., Asymptotic stability and feedback stabilization, in "*Differential Geometric Control Theory*" (eds. Brockett, Millman & Sussmann), Birkhauser, Boston, 1983.

[4] Carr, J., "*Applications of Centre Manifold Theory*", Springer Verlag, New York, 1981.

[5] Goodall, D.P., and Ryan, E.P., Controlled differential inclusions and stabilization of uncertain dynamical systems, *Proc. 25th IEEE Conference on Decision & Control (Athens)*, 1, 31–33, 1986.

[6] Ryan, E.P., and Buckingham, N.J., On asymptotically stabilizing feedback control of bilinear systems, *IEEE Trans. Automatic Control*, AC–28, 863–864, 1983.

[7] Slemrod, M., Stabilization of bilinear control systems with applications to non-conservative problems in elasticity, SIAM *J. Control & Optimiz.*, 16, 131–141, 1978.

Multivariable continuous self-adaptive control using variable structure control system design techniques

J. A. Burton A. S. I. Zinober

Abstract

In this paper a multivariable continuous self-adaptive controller is developed from variable structure control (VSC) and its continuous counterpart, smoothed variable structure control (SmVSC). The sliding mode of a VSC system is utilized along with properties within the SmVSC system to produce a continuous self-adaptive control (CSAC) system.

VSC and SmVSC are described before introducing the scalar CSAC system. This scalar CSAC system is then extended to the multivariable case. In order to extrapolate the scalar CSAC to the multivariable case a particular structure is imposed on the system. This structure is well known within VSC system design and is in fact beneficial to the design as it decouples the control actions.

With the advent of CSAC the VSC system is partitioned into three distinct phases in the design and implementation. A multivariable control system example is presented.

1 Introduction

The control of time-varying uncertain systems via Variable Structure Control theory (VSC) is well documented [9], [8], [15]. The main feature of a VSC system is the sliding mode. In this mode the discontinuous control action keeps the direction of the state trajectory towards or along the switching surface and the system becomes invariant to parameter variations and rejects a certain class of disturbance [5]. In the sliding mode infinitely fast chattering control maintains the state on the switching surface which in practice, produces an undesirable chatter motion [2]. To obviate this a continuous approximation to VSC has been developed [1], [2], [3] which gives a good approximation to ideal sliding motion and retains the basic VSC system design procedures and the associated properties.

Discontinuous self-adaptive control (SAC), developed from VSC, incorporates a time-varying switching surface. The self-adaptive scheme results in responses that are close to the time-optimal for a wide range of system parameter values without the need for the direct identification of the system parameters [13], [14], [16]. The continuous counterpart (CSAC) has been developed and shown in the scalar case to be implemented more practically than SAC [2] but possesses similar self-adaptive properties. In CSAC the position of the switching surface is altered by a small amount when a certain continuous function reaches zero. In this paper multivariable

CSAC is developed by extending the principles used in scalar CSAC. In order to specify the parameters to be adapted in the system a structure is imposed on the design. This structure is useful in easing the design of the multivariable system. Multivariable CSAC is demonstrated by an example.

2 The variable structure control system

The VSC system considered here is the linear time–invariant regulator system

$$\dot{x} = Ax + Bu \tag{2.1}$$

where x and u are of dimension n and m respectively and the VSC law is of the form

$$u_i = \begin{cases} u_i^+(x) & s_i > 0 \\ u_i^-(x) & s_i < 0 \end{cases} \quad i = 1, \ldots, m. \tag{2.2}$$

The fixed switching hyperplane is defined by

$$s(x) = Gx = 0 \tag{2.3}$$

When sliding motion occurs, the state slides on $N(s)$ the null space of the switching function. The system is designed to ensure attainment of a stable sliding mode [6], [9].

In the sliding mode

$$s(x) = Gx = 0 \text{ and } \dot{s}(x) = G\dot{x} = 0 \tag{2.4}$$

which are the equations that define the system dynamics. If GB is nonsingular, these dynamics also arise when the linear equivalent control

$$u_{eq} = -(GB)^{-1}GAx. \tag{2.5}$$

is applied. The resulting system dynamics are

$$\dot{x} = [I - B(GB)^{-1}G]Ax = A_{eq}x \tag{2.6}$$

During sliding the system behaves as an unforced system and for asymptotic stability system (2.6) should have $n - m$ stable eigenvalues, the remaining m eigenvalues being zero [7], [9].

3 The smoothed variable structure control system

One of the inherent difficulties of VSC in its application to real plant is 'chatter' motion that results about the switching hyperplane. Recently a continuous nonlinear control approximation [1], [3] has been developed. Termed smoothed variable

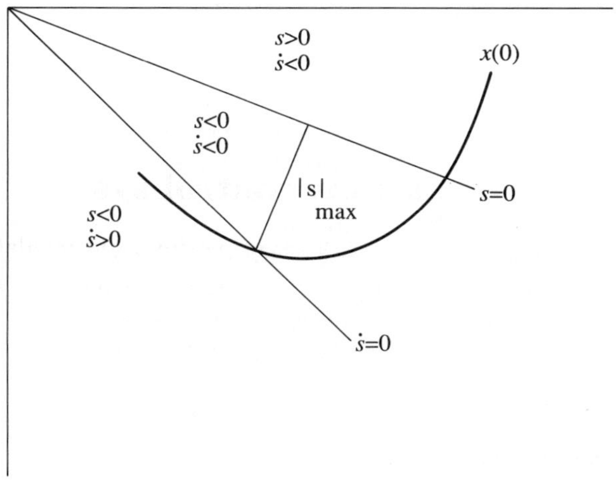

Fig. 1 Fan of derivatives in SmVSC.

structure control (SmVSC), analysis has been carried out that gives a bound on the deviation from perfect sliding, predetermines the profile of the resulting motion and proves that chatter motion is indeed eliminated [3].

The nonlinear approximation is generated by noting that for ε a small positive constant

$$u = -\delta \frac{s}{|s| + \varepsilon} \tag{3.1}$$

is a continuous control with upper and lower bounds $\pm\delta$. This control function can be approximated [3] in a small enough neighbourhood of the switching surface, to first order by

$$u = -ks \tag{3.2}$$

where $k = \delta/\varepsilon$. The approximation is essentially a high gain feedback control in the neighbourhood of $N(s)$ and the system response therefore possesses properties associated with such systems [12]. The design procedure, however, remains the VSC system design approach.

The SmVSC response is analysed by the inspection of the null spaces that correspond to zero values of the successive derivatives of a switching function. During ideal sliding motion these null spaces coincide. However, in SmVSC the null spaces are fanned out (Fig 1.) and the condition for sliding motion is only satisfied on one side of the switching surface. Results have been proved [3] showing chatter motion is eliminated since the switching hyperplane can be crossed at most n–1 times, where n is the order of the system.

4 CSAC of scalar control system

Self-adaptive control (SAC) based on VSC theory has been presented [13], [14], [16], where a time-varying 'linear' switching surface, $G(t)x = 0$, is adjusted by altering a single design parameter, $g(t)$, whenever sliding motion occurs. In SAC the design parameter has direct influence on the eigenvalues during the sliding mode which are successively shifted further into the left half plane at each adaptation step.

CSAC is derived from SmVSC. In particular, use is made of the resulting 'fan' of derivatives of the switching function (Fig 1.). CSAC defers adaptation of the design parameter until $\dot{s} = 0$, the first derivative of the switching function, is reached. This point is considered analogously to sliding motion as the sliding condition is again satisfied. The design parameter is then adjusted so that the new switching surface lies in $N(s)$.

4.1 Second order scalar systems

The CSAC function is the SmVSC function

$$u = -\delta \frac{s^{(m)}}{|s^{(m)}| + \varepsilon} \tag{4.1}$$

with the continually adapting switching line $s^{(m)}$. CSAC performs adaptation when both

$$s^{(m)} = g^{(m)} x_1 + x_2 = 0 \qquad t_m < t < t_{m+1} \tag{4.2}$$

and $N(\dot{s}^{(m)})$ have been crossed. The new switching function is then

$$s^{(m+1)} = \dot{s}^{(m)} = g^{(m)} \dot{x}_1 + \dot{x}_2 \qquad t > t_{m+1}. \tag{4.3}$$

The time t_{m+1} represents the time of attainment of the state trajectory of $N(s^{(m+1)})$. Equation (4.3) can be expressed as

$$s^{(m+1)} = g^{(m+1)} x_1 + x_2. \tag{4.4}$$

The adaptive parameter is given implicitly in (4.4) by an amount predetermined by the parameters of the SmVSC function and the system parameters.

In CSAC it is required that the time taken to reach $N(\dot{s})$ is indeed short. If this is not the case, adaptation will take place at intervals that would yield no significant advantage over SmVSC or ideal sliding motion. The adaptation steps should regularly improve the speed of response of the system by continually adapting the switching line. It is required that

$$t = t_{m+1} - t_m \simeq 0. \tag{4.5}$$

By considering the eigenvalues of the closed loop scalar SmVSC system

$$\dot{x} = (A - kBG)x = Fx \tag{4.6}$$

with G of the form $G = [g, 1]$ and the system in phase canonical form, the eigenvalues, (λ_1, λ_2), of the closed loop system are given by

$$2\lambda_i = -a_2 - k \pm k[1 + (2a_2 - 4g)/k]^{1/2} \tag{4.7}$$

where k is given in (3.2) and $a_i(i = 1, 2)$ are the elements of the system matrix A. This can be expanded to give

$$2\lambda_i = -a_2 - k \pm k(1 + \xi) \tag{4.8}$$

provided that k is sufficiently large, where

$$\xi = (a_2 - 2g)/k. \tag{4.9}$$

So from (4.8),

$$\lambda_1 = -g,\ \lambda_2 = -k \tag{4.10}$$

if the a_i are small in comparison with k. Manipulating a modal expression for the switching function it can be shown that

$$t = [\ln(\lambda_2/\lambda_1)]/(\lambda_1 - \lambda_2) \tag{4.11}$$

from which it is immediately apparent from (4.10) that

$$t \to 0 \text{ as } k \to \infty. \tag{4.12}$$

The case $\lambda_1 = \lambda_2$ is discussed later. Provided that k is large compared to the other parameters small time intervals result. This condition is also required for the approximation of VSC by SmVSC [3]. Short time internals between adaptation steps is therefore automatically satisfied by choice of a valid SmVSC function.

Resulting adaptation in CSAC can be derived by combining

$$s^{(m)} = G^{(m)}x = 0 \tag{4.13}$$
$$s^{(m+1)} = G^{(m+1)}x = 0 \tag{4.14}$$

and

$$\dot{s}^{(m)} = G^{(m)}Fx = 0 \tag{4.15}$$

to give

$$G^{(m+1)} = G^{(m)}F \tag{4.16}$$

This iterative formula allows derivation of the amount of adaptation required in the parameter $g^{(m)}$.

If A is in canonical form

$$G^{(m+1)} = [(kg^{(m)} + a_1)/(k + a_2 - g^{(m)}), 1] \tag{4.17}$$

where it can be shown by algebraic manipulation, expanding and neglecting terms of order $1/k^2$ and higher, that

$$G^{(m+1)} = [g^{(m)}(1 + \xi), 1] \tag{4.18}$$

where

$$\xi = (g^{(m)} - a_2)/k. \tag{4.19}$$

In CSAC the successive values of $g^{(m)}$ are given by discrete values predetermined by the design of the SmVSC function and the system parameters. This is in contrast to SAC, the predecessor of CSAC, where control is discontinuous and the adaptation is presented along continuous lines. For any initial stable switching line, the use of a valid SmVSC system therefore depends on the new switching line generated being stable. It is not always the case that a new switching line will be stable. There is in fact an upper limit to the adaptation procedure resulting in what is termed 'fine tuning' of the switching line to yield a maximum possible speed of response.

Fine tuning is easily illustrated for the double integrator system. From (4.17)

$$g^{(m+1)} = kg^{(m)}/(k - g^{(m)}). \tag{4.20}$$

If $g^{(m)} > k$ the adaptation step would generate $g^{(m+1)} < 0$ and hence, as the eigenvalue of the second-order system in the sliding mode is given by the negative of the adaptive parameter, an unstable switching line results. There exists then the necessary condition that $g^{(m)} > k$. The adaptation technique is halted when $g^{(m+1)}$ violates this condition. The system then continues as a SmVSC system with a response that is faster than the initial design response (since $g^{(m)} > g^{(0)}$). Closer inspection of the time interval between adaptation steps via (4.10) and (4.11) shows that at this value of $g^{(m)}$ CSAC breaks down as t becomes undefined.

4.2 Higher order scalar systems

For $n > 2$ the switching hyperplane in CSAC is also adapted by the alteration of a single design parameter; to achieve this a suitable switching hyperplane has coefficients which are the Binomial coefficients to allow easy factorization. No straightforward analysis is available to justify the scheme by generating a rigorous expression for the time interval between adaptation steps. It can be inferred that the time interval is small by comparing SmVSC with high gain feedback control and examining the fast modes of such systems forwards the switching hyperplane [12], [13].

In higher order CSAC the n dimensional switching function is expressed in the adaptable form

$$s = \sum_{i=1}^{n-1} c_i g^{n-i} x_i + x_n \tag{4.21}$$

where g is the parameter to be adapted and $c_i, (i = 1, \cdots, n-1)$ are positive constants dependent on desired initial properties. The iteration superscript of g has been dropped to ease the notation. The CSAC iteration procedure is again given by (4.16) and it can be shown [4] for a system in canonical form that

$$G^{(m)} = \left[\frac{kc_1g^{n-1} + a_1}{k + a_n - c_{n-1}g}, \frac{kc_2g^{n-2} - c_1g^{n-1} + a_2}{k + a_n - c_{n-1}g}, \cdots \right.$$
$$\left. \cdots \frac{kc_{n-1}g - c_{n-2}g^2 + a_{n-1}}{k + a_n - c_{n-1}g}, 1\right] \tag{4.22}$$

where g is the mth adaptation of the initial design parameter. Using similar methods as used in the second order CSAC the generated switching hyperplane can be written as

$$G^{(m+1)} = [c_1 g^{n-1}(1+\xi), c_2 g^{n-2}(1+\xi), \cdots, c_{n-1} g(1+\xi), 1] \tag{4.23}$$

with

$$\xi = (c_{n-1} g - a_n)/k. \tag{4.24}$$

A similar response is obtained as in the second order CSAC.

The stability again depends only on the necessity that a stable switching hyperplane is generated. For an nth–order integrator system the equivalent system matrix (2.6) is given by

$$A_{eq} = \begin{bmatrix} 0 & 1 & 0 & \cdots & \cdots & 0 \\ 0 & 0 & 1 & \cdots & \cdots & 0 \\ 0 & 0 & 0 & \cdots & 0 & 1 \\ -\frac{kc_1 g^{n-1}}{k-c_{n-1}g} & -\frac{kc_2 g^{n-2} - c_1 g^{n-1}}{k-c_{n-1}g} & \cdots & \cdots & \cdots & -1 \end{bmatrix} \tag{4.25}$$

A necessary condition for the stability of this system is that the coefficients in the bottom row are negative, therefore the necessary condition

$$k - c_{n-1} g > 0, \qquad g < \frac{k}{c_{n-1}} \tag{4.26}$$

results, giving an upper limit on the allowable fine tuning.

In higher order scalar CSAC it is possible for the derivative of the adaptive parameter to become negative and the new hyperplane to be a stable hyperplane for the system. This results in fine tuning of the system response being achieved in distinct phases. This phenomenon can be explained by examining the properties of SmVSC [3], [4].

5 CSAC of multivariable control systems

Multivariable CSAC is implemented by adapting each switching hyperplane separately (and some not at all), rather than develop a methodology for adaptation of all the switching hyperplanes concurrently. In order to extend CSAC to multivariable CSAC a structure is imposed on the parameters within the switching hyperplane matrix which has direct influence upon the eigenvalues of the sliding mode. This structure is non–restrictive and is in fact useful in making the generation of the initial controller an easier task, since the system has a form that is amenable to initial multivariable VSC design.

The point in time at which an adaptation step takes place is again when the state trajectory reaches the null space of the derivative of a switching function.

The iteration procedure for scalar CSAC is used in the analysis with the subscript j denoting any of the m switching hyperplanes in the multivariable CSAC.

Not all the coefficients of a switching hyperplane need possess the adaptive design parameter.

At some instant the nth adaptation of the jth hyperplane is given for a system in canonical form as,

$$s_j^{(m)} = \sum_{i=1}^{n_a} c_{ij} g_{a_j}^{n_a+1-i} x_i + \sum_{i=n_a}^{n} c_{ij} x_i. \tag{5.1}$$

The states have been re–ordered in the jth switching function so that the first summation represents a polynomial, the power of which is dependent upon the number n_a of key elements of that switching function in which adaptation will take place. The second summation represents those elements that have no direct influence on the eigenvalue of ideal sliding motion and hence in which no adaptation will take place. The iterative superscript has been dropped for ease of notation.

The control function used throughout is the multivariable SmVSC

$$u_i = -\delta \cdot \frac{s_i^{(m)}}{|s^{(m)}| + \varepsilon} \qquad i = 1, \cdots, m. \tag{5.2}$$

and a subspace of the state space is considered that allows the approximation

$$u_i = -ks_i \qquad i = 1, \cdots, m \tag{5.3}$$

where $k = \delta/\varepsilon$ to be valid when required [3].

Precise analytical expressions for the adaptation time interval are not derived. It is sufficient to note that the controller approximates to a high gain feedback controller in the subspace considered and in such systems it is a reasonable assumption that the time interval will be small due to the particular separation of the eigenvalues [12] into fast eigenvalues in the range space of the switching hyperplane and slow within the null space.

The CSAC system requires that the switching hyperplane matrix is expressed in a suitable form such that any adaptation of the design parameter directly increases the eigenvalue that would result if ideal sliding motion were to occur at an instance. This is termed increasing the 'slope' of the set of hyperplanes. For a multivariable control system in canonical form and matrices of the block nature given below

$$B = \begin{bmatrix} 0 \\ B_2 \end{bmatrix} \qquad G = [G_1, G_2]. \tag{5.4}$$

the response of the system in the sliding mode is that of the reduced order system [10]

$$\dot{z}_1 = (A_{11} - A_{12}G_2^{-1}G_1)z_1 \tag{5.5}$$

The multivariable CSAC system is most easily analysed by considering first the case $n - m = 1$ and then extending the approach to the case $n - m > 1$.

5.1 Multivariable CSAC: $n - m = 1$

For $n - m = 1$ (5.5) reduces to the scalar equation

$$\dot{z} = (-A_{12}G_2^{-1}G_1)z \tag{5.6}$$

where the eigenvalue of ideal sliding motion is now

$$\sigma = -A_{12}G_2^{-1}G_1. \tag{5.7}$$

In order to adapt the eigenvalue it is necessary to inspect the component $G_2^{-1}G_1$. A full description is presented in reference[2].

5.2 Multivariable CSAC: $n - m > 1$

In the general multivariable CSAC system the eigenvalues during ideal sliding motion are given by

$$\{\sigma_1, \sigma_2, \cdots, \sigma_{n-m}\} = \sigma(A_{11} - A_{12}G_2^{-1}G_1) \tag{5.8}$$

where the submatrices are as previously defined. Under the imposed constraint the eigenvalues are the roots of the polynomial

$$|\sigma I - A_{11} + A_{12}B_2G_1| = 0 \tag{5.9}$$

the solution of which has the form

$$\sigma^{n-m} + h_{n-m}\sigma^{n-m-1} + \cdots + h_2\sigma + h_1 = 0 \tag{5.10}$$

where

$$h_k = f(g_{a_1}, \cdots, g_{a_m}) \tag{5.11}$$

[2]. Adaptation of the eigenvalues in the sliding mode will depend upon adaptation of the coefficients of this polynomial being possible. Foresight into the manner and magnitude of the resulting adaptation, however, is more complex. There exist many more degrees of freedom and many more alternative strategies. Here attention is focused on forming an exact polynomial in the adaptive parameter and so adapt the eigenvalues simultaneously.

For $n - m = 2$ the quadratic equation for the eigenvalues is written in a slightly amended form

$$\sigma^2 + \Phi_2(g_a)h_2\sigma + \Phi_1(g_a)h_1 = 0 \tag{5.12}$$

where the functions $\Phi_1(g_a)$ and $\Phi_2(g_a)$ represent the adaptive element in the coefficients of the polynomial. The eigenvalues are given by

$$(\sigma_1, \sigma_2) = -\frac{\Phi_2(g_a)h_2}{2} \pm \frac{\Phi_2(g_a)h_2}{2}\left[1 - \frac{4\Phi_1(g_a)h_1}{(\Phi_2(g_a)h_2)^2}\right]^{1/2} \tag{5.13}$$

If this function possesses the correct powers of the adaptive design parameter then exact factorization is possible. If both eigenvalues are to be simultaneously increased the same adaptive parameter must be used in the switching hyperplanes; if not the degree of increase of each eigenvalue will depend on the relative value of the functions in (5.13). Detailed expressions for the magnitude of adaptation are not derived as only bounds will result.

The stability of the scheme, in both multivariable cases, is dependent only on the generation of a stable switching hyperplane matrix, since, when adaptation is not occurring, the system behaves as a multivariable SmVSC system. It is obvious, in the $n - m = 1$ case, that whilst the first parameter of any switching hyperplane is positive (for adaptation on one hyperplane) or the combination of the first parameters are positive (for adaptation on more than one hyperplane) the system will be stable. It is a requirement of the implementation of CSAC that any adaptation step is checked and if the condition is not satisfied, adaptation is stopped and the system is allowed to continue under SmVSC (improved on the initial requirement) until the state re–enters a region where adaptation can again take place.

In multivariable CSAC the switching hyperplanes are partitioned into an 'adaptable set', which influence the eigenvalue, and 'non–adaptable set' which do not. Not all adaptable hyperplanes need be adapted to improve system response; the adaptation of just one parameter would prove sufficient. It is also possible that one, or more, switching hyperplanes may not be adapted as the null space of that switching function may not be reached. This corresponds to multivariable SmVSC where a particular hyperplane is approached asymptotically. It is immediately apparent, from (5.9), that adaptation of different parameters will have varying degrees of effect on the system eigenvalues. Choice of rate of adaptation can be made through parameter inspection. Hyperplanes which are not adapted, are expressed as simple linear combinations of the states.

6 Application of the CSAC

One of the advantages of CSAC is its extreme simplicity of application. A complicated test for the occurrence of sliding motion does not need to be implemented; it is necessary to store merely the last two values of the switching functions and perform a simple test with the current value to check if the null space of the derivative of a switching hyperplane has been crossed. A further advantage is the reduction of chatter motion that is associated with SmVSC.

The case $n - m = 2$ is illustrated by the following system

$$\dot{x} = \begin{bmatrix} 0 & 1 & 0 & 0 & 0 \\ 0 & 0 & 1 & 0 & 0 \\ 0 & 0 & 0 & 1 & 0 \\ 0 & 0 & 0 & 0 & 1 \\ -1 & -5 & -10 & -10 & -5 \end{bmatrix} x + \begin{bmatrix} 0 & 0 & 0 \\ 0 & 0 & 0 \\ 1 & 0 & 1 \\ 0 & 1 & 0 \\ 0 & 0 & 1 \end{bmatrix} u \tag{6.1}$$

and initial condition $x(0) = (-0.25, 0, 0, 0, 0,)^T$. Initial design specifies eigenvalues during ideal sliding motion of $\sigma_1 = -1$, $\sigma_2 = -1$; therefore from (5.9) the hyperplane

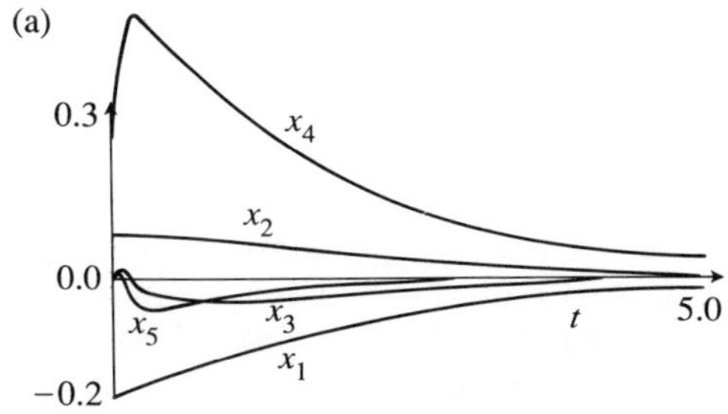

Fig. 2a SmVSC response.

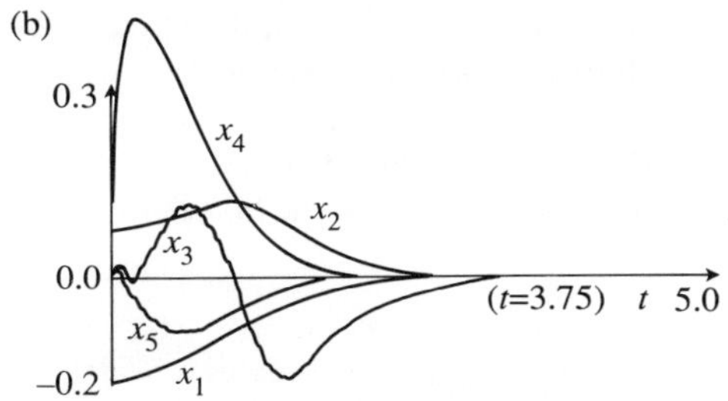

Fig. 2b CSAC response.

matrix must have the form,

$$G = \begin{bmatrix} 0.5 & 1.5 & 1 & 0 & -1 \\ 3 & 1 & 0 & 1 & 0 \\ 0.5 & 0.5 & 0 & 0 & 1 \end{bmatrix}. \tag{6.2}$$

The first two elements of the second row of the switching hyperplane matrix are chosen at random as the respective elements in B are zero. The matrix is expressed adaptively according to the general formulation (5.1) as

$$G^{(0)} = \begin{bmatrix} 0.5g_{a_1}^2 & 1.5g_{a_1} & 1 & 0 & -1 \\ 3 & 1 & 0 & 1 & 0 \\ 0.5g_{a_3}^2 & 0.5g_{a_3} & 0 & 0 & 1 \end{bmatrix} \tag{6.3}$$

i.e.

$$\begin{aligned} c_{11} &= 0.5, & c_{12} &= 1.5, \\ c_{31} &= 0.5, & c_{32} &= 0.5, \\ g_{a_1}(0) &= 1, & g_{a_3}(0) &= 1. \end{aligned} \tag{6.4}$$

No adaptive parameter is present in the second hyperplane. The system response is shown in Fig. 2a and Fig. 2b for CSAC and SmVSC. In the CSAC system each hyperplane possesses separate adaptive parameters. The adaptive parameters increase individually as each switching function reaches the null space of the first derivative associated with it.

7 Conclusion

With CSAC the control system designer can, for all orders of system, 'fine tune' an initial VSC system design and improve the speed of response of the system whilst in the sliding mode. The CSAC possesses properties of robust VSC theory . Chatter motion associated with ideal sliding motion in a VSC system is eliminated since CSAC is developed from SmVSC. The CSAC is easily implemented involving only minor adaptation to the implementation algorithm of VSC. By introducing the CSAC system the VSC system is partitioned into three phases. The attainment (hitting) of the sliding mode, the sliding mode itself and an improvement (adaptation) of the response in that sliding mode, should that be required. There are many degrees of freedom available in prescribing the rate of adaptation in the multivariable CSAC system, as compared to its scalar counterpart, and the degree greatly increases as the order of the reduced state space during the sliding mode increases. Further investigation into CSAC and the application of CSAC will provide additional interesting results.

References

[1] Ambrosino G. Celentano G. Garofalo F. 'Variable structure model reference adaptive control systems', Int. J. Control, Vol 39, 1339-1349, 1984

[2] Burton J.A., 'Adaptive Control using Variable Structure Control Theory', Ph. D. Thesis, Sheffield University, 1987

[3] Burton, J.A. Zinober A.S.I., 'Continuous approximation of variable structure control', Int. J. of Systems Sci., Vol 17, No 6, 875–885, 1986

[4] Burton, J.A., Zinober, A.S.I., ' Continuous self-adaptive control using a smoothed variable structure controller', Int. J. of Systems Sci., Vol 19, No 8, 1515–1528, 1988

[5] Draženović, B., 'The invariance condition in variable structure systems', Automatica. Vol 5, 287–295, 1969

[6] El-Ghezawi, O.M.E., Zinober, A.S.I., Billings, S.A., 'Analysis and design of variable structure systems using a geometrical approach', Int. J. Control, Vol 38, No 3, 657–671, 1983a

[7] El-Ghezawi, O.M.E., Zinober, A.S.I., Billings, S.A., 'Variable structure systems and system zeros', Proc. I.E.E., Vol 130, Pt D., No 1, 1–5, 1983b

[8] Itkis, U., 'Control Systems of Variable Structure', Keter Publishers, Jerusalem, 1977

[9] Utkin V.I. 'Variable structure systems with sliding modes', I.E.E.E. Trans. on Automatic Control, Vol AC–22, No 2, 212–222, 1977

[10] Utkin, V.I.,'Sliding Modes and Their Application in Variable Structure Systems', MIR Publishers, Moscow, 1978

[11] Wilkinson, J.H., 'The Algebraic Eigenvalue problem', Clarendon Press, Oxford, 1965

[12] Young, K-K.D., Kokotovic, P.V., Utkin, V.I., 'A singular perturbation analysis of high gain feedback systems', I.E.E.E Trans. on Automatic Control, Vol AC-22, No 6, 931-937, 1977

[13] Zinober, A.S.I., 'Adaptive relay control of second-order system', Int. J. Control, Vol 21, No 1, 81-98, 1975.

[14] Zinober, A.S.I., 'Analysis of a third order relay control system using non-linear switching surface theory', Proc. Royal Society of Edinburgh, Vol 76A, 239-254, 1977

[15] Zinober, A.S.I, 'Controller design using the theory of variable structure systems', Self Tuning and Adaptive Control, (ed. Billings S.A. and Harris C.J.), 206-299, Peter Peregrinus Publishers, London, 1981

[16] Zinober, A.S.I, 'Self-adaptive control of the time-varying nonlinear crane problem with disturbances and state constraints', Control and Computer, Vol 14, No 3, 103-109, 1986

[17] Zinober, A.S.I., 'Self adaptive control of the time varying nonlinear crane problem with disturbances and state constraints', Proc. I.A.S.T.E.D. Symp. on Automatic Control and Identification, Copenhagen, Denmark, 1983.

Asymptotic linearization of uncertain multivariable systems by sliding modes

G. Bartolini T. Zolezzi

Abstract

Asymptotic linearization of some classes of non linear multivariable control systems is obtained under deterministic uncertainty by discontinuous state feedback. Chatter reduction is discussed. A kind of switching logic is introduced to obtain finite time reachability of the (even uncertain) sliding manifold.

1 A model problem

We consider control system with deterministically uncertain dynamics described by the differential inclusions

$$\begin{matrix} \ddot{x} \in G_1(t, x, \dot{x}, y, \dot{y}, u_1, u_2) \\ \ddot{y} \in G_2(t, x, \dot{x}, y, \dot{y}, u_1, u_2) \end{matrix} \quad t \geq 0. \tag{1.1}$$

Here x, y are scalar state variables and u_1, u_2 are scalar control variables, constrained by

$$(u_1, u_2) \in U$$

a given closed subset of R^2. Motivations for considering system (1.1) come from robotics, since some dynamic equations of kinematic chains appearing in robotics may be reduced to the from (1.1). See e.g. [9] for a recent treatment.

The multifunction which describes the unknown system dynamics is given by

$$(G_1, G_2)' : [0, +\infty) \times R^4 \times U \rightrightarrows R^2.$$

We assume explicit knowledge of some upper and lower bounds of the dynamics involved. Therefore Caratheodory functions

$$g_i^+, g_i^-, \; i = 1, 2$$

are known such that

$$G_1 = [g_1^-, g_1^+], G_2 = [g_2^-, g_2^+].$$

The initial state is uncertain but bounded by some known constant. We consider a given linear time invariant (known) model

$$\ddot{w} = a_1\dot{w} + a_2 w + a_3 v_1, \qquad \ddot{z} = b_1\dot{z} + b_2 z + b_3 v_2 \tag{1.2}$$

with scalar control variables v_1, v_2, state variables w, z, $w(0)$ and $z(0)$ fixed, and arbitrarily fixed coefficients a_i, b_i $(i = 1, \cdots, 3)$.

The problem we consider is to find a state feedback control law (possibly depending on instantaneous values of v_1, v_2, w, $\dot{w}$ and z, $\dot{z}$) under which system (1.1) is asymptotically equivalent to the given linear model (1.2).

More precisely, given $\alpha > 0$ and any (v_1, v_2, w, z), we construct (in a sense explicitly) a feedback u such that every possible state (x, y) for (1.1) corresponding to it (under any uncertain dynamics $g \in (G_1, G_2)$) fulfils the model dynamics (1.2) up to an exponentially decaying error term dominated by (const.) $\exp(-\alpha t)$, $t \geq T$; moreover

$$|X(t) - W(t)| \leq (\text{const.}) \exp(-\alpha t), \;\; t \geq T$$

where

$$X = (x, \dot{x}, y, \dot{y}) = (x_1, x_2, x_3, x_4), \qquad W = (w, \dot{w}, z, \dot{z}) = (y_1, y_2, y_3, y_4),$$

and some T independent of X and explicitly estimated by known data.

Such a feedback u (in general discontinuous) is obtained by using variable structure control methods (see [11]) recently extended to non linear control systems by the authors (see [4] and [5]), provided a set of explicit inequalities is satisfied by the known bounds $g_i^{\pm}$, $i = 1, 2$, as follows. Let c_1, c_2 be such that $\alpha \leq min(c_1, c_2)$. Then put $e = x - w$,

$$s_1(e) = c_1 e_1 + e_2, \qquad s_2(e) = c_2 e_3 + e_4. \tag{1.3}$$

Consider now

$$\begin{aligned} p(x, y, v) &= c_1(y_2 - x_2) + a_1 y_2 + a_2 y_1 + a_3 v_1, \\ q(x, y, v) &= c_2(y_4 - x_4) + b_1 y_4 + b_2 y_3 + b_3 v_2. \end{aligned}$$

Then we assume existence of some $u \in U$ fulfilling

$$\begin{aligned} g_1^- &\geq p + k^2 \text{ if } s > 0; \\ g_1^+ &\leq p - k^2 \text{ if } s < 0; \\ g_2^- &\geq q + k^2 \text{ if } s > 0; \\ g_2^+ &\leq q - k^2 \text{ if } s < 0. \end{aligned} \tag{1.4}$$

These results may be generalized to higher–order control systems of the following form

$$\ddot{x}_i \in G_i(t, x_1, \dot{x}_1, \cdots, x_n, \dot{x}_n, u_1, \cdots, u_m), \; i = 1, \cdots, n.$$

It is then likely that the number of the required inequalities corresponding to (4) may be reduced by using results of [1].

The chattering effects due to the discontinuous nature of the asymptotically linearizing feedback may be reduced by appropriately combining results from [8], [2], [3] (see section 4). Moreover sufficient conditions may be obtained assuring that the feedback u is piecewise continuous (not only measurable).

An important property of the variable structure control systems (1.1), (1.3) is approximability (see [4] for the precise definition).

In essence this means that whenever some error vectors e_ε depend on disturbances described by the real parameters ε and satisfy

$$s_i(e_\varepsilon) \to 0 \text{ as } \varepsilon \to 0, \; i = 1, 2,$$

uniformly on compact intervals of $[T, +\infty)$, then $e_\varepsilon \to e_0$ in the same sense, where e_0 is uniquely defined and fulfils the sliding condition $s_i(e_0) = 0, \; i = 1, 2$. Thus approximability prevents ambiguous behaviour in the sliding mode. Sufficient conditions can be found on the available data in order to fulfil such a property.

2 A particular case

Let the uncertain control system be described by a single differential inclusion of order n with scalar control u, given by

$$\dot{x}_1 = x_2, \dot{x}_2 = x_3, \cdots, \dot{x}_{n-1} = x_n, \dot{x}_n \in G(t, x, u), \; u \in U.$$

Suppose that the model is given by

$$\dot{y}_1 = y_2, \; \dot{y}_2 = y_3, \cdots, \dot{y}_{n-1} = y_n, \dot{y}_n = \sum_{i=1}^{n} a_i y_i + bv$$

and let

$$G = [g^-, g^+].$$

Fix real number $c_1, \cdots, c_{n-1}$ such that the polynomial (in h)

$$h^{n-1} + c_{n-1} h^{n-2} + \cdots c_2 h + c_1$$

is Hurwitz. Denote by

$$e = y - x$$

the error vector and set

$$\begin{aligned} s(e) &= e_n + \sum_{1}^{n-1} c_i e_i, \\ p(x, y, v) &= bv + \sum_{i+1}^{n} a_i y_i + \sum_{i+1}^{n-1} c_i (y_{i+1} - x_{i+1}). \end{aligned}$$

The asymptotic linear behaviour (in the sense defined above) obtains provided the following holds. For any model control–state pair y, v, every $t \geq 0$ and every x we can find $u \in U$ such that

$$\begin{aligned} g^-(t, x, u) &\geq k^2 + p(x, y, v) \text{ if } s(y - x) > 0, \\ g^+(t, x, u) &\leq -k^2 + p(x, y, v) \text{ if } s(y - x) < 0, \end{aligned}$$

for some fixed constant $k \neq 0$.

Arbitrary exponential decay of the error term is obtained by a proper choice of $c_1, \cdots, c_{n-1}$. See [6], [7] for more details, examples and comparisons about this particular case.

3 More general systems

We extend the linearizing approach of section 1 to a multivariable control system described by m coupled ordinary differential inclusions of orders $n_1, \cdots, n_m$, given by

$$\dot{x}_{i1} = x_{i2}, \; \dot{x}_{i2} = x_{i3}, \cdots, \dot{x}_{in_i} \in G_i(t, X_1, \cdots, X_m, u_1, \cdots, u_m), \; i = 1, \cdots, m, \quad (3.1)$$

where X_i is the vector of components $x_{i1}, x_{i2}, \cdots, x_{in_i}$, $u = (u_1, \cdots, u_m)'$ is the control vector constrained by $u \in U$, a given control region in R^m. As before, Carathèodory functions

$$g_i^-, g_i^+, i = 1, \cdots, m$$

are known such that the uncertain dynamics (3.1) lie in $G_i = [g_i^-, g_i^+], i = 1, \cdots, m$. The initial state $X(0)$ is uncertain but bounded by a known constant.

We consider a given linear time–invariant known model

$$\dot{y}_{i1} = y_{i2}, \; \dot{y}_{i2} = y_{i3}, \cdots, \; \dot{y}_{in_i} = \sum_{j=1}^{n_i} a_{ij} y_{ij} + b_i v_i \quad (3.2)$$

with uncoupled control variables $v_1, \cdots, v_m$.

We extend to the present setting the procedure which selects some state feedback control law under which systems (3.1) will be asymptotically equivalent to the model (3.2), assuming that the system state is available. Pick real coefficients $c_{i1}, \cdots, c_{in_i-1}$ such that every polynomial

$$c_{i1} + c_{i2}\lambda + c_{i3}\lambda^2 + \cdots + c_{in_i-1}\lambda^{n_i-2} + \lambda^{n_i-1}, \; i = 1, \cdots, m \quad (3.3)$$

is Hurwitz. Consider the errors

$$e_{ij} = y_{ij} - x_{ij}, \; j = 1, \cdots, n, \; i = 1, \cdots, m$$

and put

$$s_i = e_{in_i} + \sum_{j=1}^{n_{i-1}} c_{ij} e_{ij}, \; i = 1, \cdots, m.$$

The sliding manifold is now given by

$$s_1 = 0, \; s_2 = 0, \cdots, \; s_m = 0. \quad (3.4)$$

In order to accomplish asymptotic linear behavior, we wish to send and keep the state vector on (3.4). To this purpose, let $Y_i = (y_{i1}, \cdots, y_{in_i})'$ and

$$P_i(X_i, Y_i, v) = \sum_{j=1}^{n_i} a_{ij} y_{ij} + b_i v_i + \sum_{j=1}^{n_i-1} c_{ij}(y_{ij+1} - x_{ij+1}), \quad i = 1, \cdots, m.$$

We assume that there exists some $u \in U$ and some constant $k \neq 0$ such that

$$\begin{array}{lll} s_i > 0 & \text{implies} & g_i^- \geq P_i + k^2 \\ s_j < 0 & \text{implies} & g_j^+ \leq P_j - k^2, \end{array} \tag{3.5}$$

for all i, j.

Under the assumption (3.4), it is possible to reach the sliding manifold (3.4) in finite time T (estimated from above byc/k, where c is a known bound of $|s_i(0)|$ for each i) and keep the state on it by any state feedback fulfilling the inequalities (3.5).

In this way we prove that for $t \geq T$ the system state satisfies the same model dynamics (3.2) up to an error term, which is bounded by (const.) $e^{\alpha t}$ where $\alpha < 0$ may be arbitrarily prescribed by a suitable choice of the coefficients c_{ij} in (3.3).

4 The reduction of chatter by nonlinear immersion

Let us consider the following second order system

$$\left\{ \begin{array}{lll} \dot{x}_1 & = & x_2 \\ \dot{x}_2 & = & f(x, u) \end{array} \right. \tag{4.1}$$

with f sufficiently smooth. This system may be "embedded" in the third order system

$$\left\{ \begin{array}{lll} \dot{x}_1 & = & x_2 \\ \dot{x}_2 & = & x_3 \\ \dot{x}_3 & = & f_{x_1}\dot{x}_1 + f_{x_2}\dot{x}_2 + f_u \dot{u} \end{array} \right. \tag{4.2}$$

where $x_3 = f(x, u)$.

The linearization process for system (4.1) is based on a model:

$$\left\{ \begin{array}{lll} \dot{y}_1 & = & y_2 \\ \dot{y}_2 & = & a_1 y_1 + a_2 y_2 + bv. \end{array} \right.$$

By choosing $y_3 = a_1 y_1 + a_2 y_2 + bv$ also the model may be put in the following form:

$$\left\{ \begin{array}{lll} \dot{y}_1 & = & y_2 \\ \dot{y}_2 & = & y_3 \\ \dot{y}_3 & = & a_1 \dot{y}_1 + a_2 \dot{y}_2 + b\dot{v}. \end{array} \right.$$

Following the standard procedure the error vector

$$e' = [e_1 e_2 e_3], \; e_i = y_i - x_i, \; i = 1, \cdots, 3,$$

is introduced.

The sliding manifold is chosen as

$$s(e) = e_2 + c_1 e_1 + c_2 e_2 = 0.$$

Provided on this sliding manifold the condition $\dot{s}(e) = 0$ gives rise to a stable regulator $\dot{u}_{eq} = g(t, x, e, u_{eq})$, we may obtain sliding regimes $s(e) = 0$ by means of discontinuous $\dot{u}$ but with u continuous.

If the system is perfectly known the procedure may be extended to immersions of higher order obtaining discontinuities on higher derivatives. When only approximate informations on $f(x, u)$ are available this procedure cannot be used manily because the $s(e)$ quantity depends on $x_3 = f(x, u)$ and the instants in which $s(e)$ changes its sign (and $\dot{u}$ too) are not appreciable. Then $f(x, u)$ may be estimated by the following observer:

$$\begin{aligned} \dot{\widehat{z_2}} &= \widehat{z_3} + \gamma_1 (x_2 - \widehat{z_2}) \\ \dot{\widehat{z_3}} &= \gamma_2 (x_2 - \widehat{z_2}) + \dot{u} + u_E, \end{aligned}$$

where u_E is an auxiliary control signal. The observation error

$$\begin{aligned} \dot{\widehat{e_1}} &= \widehat{e_2} - \gamma_1 \widehat{e_1} \\ \dot{\widehat{e_2}} &= -\gamma_2 \widehat{e_1} + f_{x_1} \dot{x}_1 + f_{x_2} \dot{x}_2 + f_u \dot{u} - \dot{u} - u_E \end{aligned}$$

in which $\widehat{e_1} = x_2 - \widehat{z_2}$, $\widehat{e_2} = x_3 - \widehat{z_3}$, is asymptotically stable if u_E is chosen according to: $u_E = -k(t, x, u, \dot{u})\mathrm{sgn}\{\widehat{e_1}\}$ where $k(t, x, u, \dot{u})$ is a known positive function such that $k(t, x, u, \dot{u}) \geq |f_{x_1} \dot{x}_1 + f_{x_2} \dot{x}_2 + [f_u - 1]\dot{u}|$(see [3]).

Once $\widehat{z_3}$ is available the control law $\dot{u}$ is chosen according to $\dot{\widehat{s}}(e,t)\widehat{s}(e,t) \leq -k^2 |s(e,t)|$, where $\widehat{s}(e,t) = y_3 - \widehat{z_3} + c_1 e_1 + c_2 e_2$.

With this choice of u the whole scheme obtains asymptotic linearization as in the previously considered case of known $f(x, u)$.

5 Reachability of the sliding manifold in finite time

In [1] a switching logic different from (3.5) is presented, with the aim of reducing the control effort (the norm of the control vector) required to reach the intersection of the sliding planes when uncertainties of various kind affect the system equation. Summarizing results contained in [1] a brief synthesis of this new procedure follows.

Consider a system $\dot{x} = f(t, x, u)$, $u \in U \subset R^m$ $x \in R^n$; the m–dimensional vector $s(x)$ is such that on the $(n-m)$ manifold $s'(x) = (s_1, \cdots, s_m) = 0$ the control objectives are fulfilled. Choose a simplex of constant vectors $\{w^i\}$ $i = 1, \cdots, m+1$

in R^m containing the origin, i.e. $\exists\ \lambda_i \geq 0;\ i = 1, \cdots, m$ such that $\sum_1^{m+1} \lambda_i w^i = 0$ and $\sum \lambda_i = 1$.

For any vector w^i it is possible to define a velocity vector (ds/dt) such that:

$$\left(\frac{ds}{dt}\right)^i = h(t, x) + H(t, w) w^i \tag{5.1}$$

(w^i corresponds to a vector u^i by solving an algebraic equation) ; $h(t, x)$ and $H(t, x)$ are known if $f(t, x, u)$ is known.
Perform the following transformation

$$y = H^{-1}(t, x) s, \tag{5.2}$$

$$(dy/dt)^i = g(t, x, w_i) + w^i. \tag{5.3}$$

If for every w^i

$$-g(t, x, w) = \sum_1^{m+1} \kappa_i w^i, \qquad \sum \kappa_i = a_0, \qquad \kappa_i \geq 0 \qquad a_0 \leq 1 \tag{5.4}$$

the state space y may be partitioned into regions

$$W_j, j = 1, \cdots m + 1$$

$$W_j = \left\{ y : y = \textstyle\sum_{i=1, i \neq j}^{m+1} \lambda_i w^i, \lambda_i \geq 0 \right\} \tag{5.5}$$

Theorem: [1] The control law $w = w^i$ for $y \in W_j$ ensures the asymptotic stability in the large of system 5.3 and the convergence takes place in finite time.

This theorem guarantees the reachability of the intersection of the discontinuity surfaces $s = 0$ even for uncertain system.

In (5.3) the term $y(t, x, w)$ may be uncertain without affecting the results of the theorem. It is sufficient to increase the modulus of vectors w^j to fulfil (5.4) even in presence of uncertainties.

This fact is a common feature of this approach and the unit vector method control (a comparison between the two methods based on the control effort necessary to counteract uncertainties will be made in the near future).

The peculiarity of this simplex method lies in the possibility of obtaining the reachability of the manifold $s(x) = 0$ even when uncertainties on the components of the vector $s(x)$ are considered.

Any uncertainty about vector s traduces in uncertainty about vector y by means of (5.2). The simplex of control vectors must be chosen in such a way that this uncertainty on y does not affect the possibility of recognizing the region W_j to which y belongs. This property of the proposed method, not explicitly stressed in [1], is useful when imprecision in the evaluation of the components s_i of vectors s prevent the use of the unit vector method.

References

[1] S. Baida, D. Izosimov, 'Vector methods of design of sliding motion and simplex algorithms', Autom. Telemekh. 7 (1985), 56–63.

[2] A. Balestrino, G. De Maria, L. Sciavicco: J. Dyn. Systems Meas. Control 105 (1983), 141–151.

[3] G. Bartolini, 'A note on chattering phenomenon in discontinuous control system', (in preparation).

[4] G. Bartolini, T. Zolezzi, 'Control of non linear variable structure systems,, J. Math. Anal. Appl. 118 (1986), 42–62.

[5] G. Bartolini, T. Zolezzi, 'Variable structure systems non linear in the control law', IEEE Trans. Autom. Control 30 (1985), 681–684.

[6] G. Bartolini, T. Zolezzi, 'Asymptotic linearization by variable structure control', Proc. 25th IEEE CDC (Athens, 1986), 2061–2062.

[7] G. Bartolini, T. Zolezzi, 'Asymptotic linearization of uncertain systems by variable structure control', Systems Control Lett. 10 (1988), 111–117.

[8] J.A. Burton, A.S. Zinober, 'Continuous approximation of variable structure control', Int. J. Systems Sci. 17 (1986), 875–885.

[9] B.E. Paden, S.S. Sastry, 'A calculus for computing Filippov's differential inclusion with application to the variable structure control of robot manipulators', IEEE Trans. Circuits System 34 (1987), 73–82.

[10] E.P. Ryan and M. Corless, 'Ultimate boundedness and Asymptotic Stability of a class of Uncertain Dynamical Systems via Continuous and discontinuous Feedback Control', IMA Journal of Mathematical Control & Information (1984) 1, 223–242.

[11] V.I. Utkin, 'Sliding modes and their application in variable structure systems', MIR, Moscow, 1978.

Design of discrete–time ‘adaptive’ optimal variable structure controllers

A. Parvinmehr A. J. Wilkinson R. D. Greenough

Abstract

It is well known that plants operating in a sliding regime are insensitive to parameter variations and external disturbances. The performance of these plants may not however be optimal. Guaranteed optimal performance from a mathematical view point is only achieved when the switching manifold slope is set *equal* to that if the plane's stable eigenvectors. The transient response of the system will then take place on the boundary of the regions of the sliding regime. Regimes of this type in real process control applications are susceptible to ‘break down’ because there is no guarantee that the process's gain will remain *frozen* for the duration of a transient. If the manifold slope exceeds it's admissible value the sliding regime will be violated and transients will become underdamped and may be oscillatory.

A novel approach for solving this type of control problem is looked at in this paper. The solution employed may be divided into two sections: i) plant model estimation and ii) controller design. The system's discrete–time transfer function is estimated recursively and used to design a discrete–time controller. In this way the plant model and the switching manifold slope are updated at regular intervals, compensating for any process model parameter variation. Thus once ‘hitting’ has occurred the phase plane trajectories will remain on the ‘optimum’ sliding manifold.

Simulation results of the proposed control solution with different estimated models are presented. In conclusion practical applicability of such an approach in real plants is contemplated with the view to suggest the best suitable estimation algorithm.

1 Introduction

The theory of variable structure controllers (VSC) has almost extensively been developed during the past 30 years in the U.S.S.R [1], [2]. Most of the work is based on continuous systems and very little has been done on their discrete counter parts. This paper describes the development of a novel adaptive discrete VSC algorithm which is designed to tailor the error transients in a second order discrete process with an optional zero. The adaptive behaviour is guaranteed by regular updating of the discrete process model and hence the VSC design parameters.

2 Algorithm formulation

2.1 The difference operator

Consider a first order forward difference equation of type:–

$$(x(t+1) - x(t)) + \alpha x(t) = \delta \tag{2.1}$$

α and δ are constants and $x(t)$ represents the process output. If a difference operator D is selected such that $D = Z - 1$ [3] then equation 2.1.can be expressed in terms of D, that is:–

$$(D + \alpha)x(t) = \delta \tag{2.2}$$

This requires a prediction of $x(t+1)$ which can be eliminated if the D operator is used in combination with delay in the following way:–

$$Z^{-1}D = 1 - Z^{-1} \tag{2.3}$$

Applying the $Z^{-1}D$ operator to equation 2.1

$$(D + \alpha)x(t-1) = \delta \tag{2.4}$$

The $Z^{-1}D$ operator has preserved the original dynamics and augmented a one sample delay to the process output.

Equation 2.4 may be mapped in a *difference plane* Figure 1, giving an approximate phase plane representation of the dynamics of a continuous first order system. Thus by applying the $Z^{-1}D$ operator to a discrete process the *difference plane* can be used to plot approximate *trajectory portraits.* (note : algebraic manipulation of D is only valid when all terms have the same latency!).

2.2 A discrete VSC algorithm for second order dynamic with time varying parameters and one zero

If $\sigma(t)$ represents the sliding manifold in the *phase plane* and $\dot{\sigma}(t)$ is its derivative then the existence of the *sliding motion* is guaranteed by the following inequality (Fillipov 1960 [1]):–

$$\sigma(t)\dot{\sigma}(t) < 0 \tag{2.5}$$

$$\sigma(t) = x_2 + C_t x_1 \tag{2.6}$$

C_t=piecewise sliding manifold slope. The 't' denotes nominally infrequent variation associated with any process model parameter changes. x_1 = process output or error x_2 = rate of change of x_1. The $Z^{-1}D$ operator can be used to provide a discrete analogy of the sliding criteria:–

$$\sigma(t)\dot{\sigma}(t) \rightarrow Z^{-1}(\sigma(t).D\sigma(t)) \tag{2.7}$$

Thus:–

$$\sigma(t) \overset{\Omega}{=} Z^{-1}((x_1(t+1) - x_1(t)) + C_t x_1(t)) \tag{2.8}$$

$$\overset{\Omega}{=} (D + C_t)x_1(t-1) \tag{2.9}$$

$$\text{and } \dot{\sigma}(t) \overset{\Omega}{=} Z^{-1}(\sigma(t+1) - \sigma(t))$$

$$\overset{\Omega}{=} (Z^{-1}(D + C_t) - Z^{-2}(D + C_t))x_1(t) \tag{2.10}$$

Therefore the sliding criteria 2.5 can be expressed as:–

$$(Z^{-1}(D + C_t) - Z^{-2}(D + C_t))x_1(t)\sigma(t-1) < 0 \tag{2.11}$$

$$Z^{-2}(Z(D + C_t) - (D + C_t))x_1(t)\sigma(t-1) < 0 \tag{2.12}$$

$$((D+1)(D + C_t) - (D + C_t)x_1(t-2)\sigma(t-1) < 0 \tag{2.13}$$

This reduces to:–

$$(D^2 + C_t D)x_1(t-2)\sigma(t-1) < 0 \tag{2.14}$$

Inequality 2.14 is the basic sliding criteria in *difference* form. When *hitting* occurs $\sigma(t) = 0$, from 2.9:–

$$D = -C_t \tag{2.15}$$

Figure 3 shows the general block diagram of a second order variable structure discrete process, where; $y(t)$ is the process output, $u(t)$ is the process input, $v(t)$ is the auxiliary process input, $r(t)$ is the process setpoint, $x_1(t)$ is the closed loop process error, $\Psi(t)$ is the structure switch gain, $a_{1,2}$, $b_{1,2}$, $E_{1,2}$, $c_{1,2,3}$ are Z polynomial coefficients, $A_{1,2}$, $B_{1,2}$ are the difference polynomial coefficients. This has a *difference* transfer function of the form:–

$$Z^{-2}x_1(t)(D^2 + (A_{1t} + \psi(t)B_{1t}K)D + \Psi(t)B_{2t}K + A_{2t} = 0 \tag{2.16}$$

Note that disturbance and noise are not included in the difference transfer function since VSC of the process rejects functions of this type.

Substitution can be made for D^2 and D from equations 2.16 and 2.15 in the *difference* sliding criteria 2.14, giving:–

$$((B_{2t} + B_{1t}D)\Psi(t)K - C_t^2 + A_{1t}C_t - A_{2t})x_1(t-2)\sigma(t-1) < 0 \tag{2.17}$$

Giving the sliding conditions:–

$$x_1(t-2)\sigma(t-1) < 0 \quad \text{if } (B_{2t} + B_{1t}D)\Psi(t) > E \tag{2.18}$$

$$x_1(t-2)\sigma(t-1) > 0 \quad \text{if } (B_{2t} + B_{1t}D)\Psi(t) < E$$

$$E = (C_t^2 - A_{1t}C_t + A_{2t})/K$$

$$= \text{minimum controller gain}$$

From the set of equations 2.18:–

$$D\Psi(t) = (E - \Psi(t)B_{1t})/B_{2t}, \; B_{2t} \neq 0$$

$$\Psi(t) = (E - \Psi(t)B_{1t})/B_{2t} - \Psi(t-1) \tag{2.19}$$

Equation 2.19 represents a low pass filter and to eliminate **wound up** $\Psi(t)$ must be bounded:–

$$|\Psi(t)| \leq E \tag{2.20}$$

2.3 Stability Criteria

Given parameter range;

$$\begin{array}{c} A_{1min} \leq A_{1t} \leq A_{1max} \\ A_{2min} \leq A_{2t} \leq A_{2max} \\ B_{1min} \leq B_{1t} \leq B_{1max} \\ B_{2min} \leq B_{2t} \leq B_{2max} \end{array}$$

when $\Psi(t) > 0$ the difference polynomial

$$D^2 + (A_{1t} + \Psi(t)B_{1t}K)D + \Psi(t)B_{2t}K + A_{2t} = 0 \tag{2.21}$$

must have no poles outside the '$Z-1$' disc and when $\Psi(t) < 0$ the stable pole **must** also remain within the '$Z-1$' disc.

Figure 2 shows the $Z-1$ disk defined as D. This is the unit disk centered at $(-1,0)$ and the stable region is confined to the area inside the disk.

3 Simulation examples

Simulation results on two types of plants estimated as Auto–Recursive Moving Average (ARMA) process are presented. These results demonstrate the following points:

1. Improvements in closed loop step response of a process with considerable overshoot,

2. The disturbance rejection property of variable structure controllers,

3. Improvements on the process model parameter convergence rate.

Figure 3 depicts the process closed loop block diagram. In all the simulations feedback gain is set to 0.1 and $E = 1.0$.

Process 1:–

$$y(t) = 1.6y(t-1) - 0.63y(t-2) + u(t-1) + v(t-1) - 0.2v(t-2) + c_1e(t-1) + c_2e(t-2) \tag{3.1}$$

Process 2:–

$$y(t) = 1.6y(t-1) - 0.63y(t-2) + u(t-1) - 0.5u(t-2) + v(t-1) - 0.2v(t-2) + c_1e(t-1) + c_2e(t- \tag{3.2}$$

$e(t)$ is a white noise sequence with zero mean. Figures 4 shows the setpoint changes and the auxiliary input common to both processes.

4 Conclusion and further work

The proposed control strategy and hence the discrete variable structure algorithm presented in this paper is particularly useful for process control applications. Its insensitivity to parameter changes and external disturbances provides a more robust form of process control. A unique feature of this type of controller is a fast convergent rate of process model parameter estimate due to the nature of the controlled process input. This is exceptionally useful in circumstances where the process model parameters change from their original values. This type of control strategy is most suitable for realtime process control applications since the VSC design requires very little computation time.

Other estimation algorithms are expected to behave in a similar manner to that utilised in the simulation examples. The algorithm is currently being extended to cater for systems of higher order.

References

[1] Itkis, U., 'Control Systems Of Variable Structure', John Willey & sons, 1976.

[2] Utkin, V.I, 'Variable Structure Systems with Sliding Modes', IEEE Trans. on Automat. Contr., Vol. ac–22, no. 2, pp. 212–220, 1977.

[3] Powel, G.F. and Powel, J.D., 'Digital Control of Dynamic Systems', Addison Wesley, 1980.

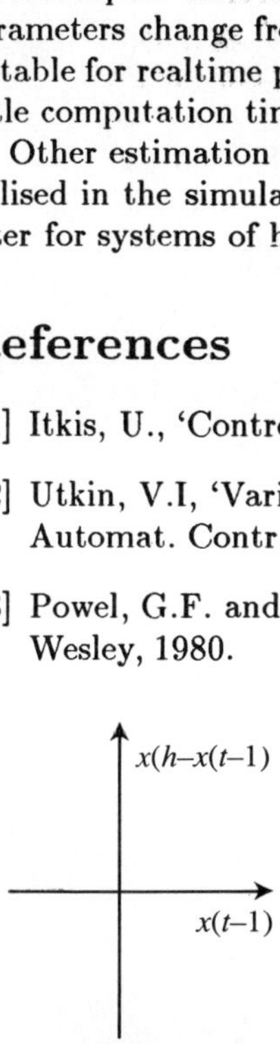

Figure 1
The difference plane.

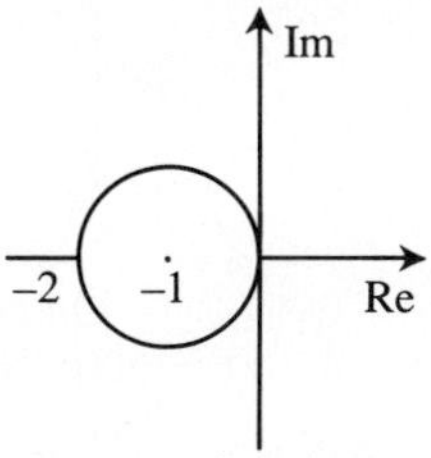

Figure 2
The Z-1 unit disk.

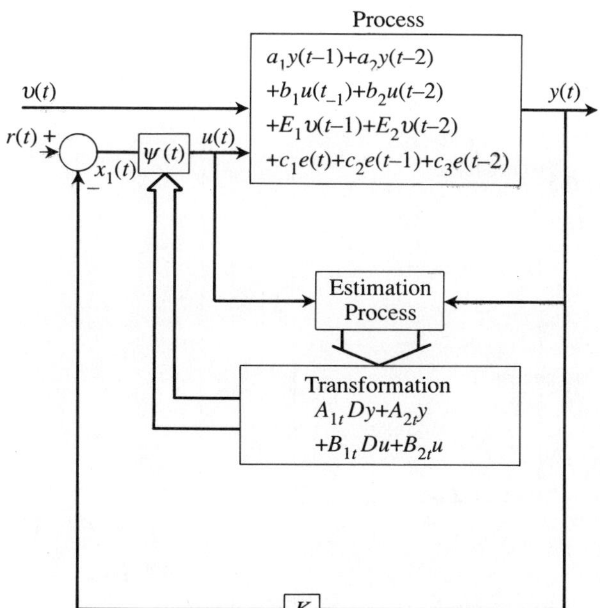

Figure 3 Adaptive variable structure control of a second order discrete process.

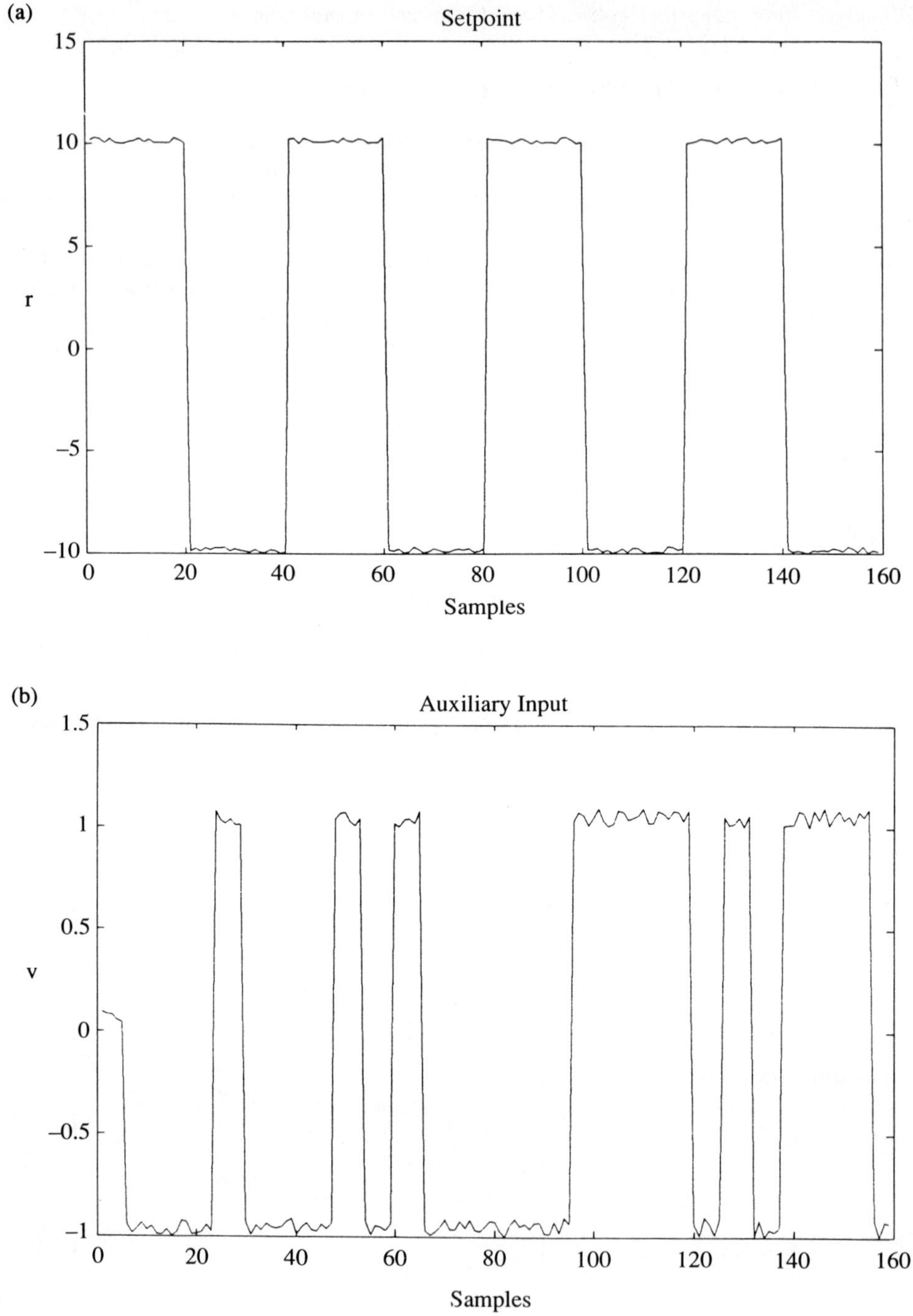

Figure 4 Process input and auxiliary input changes.

Robustness of direct adaptive control systems with slow sampling

P. A. Cook

Abstract

A model reference adaptive control scheme, with a normalised dead zone in the adaptation law, is proposed and its asymptotic behaviour analysed. It is shown that such a system, applied to a linear time–invariant plant, possesses some robustness against unmodelled dynamics and bounded disturbances, provided that the plant is asymptotically stable and the sampling rate is sufficiently low.

1 Introduction

In recent years, there has been much discussion of the robustness, if any, of adaptive control systems, against the presence of unmodelled dynamics in the plant. If, as is usual in practice, the adaptive controller is implemented in discrete–time form, i.e. using sampled data, it is plausible that the effect of any parasitic dynamics acting much faster than the sampling rate should be suppressed, thus making the system more robust. The purpose of this paper is to attempt a quantitative formulation of this process, and to consider how it is affected by various modifications of the adaptive control mechanism.

The paper is structured as follows. In section 2, the description of a uniformly sampled plant is presented as a linear discrete–time input–output relation, which is used in section 3 to formulate a fixed model–following control law, under the assumption that all relevant parameters are known. An adaptation procedure, to deal with the case where they are not, is introduced in section 4, and its asymptotic properties are analysed in section 5, with the results being used, in section 6, to investigate the robustness of the system against disturbances and unmodelled dynamics. We then illustrate the results, in section 7, by considering the particularly simple case of a plant which is sampled slowly enough for its discrete–time model to be approximated by a pure delay of one sampling interval. Finally, in section 8, we present our conclusions and discuss possible generalisations.

2 Plant description

We consider the case of a linear time–invariant plant with scalar input $u(t)$ and scalar output $y(t)$. The output is sampled at instants kT to give

$$y_k = y(kT)$$

and the input is held fixed during each sampling interval, with

$$u(t) = u_k, kT < t < (k+1)T$$

In order to express the input–output relation of the sampled system, we use the forward and backward shift operators q, q^{-1} respectively, so that

$$qy_k = y_{k+1}, \ q^{-1}u_k = u_{k-1}$$

etc. and write

$$A(q^{-1})y_k = q^{-d}B(q^{-1})u_k + \xi_k$$

where A, B are polynomials with

$$A(0) = 1, B(0) \neq 0,$$

d is a positive integer representing the delay in response, and ξ_k includes the effects of disturbances and modelling errors. We denote the degrees of A, B by n, m respectively.

3 Model–reference control

Before specifying the reference model which the controlled plant is intended to follow, we introduce another polynomial H, of degree N, whose purpose is to allow for the incorporation of integral action or other prescribed dynamics, and define new variables

$$u'_k = H(q^{-1})u_k, y'_k = H(q^{-1})y_k.$$

To accommodate N stages of integration, for example, we would have

$$H(q^{-1}) = (1 - q^{-1})^N$$

and other choices of H could be used in order to reject disturbances of known form. We next select a polynomial P, all of whose zeros lie outside the unit circle, and define a filtered tracking error

$$e_k = P(q^{-1})y_k - q^{-d}r_k$$

where r_k is a given reference signal. The object of model–reference control is then to make the tracking error vanish. In order to design a suitable control law, we first compute polynomials P', L from the identity

$$P(q^{-1}) = H(q^{-1})P'(q^{-1}) + q^{-d}L(q^{-1})$$

so that

$$e_k = P'(q^{-1})y'_k + q^{-d}\{L(q^{-1})y_k - r_k\}.$$

Here, the degrees of P, P', L are at most $N+n+d-1$, $n+d-1$, $N-1$ respectively, with L being absent if $N=0$. Next, we obtain polynomials C, α, β, whose respective degrees are at most $d-1$, $n-1$, $m+d-1$, by using

$$\begin{aligned} P'(q^{-1}) &= C(q^{-1})A(q^{-1}) + q^{-d}\alpha(q^{-1}) \\ \beta(q^{-1}) &= C(q^{-1})B(q^{-1}) \end{aligned}$$

so that

$$e_k = q^{-d}\chi_k + \eta_k$$

where

$$\begin{aligned} \chi_k &= \{\beta(q^{-1})u'_k + \alpha(q^{-1})y'_k + L(q^{-1})y_k - r_k\} \\ \eta_k &= C(q^{-1})H(q^{-1})\xi_k. \end{aligned}$$

The appropriate control law can then be obtained by setting

$$\chi_k = 0$$

which eliminates the tracking error if

$$\eta_k = 0$$

for all k.

4 Adaptation

Expanding the polynomials α, β to give

$$\begin{aligned} \alpha(q^{-1}) &= \sum_{k=0}^{n-1} \alpha_k q^{-k} \\ \beta(q^{-1}) &= \sum_{k=0}^{M-1} \beta_k q^{-k} \end{aligned}$$

where $M = m+d$, we can re–express the desired control law by defining a parameter vector

$$\overline{\theta} = (\beta_0, \cdots, \beta_{M-1}, \alpha_0, \cdots, \alpha_{n-1})^{\mathsf{T}}$$

and a regressor vector

$$\phi(k) = (u'_k, \cdots, u'_{k-M+1}, y'_k \cdot \cdots, y'_{k-n+1})^{\mathsf{T}}$$

so that

$$\chi_k = \overline{\theta}^{\mathsf{T}} \phi(k) + L(q^{-1})y_k - r_k.$$

If the components of the parameter vector are not known, we replace it by an estimate $\theta(k)$, which is to be updated through some adaptation procedure, and set

$$\theta^{\mathsf{T}}(k)\phi(k) = r_k - L(q^{-1})y_k$$

as the control law, from which u'_k will be calculated. Then, defining

$$\vartheta(k) = \theta(k) - \overline{\theta}$$

as the difference between the estimated and true parameter vectors, we have

$$e_k = \eta_k - \vartheta^{\mathsf{T}}(k-d)\phi(k-d).$$

For adaptation purposes, however, it is more convenient to use a prediction error defined as

$$\epsilon_k = P'(q^{-1})y'_k - \theta^{\mathsf{T}}(k-1)\phi(k-d)$$

which can be rewritten as

$$\epsilon_k = \eta_k - \vartheta^{\mathsf{T}}(k-1)\phi(k-d).$$

To update the estimated parameter vector, we then use a projection algorithm of the form

$$\theta(k) = \theta(k-1) + \frac{f_k D_k(\epsilon_k) R\phi(k-d)}{\phi^{\mathsf{T}}(k-d)R\phi(k-d) + \lambda_k}$$

where R is an arbitrary positive–definite symmetric matrix, λ_k is a positive scalar, D_k is a dead–zone function to be specified later, and f_k is an adjustable factor which can normally be set at unity but given a different value if the computed first component of $\theta(k)$ would otherwise vanish, as described in Goodwin, Ramadge & Caines (1980). Here, we require only that it should always be chosen to satisfy

$$\delta \leq f_k \leq 2 - \delta$$

for some constant δ such that

$$0 < \delta < 1.$$

5 Boundedness and convergence

We now investigate the properties of the adaptation algorithm, under the assumption that η_k satisfies a known bound

$$|\eta_k| \leq \Delta_k$$

so that we can define the dead–zone function

$$D_k(\epsilon_k) = \begin{cases} \epsilon_k - \Delta_k \mathrm{sgn}(\epsilon_k) & |\epsilon_k| \geq \Delta_k \\ 0 & |\epsilon_k| \leq \Delta_k \end{cases}$$

as used by, among others, Kreisselmeier & Anderson (1986), with the consequence that

$$D_k(\epsilon_k)\{D_k(\epsilon_k) - \epsilon_k + \eta_k\} \leq 0.$$

Then, defining a vector norm by

$$||\theta|| = \sqrt{\theta^\top R^{-1}\theta}$$

it follows that

$$||\vartheta(k)||^2 \leq ||\vartheta(k-1)||^2 - \left\{\frac{\delta D_k(\epsilon_k)}{\nu_k}\right\}^2$$

where

$$\nu_k = \sqrt{\phi^\top(k-d)R\phi(k-d) + \lambda_k}$$

and hence

$$\lim_{k\to\infty} \frac{D_k(\epsilon_k)}{\nu_k} = 0$$

Consequently, from the adaptation law, we obtain

$$\lim_{k\to\infty} ||\vartheta(k) - \vartheta(k-1)|| = 0$$

and so, since

$$e_k - \epsilon_k = \sum_{i=1}^{d-1}\{\vartheta(k-i) - \vartheta(k-i-1)\}^\top \phi(k-d)$$

we also have

$$\lim_{k\to\infty} \frac{e_k - \epsilon_k}{\nu_k} = 0.$$

To make further progress, we next assume that

$$\Delta_k \leq \Gamma\nu_k + \Gamma_0$$

for some non-negative constants Γ, Γ_0, so that

$$\left|\frac{\epsilon_k - D_k(\epsilon_k)}{\nu_k}\right| \leq \Gamma + \frac{\Gamma_0}{\nu_k}$$

and consequently, using the results just established,

$$\overline{\lim_{k\to\infty}} \frac{|e_k| - \Gamma_0}{\nu_k} \leq \Gamma.$$

Also, we assume that there are non-negative constants K, K_0, such that

$$\nu_k \leq K_0 + K \max_{0\leq i\leq k} |e_i|.$$

Now, if e_k becomes unbounded as k increases, then there exists an increasing sequence $j(k)$ such that

$$\lim_{k\to\infty} |e_{j(k)}| = \infty$$

and also

$$\nu_{j(k)} \leq K_0 + K|e_{j(k)}|$$

which gives a contradiction if

$$\Gamma K < 1.$$

Thus, under this condition, e_k remains bounded as k increases, and hence so does ν_k, with the result that Δ_k is bounded and also

$$\lim_{k\to\infty} (e_k - \epsilon_k) = \lim_{k\to\infty} D_k(\epsilon_k) = 0.$$

6 Robustness

In order to explore the robustness issue, we now assume that the term η_k, which includes the effects of disturbances and undermodelling, can be expressed in the form

$$\eta_k = \sum_{i=0}^{\infty} \gamma_i^{\mathsf{T}} \phi(k-d-i) + \psi_k$$

where ψ_k is bounded and the γ_i are constant vectors. Since the unmodelled dynamics are supposed to be asymptotically stable, we further assume that the magnitudes of these vectors decrease exponentially with i, so that there exists a constant μ, with

$$0 < \mu < 1$$

such that

$$\Gamma = \sqrt{\sum_{i=0}^{\infty} \mu^{-i} \gamma_i^{\mathsf{T}} R^{-1} \gamma_i} < \infty.$$

Then, if we generate the sequence of normalisation factors ν_k by setting

$$\lambda_k = \mu \nu_{k-1}^2$$

similarly to Ortega, Praly and Landau (1985), so that

$$\nu_k = \sqrt{\sum_{i=0}^{\infty} \mu^i \phi^{\mathsf{T}}(k-d-i) R \phi(k-d-i)}$$

it follows that η_k satisfies the assumed bound provided that

$$|\psi_k| \leq \Gamma_0$$

for all k. Also, from the equations

$$\begin{aligned} P(q^{-1})y'_k &= H(q^{-1})\{e_k + q^{-d}r_k\} \\ q^{-d}B(q^{-1})u'_k &= A(q^{-1})y'_k - H(q^{-1})\xi_k \end{aligned}$$

it follows, provided the plant is stably invertible and the disturbances and reference signals are bounded, that ν_k satisfies a bound of the required form. Consequently, all we now require is that μ can be chosen so that the condition for boundedness of the tracking error is satisfied. In order to find out if this is possible, we need to examine the dependence of Γ and K on μ. It is clear that the system does possess some robustness since, with K being bounded, the condition will hold if Γ is small enough. However, as the details are in general complicated and model-dependent, we now concentrate on the case where the sampling interval is long compared with the timescale of the dynamics, so that the sampled plant can be approximately modelled as a pure one-step delay.

7 Slow sampling

Let us assume that the plant is represented, in continuous time, by an impulse response $g(t)$, with corresponding transfer function

$$G(s) = \int_0^\infty g(t)\exp(-st)dt$$

so that, ignoring disturbances and initial condition effects, and taking

$$u(t) = 0(t < 0)$$

we have

$$y(t) = \int_0^t g(t-\tau)u(\tau)d\tau.$$

The input-output relation for the sampled plant is then

$$y_k = \sum_{i=0}^{k-1} w_i u_{k-1-i}$$

where

$$w_i = \int_{iT}^{(i+1)T} g(t)dt.$$

Hence, if the plant is asymptotically stable, we have

$$\lim_{T\to\infty} w_0 = G(0), \quad \lim_{T\to\infty} w_i = 0(i > 0)$$

and so, provided that

$$G(0) \neq 0$$

we can, for sufficiently large T, represent the sampled system in the assumed form, with

$$A(q^{-1}) = 1, B(q^{-1}) = w_0, d = 1$$

and

$$\xi_k = \sum_{i=1}^{k-1} w_i u_{k-1-i}.$$

We consider first the case where there is no integral action, and set

$$P(q^{-1}) = H(q^{-1}) = 1$$

so that the reference model is a one–step delay with unit gain. The regressor vector in this case reduces to a scalar

$$\phi(k) = u_k$$

since there is only one unknown parameter in the control law, and we can take $R = 1$. Also,

$$\eta_k = \xi_k$$

and hence

$$\Gamma = \sqrt{\sum_{i=1}^{\infty} \mu^{-i} w_i^2}$$

assuming μ to be chosen so that the summation converges, for which purpose it is sufficient that there exists a positive constant σ such that

$$\lim_{t\to\infty} g(t)\exp(\sigma t) = 0$$

and that

$$\mu > \exp(-2\sigma T).$$

Further, by considering the inverse of the plant, we obtain

$$|\mu_{k-1}| \le \kappa \max_{i\le k} |y_i|$$

where

$$\kappa = \frac{1}{|w_0| - \sum_{i=1}^{\infty} |w_i|}$$

provided that the denominator is positive, as it will be for large enough T, and then, since

$$\nu_k = \sqrt{\sum_{i=0}^{\infty} \mu^i u_{k-1-i}^2}$$

we can set

$$K = \frac{\kappa}{\sqrt{1-\mu}}.$$

Similarly, if integral action is present, we obtain the same value for Γ, since

$$y'_k = \sum_{i=0}^{k-1} w_i u'_{k-i-1}$$

though the value of K will in general be different, and dependent on the reference model. In either case, however, we have

$$\lim_{T \to \infty} \Gamma K = 0$$

so that, if the sampling interval is sufficiently long, the condition $\Gamma K < 1$ is satisfied and hence all signals in the system are bounded.

8 Discussion and conclusions

We have shown the existence of some robustness against disturbances and under-modelling in a certain type of adaptive control scheme, provided that the sampling rate is low enough. The spirit of this work is similar to that of Owens & Wang (1987), who use a pole–assignment control law, in conjunction with an estimator containing a dead zone, and rely on the asymptotically vanishing rate of change of the parameter estimates, together with some other assumptions, to establish robust stability. Here, we take a model–following approach, using a projection–type algorithm, modified by including a relative dead zone depending on a normalisation factor, for the adaptation, and invoke the stability of the plant inverse in the robustness conditions. This assumption of stable invertibility, being essentially what restricts us to the case of slow sampling, should preferably be avoided if possible, perhaps by modifying the model–reference formulation into a self–tuning approach as in Gawthrop & Lim (1982). Other avenues of generalisation would be to employ alternative adaptation algorithms, such as recursive least squares, and to consider the extension of our results to the cases of nonlinear, time–varying and multivariable systems. It is intended that all these possibilities will be explored later.

References

[1] P.J. Gawthrop and K.W.Lim (1982). 'Robustness of self–tuning controllers', *IEE Proc, Pt. D*, 129, pp. 21–29.

[2] G.C.Goodwin, P.J.Ramadge and P.E.Caines (1980). 'Discrete–time multivariable adaptive control',*IEEE Trans. Autom. Control*, AC–25, pp. 449–456.

[3] G.Kreisselmeier and B.D.O.Anderson (1986). 'Robust model reference adaptive control',*IEEE Trans. Autom. Control*, AC–31, pp. 127–132.

[4] R.Ortega, L.Praly and I.D.Landau (1985). 'Robustness of discrete–time direct adaptive controllers', *IEEE Trans. Autom. Control*, AC–30, pp. 1179–1187.

[5] D.H.Owens and L.Wang(1987). 'Robust adaptive controller using first order models under slow sampling speed',*Proc. 26th IEEE Conf. on Decision and Control.* pp. 861–862.

Multivariable reduced–variance pole–assignment self–tuners

M. B. Zarrop R. Davies

Abstract

The pole–assignment approach to self–tuning regulation and control for both SISO and MIMO systems is well known [5] [6] [7] and is inspired by a classical rather than an optimization approach to controller synthesis. This approach, however, can lead to large signal variances when disturbances are present so that, in achieving the specified closed–loop characteristics, the noise rejecting properties of the control system are impaired. For SISO self–tuning systems it has been shown [8] [9] that controller overparametrization can be used to introduce extra degrees of freedom in order to reduce signal variances while preserving the required closed loop pole set. This paper presents preliminary results extending the SISO results to square multivariable systems represented by polynomial matrix models. It is shown that the 'covariance reduction' operation is equivalent to employing a suitably constructed feedforward signal. The self–tuning property is investigated in the regulator case and a simulation example presented in the servo case.

1 Introduction

This paper extends to square multivariable stochastic systems the reduced–variance pole–assignment framework developed in [8] [9] for the SISO case. Polynomial matrix function descriptions [3] [5] [6] are employed (Section 2). Controller overparametrization is used to provide extra degrees of freedom in order to reduce signal variances while preserving the required closed–loop pole set (section 3). It is shown (section 4) that these controllers are equivalent to standard controllers augmented by a feedforward signal of filtered reconstructed noise. Feedforward for disturbance reduction is discussed by Astrom [1] and Clarke [2]. Astrom ([1,p.242]) also refers to the possible use of the arbitrariness in the solution of the pole–assignment identity to reduce sensitivity and disturbance effects.

In certain cases, the optimal reduced–variance pole–assignment regulator coincides with the minimum variance regulator (section [5]). As with the standard pole–assignment controllers, every reduced–variance version can be placed within an explicit self–tuning framework. In the MIMO regulator case (section 6), the validity of the 'self–tuning property' is discussed and it is shown that the property holds for the lowest degree of overparametrization. For some higher degrees of overparametrization the problem is unresolved, but it can be established that beyond a well–defined limit the property cannot hold.

A simulated example is presented of the incremental servo controller.

2 Multivariable pole–assignment

A brief review of pole–assignment for known multivariable systems is presented in order to fix the notation and formulation. Most of the material is well known [5][6]. Systems and controllers are represented by discrete time polynomial matrix models.

The system to be controlled is represented by the linear multivariable difference equation

$$A(z^{-1})\underline{y}(t) = B(z^{-1})\underline{u}(t) + C(z^{-1})\underline{e}(t) \tag{2.1}$$

where $\underline{y}(t)$, $\underline{u}(t)$, $\underline{e}(t)$ are m–vectors of outputs, inputs and disturbances respectively. The disturbance $\underline{e}(t)$ is zero mean white noise with covariance matrix Ω.

The matrices A, B, C are $m \times m$ polynomial matrices in the unit delay operator z^{-1}. All polynomial matrices have the form

$$X = X_0 + X_1 z^{-1} + \cdots + X_{n_X} Z^{-n_X}$$

where n_X is the degree of X (the maximum degree of its polynomial elements). The polynomial matrices A, B, C satisfy the conditions:

1. $A_0 = I = C_0$
2. A, B left coprime
3. $\det C(z) = 0 \rightarrow |z| > 1$

If the minimum time delay in any loop is $k(\geq 0)$ then B_i is the zero matrix for $0 \leq i \leq k-1$ and $B_k \neq 0$. Exact knowledge of k is not crucial when a pole assignment approach is taken to system control and k is usually set to unity within a self–tuning framework [5].

A linear digital controller structure for (2.1) is

$$F(z^{-1})\underline{u}(t) + G(z^{-1})\underline{y}(t) = H(z^{-1})\underline{r}(t) \tag{2.2}$$

where $\underline{r}(t)$ is a given m–vector of set points and F, G, H are $m \times m$ polynomial matrices ($F_0 = I$) chosen so that the resulting closed–loop poles are suitably allocated and there is zero steady state output tracking error.

In order to obtain the closed–loop system equation corresponding to (2.1) and (2.2) standard pseudo–commutativity relations are used of the form:

$$X\tilde{Y} = Y\tilde{X} \tag{2.3a}$$

where (X, Y) are left coprime $m \times m$ polynomial matrices and $(\tilde{X}, \tilde{Y})$ are the associated right coprime $m \times m$ polynomial matrices. Further,

$$deg\tilde{X} = degX; deg\tilde{Y} = degY \tag{2.3b}$$

$$det\tilde{X} = detX; det\tilde{Y} = detY \tag{2.3c}$$

In the following it is assumed that the required matrix inverses exist. Let A, B, $\tilde{A}$, $\tilde{B}$ and also F, G, $\tilde{F}$, $\tilde{G}$ satisfy (2.3) and define the polynomial matrices $T_{(L)}$, $T_{(R)}$ by:

$$T_{(L)} = A\tilde{F} + B\tilde{G} \tag{2.4a}$$
$$T_{(R)} = F\tilde{A} + G\tilde{B} \tag{2.4b}$$
$$degT_{(R)} = degT_{(L)} \tag{2.4c}$$

where $\tilde{F}_0 = I$, $\tilde{B}_0 = \tilde{B}_1 = \cdots = \tilde{B}_{k-1} = 0$ and the argument z^{-1} is omitted for brevity.

From (2.1)–(2.4) the closed–loop equations take either the left matrix function description (LMFD) form

$$\underline{y}(t) = \tilde{F}T_{(L)}^{-1}BF^{-1}H\underline{r}(t) + \tilde{F}T_{(L)}^{-1}C\underline{e}(t) \tag{2.5a}$$
$$\underline{u}(t) = \tilde{G}T_{(L)}^{-1}AG^{-1}H\underline{r}(t) - \underline{G}T_{(L)}^{-1}C\underline{e}(t) \tag{2.5b}$$

or the right matrix function description (RMFD) form:

$$\underline{y}(t) = \tilde{B}T_{(R)}^{-1}H\underline{r}(t) + \tilde{B}T_{(R)}^{-1}FB^{-1}C\underline{e}(t) \tag{2.6a}$$
$$\underline{u}(t) = \tilde{A}T_{(R)}^{-1}H\underline{r}(t) - \tilde{A}T_{(R)}^{-1}GA^{-1}C\underline{e}(t) \tag{2.6b}$$

Provided (2.3) and (2.4) hold, the two forms may be mixed if required.

From (2.5) and (2.6)

$$det(\tilde{F}_{(L)}^{-1}BF^{-1}H) = det(\tilde{B}T_{(R)}^{-1}H) = detB.detH/T \tag{2.7a}$$

where

$$T = detT_{(L)} = detT_{(R)} \tag{2.7b}$$

Hence the closed–loop poles coincide with the zeros of T and may be specified through selecting $T_{(L)}$ or $T_{(R)}$. For example, given A, B, C, $T_{(L)}$, the diophantine equation (2.4a) is solved for $\tilde{F}$, $\tilde{G}$ and then, F, G are obtained from (2.3). Note, however, that there are an infinity of solutions F, G satisfying (2.4a), each giving rise to a $T_{(R)}$ satisfying (2.7b). From (2.5) and (2.6) each solution gives rise to particular interaction behaviour and disturbance structure. This is exploited below.

3 Reduced–variance pole–assignment

The diophantine equation (2.4a) yields a unique solution $\tilde{\mathcal{F}}$, $\tilde{\mathcal{G}}$ (and thence $\mathcal{F}$, $\mathcal{G}$) if

$$deg\mathcal{F} = n_b - 1 = deg\tilde{\mathcal{F}} \tag{3.1a}$$
$$deg\mathcal{G} = n_a - 1 = deg\tilde{\mathcal{G}} \tag{3.1b}$$
$$degT_{(L)} \leq n_a + n_b - 1 \tag{3.1c}$$

As in the SISO case [8][9], this solution may give rise to large output and/or input variances.

The general solution of (2.4a) is

$$\tilde{F} = \tilde{\mathcal{F}} + \tilde{B}P; \tilde{G} = \tilde{\mathcal{G}} - \tilde{A}P \tag{3.2}$$

where P is an arbitrary $m \times m$ polynomial matrix [4, p.45]. Hence the cost function

$$J_1(P) = trace[Q_1 cov\underline{y}(t) + Q_2 cov\underline{u}(t)] \tag{3.3}$$

where Q_1, Q_2 are $m \times m$ symmetric positive semi definite weighting matrices for the output and input covariances respectively, is a quadratic function of the $m^2(n_p+1)$ elements of P. Specifically,

$$\begin{aligned} J_1(P) = E[&(\tilde{F}T_{(L)}^{-1}C\underline{e}(t))^T Q_1(\tilde{F}T_{(L)}^{-1}C\underline{e}(t)) \\ &+(\tilde{G}T_{(L)}^{-1}C\underline{e}(t))^T Q_2(\tilde{G}T_{(L)}^{-1}C\underline{e}(t))] \end{aligned} \tag{3.4}$$

with $\tilde{F}$, $\tilde{G}$ given by (3.2).

Note that

$$\deg F = n_b + n_p = \det\tilde{F} \tag{3.5a}$$

$$\deg G = n_a + n_p = \det\tilde{G} \tag{3.5b}$$

so that the standard case can be considered as corresponding to $n_p = -1$. For $n_p = 0, 1$ there are m^2, $2m^2$ degrees of freedom respectively and so on.

From (3.4), the stationarity conditions for the minimizing P are the set of linear equations

$$\begin{aligned} &E[(z^{-s}\tilde{B}N_{ij}T_{(L)}^{-1}C\underline{e}(t))^T Q_1(\tilde{F}T_{(L)}^{-1}C\underline{e}(t)) \\ &- \quad (z^{-s}\tilde{A}N_{ij}T_{(L)}^{-1}C\underline{e}(t))^T Q_2(\tilde{G}T_{(L)}^{-1}C\underline{e}(t))] = 0 \\ &\quad (i,j = 1,\cdots,m; s = 0,\cdots n_p) \end{aligned} \tag{3.6}$$

where N_{ij} is the $m \times m$ matrix with unity as the (i,j) entry and zeros elsewhere. As in the SISO case, the solution involves the inversion of a $m^2(n_p+1)$–order matrix whose elements are complex integrals.

4 A feedforward interpretation

The overparametrized controller polynomial matrices in (3.2) lead to a feedforward interpretation. Given $T_{(L)}$, P the matrices F, G, $T_{(R)}$ can be derived. Form (2.4b), for some P'

$$\begin{aligned} F &= F' + P'B; \\ G &= G' - P'A \end{aligned} \tag{4.1}$$

where F', G' is the minimum degree solution of (2.4b). In general, F' and $\mathcal{F}$ are not equal. From (2.2) and (4.1) using (2.1)

$$\begin{aligned} F'\underline{u}(t) + G'\underline{y}(t) &= H\underline{r}(t) + P'(A\underline{y}(t) - B\underline{u}(t)) \\ &= H\underline{r}(t) + P'C\underline{e}(t) \end{aligned} \tag{4.2}$$

The role of P' is seen as filtering a reconstructed disturbance term in feedforward to the closed–loop system. Using measured/estimated disturbance feedforward for reducing the effect of disturbances is a standard technique in control engineering [1, pp. 123,213] [2, pp. 36,60].

5 Minimum variance regulator

Consider the standard pole assignment regulator case derived by assuming

$$\begin{aligned} \underline{r}(t) &= 0 \\ T_{(L)} &= CT' \end{aligned} \tag{5.1}$$

From (2.5a)

$$\underline{y}(t) = \tilde{F}T'^{-1}\underline{e}(t) \tag{5.2}$$

where

$$CT' = A\tilde{F} + B\tilde{G} \tag{5.3}$$

If $Q_2 = 0$, the stationarity conditions (3.6) become:

$$\begin{gathered} E[(z^{-s}\tilde{B}N_{ij}T'^{-1}\underline{e}(t))^T Q_1(\tilde{F}T'^{-1}\underline{e}(t))] = 0 \\ (i,j = 1,\cdots,m; s = 0,\cdots,n_p) \end{gathered} \tag{5.4}$$

These conditions are satisfied if $B = z^{-k}B'$ ($B'_0 \neq 0$) and there exists an $m \times m$ polynomial matrix R such that

$$\tilde{F} = RT' \tag{5.5a}$$

$$n_r \leq k - 1 \tag{5.5b}$$

From (3.2), (5.5a)

$$z^{-k}B'P = RT' - \mathcal{F} \tag{5.6}$$

In order to be able to select a suitable R satisfying (5.5b), (5.6) and yielding P:

$$n_b = k \quad (i.e. B = z^{-k}B_k) \tag{5.7a}$$

$$n_r = k - 1 \tag{5.7b}$$

so that

$$n_p = n_{t'} - 1 \tag{5.8}$$

From (5.3), (5.5a), (5.7a):

$$C = AR + z^{-k}G' \tag{5.9a}$$

where

$$B_k\tilde{G} = G'T' \tag{5.9b}$$

This is the standard partition for the minimum variance regulator [2,p.72] and leads to the following MIMO version of the SISO result in [8]:
THEOREM: If

$$\begin{aligned} B(z^{-1}) &= z^{-k}B_k, \\ n_p &\geq n_{t'} - 1 \end{aligned} \tag{5.10}$$

then the reduced–variance pole–assignment regulator coincides with the minimum variance regulator and the optimal P_i are zero for $n_{t'} \leq i \leq n_p$.

6 Self–tuning regulation

The standard MIMO pole–assignment self–tuning regulator [5] is easily extended to include the reduced variance facility. A possible on–line algorithm is based on the linear model

$$\mathbf{A}\underline{y}(t) = \mathbf{B}\underline{u}(t) + \underline{\varepsilon}(t) \tag{6.1}$$

where $\mathbf{A}_0 = I$ and only $\mathbf{B}_0$ is assumed to be zero.

Algorithm Set $n_A(=n_a)$, $n_B(=n_b)$, n_p, Q_1, Q_2, T'. At each sample time:

1. Estimate $\hat{\mathbf{A}}$, $\hat{\mathbf{B}}$ using RLS on the model (6.1) to minimize $\underline{\varepsilon}(t)^T\underline{\varepsilon}(t)$

2. Solve for the standard $\hat{\tilde{\mathbf{F}}}$, $\hat{\tilde{\mathbf{G}}}$ from the identity

$$T' = \hat{\mathbf{A}}\hat{\tilde{\mathbf{F}}} + \hat{\mathbf{B}}\hat{\tilde{\mathbf{G}}} \tag{6.2}$$

   If $n_p = -1$, go to (6).

3. Solve the pseudo commutativity conditions (2.2) for $\hat{\tilde{\mathbf{A}}}$, $\hat{\tilde{\mathbf{B}}}$.

4. Solve for P from the stationarity conditions

$$\begin{aligned} E\Big[&\left(z^{-s}\hat{\tilde{\mathbf{B}}}N_{ij}T'^{-1}\underline{w}(t)\right)^T Q_1 \left(\hat{\tilde{F}}T'^{-1}\underline{w}(t)\right) \\ &- \left(z^{-s}\hat{\tilde{\mathbf{A}}}N_{ij}T'^{-1}\underline{w}(t)\right)^T Q_2 \left(\hat{\tilde{G}}T'^{-1}\underline{w}(t)\right)\Big] = 0 \\ &(i, j = 1, \cdots, m;\ s = 0, \cdots, n_p) \end{aligned} \tag{6.3a}$$

   where

$$\hat{\tilde{F}} = \hat{\tilde{\mathbf{F}}} + \hat{\tilde{\mathbf{B}}}\hat{P},\ \hat{\tilde{G}} = \hat{\tilde{\mathbf{G}}} - \hat{\tilde{\mathbf{A}}}\hat{P} \tag{6.3b}$$

   and the m–vector $\underline{w}(t)$ is white noise, covariance I.

5. Calculate $\hat{\tilde{F}}$, $\hat{\tilde{G}}$ from (6.3b).

6. Calculate $\hat{F}$, $\hat{G}$ from (2.3).

7. Compute $\underline{u}(t)$ from the control law

$$\hat{F}\underline{u}(t) + \hat{G}\underline{y}(t) = 0 \tag{6.4}$$

8. Input $\underline{u}(t)$ to the system (2.1). Observe $\underline{y}(t+1)$ and return to (1).

Note that RLS is used even though $C = I$ in general. For $n_p = 0$, the standard proof of the *self-tuning property* [5] shows that, if convergence takes place, then, under fairly weak conditions, the estimated regulator polynomial matrices converge to the solution of

$$CT' = A\tilde{\mathbf{F}} + B\tilde{\mathbf{G}} \tag{6.5}$$

if

$$n_{t'} \leq n_a + n_b - n_0 - 1 \tag{6.6}$$

and (3.1a,b) are satisfied.

As in the standard MIMO case [5], *the same proof goes through in the MIMO case for $n_p = 0$ where P_0 is arbitrary.*

For $n_p > 0$, the standard proof breaks down but simulation indicates that the self-tuning property still holds provided the stationarity conditions (6.3) are satisfied on-line. If, however,

$$n_p \geq n_{t'} + n_c \tag{6.7}$$

the self-tuning property *cannot* hold. This is easily shown in the SISO case, using (6.5) and (6.2) at convergence. Then

$$\frac{\tilde{G}}{\tilde{F}} = \frac{CA - A}{B - CB} \tag{6.8}$$

Cancelling z^{-1} from the RHS numerator and denominator, if (6.7) holds then $\tilde{G}$, $\tilde{F}$ must have more than $n_{t'}$ zeros in common. This is ruled out by (6.2). Hence, if (6.7) holds, it may be necessary to estimate C.

7 Incremental control

Integral action may be incorporated into the regulator or controller in order to reject slowly varying load disturbances and ensure zero steady state tracking error. This corresponds to a control law of the form

$$\Delta F\underline{u}(t) + G\underline{y}(t) = H\underline{r}(t) \tag{7.1a}$$

where

$$\Delta = 1 - z^{-1} \tag{7.1b}$$

$$H(1) = F(1)B(1)^{-1}T_{(L)}(1)\tilde{F}(1)^{-1} \tag{7.1c}$$

The only difference to the previous development in Sections 2,3 is that F is replaced by ΔF, requiring that n_g is increased by unity. The details are omitted here.

The following example illustrates the effect of incremental reduced variance control on a two-input, two-output system.

Example 7.1 Consider the case $m = 2$:

$$A = \begin{bmatrix} 1 - 0.085z^{-1} & 0.05z^{-1} \\ 0.05z^{-1} & 1 - 0.07z^{-1} \end{bmatrix}, \quad B = \begin{bmatrix} z^{-1} - 0.1z^{-2} & 0.005z^{-2} \\ 0.005z^{-2} & z^{-1} - 0.1z^{-2} \end{bmatrix}$$

$C = diag(1 + 0.8z^{-1}, 1 + 0.5z^{-1})$, $\Omega = diag(0.01, 0.01)$, $T_{(L)} = CT'$
where $T' = diag(1 - 0.91z^{-1}, 1 - 0.5z^{-1})$.
The standard solution ($n_p = -1$) yields the covariance values:

$$\text{cov}\underline{y} = \begin{bmatrix} 0.0389 & -0.0370 \\ -0.0370 & 0.0591 \end{bmatrix}, \quad \text{cov}\underline{u} = \begin{bmatrix} 0.0601 & -0.0554 \\ -0.0554 & 0.0641 \end{bmatrix} \tag{7.2}$$

Selecting $Q_1 = I$, $Q_2 = 0$, $n_p = 0$ yields:

$$P_0 = \begin{bmatrix} 1.310 & -0.629 \\ -1.950 & 0.509 \end{bmatrix} \tag{7.3a}$$

$$\text{cov}\underline{y} = \begin{bmatrix} 0.0103 & -0.0018 \\ -0.0018 & 0.0130 \end{bmatrix}, \quad \text{cov}\underline{u} = \begin{bmatrix} 0.0107 & -0.0013 \\ -0.0013 & 0.0121 \end{bmatrix} \tag{7.3b}$$

Note that, in this case, reducing the output variances leads to a reduction in the input variances without penalizing the latter in the cost function. This does not happen in general.

For the input–output behaviour of the closed–loop system, the square wave reference signals ($t \geq 0$) are given by:

$$r_1(t) \quad \begin{aligned} &= 1 \quad && 100k \leq t < 100(k+1) \\ &= -1 && \text{otherwise} \end{aligned}$$

$$r_2(t) \quad \begin{aligned} &= 1 \quad && 100(k+1/2) \leq t < 100(k+3/2) \\ &= -1 && \text{otherwise} \end{aligned}$$

for $k = 0, 2, 4, \cdots$ The standard pole–assignment incremental controller is used initially and an optimal constant P(i.e.$n_p = 0$) is used from $t = 200$. The choice of H is the constant matrix (7.1c).

The simulation shows clearly the reduction in the output variances. There is also little output coupling in this case. This is purely fortuitous and in general difference choices of n_p will lead to different (possibly adverse) interaction effects. In such cases it may be useful to choose a nondiagonal $T_{(L)}$, consistent with the required pole set, and a nonconstant H to minimize interaction according to appropriate criteria. In [6], one such method is given for selecting H with $T_{(L)}$ diagonal. This is not pursued further here.

8 Concluding remarks

The reduced variance pole assignment framework developed in [8] [9] for SISO system carries over in a straightforward way to the control (either incremental or non-incremental) of MIMO systems. It has been shown that the synthesized controller is equivalent to a standard pole assignment controller augmented by a feedforward signal. In the case of the self-tuning regulator, the self-tuning property holds for the minimal degree of overparametrization but fails if the degree of overparametrization is too high. Intermediate cases remain unresolved. In the combined servo/regulator case, the *reduced variance* strategy affects the closed-loop output interaction and this should be taken into account when designing the precompensator.

References

[1] Astrom, K.J. and Wittenmark, B., "Computer Controlled Systems". Prentice-Hall, 1984.

[2] Harris, C.J. and Billings, S.A. (eds.), "Self-Tuning and Adaptive Control: Theory and Applications". Peter Peregrinus, 1981.

[3] Kailath, T., "Linear Systems". Prentice-Hall, 1980.

[4] Kucera, V., "Discrete Linear Control". Academia Praha, 1979.

[5] Prager, D. and Wellstead, P.E., Multivariable Pole-Assignment Self-Tuning Regulators. IEE Proc., vol. 128, Part D, No. 1, 9-18, 1980.

[6] Wahab, W. and Wellstead, P.E., Multivariable Self-Tuning for Robotic Servo Applications. Proc. 2nd IFAC Workshop on Adaptive Systems in Control and Signal Processing, Lund, Sweden, 1986.

[7] Wellstead, P.E., Prager, D. & Zanker, P., Self-Tuning Pole/Zero Assignment Regulators. Int.J.Control, 30, 1-26, 1971.

[8] Zarrop, M.B. and Fischer, M., Reduced-Varianced Pole-Assignment Self-Tuning Regulation. Int.J.Control, 42,5, 1013-1033, 1985.

[9] Zarrop, M.B. & Karafakioglu, A., Reduced-Variance Servo Self-Tuning. Proc. 2nd IFAC Workshop on Adaptive Systems in Control & Signal Processing, Lund, Sweden, 1986.

Predictive control with costed inputs

K. Warwick D.H. Owens

Abstract

Predictive control techniques are generally founded on the assumption that future control inputs are well classified. A finite output horizon is specified and a prediction of future outputs is made within this horizon. By suitable selection of the future control inputs, it is possible to tailor the output signal such that it meets pre-specified performance objectives.

An end result of the employment of a predictive controller is that at each sampling time instant, a set of predictions of future control actions to be taken is available. This means that at any particular time instant a group of predicted values for the control input at that time instant are accessible, these predictions having been found at previous time instants. The control input actually applied can therefore be found in terms of a costed summation of the predicted input signal values.

In this paper the Generalised Predictive Control method is employed as a basic design scheme and a cost function is derived which is founded on the use of combinations of predicted control values to form a new control input signal. It is shown that the optimal solution to this cost function is a convex combination of the most recently calculated input signal value along with the previously obtained predictions.

The optimal solution therefore reduces to the original Generalised Predictive Control solution when the costing on the predicted control actions is set to zero.

1 Introduction

A considerable number of advances have recently been made in the area of predictive control techniques, when applied to the adaptive discrete–time control of systems [2], such that systems with variable dead–time, open–loop unstable or non minimum–phase characteristics can be readily dealt with. With a basic assumption that future input signals are well classified, predictive controllers, based on a finite horizon output prediction, can cope with either univariate or multivariate systems [7]. In fact, many of the predictive control schemes which do exist, e.g. [9, 13], can be seen to be simply related, in many cases the only differences which do exist are due to output horizon selection and control classification.

When required to operated within an adaptive, self–tuning control scheme [1], it is necessary that the parameters actually employed within the predictive controller are periodically updated in response to variations in the system characteristics. To this end an autoregressive plant model, based on integrated plant input/output signal values, is recursively estimated such that the coefficients in this model are used as a basis for controller parameter updating. An advantage in the use of integrated data is that offsets, caused by undesirable disturbances, are both more adequately modelled and more easily dealt with. The controller described in this paper is based on the use of an integrated data plant model, and although only univariate systems are considered, the controller is just as applicable in a multivariate environment, an explanation is rather more detailed however.

In Clarke and Zhang [3] it was shown that self–tuning predictive controllers e.g. [2, 4] can exhibit improved robustness properties, when compared with such as pole placement [6], minimum–variance [1] or deadbeat [10] self–tuning controllers. With predictive controllers such as these, the control action can be specified to ensure that no variations occur outside the selected horizon. A positive result is then that load disturbance effects are automatically removed by the inherent integral action.

It is a consequence of the use of a predictive controller that at each sampling instant a set of present and future predicted control inputs is calculated, the size of the set depending directly on the control horizon. So, at any one time instant not only is the most recently calculated control signal available, but also available is a number of predictions of the control input, at that instant, each calculated at a different sampling instant in the past. The control input to be actually applied at any particular instant can therefore be found in terms of a summation of weighted control input predictions [8, 12].

The controller described in this paper is founded on a generalised predictive control scheme [2], however control input predictions made at different time instants are weighted and combined to form the actual applied control input. This results in the provision of a filtering action on the input signals and improves robustness properties by reducing the effects of temporarily unmeasured dynamics, poor parameter estimates or spurious control signals. Further, due to the choice of weightings used, the controller can be made plant specific, in order to fit in more closely with likely plant behaviour. Also, because the emphasis is no longer placed on a direct output to input feedback path every sampling instant, the technique is applicable for use within a fault tolerant control design.

2 System definition

The single input–single output linear system under control is modelled by means of the relationship

$$Ay(t) = Bu(t-1) + Ce(t) \tag{2.1}$$

where $\{u(t) : t \in \tau\}$ and $\{y(t) : t \in \tau\}$ are the system input and output sequences respectively, and $\{e(t) : t \in \tau\}$ is a disturbance signal. The disturbance is modelled as $\delta e(t) = \varepsilon(t)$, in which $\delta = 1 - z^{-1}$, such that the zero mean white noise signal, $\varepsilon(t)$, can contain an element of bias [2].

The polynomials A, B and C are defined as:

$$\begin{aligned} A &= 1 + a_1 z^{-1} + a_2 z^{-2} + \cdots + a_{n_a} z^{-n_a} \\ B &= b_0 + b_1 z^{-1} + b_2 z^{-2} + \cdots + b_{n_a} z^{-n_b} \end{aligned} \tag{2.2}$$

and C is, in its general form, similar to A, however in order to simplify the controller description, an assumption has been made that $C = 1$. A more general system, with $C \neq 1$ can be readily taken into account in the controller to be described though. Also z^{-1} is the backward shift operator, such that $z^{-i}y(t) = y(t-i)$.

The system equation (2.1) can be reconsidered in terms of a power series expansion, rather than a transfer function. This can be seen if both sides of (2.1) are divided by the polynomial A, then it follows that

$$y(t) = \delta H u(t-1) + A^{-1}e(t) \tag{2.3}$$

in which

$$H = (\delta A)^{-1} B = h_0 + h_1 z^{-1} + \cdots \tag{2.4}$$

In the controller description which immediately follows, it is assumed that the parameters a_i, b_i and therefore h_i are known time invariant values. The possibility of an adaptive controller being formed in order for the controller response to be modified, in response to time varying system characteristics, is discussed at a later stage.

3 General predictive control form

Consider the polynomials:

$$E_k = 1 + e_1 z^{-1} + e_2 z^{-2} + \cdots + e_{n_e} z^{-n_e} \tag{3.1}$$

and

$$F_k = f_0 + f_1 z^{-1} + f_2 z^{-2} + \cdots + f_{n_f} z^{-n_f} \tag{3.2}$$

where $k \geq 1$ is a specified integer value, such that E_k and F_k are unique polynomials, for any particular k and A; and are found from the Diophantine equation

$$\delta E_k A + z^{-k} F_k = 1 \tag{3.3}$$

In (3.3) we have that, for a unique solution to be found:

$$n_e = k - 1 \text{ and } n_f = n_a \tag{3.4}$$

By multiplying (2.1) throughout by $z^{+k\delta}E_k$, it follows that:

$$\delta E_k A y(t+k) = B \delta E_k u(t+k-1) + E_k \varepsilon(t+k) \tag{3.5}$$

remembering that $C = 1$.

Once the Diophantine equation (3.3) is applied to (3.5) the system equation becomes

$$y(t+k) = BE_k\delta u(t+k-1) + F_k y(t) + E_k\varepsilon(t+k) \tag{3.6}$$

which means that all of the disturbance terms are future values, due to $n_e = k-1$.

If a prediction is made, at the present time instant, t, of what the output signal will be at time instant $t+k$, the optimal solution is found to be (3.6) with the disturbances removed, that is

$$\hat{y}(t+k/t) = BE_k\delta u(t+k-1/t) + F_k y(t) \tag{3.7}$$

The output prediction for time $t+k$ is based only on *a priori* known output signal values at the present time instant, t. Also, the prediction is dependent on absolute knowledge of the future input signal values, up to time $t+k-1$.

It is worth noting that by multiplying the Diophantine equation (3.3) throughout by H, we have that

$$E_k B + z^{-k}F_k H = H \tag{3.8}$$

such that on expansion of the prediction equation (3.7) this leads to:

$$\begin{aligned}\hat{y}(t+k/t) \quad = \quad & F_k y(t) + [BE_k - (h_0 + h_1 z^{-1} + \cdots + h_{k-1}z^{-(k-1)})\delta u(t+k-1) \\ & +[h_0 + h_1 z^{-1} + \cdots + h_{k-1}z^{-(k-1)}]\delta u(t+k-1)\end{aligned} \tag{3.9}$$

in which $[h_0+h_1z^{-1}+\cdots+h_{k-1}z^{-(k-1)}]\delta u(t+k-1)$ is a summation of future terms, whereas $Fy(t) + [BE_k - (h_0 + h_1 z^{-1} + \cdots + h_{k-1}z^{-(k-1)})\delta u(t+k-1)] = f(t+k)$ is a summation of known values at time instant t. It follows then that:

$$y(t+k) = \hat{y}(t+k/t) + E_k\varepsilon(t+k) \tag{3.10}$$

in which the output prediction is given by:

$$\begin{aligned}\hat{y}(t+k/t) = {} & [h_0 + h_1 z^{-1} + \cdots + h_{k-1}z^{-(k-1)}] \\ & \delta u(t+k-1) + f(t+k)\end{aligned} \tag{3.11}$$

The value $\hat{y}(t+k/t)$ is a prediction of the output signal at time $t+k$, made at time t by making use of information available up to and including that at time t. So, a k step-ahead prediction is made, where k is a general integer value. Let a maximum value of $N \geq 1$ be set for k, such that at any time t, a set of predictions can be found over the finite horizon $1 \leq k \leq N$. It is then possible to rewrite (3.10) and (3.11) in terms of the set of predictions made for $k = 1$ to N, i.e.

$$\underline{y} = G\underline{\overline{u}} + \underline{f} + \underline{\varepsilon} \tag{3.12}$$

where

$$\begin{aligned}\underline{y}^{\mathsf{T}} &= [y(t+1), \cdots, y(t+N)] \\ \underline{\overline{u}}^{\mathsf{T}} &= [\delta u(t), \cdots, \delta u(t+N-1)] \\ \underline{f}^{\mathsf{T}} &= [f(t+1), \cdots, f(t+N)] \\ \underline{\varepsilon}^{\mathsf{T}} &= [E_1\varepsilon(t+1), \cdots, E_N\varepsilon(t+N)]\end{aligned} \tag{3.13}$$

in which the matrix G is of dimension $N \times N$ and has rows defined, in terms of the ith row from the top, as

$$G_i = [g_{i-1}, g_{i-2}, \cdots, g_0, 0, \cdots, 0] \quad (3.14)$$

$$g_j = h_j \quad (3.15)$$

Having written the set of output predictions in a form which is appropriate, i.e. as in (3.12), the general predictive controller can be considered in terms of the relevant control objective, which entails finding a minimum value of the cost function

$$J = \frac{1}{2}[||\underline{y} - \underline{v}||_Q^2 + ||\overline{\underline{u}}||_R^2] \quad (3.16)$$

where $\underline{v}$ is a vector containing future, specified, reference input signal values.

$$\underline{v}^{\mathsf{T}} = [v(t+1), \cdots, v(t+N)] \quad (3.17)$$

By differentiating J, it is found that a cost function minimum is obtained when

$$\overline{\underline{u}} = (G^{\mathsf{T}}QG + R)^{-1} G^{\mathsf{T}} Q(\underline{v} - \underline{f}) \quad (3.18)$$

By putting equation weight on all of the output predictions from $y(t+1)$ to $y(t+N)$, $Q = 1$, and further if $r(k)$ is a costing applied to the control inputs, such that $R = \text{diag}\{r(k)\}$, then

$$\overline{\underline{u}} = (G^{\mathsf{T}}G + R)^{-1} G^{\mathsf{T}} (\underline{v} - \underline{f}) \quad (3.19)$$

It can be seen from (3.13) and (3.19) that as $\overline{\underline{u}}$ contains a common δ term, so the control input predictions found each have a direct integrating action.

The standard general predictive controller [2] actually applied at time instant t, given information available up to and including that at time t, is

$$\delta u_{GPC}(t/t) = e_1^{\mathsf{T}} \overline{\underline{u}} \quad (3.20)$$

where

$$e_1^{\mathsf{T}} = [1, 0, \cdots, 0] \quad (3.21)$$

which means that only the first row of $(G^{\mathsf{T}}G + R)^{-1}G^{\mathsf{T}}$ is employed, despite the fact that the whole of $(G^{\mathsf{T}}G + R)$ must be inverted.

4 Input costing

It can be seen from (3.19) that a vector of present and future control inputs is available at each time instant t. However the standard general predictive controller only makes use of one of the elements in the $\overline{\underline{u}}$ vector.

The vast majority of the computational effort involved in finding $u_{GPC}(t/t)$ is that due to the inversion of the $N \times N$ matrix $(G^{\mathsf{T}}G + R)$. For only a small proportional increase in computational effort it is though possible to find the complete vector $\underline{u}$, such that at time instant t, a set of control input signals is found, i.e.

$$\{u(t+k-1/t) : \ k = 1, \cdots, N\}. \tag{4.1}$$

As an example of the elements in this set, when $k = 2$, $u(t+1/t)$ denotes a prediction of the control input to be applied at time instant $t+1$, which is made at time instant t.

It follows directly that at a particular time instant t, not only is the most recently calculated value to be applied at time t available, but also available are the predictions of the value to be applied at time t, which were made in the past.

The control input actually applied at time instant t can be formed from the summation of weighted versions of the control input suggestions for that time instant, which results in an applied signal in which the effect of one–off signal values is much reduced. If the actual control input signal to be applied at time t is denoted by $u'(t)$, then

$$u'(t) = \frac{\sum_{k=1}^{N} \lambda(t) \cdot u(t/t-k+1)}{\sum_{k=1}^{N} \lambda(k)} \tag{4.2}$$

in which $\lambda(k)$ is an operator specified input signal weighting and $u(t/t-k+1)$ denotes the value of control input signal predicted for time instant t, the prediction being made at time instant $t-k+1$.

Employment of the weighting function $\lambda(k)$ can be seen as modifying the original general predictive controller by taking into account previously predicted control input values to be applied at that particular time. In fact the standard general predictive control algorithm [2] is a special case of (4.1) in which $\lambda(k) = 1$ for $k = 1$, whereas $\lambda(k) = 0$ for $k = 2, \cdots, N$.

The type of control input requested, with costed inputs taken on board, can be regarded as requiring a modification to the cost function considered, such that now

$$\begin{aligned} J &= \frac{1}{2}[\sum_{k=1}^{N}\{y(t+k/t) - v(t+k)\}^2 + \sum_{k=1}^{N} r(k)\{\delta u(t+k-1/t)\}^2 \\ &+ \sum_{k=1}^{N-1} \mu(k)\{\delta u(t/t) - \delta u(t/t-k)\}^2] \end{aligned} \tag{4.3}$$

where it is assumed that the weighting $\mu(k)$ is applied to control increments, with no loss of generality, the weighting $\lambda(k)$ having been applied to actual signal values. Also $Q = I$ and $R = \mathrm{diag}\{r(k)\}$.

By defining

$$W = (G^{\mathsf{T}}G + R)^{-1} \tag{4.4}$$

the minimum for (4.3) is found when

$$\delta u(t) = e_1^{\mathsf{T}}(I + W\mu_0 e_1 e_1^{\mathsf{T}})^{-1}\overline{\underline{u}}$$

$$+ \quad e_1^\mathsf{T}(I + W\mu_0 e_1 e_1^\mathsf{T})^{-1} e_1 \sum_{k=1}^{N-1} \mu(k) e_1^\mathsf{T} d_k \tag{4.5}$$

in which d_k is chosen to satisfy

$$e_1^\mathsf{T} d_k = \delta u(t/t-k) \tag{4.6}$$

and

$$\mu_0 = \sum_{k=1}^{N-1} \mu(k) \tag{4.7}$$

It can be noted that the control input found in (4.5) reduces to the general predictive control input when $\mu(k) = 0$ for all k.

But we have also that:

$$(I + W\mu_0 e_1 e_1^\mathsf{T})^{-1} = I - \frac{1}{1 + \mu_0 e_1^\mathsf{T} W e_1} W\mu_0 e_1 e_1^\mathsf{T} \tag{4.8}$$

such that

$$e_1^\mathsf{T}(I + W\mu_0 e_1 e_1^\mathsf{T})^{-1} = \frac{1}{1 + \mu_0 e_1^\mathsf{T} W e_1} e_1^\mathsf{T} \tag{4.9}$$

The optimal increment in control is therefore

$$\begin{aligned} \delta u'(t) \quad &= \quad \delta u(t/t) = \frac{1}{1 + \mu_0 e_1^\mathsf{T} W e_1} \delta u_{GPC}(t/t) \\ &+ \quad \sum_{k=1}^{N-1} \frac{\mu(k) e_1^\mathsf{T} W e_1}{1 + \mu_0 e_1^\mathsf{T} W e_1} \delta u(t/t-k) \end{aligned} \tag{4.10}$$

and this is a convex combination of $\delta u_{GPC}(t/t)$ and the previous predictions made for the control input to be applied at time instant t, because

$$\frac{1}{1 + \mu_0 e_1^\mathsf{T} W e_1} + \sum_{k=1}^{N-1} \frac{\mu(k) e_1^\mathsf{T} W e_1}{1 + \mu_0 e_1^\mathsf{T} W e_1} = 1 \tag{4.11}$$

The control input signal of (4.2) can be applied within a self–tuning control format [8], and is particularly suited for use with a simplified, low computation, recursive parameter estimation scheme [5], as detailed in [11]. Some implementation results, particularly a trial comparison with a well tuned PID controller, are given in Owens and Warwick [8].

5 Conclusions

The predictive control method described in this paper can in fact be used on a variety of basic predictive control algorithms, e.g. dynamic matrix control. Its use on a generalised predictive controller in this paper was merely one example, out of those possible.

As with any operator selection procedure, immediate questions arise as to how the controller operator decides on particular controller values. In this case the choice of weighting values poses a problem. Some obvious simple solutions exist, such as to provide a low pass filtering action on the control input signal [12] or to include a fault tolerant emphasis on the controller structure by allowing for a hold back time on spurious signals.

The method shown in this paper gives additional robustness and reliability characteristics to standard predictive control schemes, particularly when employed within an adaptive/self-tuning control framework, as the effect of wayward control signals can be significantly reduced.

References

[1] Astrom, K.J. and Wittenmark, B., On self-tuning regulators, Automatica, 9, 185–199, 1973.

[2] Clarke, D.W., Mohtadi, C. and Tuffs, P.S., Generalised predictive control, Parts I and II, Automatica, 23, 137–160, 1987.

[3] Clarke, D.W. and Zhang, L., Does long range predictive control work?, Proc. IEE Int. Conference Control 85, 13–18, 1985.

[4] Cutler, C.R. and Ramaker, B.L., Dynamic matrix control–a computer control algorithm, Proc. JACC, San Francisco, 1980.

[5] Farsi, M., Karam, K.Z. and Warwick, K., Simplified recursive identifier for ARMA processes, Electronics Letters, 20, 913–915, 1984.

[6] Karam, K.Z., Warwick, K. and Farsi, M., An implicit pole assignment self-tuning regulator, Journal of Optimal Control: Applications and Methods, 7, 127–134, 1986.

[7] Mohtadi, C., Shah, S.L. and Clarke, D.W., Generalised predictive control of multivariable systems, Oxford University Engineering Labs. Report No. 1640/86, 1986.

[8] Owens, D.H. and Warwick, K., Extended predictive control, Proc. IEE Int. Conference Control 88, 622–627, 1988.

[9] Peterka, V., Predictor-based self-tuning control, Automatica, 20, 39–50, 1984.

[10] Warwick, K., Adaptive deadbeat control of stochastic systems, Int. J of Control, 44, 651–663, 1986.

[11] Warwick, K., System identification, Chapter 6 in Industrial Digital Control Systems, K. Warwick and D. Rees (eds.), comprehensively revised 2nd ed., Peter Peregrinus Ltd., 1988.

[12] Warwick, K. and Clarke, D.W. Weighted input predictive controller, Proc. IEE, Part D, 135, 16–20, 1988.

[13] Ydstie, B.E., Extended horizon adaptive control, Proc. IFAC 9th World Congress, Budapest, Hungary, 1986.

PID self–tuning control via optimal control methods

R. W. Jones R. P. D. Baker

Abstract

The PID controller is still by far the most popular algorithm used for industrial process control. The tuning procedure is, in many cases, difficult and extremely time consuming with values of the parameters which give a desired closed–loop behaviour being very hard to find. Automatic tuning of the controller is therefore of great interest. One possibility is the use of a self–tuning approach.

In this paper an optimal self–tuning PID controller will be obtained from a LQG control law based upon the polynomial equation approach ([11]). Two cases will be examined. Initially the optimal PID self–tuning in developed for a second–order discrete process model with a delay of one sample, then the use of a predictor in the algorithm is investigated to improve the performance of the PID self–tuner in the presence of time–delay processes.

1 Introduction

The coming of digital process control computers has seen very little change occurring in the types of control algorithm used for industrial process control. In other words the revolution in control hardware digital computer–based systems has not been accompanied by a parallel revolution in the types of control algorithm used. The continuous proportional–integral–derivative (PID) controller is very well entrenched and so to a large extent most digital computer systems and programmable controller units are set up with digital PID algorithms utilizing very short control intervals in order to mimic their continuous counterparts. The two main reasons why PID control is so popular are:

1. it is robust, in a practical sense, and remarkably effective for a wide range of processes, and,

2. the unfamiliarity of control engineers with modern control methods which would (in some cases) require extensive retraining.

For the PID controller three parameters can be tuned usually in the closed–loop, using one of the many tuning schemes (e.g. [23], [14] and [3]) or by trial and error. The tuning procedure is, in many cases, difficult and extremely time consuming with values of the parameters which give a desired closed–loop behaviour being very hard to find. Automatic tuning of the controller is therefore of great interest. One possibility is the use of a self-tuning approach.

Self-tuners will generally have considerably more parameters than a conventional PID controller and usually have a quite different control structure. Various researchers though have investigated the performance of PID self-tuners based upon, for example, a generalised minimum variance control law ([5]) and pole-placement control laws ([21] and [4]). To obtain the correct form of control law though assumptions have to be made about the process model complexity used in the self-tuner, the discrete representation usually being second-order with a discrete delay of one sample. Most process dynamics can be approximated as second-order but the delay restriction makes these self-tuners sensitive to a time-delay process unless the sample time is set equal to the delay. This is generally not a practical proposition.

In this paper an optimal self-tuning PID controller will be obtained from an LQG control law based upon the polynomial equation approach([11]). Two cases will be examined. The optimal PID self-tuner is initially developed for a second order discrete process model with a delay of one sample, then the use of a predictor is incorporated in the algorithm to improve the performance of the PID self-tuner in the presence of larger sample time-delay processes.

The LQG controller expressions to be utilized in this paper are based upon the general polynomial equation approach developed by Peterka ([19]) and Kucera ([12]). The control calculations normally involve the solution of two diophantine equations and one polynomial equation which must be spectrally factored. However the algorithm here, which is used as the basis for the PID self-tuner, involves the solution of only one diophantine equation and a spectral factorisation by making some simplifying assumptions about the system. This approach has been successfully implemented in several self-tuning algorithms ([22] and [9]).

2 LQG design

Consider a single-input single-output linear discrete time randomly disturbed model described by

$$A(z^{-1})y(t) = B(z^{-1})u(t) + C(z^{-1})\xi(t) \tag{2.1}$$

where z^{-1} is interpreted as the backward shift operator $z^{-1}x(t) = x(t-1)$, $y(t)$ and $u(t)$ are the output and input sequences and $\xi(t)$ is a zero-mean white noise disturbance sequence. The system polynomials A, B, C are defined by:

$$A(z^{-1}) = 1 + a_1 z^{-1} + \cdots + a_n z^{-n} \tag{2.2}$$
$$B(z^{-1}) = z^{-k}(b_0 + b_1 z^{-1} + \cdots + b_r z^{-r}) \tag{2.3}$$
$$C(z^{-1}) = 1 + c_1 z^{-1} + \cdots + c_m z^{-m} \tag{2.4}$$

where k is the time-delay and r is chosen so that all values of the possibly varying dead-time can be accommodated.

The LQG control laws under consideration are designed to minimise a performance criterion:

$$J = E\left[(H_q e)^2(t) + (H_r u)^2(t)\right] \tag{2.5}$$

where the tracking error $e(t)$ is defined by:

$$e(t) \triangleq w(t) - y(t)$$

Re–writing this in the frequency domain gives:

$$J = \frac{1}{2\pi j} \oint_{|z|=1} \left[Q_c(z^{-1})\phi_{ee} + R_c(z^{-1})\phi_{uu}\right] \frac{dz}{z} \tag{2.6}$$

where $Q_c \triangleq H_q^* H_q$ and $R_c \triangleq H_r^* H_r$, and ϕ_{ee} and ϕ_{uu} denote the spectral densities of the signals e and u, respectively. The weighting elements Q_c and R_c are positive constants or strictly positive dynamical operators on $|z| = 1$:

$$Q_c \triangleq \frac{B_q^* B_q}{A_q^* A_q}, R_c \triangleq \frac{A_r^* B_r}{A_r^* A_r} \tag{2.7}$$

where A_q and A_r are strictly Hurwitz polynomials, and $A_q(0) = A_r(0) = 1$. The adjoint of a polynomial $X(z^{-1})$ has been denoted by $X^*(z^{-1})$ such that $X^*(z^{-1}) = X(z)$. The arguments of the polynomials will now be omitted for notational simplicity.

The effect of the dynamic (frequency–dependent) cost function weights is to allow the error and control variations to be penalised in different frequency ranges. These are the main design parameters which must be chosen by the system user. There are many ways to choose these weighting elements, however the simplest, which will be used in this paper, is to define them as

$$Q_c = \frac{1}{(1-z)(1-z^{-1})}, R_c = R_0 \tag{2.8}$$

Choosing A_q as an integrator allows the introduction of integral action into the control while R_0 can be interpreted as an online tuning knob which varies the relative magnitude of the error and control variances.

In the closed–loop system a two–degrees–of–freedom control structure is used. This structure allows the systems servo response to be shaped independently of the feedback properties of the system and gives a lower optimal cost for the controller ([7]). The control law is given by:

$$A_q(z^{-1})H(z^{-1})u(t) = S(z^{-1})w(t) - Gz^{-1})y(t) \tag{2.9}$$

where $w(t)$ is the desired set point. The approach taken here will be to design the $G(z^{-1})$ and $H(z^{-1})$ polynomials to give optimal (in the LQG sense) regulation, and then to design $S(z^{-1})$ so that in the steady–state there is unity gain between output and setpoint. The polynomials G and H are determined by the following two–step procedure.

The characteristic polynomial of the optimal closed–loop system is first obtained as a solution to the spectral factorization problem:

$$D_c^* D_c = B^* B + A^* A_q^* R_0 A_q A \tag{2.10}$$

The second step involves the solution of the following coupled diophantine equations in terms of the unknown G, H and F to provide the unique particular solution with minimal degree with respect to F

$$D_c^* G z^g + Y A A_q = B^* C z^{-g} \tag{2.11}$$
$$D_c^* H z^{-g} - Y B = A^* A_q^* R_0 C z^{-g} \tag{2.12}$$

where $g \triangleq \max(ndc, nb + k, na + naq)$. These equations can be combined to obtain the 'implied' equation

$$A A_q H + B G = D_c C \tag{2.13}$$

By making the assumptions that A, B and C are relatively prime and that their orders satisfy $n > m - k$ and $r > m$ the controller calculation can be simplified. It has been shown by Jezek ([10]) that in this case the solution of the implied diophantine equation is unique thus the controller can be found directly from eqn. (2.13).

Solving eqns. (2.1) and (2.9) for the output $y(t)$ gives:

$$y(t) = \frac{BS}{(AA_q H + BG)} w(t) + \frac{C}{(AA_q H + BG)} \xi(t) \tag{2.14}$$

From this it can be seen that the servo response is defined as BS/CD_c. The polynomial $S(z^{-1})$ can then be taken to be γC where γ is chosen so that there is unity steady–state gain between $y(t)$ and $w(t)$ i.e.:

$$\gamma = \frac{D_c(1)}{B(1)} \tag{2.15}$$

3 PID controller structure

The conventional digital PID–controller (e.g. [2]) can be derived from the continuous–time form as:

$$u(t) = K_P \left[+ \frac{K_I z^{-1}}{(1 - z^{-1})} + \frac{K_D (1 - z^{-1})}{(1 + \alpha z^{-1})} \right] (w(t) - y(t)) \tag{3.1}$$

where K_P, K_I and K_D are, respectively, the proportional, integral, and derivative gains, and α is the derivative filter constant $(-1 < \alpha < 0)$. The delay in the integral part can be removed by using a backward–difference approximation.

PID controllers can be implemented and parameterised in many different ways ([21]). One alternative to the "standard" form shown above is to include the set–point signal $w(t)$ only in the integral part to avoid setpoint and derivative kick (sudden changes in the manipulated variable when the set point is changed) ([3]).

Removing the derivative kick and the delay in the integral term in eqn. (3.1) gives upon rearrangement:

$$\begin{aligned}(1-z^{-1})(1+\alpha z^{-1})u(t) &= \\ &\left[K_P(1+K_I)+K_P(K_I\alpha-(1-\alpha))z^{-1}-\alpha K_P z^{-2}\right](w(t)-y(t)) \\ + &\left[-K_P K_D+2K_P K_D z^{-1}-K_P K_D z^{-2}\right]y(t) \end{aligned} \tag{3.2}$$

which is a two–degree of freedom controller structure due to the exclusion of the derivative kick. If the derivative term is removed ($K_D = \alpha = 0$) then the control law becomes:

$$(1-z^{-1})u(t) = \left[K_P(1+KI)-K_P z^{-1}\right](w(t)-y(t)) \tag{3.3}$$

and the controller structure reverts to the usual error driven (single–degree of free-dom) form.

4 PID form LQG controller

4.1 Discrete delay of one sample

In the previous section a PID controller without derivative kick was shown to have a tow-degree of freedom control structure. If we equate this eqn. (3.2) to the previously defined LQG control law, (eqn.(2.9)), we obtain the following relationships:

$$\begin{aligned} g_0 &= K_P(1+K_I+K_D) \\ g_1 &= K_P(K_I\alpha-(1-\alpha)-2K_D) \\ g_2 &= K_P(K_D-\alpha) \end{aligned} \tag{4.1}$$

$$\begin{aligned} h_0 &= 1.0 \\ h_1 &= \alpha \end{aligned} \tag{4.2}$$

$$\begin{aligned} s_0 &= K_P(1+K_I) \\ s_1 &= K_P(K_I\alpha-(1-\alpha)) \\ s_2 &= \alpha K_P \end{aligned} \tag{4.3}$$

thus it can be seen that the PID and LQG controllers are equivalent when the polynomial degrees $nh = 1$, $ng = 2$ and $ns = 2$.

The model–order which produces this equivalence of controller parameters is, as indicated in the introduction, second order ($na = nb + k = 2, nc = 1$) with a discrete delay of one sample. From the spectral factorisation, eqn. (2.10) $nd_c = 3$ as $\max(nb, naq + na) = 3$ and then in the 'implied' diophantine, eqn. (2.13), $nh = nb + k - 1$ and $ng = na + na_q - 1$. The order of the right–hand side of the diophantine, CD_c, has to be fourth order and as $nd_c = 3$ this gives the requirement that $nc = 1$ in the process model. Normally $na = nc$ but the significance of this apparent underparameterisation of the noise polynomial and its effect on the self–tuning algorithm will be discussed in section 5. From section 2. we can see that the design of $S(z^{-1}) = \gamma C(z^{-1})$ is inadequate in terms of model order (ns should be 2).

The $S(z^{-1})$ polynomial will, therefore, be defined from the values of K_P, K_I, K_D and α (i.e. h_1). These values can be found from eqns. (4.1–4.3) by solving in terms of g_0, g_1, g_2 and α to give:

$$K_P = \frac{\alpha g_0 - g_1 - g_2(2+\alpha)}{(1+2\alpha+\alpha^2)} \tag{4.4}$$

$$K_D = \frac{g_2 + \alpha K_P}{K_P} \tag{4.5}$$

$$K_I = \frac{g_0 - K_P K_D - K_P}{K_P} \tag{4.6}$$

It can also be seen from the definitions of eqns. (4.1–4.3) that in the steady state:

$$G(1) = S(1) = K_P K_I (1+\alpha) \tag{4.7}$$

Thus given these definitions, for a value of control weighting R_0 in the LQG cost function the equivalent PID gains K_P, K_I and K_D can be calculated.

Different formulations of the controller are of course possible depending upon the choice of model–order and A_q. For example, if the model is first order with a delay of one sample and $A_q = (1 - z^{-1})$ then the controller is of a PI form but this model parameterisation is not very practical for self–tuning as very few systems are adequately described by such a model. Another formulation that perhaps could be of some use is when, for example, the simplified input–output relationship between the rudder angle and the ships heading is given by the transfer function:

$$G(s) = \frac{Ke^{-s\tau}}{s(s+\delta)}$$

where the time delay τ approximates the effects of high order dynamics and the dynamics of the steering engine. Discretisation of the above, assuming that the sample time $\gg \tau$, gives a second–order model with a single sample delay. As integral action occurs naturally in the process, $A_q = 1.0$ which gives a controller where $ng = 1$ and $nh = 1$, i.e., a type of PD control law.

In this paper only one PID form has been considered. The form chosen is sometimes application specific, for example, McInnis et al ([16]) derived a pole–placement type self–tuning PID algorithm for a biomedical application. Because in this application, overshoot could have a disastrous (i.e. fatal) effect their PID algorithm has both proportional and derivative kicks removed to damp the response as much as possible to changes in setpoint.

4.2 Discrete delays greater than one sample

From the implied diophantine equation it can be seen, for a second order model, that as the discrete delay in the model increases so does the order of the $H(z^{-1})$ polynomial (as $nh = nb + k - 1$). Therefore, the equivalence between the LQG and PID control laws disappears when $k > 1$.

By incorporating a predictor into the controller this problem can be overcome. This compensates for the dead-time in the process, therefore, ensuring that $k = 1$ in the characteristic polynomial and the diophantine equation. In this paper to demonstrate this the discrete form of the well-known Smith Predictor ([15]) will be utilized. It is given as

$$y^s(t + k(t)) = y(t) + (1 - z^{-k})\frac{B}{A}u(t) \tag{4.8}$$

where $y^s(t + k/t)$ is the predicted output k samples into the future. Algebraic manipulation of the system yields the closed-loop transfer-functions:

$$\frac{y(t)}{w(t)} = \frac{z^{-k}BSG}{AA_qH + z^{-1}(b_0 + b_1z^{-1})G} \tag{4.9}$$

$$\frac{u(t)}{e(t)} = \frac{AG}{AA_qH + z^{-1}GB(1 - z^{-k+1})} \tag{4.10}$$

Thus the system delay has been replaced in the characteristic equation by $k = 1$.

This particular prediction method has, however, some disadvantages: it cannot be used with open-loop unstable systems and it is sensitive to model mismatch. The second problem can be compensated for by using an adaptive version which is the intention in this contribution.

If the process is likely to be open-loop unstable then a stochastic k-incremental predictor could be used instead. This has the form:

$$y^*(t + k/t) = y(t) - E\xi(t) + F\Delta_k y(t) + G\Delta u(t) \tag{4.11}$$

where $\Delta_k = (1 - z^{-k})$ and $E(z^{-1})$ and $F(z^{-1})$ are polynomials of degrees $(k - 1)$ and $(na - 1)$ found from the diophantine equation:

$$C = EA + z^{-k}F$$

and $G = EB$. The increased computational complexity (due to the need to solve eqn.(4.12)) is the price paid for the ability to predict the output of open-loop unstable systems.

The explicit generation of the predictor (whichever one it may be) allows the inclusion of a facility to compensate for unknown or time-varying delays. Appropriate methods for estimating K are discussed later.

5 The self–tuning algorithm

Implementation of the controller design requires estimates of the $A(z^{-1}), B(z^{-1})$ and $C(z^{-1})$ polynomials. Extended least squares (ELS) ([18]) identification will provide unbiased estimates of $A(z^{-1}), B(z^{-1})$ and $C(z^{-1})$. Combined with this a forgetting factor technique (e.g. [8]) should be utilized to estimate time–varying parameters and to ensure that the adaptive predictor operates successfully. Earlier it was indicated that there is an apparent underparameterisation of the $C(z^{-1})$ polynomial so that a PID structure is obtained for the LQG control law. This could possibly have implications for the control and identification as $\hat{C}(z^{-1})$, the estimate of $C(z^{-1})$ should be inverse stable, and is assumed so by the representation theorem ([1]). If $\hat{C}(z^{-1})$ does become inverse unstable then a stable representation of $\hat{C}(z^{-1})$ (i.e. $\tilde{C}(z^{-1})$) can be found, and used, from the following spectral factorisation:

$$\tilde{C}^*(z^{-1})\tilde{C}(z^{-1}) = \hat{C}^*(z^{-1})\hat{C}(z^{-1}) \tag{5.1}$$

5.1 Calculation of $G(z^{-1})$ and $H(z^{-1})$

To solve the design problem it is necessary to calculate a stable spectral factor, D_c, eqn. (2.10) and then to solve the implied diophantine equation, eqn. (2.13). The spectral factorization problem can be solved by the method proposed by Gable, Neuvo and Rudin ([6]). This algorithm is iterative and has the advantage that the polynomial $D_c(z^{-1})$ obtained at each step is guaranteed to be stable. In practice only a few iterations are required and the spectral factor from the previous time step can be used as the first iterate.

For the implied diophantine:

$$AA_qH + BG = T \tag{5.2}$$

where $T \triangleq D_cC$, as the order of the equation is never greater than 4th order it is trivial to generate direct relationships for the coefficients of the unknown polynomials $G(z^{-1})$ and $H(z^{-1})$ to $T(z^{-1})$. Equating coefficients gives:

$$H_1 = \frac{b_1\left[b_1^2(t_1 - a_1 + 1) + b_1b_0(a_2 - t_2 - a_1) + b_0^2(t_3 + a_2)\right] - b_0^3t_4}{\left[b_1^3 + b_1^2b_0(1 - a_1) + b_1b_0^2(a_2 - a_1) + b_0^3a_2\right]} \tag{5.3}$$

$$g_0 = \frac{t_1 - a_1 + 1 - h_1}{b_0} \tag{5.4}$$

$$g_1 = \frac{t_2 - b_1g_0 + h_1(1 - a_1) + a_1 - a_2}{b_0} \tag{5.5}$$

$$g_2 = \frac{t_4 + h_1a_2}{b_1} \tag{5.6}$$

5.2 Unknown and time–varying delays

Kurz and Goedeche ([13]) suggested a method of estimation k by fitting the impulse response of a model with known model order. This procedure is very time consuming especially when the time delay is large, but has a major advantage because its time-delay estimation procedure uses a criterion function to decide when b_i coefficients are zero. It is appropriate to the work of this paper and is to be adopted as a future development.

5.3 The effect of common factors

It was indicated earlier that in most practical cases it is sufficient to solve the computationally simpler implied diophantine equation to calculate the optimal feedback controller. If the discrete model, eqn. (2.1), is overparameterised, i.e. the physical process is first-order, then common factors occur between the $A(z^{-1})$ and $B(z^{-1})$ polynomials. If these are stable then the coupled diophantine equations (2.11 and 2.12) must be used to calculate the optimal feedback controller, unfortunately the correct orders of $G(z^{-1})$ and $H(z^{-1})$ polynomials are unlikely to be obtained to generate the PID control law. It can be seen from the spectral factorisation that the common factors will also be in $D_c(z^{-1})$. Thus if the implied diophantine equation is used in all cases, which has been suggested, then with stable common factors in $A(z^{-1})$ and $B(z^{-1})$ the equation is still solvable. The effect on the controller is that a closed-loop feedback system will be calculated with optimal pole positions but sub-optimal zero positions.

These properties of the optimal design can be constructed with standard pole-placement algorithms (e.g. [20]) where a similar diophantine must be chosen by the user, a solution cannot be found when common factors occur between $A(z^{-1})$ and $B(z^{-1})$.

5.4 The implemented algorithm

The calculation steps for the two PID self-tuners whose basic algorithms are discussed in sections 4.1 and 4.2 are presented here.

Data: Control Weighting R_0

Step 1: Estimate the model polynomials $A(z^{-1})$, $B(z^{-1})$ and $C(z^{-1})$ using ELS.

Step 2: Calculate the stable spectral factor, $D_c(z^{-1})$.

Step 3: Calculate $G(z^{-1})$ and $H(z^{-1})$ from the implied diophantine, eqn. (2.13).

Step 4: Caculate the equivalent PID gains K_P, K_I and K_D and therefore calculate $S(z^{-1})$ from eqn. (4.3).

Step 5: If $k > 1$ generate the prediction using equation (4.8).

Step 6: Calculate the control ($k = 1$ or $k > 1$).

6 Conclusions

The problem of constructing an adaptive PID controller has been approached along an optimal control path. This approach has the advantage over pole placement based PID self–tuners that choice of design parameters is more natural–control weighting as against pole locations. In the time–delay case optimal control methods have been combined with a discrete Smith Predictor to give an algorithm with an identical structure from the users viewpoint.

7 Acknowledgements

The authors would like to thank Miss Yvonne Fleming for typing the paper and also Dr. Reza Katebi who kindly presented the paper at the IMA conference.

References

[1] Astrom. K.J., "Introduction to stochastic control theory", Academic Press, New York, 1970.

[2] Astrom, K.J. and Wittenmark, B., "Computer Controlled Systems: Theory and Design", Prentice–Hall International, London, 1984.

[3] Auslander, D.M., Takahashi, Y. and Tomizaka, M., "Direct digital process control: Practice and algorithm for microprocessor applications", Proc. IEEE., 66, 2, 199–208, 1978.

[4] Banyasz, C. and Keviczky, L., "Direct Methods for Self–tuning PID regulators, 6th IFAC Symp. on Ident. and System Parameter Est., 1249, 1982.

[5] Cameron, F. and Seborg, D.E., "A Self–tuning Controller with a PID structure, Int. J. Control, 38, 401, 1983.

[6] Gabel, R.A., Neuvo, Y. and Rudin, H., Iterative Spectral Factorization Algorithms for digital filters, In: "Digital Signal Processing", Ed. Cappellini, V. and Constantini, A.G., Academic Press, 1980.

[7] Gawthrop, P.J., Developments in optimal and self–tuning control theory, D. Phil thesis, University of Oxford, OUEL report 1239/78, 1978

[8] Hagglund, T., New estimation techniques for adaptive control, Ph.D. thesis, Lund Institute of Technology, 1983.

[9] Hunt, K.J. and Jones, R.W., The Application of LQG Self–tuning control to the Hunterston 'B' Simulator, Industrial Control Unit Report, University of Strathclyde, 1987.

[10] Jezek, J., New Algorithm for minimal solution of linear polynomial equations, Kybernetika, 18, 505–516, 1982.

[11] Jones, R.W., Hunt, K.J. and Grimble, M.J., Principles of LQG Self-tuning Systems, In: "Control System Design: Robustness and Polynomial Methods". Ed. M.A. Johnson, 1986.

[12] Kucera, V., "Discrete Linear Control: The Polynomial Equation Approach". Wiley, New York, 1979.

[13] Kurz, H. and Goedecke, W., Digital parameter–adaptive control of processes with unknown dead–time. Automatica, 17, 245–252, 1981.

[14] Lopez, A.M., Miller, J.A., Smith, C.L. and Murrill, P.W., Tuning Controllers with Error–Integral Criteria, Inst. Tech., November, 1967.

[15] Marshall, J.E., "Control of Time–Delay Systems", Peter Peregrinus, Stevenage, 1979.

[16] McInnis, B.C., Guo, Z.W., Lu, P.C. and Wang, J.C., Adaptive control of left ventricular bypass assist devices, IEEE Trans. Auto. Contr., AC–30, 322–329, 1985.

[17] Nomoto, K., Taguchi, T., Honda, K. and Hirano, S., On the steering qualities of ships, Int. Shipbuilding Progress, 4, 1957.

[18] Panuska, V., A stochastic approximation method for identification of linear systems using adaptive filtering, Joint Automatic Control Conference, Ann Arbor, 1968.

[19] Peterka, V., On steady state minimum variance control strategy, Kybernetica, 8, 219–232, 1972.

[20] Wellstead, P.E., Prager, D and Zanker, P., Pole assignment self–tuning regulator, Proc. IEE., Pt. D., 126, 781–787, 1979.

[21] Wittenmark, B., Self–tuning PID–controllers based on pole–placement, Department of Automatic Control, Lund Institute of Technology Report, CODEN: LUTFD2/(TFRT–7179)/1–037/, 1979.

[22] Zhao–Ying, Z. and Astrom, K.J., A microprocessor implementation of an LQG self–tuner, Department of Automatic Control, Lund Institute of Technology, 1981.

[23] Ziegler, J.G. and Nichols, N.B., Optimum settings for automatic controllers, Trans ASME, 64, 759–768, 1942.

Robustness of adaptive control systems

L. Wang D. H. Owens

Abstract

The problem of the robustness of adaptive control systems will be examined and discussed in this paper. By connecting point-wise stability with the stability criteria given by frequency domain conditions a general robust stabilization theorem is derived for those adaptive systems designed according to the certainty equivalence principle.

1 Introduction

As we know, most physical system structures are very complicated and high order, control system design methods without robustness against unmodelled dynamics would have severe limitations in their applications. Convincing evidence given by Rohrs. et. al (1982) shows the basic adaptive control algorithms can indeed lead to instability of closed–loop adaptive system when they are applied to the systems having unmodelled dynamics or disturbance, even if they are very small. However, it has been shown in Astrom (1980) that a 7th order process plant can be successfully controlled respectively by 1st order, second order and third order adaptive controllers, and the closed–loop stability is maintained. These facts point out that the adaptive control systems are conditionally stable in the presence of unmodelled dynamics.

The basic task of this paper is to derive robustness conditions for the adaptive control systems designed according to the certainty equivalence principle. The interpretation of the robustness conditions is given as consequence of local stabilization conditions and as indications of possible methods to improve their robustness. The structure of this paper is outlined as follows: in section 2, the adaptive control systems are described by linear time varying system, and their stability conditions are examined in terms of point–wise stability and slow variation assumptions. Based on the linear time varying system theory, a robust stabilization theorem is derived for the adaptive control systems designed according to certainty equivalence principle. In section 3, the robust stabilization theory is interpreted as local stabilization neighbourhood theory.

2 Robustness conditions for adaptive control systems

The overall adaptive system is nonlinear and time varying, because of the involvement of the adaptive mechanisms and the variation of the parameters in adaptive controllers. However, the examination of the stability of nonlinear systems can be converted to an examination of a linear time varying system, This is reasoned as follows: consider adaptive systems described by the nonlinear difference equation

$$\begin{aligned} x(k+1) &= A(\Phi(x(k)))x(k) + B(\Psi(x(k)))\gamma(k) \\ y(k) &= C(\xi(x(k)))x(k) \end{aligned} \tag{2.1}$$

where Φ, Ψ, and ξ are state dependent functions and representations of parameter variations.

Let $z(k) = x(k)$, $z(0) = x(0)$, the linear time varying difference equation is formed as

$$\begin{aligned} z(k+1) &= A(\Phi(x(k)))z(k) + B(\Psi(x(k)))\gamma(k) \\ \tilde{y}(k) &= C(\xi(x(k)))z(k) \end{aligned} \tag{2.2}$$

It is easy to see that these two equations have the same solution $x(k) = z(k)$, $k \geq 0$. Therefore, the nonlinear equation (2.1) is stable from $x(0)$ if the linear system (2.2) is stable from $z(0)$. Regarding $A(\Phi(x(k))) = \Phi(k)$, $B(\Psi(x(k))) = \Delta(k)$, and $C(\xi(x(k))) = C(k)$ simply as time varying matrices we have

$$\begin{aligned} z(k+1) &= \Phi(k)z(k) + \Delta(k)\gamma(k) \\ y(k) &= C(k)z(k) \end{aligned} \tag{2.3}$$

Hence the theory of linear time varying system can be applied to analyze the stability of adaptive systems. Some stability conditions for unforced system of (2.3) with $\gamma(k) \equiv 0$ were derived by Desoer (1970). The case of bounded input and bounded output stability conditions will be given by the following theorem which generalised Desoer's result.

Theorem 2.1: The adaptive control system represented by linear time varying difference equation (2.3) will be stable in the sense that for a given bounded input $\gamma \in \ell^\infty \triangleq \{\gamma : \ ||\gamma|| = \sup_k |\gamma_k| \leq M, \ M < \infty\}$ the output $y \in \ell^\infty$ if the following conditions hold true for sufficiently small ξ

1.
$$\begin{aligned} \sup_k \quad & ||\Phi(k)|| \leq a_m < \infty \\ \sup_k \quad & ||\Delta(k)|| \leq b_m < \infty \\ \sup_k \quad & ||C(k)|| \leq C_m < \infty \end{aligned}$$

2.
$$\sup_k \sup_j |\lambda_j(\Phi(k))|| < 1 - 2\epsilon, \ \epsilon > 0, \tag{2.4}$$

where $\lambda_j, 1 \leq j \leq m$, are the eigenvalues of $\Phi(k)$ and

3. $\sup_k ||\Phi(k) - \Phi(k-1)|| \leq \xi$, ξ is dependent on ϵ, a_m.

Proof: According to Desoer (1970), when $\gamma(\cdot) = 0$, under the conditions given in the theorem, the constructed Lyapunov function $V(z(k))$ has the following properties

$$0 \leq V(z(k)) \leq \frac{m^2}{1-\rho^2}\|z(k)\|^2 \tag{2.5}$$

and

$$V(z(k+1)) - V(z(k)) \leq -\eta\|z(k)\|^2, \ 1 > \eta > 0 \tag{2.6}$$

where $m \triangleq \frac{(1-\epsilon)+a_m^{n-1}}{\epsilon^n}$, $\rho = 1-\epsilon$, $0 < \epsilon < 1$, and n is the dimension number of matrix $\Phi(k)$.

Equation (2.5) and (2.6) yield

$$V(z(k+1)) - V(z(k)) \leq -\frac{\eta}{m^2}\left(1-\rho^2\right) V(z(k)) \tag{2.7}$$

Equation (2.7) gives

$$\begin{aligned} V(z(k+1)) &\leq \left(1 - \frac{\eta}{m^2}\left(1-\rho^2\right)\right) V(z(k)) \\ &= \lambda^2 V(z(k)) \end{aligned} \tag{2.8}$$

Now that since $\epsilon < 1, m > 1$, therefore, $\lambda^2 < 1$. This leads to $\lambda < 1$. Hence, it is obtained that

$$V(z(k)) \leq \lambda^{2k} V(z(0)), \ \lambda < 1 \tag{2.9}$$

That is

$$\|z(k+1)\| \leq \sqrt{\frac{m^2}{1-\rho^2}}\lambda^k\|z(0)\|, \ \lambda < 1 \tag{2.10}$$

Therefore

$$\|\Phi(k)\Phi(k-1)\cdots\Phi(0)\| \leq \sqrt{\frac{m^2}{1-\rho^2}}\lambda^k \tag{2.11}$$

When $\gamma(\cdot) \neq 0$, the solution of the forced non-stationary difference equation is bounded by

$$\begin{aligned} \|z(k+1)\| &\leq \|\prod_{i=0}^{k}\Phi(k-1)z(0)\| + \prod_{i=0}^{k-i}\Phi(k-i)\Delta(0)\gamma(0)\| \\ &\quad + \cdots + \|\Phi(k)\Delta(k-1)\gamma(k-1)\| + \|\Delta(k)\gamma(k)\| \\ &\leq \sqrt{\frac{m^2}{1-\rho^2}}\|z(0)\| + b_m\|r\| + \frac{\lambda b_m}{1-\lambda}\sqrt{\frac{m^2}{1-\rho^2}}\|\gamma\| \end{aligned} \tag{2.12}$$

Hence

$$\|y(k)\| \leq \sup_k \|C(k)\| \sup_k \|z(k)\| < \infty$$

is obtained as required.

This theorem examines the stability of adaptive system from point–wise stability and slow time variation point of view. It sets the basis for further analysis of the robustness of adaptive control systems.

In the following, by connecting the point–wise stability with the stability criteria given by frequency domain conditions a general robust stabilization theorem will be derived for those adaptive systems designed according to the certainty equivalence principle (Bar–shalom, 1974).

Suppose that there exist unmodelled dynamics and unmodelled disturbance in the adaptive control system design. In general, the unmodelled factors will prevent the identified model from converging to the mathematical model of the plant, and lead to the existence of the modelling error between the identified model and the plant. Let $\Delta G(k,z)$ denote the modelling error at time k, defined as an additive error

$$\Delta G(k,z) = G(z) - G_A(k,z) \tag{2.13}$$

where $G(z)$ represents the unknown system, $G_A(k,z)$ represents the estimated model at time k. The robustness of the adaptive control system with respect to the additive modelling error $\Delta G(k,z)$ described by the following theorem.

Theorem 2.2: Suppose that the adaptive controller described by $K(k,z)$ at time k, is designed to stabilize the estimated model $G_A(k,z)$ point–wisely. If the parameters in the adaptive controller are uniformly bounded over all time k, then the closed–loop adaptive control system will be ℓ^∞ sense stable provided that there exists $k*$ such that for all $k \geq k*$ the following conditions hold true for sufficiently small ϵ_1

1. $\Delta G(k,z)$ is point–wisely stable
2. $\sup_{|z|\geq 1} |(1 + G_A(k,z)K(k,z))^{-1}K(k,z))||\Delta G(k,z)|| < 1$
3. $||\Phi(k) - \Phi(k-1)|| \leq \epsilon_1, \ \epsilon_1 > 0$

where $\Phi(k)$ is the state space matrix of the closed–loop adaptive system.

Proof: Condition (iii) in the theorem ensures slow time variation in the adaptive control system. hence the concentration is laid on showing that under the conditions given in the theorem, there exists $\epsilon > 0$ such that

$$|\lambda_j(\Phi(k))| < 1 - \epsilon, \ \forall j, \ \forall k.$$

Before the proof is given, a lemma is introduced as follows.

Lemma 2.1: For non–stationary discrete time system given by

$$x(k+1) = A(k)x(k), \ A(k) : z^+ \rightarrow R^{n\times n} \tag{2.14}$$

if

1. $\sup_k ||A(k)|| \leq a_m < \infty$ and
2. $\inf_k |\mathrm{Det}[zI - A(k)]| > 0, \ \forall |z| \geq 1, \ \forall k$

where $\det(M)$ denotes the determinant of matrix M, then there exists $\epsilon > 0$ such that

$$\sup_k \sup_i |\lambda_i(A(k))| < 1 - \epsilon, \; i = 1, \cdots, n,$$

where $\lambda_i(A(k))$ denotes the ith eigenvalue of $A(k)$.

Proof: (See appendix).

Direct calculation yields that

$$\begin{aligned} |\det[zI - \Phi(k)]| &= |\det[zI - \tilde{\Phi}(k)] \cdot \frac{\det[zI_n - A_p]}{\det[zI_{\tilde{n}} - \tilde{A}_p(k)]} \times \\ &\quad \{1 + (1 + G_A(k,z)K(k,z))^{-1}K(k,z)\Delta G(k,z)\}| \end{aligned}$$

where $\tilde{\Phi}$ denotes the state–space matrix of closed–loop system generated by the estimated model $G_A(k,z)$ and the adaptive controller $K(k,z)$. A_p and $\tilde{A}_p(k)$ are the state space matrices given by plant G and the estimated model $G_A(k,z)$ respectively.

Since the adaptive controller $K(k,z)$ has been designed to stabilise the estimated model $G_A(k,z)$ point–wisely, we have

$$\det[zI - \tilde{\Phi}(k)] \neq 0, \; \forall k, \; |z| \geq 1.$$

and the point–wise stability of the perturbation $\Delta G(k,z)$ ensures that

$$\frac{\det[zI_n - A_p]}{\det[zI_{\tilde{n}} - \tilde{A}_p(k)]} \neq 0, \; \forall k \geq k^*, \; \forall |z| \geq 1.$$

Therefore, $|\det[zI - \Phi(k)]| > 0$, if $|[1 + (1 + G_A(k,z)K(k,z))^{-1}K(k,z)\Delta G(k,z)]| > 0$, for $|z| \geq 1$. This is certainly true if

$$|(1 + G_A(k,z)K(k,z))^{-1}K(k,z)\Delta G(k,z)| < 1, \; |z| \geq 1, \; \forall k \geq k^*,$$

By applying lemma (2.1), and noting that $\Phi(k)$ is bounded it can be concluded that there exists $\epsilon > 0$ such that

$$|\lambda_j \Phi(k)]| < 1 - \epsilon \; \forall j, \; \forall k \geq k^*$$

The ℓ^∞ stability of the adaptive control system is hence concluded by applying theorem (2.1)

Remarks:

1. The self–tuning controllers and adaptive controllers based on the certainty equivalence principle (Bar–shalom, 1974), almost without exception, are designed to stabilize the estimated model G_A point–wisely. For example, in the pole–assignment adaptive controller, if the closed–loop polynomial $T(z)$ is chosen to be stable and the Sylvester matrix is non–singular the pole–assignment adaptive controller stabilises the estimated model point–wisely.

2. Condition (i) and (ii) are not sufficient to ensure the stability of the adaptive system because they only guarantee the point–wise stability of the adaptive system. Examples are given in Rosenbrock (1963) to show that instability of linear time varying system can occur even if it is point–wisely stable. This kind of instability, if it happens in an adaptive system, can be interpreted as fast adaptation instability.

If the plant is known, the conditions for robust stability of adaptive control system can be checked. However, if the plant is unknown, an extra identifier with assumed correct structure is suggested (Owens and Wang, 1988) to obtain the mathematical model of the plant, and the robustness conditions can be checked.

3 Interpretation of the robustness theory

The robustness theory derived suggested two fundamental aspects in ensuring closed–loop stability of adaptive system. The first is the satisfaction of the usual robust criteria (Owens and Chotai, 1983) in point–wise form; the second is the satisfaction of the slow variation assumption. However, the slow variation and parameter boundedness assumption can be satisfied if the robust estimation algorithm with deadzone (Godwin and Colleagues, 1985) is used to cope with the unmodelled dynamics in adaptive control systems. Our analysis in the following will be focussed on the robust criteria. The derived robustness theory for adaptive control system can be interpreted as a local theory. After slight change, the condition (ii) theorem (2.2) could be replaced by

$$\inf_{|z|\geq 1} |1 + (K(k,z)G_A(k,z))^{-1}| > \sup_{|z|\geq 1} |G_A^{-1}(k,z)\Delta G(k,z)| \ \forall k \geq k^*$$

Let us examine the inequality. On the right hand side, for each k, it can be regarded to represent the distance of the on–line identification model to the real plant in the frequency domain. The left hand side can be regarded to represent the permissible distance such that within this distance the adaptive controller designed to stabilize the estimated model will stabilise the plant point–wisely. The theory tells us that if the identified model is sufficiently close to the plant in the sense that the above inequality holds true and the closed–loop adaptive system is slowly time varying, the the adaptive control system is robustly stable. Generally speaking, lower gain adaptive controller can tolerate larger modelling error and higher gain adaptive controller requires a more accurate model.

There are two options open to the designers for robustness of adaptive control system. The first is to reduce the modelling error between the identified model and the plant by using more complicated model structure and better identification strategy, the second is to reduce the gain of the adaptive controller by choosing the desired performance corresponding to slower dynamic responses.

4 Conclusions

Robust stability theorems have been derived for adaptive systems designed according to the certainty equivalence principle. The robustness conditions are given in terms of the robust control criteria in the frequency domain and slow time variation assumptions.

5 Appendix: proof of lemma 2.1

First, assume that there exists an eigenvalue of $A(k)$, $\lambda_i(k)$ such that

$$|\lambda_i(k)| = |\lambda_i[A(k)]| > 1$$

then there must have $z = \lambda_i(k)$ such that

$$\inf_k |\det[zI - A(k)]| = \inf_k |\prod_{i=1}^{\eta}(z - \lambda_i(k))| = 0$$

which is a contradiction to the assumption (2) given in the Lemma. Therefore for assumption (2) to hold true we only have the possibility that

$$\sup_k \sup_i |\lambda_i[A(k)]| \leq 1$$

we assume that

$$\lambda_i(k) = \alpha_i(k)e^{j\omega}, \; 0 \leq \omega \leq 2\pi$$

and $\alpha_i(k) = 1 - \delta(k)$, $\lim_{k \to k_0} \delta(k) = 0$

Since

$$\begin{aligned} \inf_k |\det(zI - A(k))| &= \inf_k |\prod_{j=1}^{n}(z - \lambda_i(k))| \\ &= \inf_k |(z - \lambda_j(k))\prod_{\substack{j=1 \\ j \neq i}}^{n}(z - \lambda_j(k))| \end{aligned}$$

Let $z = e^{j\omega}$

$$\begin{aligned} \inf_k |\mathrm{Det}(zI - A(k))| &= \inf_k |\delta(k)\prod_{\substack{j=1 \\ j \neq i}}^{n}(e^{j\omega} - \lambda_j(k))| \\ &\leq \inf_k |\delta(k)\prod_{\substack{j=1 \\ j \neq i}}^{n}(1 + |\lambda_i(k)|). \end{aligned}$$

Since $\sup_k |\lambda_j(k)| \leq \sup_k \|A(k)\| \leq a_m$
thus

$$\inf_k |\det(zI - A(k))| \leq \delta(k)(1 + a_m)^{n-1}$$

if

$$\lim_{k \to k_0} \delta(k) = 0, \text{ then } \inf_k |\det(zI - A(k))| = 0.$$

which is contradict to condition (2).

Therefore we are able to conclude $\varepsilon = \inf_k \delta(k) > 0$ and hence we must have

$$\sup_k \sup_i |\lambda_i[A(k)]| \leq 1 - \varepsilon.$$

References

[1] Bar–shalom, Y. and E. Tse (1974): Dual effect, certainty equivalence and separation in stochastic control. IEEE Trans. Automat. Contr. Vol–19, No. 5.

[2] Desoer, C.A. (1970): Slowly varying discrete system $X_{i+1} = A_i X_i$ Electronics letters, 6, 11, 339–440, May.

[3] Doyle, J.C. and G. Stein (1981): Multivariable feedback design: concepts for a classical/Modern Synthesis, IEEE Trans. Automatic Control Vol. AC–26, No. 1.

[4] Goodwin, G.C., P.J. Ramadge, P.E Caines (1980): Discrete time Multivariable adaptive control. IEEE Trans. Automat. Contr. Vol. AC–25, pp. 449–455.

[5] Goodwin , G.C. and Sin, K.S. (1984): Adaptive filtering prediction and control, Prentice–Hall, Inc. Inglewood Cliffs.

[6] Hill, D.J. Middleton, R.H., Goodwin G.C., (1986): A class of robust adaptive control algorithms. IFAC, Adaptive control signal process, Lund Institute, pp. 25–30.

[7] Owens, D.H., A. Chotai (1983): Robust controller design for linear dynamic systems using approximate models. Proc. IEE, 130(0), pp.45–57.

[8] Owens, D.H., A. Chotai (1984): Stability and performance deterioration due to modelling errors in 'Nonlinear systems design'(Ed. J.O. Gray, S.A. Billings, D.H. Owens), Peter Peregrinus pp.82–89.

[9] Riedle, B.D. and Kokotovic P.V., (1985): Stability analysis of an adaptive system with unmodelled dynamics. Int. J. Control, Vol. 39, pp.389–402.

[10] Rohrs, C., Valavani L., Athans M. and G. Stein (1982): Robustness of continuous–time adaptive control algorithms in the presence of unmodelled dynamics, Proc. 21st IEEE Conf. Decision Contr., Orlando, pp.3–11.

[11] Rosenbrock, H.H. (1963): The stability of linear time–dependent control systems. J. Electronics and control. Vol. 15, pp.73–80.

[12] Wellstead, P.E., Prager D. and Zanker P. (1979): Pole–assignment self–tuning regulator, Proc. IEE, Vol. 126, No. 8, pp.781–787.

[13] Owens, D.H. and L. Wang (1987): Robust adaptive controllers for first order models under slow sampling conditions. 26th conference on Decision and Control, U.S.A.

[14] Owens, D.H. and L. Wang (1988): Switch off/switch on algorithms for robust adaptive control. 8th IFAC symposium on Identification and parameter Estimation. Beijing, China.

[15] Wang, L. and D.H. Owens (1988): Robust pole–assignment adaptive controllers with on–line tuning parameters. IEE control conference, Oxford, U.K.

[16] Wang, L. (1988): Robustness of adaptive control systems. Ph.D thesis, University of Sheffield.

[17] Goodwin, G.C., D.J. Hill, D.Q. Mayne and R.H. Middleton (1985): Adaptive robust control (convergence, stability and performance). Technical report, Dept. of Electrical and Computer Engineering, University of Newcastle.

Observer based design for robust stabilization of nonlinear systems

W. Dayawansa G. Knowles C. Martin

1 Introduction

In this paper we discuss a new approach and present a number of intriguing results for the first time. For the purpose of the discussion, in this section we consider a very general configuration which will include parameter uncertainties and bounded disturbances. Our main result is that this configuration allows the design of *robust controllers of a linear type* which will simultaneously handle these unknown *variations in the parameters and certain bounded disturbances.*

Consider the following system

$$x^{(n)} = f(x, x^{(1)}, \cdots, x^{(n-1)}) + u(t) \tag{1.1}$$

where for simplicity of exposition we let $x(t) \in R^n$, $u(t) \in R^1$ [1] and as an output simply $x(t)$.[2] Our standing assumptions in this section will be that $\partial/\partial x^{(j)} f(x, x^{(1)}, \cdots, x^{(n-1)})$ is bounded on R^n for $j = 0, 1, 2, \cdots, n-1$ and that the uncertain parameters are also bounded. Allowing there to be the possibility of even "large" uncertainties seems extremely realistic as certain describing parameters such as hinge friction levels are dependent upon temperature, and viscoelastic and material damping may have long–term transient effects.

Using the mean value theorem, we may consider the following system of nonlinear equations

$$x^{(n)} = \sum_{j=1}^{n-1} a_j(x, x^{(1)}, \cdots, x^{(n-1)}) x^{(j)} + u(t) \tag{1.2}$$

where the a_j's are C^1 functions which are bounded on R^n. The corresponding matrix representation can be given as:

$$\begin{pmatrix} \dot{z}_1 \\ \vdots \\ \dot{z}_{n-2} \\ \dot{z}_{n-1} \end{pmatrix} = \begin{pmatrix} z_2 \\ \vdots \\ z_{n-1} \\ a_1(z)z_1 + \cdots + a_n(z)z_n \end{pmatrix} + \begin{pmatrix} 0 \\ 0 \\ \cdots \\ 0 \\ 1 \end{pmatrix} u. \tag{1.3}$$

[1] Our results have partial extensions to the MIMO situation utilizing Brunovsky canonical forms and the full feedback group.

[2] We can consider much more general outputs $y = g(x, x^{(1)}, x^{(2)}, \cdots, x^{(n-1)})$ subject to some growth constraints.

One approach for stabilization (which we do not adopt here) is to use nonlinear state feedback

$$u = -\sum_{i=1}^{n} a_i(z) z_i + K(z)$$

to first linearize the system and then apply linear state feedback. The difficulty of this approach when faced with possible parameter uncertainty is that the method depends upon *precise knowledge* of the coefficients $a_j(z)$ and the need to measure *all* states. Instead we will employ only input and output measurements to construct the feedback controller.

The following design has the unexpected additional advantage of allowing the implementation of a *linear robust controller design* which does not rely upon any adaptive parameter update mechanisms. For purposes of exposition, we consider only the output equation of the form

$$y = z_1.$$

The linear observer will be given by

$$\dot{w} = Fw + bu + cy, \tag{1.4}$$

where F, b, c are constant. We require a technical lemma which will state that there exists a choice of F, b, c, K such that the composite system is globally asymptotically stable independent of the parameter uncertainties and bounded disturbances.

Consider the following family of linear systems:

$$\begin{pmatrix} \dot{z}_1 \\ \vdots \\ \vdots \\ \vdots \\ \dot{z}_n \end{pmatrix} = \begin{pmatrix} 0 & 1 & & & \\ & 0 & 1 & & \\ & & & \ddots & \\ & & & & 1 \\ a_n(\sigma) & & \cdots & & a_1(\sigma) \end{pmatrix} \begin{pmatrix} z_1 \\ \vdots \\ \vdots \\ \vdots \\ z_n \end{pmatrix} + \begin{pmatrix} 0 \\ \vdots \\ \vdots \\ 0 \\ 1 \end{pmatrix} u \tag{1.5}$$

$$y = z_1$$

Here $a_1(\sigma), a_2(\sigma), \cdots, a_n(\sigma)$ are bounded functions of some (unknown) parameter σ. Then there exists an "observer"

$$\begin{aligned} \dot{w} &= Fw - Ey + bu, \\ u &= -Kw + v \end{aligned} \tag{1.6}$$

with F, E, K fixed, such that the composite system below is globally asymptotically stable for all values of the parameter σ. Assuming this lemma, we will prove the robust stabilizability (and partial control) of the nonlinear system

$$x^{(n)} = f(x, x^{(1)}, x^{(2)}, \cdots, x^{(n-1)}) + u(t) \tag{1.7}$$

subject to bounded parameter uncertainty [3,6] and/or bounded disturbances [2,4].

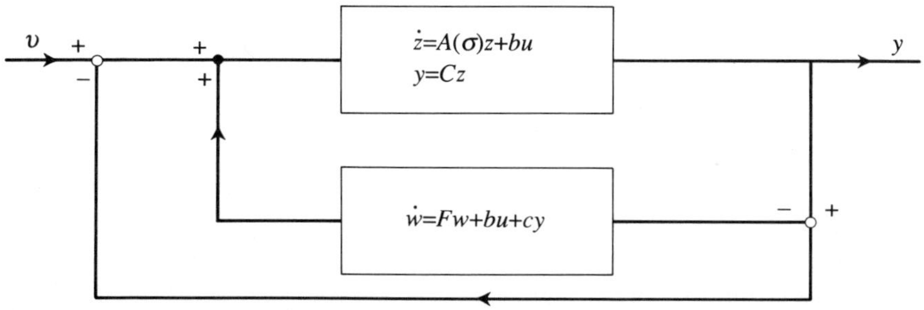

Our results seem to indicate the existence of some very general underlying theory concerning robust stabilization of nonlinear systems utilizing this observer configuration. There are a number of issues which we would suggest require much further investigations. The role of non–minimum phase zero structure is not well understood, the relationship to (A, B)–invariant subspaces and the relationship to structured perturbations [1,2,4,5].

2 Observer design

Consider a nonlinear system of the form

$$\frac{d^n x(t)}{dt^n} = f(x(t), x^{(n)}(t), \cdots, x^{(n-1)}(t)) + u(t). \tag{2.1}$$

Here $x \in R^n$ and $x^{(j)}(t)$ is the j–th derivative of $x(t)$ in time. We will assume for purposes of discussion that zero is an equilibrium point for (2.1). Also, for simplicity we take the measured output to be $x(t)$; although we may, in fact, allow for more complex outputs.

For the purposes of this paper, we will assume that all partial derivatives $\partial f/\partial x^{(i)}$ are bounded functions on R^n, and that $f : R^n \to R$ is C^2.

We shall now show that the system (2.1) can in fact be stabilized using a linear controller which uses the input u and the output y but does not require the complete state $(x, x^{(1)}, \cdots, x^{(n-1)})$.

Let us first put (2.1) in state space form:

$$\left.\begin{aligned} \frac{d}{dt}\begin{pmatrix} z_1 \\ \vdots \\ \vdots \\ z_n \end{pmatrix} &= \begin{pmatrix} z_2 \\ \vdots \\ \vdots \\ f(z_1, \cdots, z_n) \end{pmatrix} + \begin{pmatrix} 0 \\ \vdots \\ 0 \\ 1 \end{pmatrix} u \\ y &= z_1 \end{aligned}\right\} \tag{2.2}$$

Remark [2.1]

One approach which has been proposed in the literature to stabilize (2.2) is to use nonlinear state feedback to cancel out the term $f(z_1, z_2, \cdots, z_n)$. Then they apply linear state feedback to stabilize the system. Such an approach depends to a large extent on precise knowledge of f and the state vector $[z_1, z_2, \cdots, z_n]^t$. We do not adopt this approach here.

Our solution requires the bounds on $\partial f/\partial z_i$ and the input and output measurements, and hence will work when some parameters of $f(z)$ are uncertain.

Among the applications of this methodology, one can use feedback control directly or one could first apply nonlinear state feedback first and then use our feedback controller to account for inaccuracies of the state estimation and the uncertainty of f.

Given $z = (z_1, z_2, \cdots, z_n) \in R^n$, we have

$$\left.\begin{aligned} f(z) &= f(0) + \textstyle\int_0^1 \frac{d}{dt} f(tz)dt \\ &= \textstyle\sum_{i=1}^n \left(\int_0^1 \frac{\partial f}{\partial z_i}(tz)dt \right) z_i \end{aligned}\right\} \tag{2.3}$$

Let $g_i(z)$ be defined as

$$\int_0^1 \frac{\partial f}{\partial z_i}(tz)dt; i = 1, \cdots, n; z \in R^n. \tag{2.4}$$

Then $g_i : R^n \to R$ is C^1 and bounded since $(\partial f/\partial z_i)(z)$ is bounded. We can now express (2.2) as

$$\frac{d}{dt}\begin{pmatrix} z_1 \\ \vdots \\ z_{n-1} \\ z_n \end{pmatrix} = \begin{pmatrix} z_2 \\ \vdots \\ z_n \\ \sum g_i(z) z_i \end{pmatrix} + \begin{pmatrix} 0 \\ \vdots \\ 0 \\ 1 \end{pmatrix} u. \tag{2.5}$$

An observer for this system can be expressed as

$$\frac{d}{dt}\begin{pmatrix} w_1 \\ \vdots \\ w_{n-1} \\ w_n \end{pmatrix} = \begin{pmatrix} w_2 \\ \vdots \\ w_n \\ 0 \end{pmatrix} - \begin{pmatrix} D_1 \\ \vdots \\ D_{n-1} \\ D_n \end{pmatrix} w_1 + \begin{pmatrix} D_1 \\ \vdots \\ D_{n-1} \\ D_n \end{pmatrix} y + \begin{pmatrix} 0 \\ \vdots \\ 0 \\ 1 \end{pmatrix} u. \tag{2.6}$$

and feedback from the "observed state" w,

$$u = -Kw = -[K_1, K_2, \cdots, K_n]w. \tag{2.7}$$

The composite system (2.5), (2.6), and (2.7) takes the following form:

$$\frac{d}{dt}\begin{pmatrix} z_1 \\ \vdots \\ z_{n-1} \\ z_n \\ \eta_1 \\ \vdots \\ \eta_{n-1} \\ \eta_n \end{pmatrix} = \begin{pmatrix} 0 & 1 & & & & & & 0 \\ & & \ddots & & & & & \\ & & & 1 & & & & \\ g_1(z)-K_1 & \cdots & & g_n(z)-K_n & K_1 & \cdots & \cdots & K_n \\ & & & & -D_1 & 1 & & \\ & 0 & & & \vdots & & \ddots & \\ & & & & & & & 1 \\ g_1(z) & & & g_n(z) & -D_n & & & 0 \end{pmatrix} \begin{pmatrix} z_1 \\ \vdots \\ z_{n-1} \\ z_n \\ \eta_1 \\ \vdots \\ \eta_{n-1} \\ \eta_n \end{pmatrix} \quad (2.8)$$

where $\eta_i = (z_i - w_i)$ is the error in the "observed state."

Theorem [2.2]

There exists a choice of an observer based feedback controller (2.6) and (2.7) such that the composite system (2.8) is globally asymptotically stable. Moreover, the controller can be designed by using the knowledge of the bounds on $\partial f/\partial z_i(z)$ (for $j = 1, 2, \cdots, n$) only. This feedback mechanism will stabilize systems of the type (2.1) for which $\partial f/\partial z_i(z)$ (for $j = 1, 2, \cdots, n$).

Remark [2.3]

Before the proof is undertaken, we observe that the bounds on $(\partial f/\partial z_i)(z)$ may be taken to be large. In this case the stability of the overall system is very robust.

The above statements can be applied to certain MIMO systems, as previously indicated, of the type

$$\left.\begin{array}{c} \frac{d^{r_1}}{dt^{r_1}} x_1(t) = f_1(x_1, \cdots, x^{(r_1-1)}, x_2^{(r_2-2)}, \cdots, x_m^{(r_m-1)}) + u_1(t) \\ \vdots \\ \frac{d^{r_m}}{dt^{r_m}} x_m(t) = f_m(x_1, \cdots, x^{(r_1-1)}, x_2^{(r_2-2)}, \cdots, x_m^{(r_m-1)}) + u_m(t) \end{array}\right\} \quad (2.9)$$

We first prove this result on simultaneous stabilization for a family of linear systems. As far as we are aware, this is in itself new in terms of this approach.[3]

Consider a parameterized family of linear systems (possibly after feedback) given by

$$\frac{d}{dt}\begin{pmatrix} z_1 \\ \vdots \\ z_{n-1} \\ z_n \end{pmatrix} = \begin{pmatrix} 0 & 1 & & \\ & & \ddots & \\ & & & 1 \\ a_{1\sigma} & & \cdots & a_{n\sigma} \end{pmatrix} \begin{pmatrix} z_1 \\ \vdots \\ z_{n-1} \\ z_n \end{pmatrix} + \begin{pmatrix} 0 \\ \vdots \\ 0 \\ 1 \end{pmatrix} u \quad (2.10)$$

$\sigma \in I$ some index set for the family.

[3]The authors would like to draw attention to some recent related developments in the work of Petesen, I.R. and Hollot, C.V. [6] and the work of Khalil, H.K. and Saberi, A. [3]

Assumption [2.4]

There exists an L with $|a_{i\sigma}| < L$ for $\sigma \in I$ and $i = 1, 2, \cdots, n$. Now consider the observed based feedback design configuration for (2.10) as in (2.6) and (2.7). A straightforward computation shows that the composite system takes the form

$$\frac{d}{dt}\begin{pmatrix} z_1 \\ \vdots \\ z_{n-1} \\ z_n \\ \eta_1 \\ \vdots \\ \eta_{n-1} \\ \eta_n \end{pmatrix} = \begin{pmatrix} 0 & 1 & & & & & 0 & \\ & & \ddots & & & & & \\ & & & 1 & & & & \\ a_1(z)-k_1 & \cdots & & a_n(z)-k_n & k_1 & \cdots & \cdots & k_n \\ & & & & -D_1 & 1 & & \\ & 0 & & & \vdots & & \ddots & \\ & & & & & & & 1 \\ a_{1\sigma} & & & a_{n\sigma} & -D_n & & & 0 \end{pmatrix}\begin{pmatrix} z_1 \\ \vdots \\ z_{n-1} \\ z_n \\ \eta_1 \\ \vdots \\ \eta_{n-1} \\ \eta_n \end{pmatrix} \tag{2.11}$$

where $\eta = z - w$.

Theorem [2.5]

1. With the above boundedness assumption there exists a fixed controller given by (2.6) and (2.7) such that (2.11) is asymptotically stable independent of σ in I.

2. There exists a positive definite matrix P such that $V(x) = x^t P x$ is a Lyapunov function for systems described by (2.11) independent of choices of σ in I.

   Proof:

   Let us introduce

$$A^{(\sigma,k,d)} = \begin{pmatrix} 0 & 1 & & & & & 0 & \\ & & \ddots & & & & & \\ & & & 1 & & & & \\ a_1(z)-k_1 & \cdots & & a_n(z)-k_n & k_1 & \cdots & \cdots & k_n \\ & & & & -D_1 & 1 & & \\ & 0 & & & \vdots & & \ddots & \\ & & & & & & & 1 \\ a_{1\sigma} & & & a_{n\sigma} & -D_n & & & 0 \end{pmatrix} \tag{2.12}$$

Without loss of generality, we assume that for a certain value σ_0 of the parameter σ

$$a_{1\sigma_0} = a_{2\sigma_0} = a_{n\sigma_0} = 0. \tag{2.13}$$

Pick $k = (k_1, k_2, \cdots, k_n)$ and $d = (d_1, d_2, \cdots, d_n)$ so that the roots of $s^n + k_n s^{n-1} + \cdots + k_1 = 0$ are $(-1, -2, \cdots, -n)$ and the roots of $s^n + d_1 s^{n-1} + \cdots + d_n$ are $(-n-1, -n-2, \cdots, -2n)$, respectively. Fix this k and d at these values for the remainder of this paper.

For each positive real r, define

$$\begin{aligned} k^r &= (r^n k_1, r^{n-1}k_2, \cdots, rk_n) \\ d^r &= (rd_1, r^2 d_2, \cdots, r^n d_n) \end{aligned} \tag{2.14}$$

The desired compensator will be obtained for a certain range of r whenever $K = k^r$ and $D = d^r$. By our choice of k and d,

$$A^{(\sigma_0,k,d)} = \begin{pmatrix} 0 & 1 & & & & 0 & & \\ & & \ddots & & & & & \\ & & & 1 & & & & \\ -k_1 & \cdots & & k_n & k_1 & \cdots & \cdots & k_n \\ & & & & -d_1 & 1 & & \\ & 0 & & & \vdots & & \ddots & \\ & & & & & & & 1 \\ & & & & -d_n & & & 0 \end{pmatrix} \tag{2.15}$$

It immediately follows that $A^{(\sigma_0,k,d)}$ will have as eigenvalues $(-1, -2, \cdots, -2n)$. Let T_0 be the $2n \times 2n$ matrix whose columns consist of eigenvectors of $A^{(\sigma_0,k,d)}$ and set

$$P_0 = T_0^{-t} T^{-1}. \tag{2.16}$$

It follows that $P_0 A^{(\sigma_0,k,d)} + [A^{(\sigma_0,k,d)}]^t P_0$ will be negative definite.

Let ζ_{2n} denote the vector space of real symmetric $2n \times 2n$ matrices. Define $F : R^{2n\times 2n} \to \zeta^{2n}$ by $F(A) = P_0 + A^t P_0$. Clearly F is continuous. Therefore there will exist some neighbourhood U of $A^{(\sigma_0,k,d)}$ in $R^{2n\times 2n}$ with $P_0 A + A^t P_0$ negative definite for $A \in U$. By choosing U smaller if necessary, we can suppose that every A in U has distinct eigenvalues which lie on the negative real axis.

Choose $\delta > 0$ such that for $|\mathrm{k}_i - k_i| < \zeta$ and $|e_i| < \delta$, $i = 1, 2, \cdots, n$ the corresponding matrix $A(\mathrm{k}, k, d, e)$ is in U.

Here $A(\mathrm{k}, k, d, e)$ is given by

$$A(\mathrm{k}, k, d, e) = \begin{pmatrix} 0 & 1 & & & & & 0 & \\ & & \ddots & & & & & \\ & & & & 1 & & & \\ \mathrm{k}_1 & \cdots & & \mathrm{k}_n & k_1 & \cdots & & k_n \\ & & & & d_1 & 1 & & \\ & 0 & & & \vdots & & \ddots & \\ & & & & & & & 1 \\ e_1 & \cdots & & e_n & d_n & & & 0 \end{pmatrix} \tag{2.17}$$

With k, d, e as above and for $r > 0$, define $\mathrm{k}^r, k^r, d^r, e^r$ as in (2.14). Introduce

$$\Phi_r = \begin{pmatrix} 1 & & & & & & & \\ & r & & & & & 0 & \\ & & \ddots & & & & & \\ & & & r^{n-1} & & & & \\ & & & & 1 & & & \\ & & & & & r & & \\ & 0 & & & & & \ddots & \\ & & & & & & & r^{n-1} \end{pmatrix} \tag{2.18}$$

The eigenvalues of $A(\mathrm{k}, k, d, e)$ are denoted by $(\lambda_1, \lambda_2, \cdots, \lambda_{2n})$. Let T be the $2n \times 2n$ matrix whose columns consist of the eigenvectors of A. (From our assumptions on U, $\lambda_1, \lambda_2, \cdots, \lambda_{2n}$, there will be such a real nonsingular T.)

Proposition [2.6]

With the above notation, the following hold.

$$A(\mathrm{d}^r, k^r, d^r, e^r)\Phi_r = r\Phi_r A(\mathrm{k}, k, d, e) \tag{2.19}$$

and

$$A(\mathrm{k}^r, k^r, d^r, e^r) \tag{2.20}$$

has eigenvalues $(r\lambda_1, r\lambda_2, \cdots, \lambda_{r2n})$ and a corresponding matrix $\Phi_r T$ of eigenvectors.

The identity (2.19) can be found by direct calculation. (2.20) now follows almost immediately from (2.19).

Since $a_{i\sigma}$ is bounded on I, $i = 1, 2, \cdots, n$, it is possible to choose r to satisfy

$$\left|\frac{a_{i\sigma}}{r^{n-i+1}}\right| < \delta \tag{2.21}$$

for all $\sigma \in I$; $i = 1, 2, \cdots, n$.

If we let $a_\sigma(r) := ((1/r^n)a_{1\sigma}, (1/r^{n-1})a_{2\sigma}, \cdots, (1/r)a_{n\sigma})$.

$$A(\mathrm{k} - a_\sigma(r), k, d, a_\sigma(r)) \in U, \ \forall \sigma \in I. \tag{2.22}$$

From this and (2.20), we can conclude that $A((k - a_\sigma(r))^r, k^r, d^r, (a_\sigma(r)^r)$ will have eigenvalues on the negative real axis. From Proposition [2.6] we have that

$$A((\mathrm{k} - a_\sigma(r)^r, k^r, d^r, a_\sigma(r)^r) = A(\sigma, k^r, d^r), \tag{2.23}$$

which now shows that we may simultaneously stabilize the family (2.10) by choosing $K = k^r$ and $D = d^r$.

To prove part (2) of Theorem [2.5], we consider the parameterized family

$$\dot{X} = A(\sigma^r, k^r, d^r)X; \ \sigma \in I. \tag{2.24}$$

Choose T_0 as before to be the matrix of eigenvectors of $A(\sigma, k, d)$ and $P_0 = T_0^{-t}T_0^{-1}$. If we set

$$P = \Phi_r^{-1}P_0\Phi_r^{-1}, \tag{2.25}$$

then P is positive definite. Moreover, by differentiating X^tPX along an arbitrary solution of (2.24), we obtain

$$(d/dt)\left(X^TPX\right) = 2X^t\Phi_r^{-1}A(\sigma, k^r, d^r)X$$

An application of this becomes

$$= 2X^t\Phi_r^{-1}P_0(\nu A(\mathrm{k} - a_\sigma(r), k, d, a_\sigma(r))).$$

Or, rewriting, we get

$$(d/dt)(X^tPX) = 2r(\Phi_r^{-1}X)^tP_0A(\mathrm{k} - a_\sigma(\nu), k, d, a_\sigma\nu)\Phi_r^{-1}X.$$

By construction, $A(\mathrm{k} - a_\sigma(r), k, d, a_\sigma(r))$ is always in U. By inspection we see that $(d/dt)(X^tPX) < 0$ for every $X \neq 0$. Hence $V(x) = x^tPx$ is a Lyapunov function for every σ, as claimed.

Proof of Theorem [2.2]

In the proof of Proposition [2.6] above, P_0, k, d were fixed but r is allowed to take on any value provided only that it satisfies (2.21). That is, it depends only upon the bound, say I_o, of $|a_{i\sigma}|$ for $\sigma \in I$, $i = 1, 2, \cdots, n$. With this in mind, let us choose L as follows:

$$L = \sup(|\partial f_i/\partial z_i)(z)| : \ i = 1, 2, \cdots, n; \ z \in R^n) \tag{2.26}$$

and now choose r, as before, to correspond to this value of L. Consider the following modification of (2.8).

$$\frac{d}{dt}\begin{pmatrix} z_1 \\ \vdots \\ z_{n-1} \\ z_n \\ \eta_1 \\ \vdots \\ \eta_n \end{pmatrix} = \begin{pmatrix} 0 & 1 & & & 0 & & \\ & & \ddots & & & & \\ & & & & 1 & & \\ g_1(z) - r^nk_1 & \cdots & g_n(z) - rk_n & & r^nk_1 & \cdots & rk_n \\ & & & & -rd_1 & 1 & \\ & 0 & & & \vdots & \ddots & \\ & & & & & & 1 \\ g_1(z) & & g_n(z) & & -rd_n & \cdots & 0 \end{pmatrix}\begin{pmatrix} z_1 \\ \vdots \\ z_{n-1} \\ z_n \\ \eta_1 \\ \vdots \\ \eta_n \end{pmatrix} \tag{2.27}$$

which, for simplicity , we write as:

$$(d/dy)X = A(X, k^r, d^r)X \tag{2.28}$$

Now consider the solutions of (2.27). Along any trajectory of (2.27) we have

$$(d/dt)(X^tPX) = 2X^tPX = 2X^tPA(X, k^r, d^r)X. \tag{2.29}$$

According to Theorem [2.5], $V(x) = x^t P x$ is a Lyapunov function for any system (2.24) for any chosen value of $\sigma \in I$. As such, the system (2.27) is stable.

Remark [2.7]

There are some easy extensions of the above methodology which allows for some zero structure to appear in (2.10). However, the overall picture invites much further investigation.

References

[1] Barmish, B.R., and Wei, K.H., "Simultaneous Stabilizability of Single Input–Single Output Systems," Modelling, Identification and Robust Control. Eds, E.I. Byrnes, and A. Lindquist, Elsevier (North–Holland), 1986, pp. 431–443.

[2] Bhattacharyya, S.P., "Robust Stabilization Against Structured Perturbations," Springer–Verlag: Lecture Notes in Control and Information Science, vol. 99, 1987.

[3] Khalil, H.K. and Saberi, A., "Adaptive Stabilization of a Class of Nonlinear Systems Using High–Gain Feedback," IEEE Trans. Aut. Contr., vol. AC–32, no. 11, Nov. 1987

[4] Kimura, H., "Robust Stabilizability for a Class of Transfer Functions," IEEE Trans. Aut. Contro., vol. AC–29, no. 9, Sept. 1984.

[5] Kwakernaak, H., "A Condition for Robust Stabilizability," Systems and Control Letters, vol. 2, no. 1, July 1982.

[6] Petersen, I.R. and Hollot, C.V., "High Gain Observers Applied to Problems in the Stabilization of Uncertain Linear Systems, Disturbance Attenuation and H^∞–Optimization," Proc. 1988 American Control Conference, Atlanta, Georgia.

Evaluation of robustness for systems with structured perturbations

G. Celentano F. Garofalo L. Glielmo

Abstract

The question of the stbility of a system affected by parameter uncertainties is dealt with. The system which has been considered is made up of nominal part, linear, time-invariant and asymptotically stable, and an uncertain part, which is time-varying and structured, i.e., given as a multiplication of known matrix by an uncertain one. Using Lyapunov second method as a stability test, various rectangular *regions of stability* are obtained in the parameter space; their form can be a *priori* fixed to better fit the relative ranges of variation of the uncertain parameters. A further, less pessimistic evaluation of the region of stability is given, which generally determines an asymmetrical region. Numerical examples and comparisons with recent results are presented.

1 Introduction

Main aim of the control engineer is to design a system which is able to satisfy certain specifications regarding its input-output behaviour. The part of the overall system to be controlled, the so called "plant", is generally designed taking in mind a variety of possible applications and a set of technological and economical constraints, so that it is quite uncommon that no control system is necessary. In certain cases the plant is completely *known*, at least in an engineering sense, and what is to be modified is, for example, the velocity of its response to certain commands, the shape of that response, and so on. On other occasions, even though we can give a mathematical description of the plant, some parameters in it are not exactly known, as happens for the current gain of a transistor or the torque load of an electrical motor. In classical control theory such concepts as the *gain and phase margins* were developed just to quantify the robustness of the system to certain kinds of uncertainties.

Roughly speaking we can say that now two main approaches exist. One is the *Probabilistic* approach: in it the uncertainties affecting the system are viewed as realizations of stochastic processes, so that typically the control is mainly interested on the average values of the variables, and the degree of freedom which may be left are exploited to minimize some indices related to the probabilistic descriptions, as the variance of the error. Unfortunately this kind of analysis and synthesis methods may become very difficult and practically unusable when certain crucial hypotheses (Gaussian distributions, ergodicty, stochastic independence, linearity of the systems, and so on) are not met.

On the other hand, the *deterministic* approach tries to ensure the behaviour of the system regardless of the particular realizations and the stochastic properties of the disturbances, provided that their values range in a *priori* known sets. Strong conditions on the disturbances can thus be avoided. Examples of this technique are above-mentioned gain and phase margins and, more recently, the results which have been derived by *Kharitonov's theorem* [6] for time-invariant systems.

In this paper we refer to the deterministic approach and use Lyapunov's second method [4] for guaranteeing the stability of a linear system affected by time-varying perturbations. Similar analyses have been carried out in [3; 5; 7-10].

2 Statement of the problem

In many practical situation a control system is made up of a linear time-invariant controller and a plant which can be seen as linear time-invariant system affected by (time-varying) uncertainties in the form of modeling errors and parameter variations. The resulting system can be written as follows:

$$\dot{x}(t) = Ax(t) + f(x(t), t), \tag{1a}$$

$$x(t_0) = x_0, \tag{1b}$$

where $x \in \mathbf{R}^n$ is the state, $A \in \mathbf{R}^{n\times n}$ is a known matrix, and $f(\cdot) : \mathbf{R}^n \times \mathbf{R} \mapsto \mathbf{R}^n$ represents the various perturbations acting on the system and satisfies certain structural relations. In particular the structural relations we consider in this paper are

$$f(x, t) = E(x, t)x, \tag{2}$$

with

1. $E(x,t) = GF(x,t)$, where $G \in \mathbf{R}^{n\times m}$ is a known matrix and $F(\cdot) : \mathbf{R}^n \times \mathbf{R} \mapsto \mathbf{R}^{m\times n}$;
   or

2. $E(x,t) = \Sigma_{i=1}^{\nu} H_i p_i(x,t)$, where $H_i \in \mathbf{R}^{n\times n}$ for $i = 1, 2, \cdots, \nu$, are known matrices and $p_i(\cdot) : \mathbf{R}^n \times \mathbf{R} \mapsto \mathbf{R}$, for $i = 1, 2, \cdots, \nu$.

In the following we will refer to the above cases as *Structure a)* or *Structure b)* respectively. We will drop the arguments of the functions when no confusion is likely to occur. The vector norm which will be used is the Euclidean one, the matrix norm is the induced one, i.e.,

$$||D|| \triangleq \lambda_{\max}^{1/2}[D^T D]$$

where $\lambda_{\max}[\cdot]$ represents the operation of taking the maximum eigenvalue of the argument. Similarly $\lambda_{\min}[\cdot]$ will stand for the minimum eigenvalue.

Following the classification introduced by Barrett [2], the perturbation in eq. (1) is *structured*, while in the form (2) it can be considered either weakly *structured* or *highly structured* according as the bound on either the norm or the individual elements of matrix $E(x,t)$ is known. In [7] the unstructured problem was addressed and results for the structured cases were obtained as particularizations. The highly structured problem (2) was explicitly considered in [9-10]. Further levels of structure are added in *Structure a)* and *Structure b)*, which respectively represent weakly and highly structured perturbations. *Structure b)* has been recently taken into consideration in [5]. It should be noted that both the highly structured case of eq. (2) and *Structure a)* can be viewed, by a proper choice of matrices H_i, as special cases of *Structure b)*.

Assumption 1. The nominal dynamical system

$$\dot{x} = Ax, \tag{3}$$

is asymptotically stable.

The problem is to determine upper bounds for the norm of the disturbances so as the asymptotic stability of the system (1) is assured regardless of the particular form of the disturbance functions and the particular realizations of them. In the space of the uncertain parameters these bounds will describe a *stability region* which is to be compared with the *feasible region* of those parameters.

Here we wish to point out that quite conservative results are obtained if the two regions are very dissimilar. For example, let us suppose that, in the framework of *Structure a)*, $m = 1$ and then

$$F = (f_{11}\ f_{12} \cdots f_{1n}), \tag{4}$$

with $|f_{1i}| < s_i$, for $i = 1, 2, \cdots, n$, and $\underline{s} \triangleq \min_i s_i \ll \bar{s} \triangleq \max_i s_i$. Then stability bounds as

$$\|F\| = \sqrt{f_{11}^2 + f_{12}^2 + \cdots + f_{1n}^2} < \rho$$

or

$$|f_{1i}| < \rho_i = \varepsilon_i s_i, \quad \text{for } i = 1, 2, \cdots, n,$$

with $\max_i \varepsilon_i / \min_i \varepsilon_i \gg 1$, are not appropriate because the stability regions they establish are too small (see Figure 1).

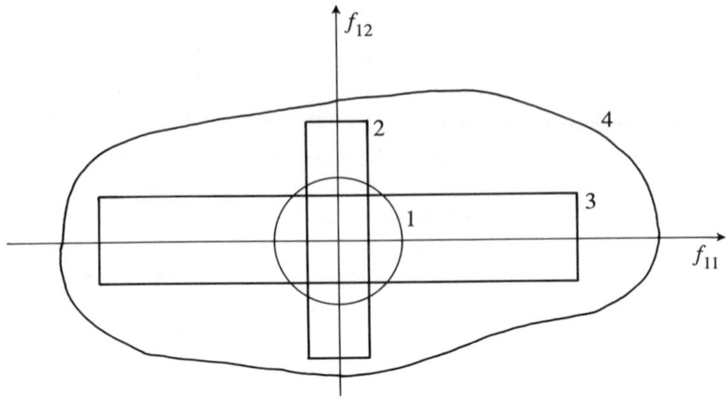

Figure 1. Stability regions vs. feasible region.

1. $\|F\| < \rho$;
2. $f_{1i} < \rho_i$;
3. feasible region;
4. actual stability region.

3 Main results

As already said, the stability test which will be used in this paper is based on Lyapunov second method (see for example, [4]). Using this approach Patel and Toda [7] showed that, selecting as Lyapunov function $V(x) \triangleq x^T P x$, with P solution of

$$A^T P + PA = -2Q, \quad Q \text{ p.d.}, \tag{5}$$

then the stability region of system eqs. (1) and (2) is described by

$$\|E(x,t)\| < \rho \triangleq \frac{\lambda \min[Q]}{\lambda \max[P]}, \quad \text{for any } x \text{ and } t. \tag{6}$$

Moreover they showed that the best result, that is the maximum ρ, is obtained by choosing matrix Q equal to the identity matrix I_n. Similar results were obtained in [1], in the framework of ultimate boundedness control of uncertain systems.

The above-mentioned bound may turn out to be rather conservative in case of *Structure a)*. In this respect we give the following

Theorem 1.
If the uncertainties satisfy Structure a) *and*

$$\|F(x,t)\| < \frac{\lambda \min[Q]}{\|PG\|}, \quad \textit{for any } x \textit{ and } t, \tag{7}$$

where P and Q derive from eq. (5), *then system* (1) *is globally asymptotically stable.*

Proof.
Choosing as Lyapunov function candidate

$$V(x) \triangleq x^T P x, \tag{8}$$

with P solution of eq. (5), its derivative along the solutions of system (1) under *Structure a)* uncertainties is given by

$$\begin{aligned} \dot{V}(x,t) &\triangleq \nabla_x V(x)(Ax + GF(x,t)x) \\ &= -2x^T Qx + 2x^T PGF(x,t)x \\ &\leq -2\lambda_{\min}[Q]\|x\|^2 + 2\|PG\|\,\|F(x,t)\|\,\|x\|^2. \end{aligned} \tag{9}$$

If eq. (7) applies, $\dot{V}(x,t) < 0$ and the thesis follows. □

It should be noted that, since $\|PG\| \leq \|P\|\,\|G\|$, the stability region characterized by eq. (7) is larger than that one derived by eq. (6). Unfortunately, but expectedly, the largest region is not usually attained by $Q = I_n$. This is pointed out in the following simple

Example.
Let us consider the system

$$\dot{x}_1 = -x_1 + 100x_2 + f_{11}(x_1, x_2, t), \tag{10a}$$

$$\dot{x}_2 = -0.1x_1 - 2x_2. \tag{10b}$$

It follows

$$\begin{aligned} A &= \begin{pmatrix} -1 & 100 \\ -0.1 & -2 \end{pmatrix}, \\ G &= \begin{pmatrix} 1 \\ 0 \end{pmatrix}, \\ F &= (f_{11}\ 0). \end{aligned} \tag{11}$$

Solving the Lyapunov equation (5) for $Q = I_2$, we have, using eq. (6),

$$|f_{11}| < \frac{1}{278.25},$$

while, using eq. (7), we have

$$|f_{11}| < \frac{1}{5.573}.$$

Finally, optimizing the bound (7) with respect to a diagonal Q matrix, one obtains

$$|f_{11}| < \frac{1}{1.015}.$$

□

Theorem 1, like Patel and Toda's result, consider the disturbance matrix $F(x,t)$ as a whole since the bound is given on the norm of that matrix. Nevertheless, in some situations and typically in the design of robust linear compensators, *Structure a)* may arise with matrix F partitioned in m rows which refer to different kinds of uncertainties of the plant. It results that different amplitudes of variations are possible and a unique overall stability margin is bound to be too much conservative.

Thus let

$$F(x,t) = \begin{pmatrix} f_1^T(x,t) \\ f_2^T(x,t) \\ \vdots \\ f_m^T(x,t) \end{pmatrix}, \tag{12}$$

with $f_i(\cdot) : \mathbf{R}^n \times \mathbf{R} \mapsto \mathbf{R}^n$, for $i = 1, 2, \cdots, m$. Suppose that the *relative* magnitudes of the norms $||f_i||$, for $i = 1, 2, \cdots, m$, are given, i.e., the form (which we desire) of the stability hyperrectangle in the space $(||f_1||, ||f_2||, \cdots, ||f_m||)$ is established by means of its relative dimensions $\varphi_1, \varphi_2, \cdots, \varphi_m$ (see Figure 2). Moreover let the matrix G be the column-partitioned

$$G = (g_1 \; g_2 \; \cdots g_m). \tag{13}$$

Then the following theorem holds true.

Theorem 2.
If P and Q are p.d. matrices which satisfy Lyapunov equation (5) *and*

$$||f_i(x,t)|| < \frac{\varphi_i}{\sum_{k=1}^m ||Pg_k||\varphi_k}\lambda_{\min}[Q], \quad \begin{cases} \textit{for any } x \textit{ and } t, \\ \textit{for } i = 1, 2, \cdots, m, \end{cases} \tag{14}$$

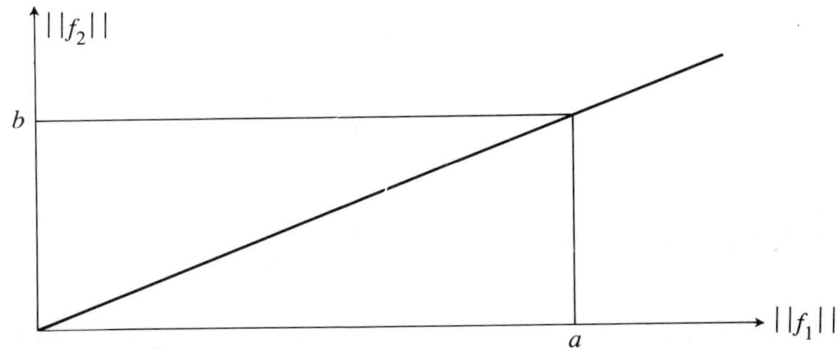

Figure 2. Relative dimensions of the hyperrectangle: $\varphi_1 : \varphi_2 = a : b$.

then system (1) *under* Structure a) *uncertainties is globally asymptotically stable.*

Proof.
Choosing the Lyapunov function candidate $V(x) \triangleq x^T P x$, with P p.d. solution of eq. (5), the derivative along the solutions of system (1) is given by

$$\begin{aligned} \dot{V}(x,t) &= -2x^T Q x + 2x^T PGFx \\ &\leq -2\lambda_{\min}[Q]||x||^2 + 2\sum_{k=1}^{m} x^T P g_k f_k^T x. \end{aligned}$$

Since

$$\sum_{k=1}^{m} x^T P g_k f_k^T x \leq \sum_{k=1}^{m} |x^T P g_k|\, |f_k^T x| \leq \sum_{k=1}^{m} ||Pg_k|| ||f_k|| ||x||^2,$$

we have that $\dot{V}(x,t) < 0$ if $\sum_{k=1}^{m} ||Pg_k|| ||f_k|| < \lambda_{\min}[Q]$, which in turn is verified if eq. (14) applies. □

A result similar to that one of Theorem 2 holds true if uncertainties are under *Structure b)*. In this hypothesis, let us suppose that the relative dimensions of the desired hyperrectangle in the space $(p_1, p_2, \cdots, p_\nu)$ are given by means of ν positive real numbers $\psi_1, \psi_2, \cdots, \psi_\nu$. We can state the following

Theorem 3. *If P and Q are p.d. matrices which satisfy Lyapunov equation (5) and*

$$|p_i(x,t)| < \frac{\psi_i}{\sum_{k=1}^{\nu} \|PH_k\|\psi_k} \lambda_{\min}[Q], \quad \begin{cases} \text{for any } x \text{ and } t, \\ \text{for } i = 1, 2, \cdots, \nu, \end{cases} \tag{15}$$

then system (1) *under* Structure b) *uncertainties is globally asymptotically stable.*

Proof.
It is similar to that one of Theorem 2. □

Remark 1:
In case matrices H_k, for $k = 1, 2, \cdots, \nu$, are of unitary rank, i.e., of the form $H - k = h_k \chi_k^T$, then the bound eq. (15) can be written as

$$|p_i(x,t)| < \frac{\psi_i}{\sum_{k=1}^{\nu} \|Ph_k\|\|\chi_k\|\psi_k} \lambda_{\min}[Q], \quad \begin{cases} \text{for any } x \text{ and } t, \\ \text{for } i = 1, 2, \cdots, \nu, \end{cases} \tag{16}$$

Remark 2:
No computation of eigenvalues is necessary in order to evaluate the denominators in eqns (14) and (16). This may be useful if one desires to optimize the above bounds with respect to matrix Q.

Remark 3:
In order to enlarge the stability regions, the above theorems suggest three optimization problems which can be solved by means of ordinary numerical optimization methods for constrained problems. They are

Problem 1.

$$\min_{Q} \|PG\|$$

subject to: Q symmetric;
$\lambda_{\min}[Q] = 1$;
P solution of eq. (5).

Problem 2.

$$\min_{Q} \sum_{k=1}^{m} \|Pg_k\|\varphi_k$$

subject to: Q symmetric;
$\lambda_{\min}[Q] = 1$;
P solution of eq. (5).

Problem 3.

$$\min_Q \sum_{k=1}^{m} ||PH_k||\varphi_k$$

subject to: Q symmetric;
$\lambda_{\min}[Q] = 1$;
P solution of eq. (5).

The constraint on $\lambda_{\min}[Q]$ is easily explained in virtue of the fact that the ratios in eqs. (7), (14), (15) and (16), do not change if Q is multiplied by a scalar [7]. That constraint can be avoided by dividing the above objective functions by $\lambda_{\min}[Q]$ and adding the condition that Q is p.d. It must be added that, for a problem similar to Problem 3, a gradient based descent procedure has been proposed in [5].

Theorems 1, 2 and 3, like all the results in the quoted literature, give stability regions which are symmetrical to the origin of the respective spaces. This is reasonable, since generally the *nominal* system is determined by the average values of the parameters. Nevertheless in certain situations it may be interesting to determine less conservative regions, even though not centered around the origin.

In this respect, let us consider again system (1) under *Structure b)* uncertainties. Define

$$\Gamma^+(S) = \max_i\{0, \lambda_i(S)\},$$

$$\Gamma^-(S) = |\min_i\{0, \lambda_i(S)\}|,$$

where S is a symmetric matrix, and let $(D)_s$ indicate the symmetric part of a square matrix D. Moreover partition the space $(p_1, p_2, \cdots, p_\nu)$ in the 2^ν subsets $\mathcal{S}(s_1, s_2, \cdots, s_\nu)$, when s_i, for $i = 1, 2, \cdots, \nu$ is "+" or "−" according as p_i ranges over positive or negative values.

We have the following

Theorem 4.
Let P and Q be p.d. matrices which satisfy Lyapunov equation (5). *In every subset $\mathcal{S}(s_1, s_2, \cdots, s_\nu)$, with $s_i = +|-$ for $i = 1, 2, \cdots, \nu$, define the sets*

$$\mathcal{R}(s_1, s_2, \cdots, s_\nu) \triangleq \{(p_1, p_2, \cdots, p_\nu) \in \mathcal{S}(s_1, s_2, \cdots, s_\nu) : \\ |p_i| < \frac{\psi_i}{\sum_{k=1}^{\nu} \Gamma^{s_k}[(PH_k)_s]\psi_k} \lambda_{\min}[Q], \text{ for } i = 1, 2, \cdots, \nu\}, \tag{17}$$

and their union

$$\mathcal{R} \triangleq \bigcup_{\text{all } (s_1, s_2, \cdots, s_\nu)} \mathcal{R}(s_1, s_2, \cdots, s_\nu) \tag{18}$$

If

$$(p_1(x,t), p_2(x,t), \cdots, p_\nu(x,t)) \in \mathcal{R}, \quad \textit{for any } x \textit{ and } t, \tag{19}$$

then system (1) *under* Structure b) *uncertainties is globally asymptotically stable.*

Proof.
Choosing the Lyapunov function candidate $V(x) \triangleq x^T P x$, with P p.d. solution of eq. (5), the derivative along the solutions of system (1) satisfies the inequality

$$\dot{V}(x,t) \leq -2\lambda_{\min}[Q]||x||^2 + 2\sum_{k=1}^{\nu}[x^T(PH_k)_s x]p_k(x,t). \tag{20}$$

If now we consider that

$$\lambda_{\min}[(PH_k)_s]||x||^2 \leq x^T(PH_k)_s x \leq \lambda_{\max}[(PH_k)_s]||x||^2,$$

and in every region $\mathcal{R}(s_1, s_2, \cdots, s_\nu)$ the worst-case situation for the second term on the RHS of eq. (20) occurs when the quadratic form $x^T(PH_k)_s x$ is positive, if $s_k = +$, or is negative, if $s_k = -$, for $k = 1, 2, \cdots, \nu$, the thesis follows. □

Remark 4:
The regions $\mathcal{R}(s_1, s_2, \cdots, s_\nu)$ are hyperrectangles whose sides satisfy the desired ratios $\psi_1 : \psi_2, \cdots, \psi_\nu$.

Remark 5:
The asymmetrical region $\mathcal{R}$ defined in eq. (18) is a closer approximation of the generally asymmetrical region in which negative definiteness of $\dot{V}(x,t)$ is *actually* guaranteed. It is, of course, always possible to determine the largest symmetrical region included in it.

Remark 6:
It is apparently more difficult to state a unique optimization problem related to Theorem 4. For example one could maximize the "volume" of region $\mathcal{R}$ or of the largest symmetrical region included.

4 An example

Consider a system whose nominal dynamical matrix A is given by

$$A = \begin{pmatrix} 0 & -2 \\ 0.5 & -50 \end{pmatrix},$$

and whose disturbance matrix E is

$$E = \begin{pmatrix} 0 & 0 \\ p_1(x,t) & 0 \end{pmatrix} = \begin{cases} \begin{pmatrix} 0 \\ 1 \end{pmatrix} (p_1(x,t)\ 0), & \textit{Structure a),} \\ \begin{pmatrix} 0 & 0 \\ 1 & 0 \end{pmatrix} p_1(x,t), & \textit{Structure b),} \end{cases}$$

so that

$$g_1 = \begin{pmatrix} 0 \\ 1 \end{pmatrix},$$

$$H_1 = \begin{pmatrix} 0 & 0 \\ 1 & 0 \end{pmatrix}.$$

The solution P of the Lyapunov equation (5), with $Q = I_2$, is

$$P = \begin{pmatrix} 50.025 & -2 \\ -2 & 0.1 \end{pmatrix}.$$

The bound obtained by application of Theorems (2) and (3) are, in this simple example, both equal to

$$|p_1(x,t)| < \frac{1}{||Pg_1||} = \frac{1}{||PH_1||} = \frac{1}{2.003}. \tag{21}$$

Using the result in Theorem 4, we have

$$\frac{1}{\lambda_{\min}[(PH_1)_s]} = -\frac{1}{2.001} < p_1(x,t) < \frac{1}{\lambda_{\max}[(PH_1)_s]} = \frac{1}{0.00125}. \tag{22}$$

The symmetric region obtained from eq. (22)

$$|p_1(x,t)| < \frac{1}{2.001}$$

is equal too that one which is determined by Yedavalli's bound [9] given by

$$|p_1(x,t)| < \frac{1}{||(P_m H_1)_s||}$$

where P_m indicates the matrix whose entries are the absolute values of the entries of P.

Another uncertainty on the (1, 2) entry of the matrix A, say $p_2(x,t)$, can be represented as

$$\begin{pmatrix} 0 & p_2(x,t) \\ 0 & 0 \end{pmatrix} = \begin{cases} \begin{pmatrix} 1 \\ 0 \end{pmatrix} (0 \;\; p_2(x,t)), & \textit{Structure a),} \\ \begin{pmatrix} 0 & 1 \\ 0 & 0 \end{pmatrix} p_2(x,t), & \textit{Structure b).} \end{cases}$$

Supposing that the desired hyperrectangle sides are in the ratio 1 : 1, Theorems (2) and (3) give the bound

$$|p_{1,2}(x,t)| < \frac{1}{\|Pg_1\| + \|Pg_2\|} = \frac{1}{52.06}. \tag{23}$$

Theorem (4) gives

$$|p_{1,2}(x,t)| < \begin{cases} 1/24.03, & \text{in region } \mathcal{R}(+,+); \\ 1/26.03, & \text{in region } \mathcal{R}(+,-); \\ 1/28.03, & \text{in region } \mathcal{R}(-,-); \\ 1/26.03, & \text{in region } \mathcal{R}(-,+). \end{cases} \tag{24}$$

The largest symmetrical region included is given by

$$|p_{1,2}(x,t)| < \frac{1}{28.03}, \tag{25}$$

which is slightly smaller than that one characterized by Yedavalli's bound

$$p_{1,2}(x,t) < \frac{1}{\|(P_m(H_1+H_2))_s\|} = \frac{1}{27.06}. \tag{26}$$

5 Conclusions

Some results, aimed at quantifying the stability robustness of a linear time-invariant system subject to time-varying structured perturbations, have been presented. Even though generally they are not so good as Yedavalli's ([9]), they easily lend themselves to optimization procedures. The last result determines an *asymmetrical* region of stability in the uncertain parameter space, thus increasing the information on the system robustness. An emphasis has been given to the problem of the *form* of the stability region.

References

[1] Barmish, B. R., and Leitmann, G., " On Ultimate Boundedness Control of Uncertain Systems in the Absence of Matching Conditions," *IEEE Trans. Automat. Contr.,* **AC-27**, 153, 1982.

[2] Barrett, M. F., " Conservatism with Robustness Tests for Linear Feedback Control Systems," *Proc. 19th Conf. Decision Contr.*, 885-890, 1980.

[3] Corless, M., and Da, D., " New Criteria for Robust Stability," presented at *International Workshop on Robustness in Identification and Control*, Torino, 1988.

[4] Kalman, R. E., and Bertram, J. E., "Control System Analysis and Design via the Second Method of Lyapunov," *ASME Journal of Basic Engineering*, **82**, No. 2, 1960.

[5] Keel, L. H., Bhattacharyya, S. P., and Howze, J. W., "Robust Control with Structured Perturbation," *IEEE Trans. Automat. Contr.*, **AC-33**, 68-78, 1988.

[6] Kharitonov, V. L., "Asymptotic Stability of an Equilibrium Position of a Family of Systems of Linear Differential Equations," *Differential Uravnen*, **14**, 2086-2088, 1978.

[7] Patel, R. V., and Toda, M., "Quantitative Measures of Robustness for Multivariable Systems," *Proc. Joint Automat. Contr.*, San Francisco, TP8-A, 1980.

[8] Petkovski, D. B., "On the Robust Stability of Linear State Space Models," *Proc. of IEE Conf. Control88*, Oxford, 80-82, 1988.

[9] Yedavalli, R. K., "Improved Measures of Stability Robustness for Linear State Sapce Models," *IEEE Trans. Automat. Contr.,* **AC-30**, 577-579, 1985.

[10] Yedavalli, R. K., and Liang, Z., "Reduced Conservatism in Stability Robustness Bounds by State Transformation," *IEEE Trans. Automat. Contr.,* **AC-31**, 863-866, 1986.

A stability radius for time-varying linear systems

D. Hinrichsen A. Ilchmann A. J. Pritchard

Notation

$\mathbb{R}_+ = \{z \in \mathbb{R} | z \geq 0\}$

$\mathbb{C}_- = \{z \in \mathbb{C} | Re z < 0\}$

$\sigma(A)$ spectrum of $A \in \mathbb{C}^{n \times n}$

$GL_n(\mathbb{C})$ the set of all invertible matrices $T \in \mathbb{C}^{n \times n}$

$||x||$ Euclidean norm of $x \in \mathbb{C}^n$

$||D||$ induced operator norm for $D \in \mathbb{C}^{m \times p}$

$||D(\cdot)||_{L_\infty} = \sup_{t_0 < t < t_1} \{||D(t)||\}$ for $D(\cdot) \in PC\left((t_0, t_1); \mathbb{C}^{m \times p}\right)$

$L_2(t_0, t_1; \mathbb{C}^m)$	space of functions $u : (t_0, t_1) \to \mathbb{C}^m$ s. t. $t \mapsto ||u(t)||^2$ is integrable over (t_0, t_1)
$PC((t_0, t_1); \mathbb{C}^{n \times m})$	set of piecewise continuous matrix functions $D(\cdot) : (t_0, t_1) \to \mathbb{C}^{n \times m}$
$PC_b((t_0, t_1); \mathbb{C}^{n \times m})$	set of all bounded matrix functions in $PC((t_0, t_1); \mathbb{C}^{n \times m})$
$PC^1((t_0, t_1); GL_n(\mathbb{C}))$	set of all piecewise continuously differentiable functions $D(\cdot) : (t_0, t_1) \to GL_n(\mathbb{C})$
$C^1(t_0, t_1; \mathbb{C}^{n \times m})$	set of all continuously differentiable $D(\cdot) : (t_0, t_1) \to \mathbb{C}^{n \times m}$

1 Introduction

In recent years problems of robust stability have received a good deal of attention. Most of the work on time-invariant linear systems – including the successful H^∞-approach (see [4], [12]) – is based on transform techniques. However, in [7], [8] a state space approach via the concept of ***stability radius*** is proposed. In the present paper this approach is extended to a time-varying setting.

Consider a nominal system of the form

$$\dot{x}(t) = A(t)x(t), \quad t \geq 0 \tag{1.1}$$

where $A(\cdot) \in PC(\mathbb{R}_+, \mathbb{C}^{n\times n})$. Assume that (1.1) is ***exponentially stable***, i.e. there exist M , $w > 0$ so that

$$\|\phi(t,s)\| \leq Me^{-w(t-s)} \quad \text{for all } t \geq s \geq 0 \tag{1.2}$$

where $\phi(t,s)$ denotes the transition matrix of (1.1). Many authors (see [1], [2], [3], [5], [10]) have determined bounds $\delta > 0$ so that exponential stability of the disturbed system

$$\dot{x}(t) = [A(t) + D(t)]x(t) \quad ,t \geq 0 \tag{1.3}$$

is preserved whenever

$$\|D(\cdot)\|_{L_\infty} < \delta \qquad \text{for } D(\cdot) \in PC(\mathbb{R}_+, \mathbb{C}^{n\times n}). \tag{1.4}$$

These bounds are conservative. Our problem is to determine a ***sharp*** upper bound. We call this bound the (complex)[1] ***stability radius*** and define it by

$$r_{\mathbb{C}}(A) = \inf\{\|D(\cdot)\|_{L_\infty} | D \in PC_b(\mathbb{R}_+, \mathbb{C}^{n\times n}) \text{ and (1.3) is not exponentially stable }\} \tag{1.5}$$

We also consider the case where A is subjected to ***structured*** pertubations, so that the perturbed system is

$$\dot{x}(t) \;=\; [A(t) + B(t)D(t)C(t)]\, x(t), \quad t \geq 0 \tag{1.6}$$

where $D(\cdot) \in PC_b(\mathbb{R}_+, \mathbb{C}^{m\times p})$ is an ***unknown*** bounded time-varying disturbance matrix and $B(\cdot) \in PC(\mathbb{R}_+, \mathbb{C}^{n\times m})$, $C(\cdot) \in PC(\mathbb{R}_+, \mathbb{C}^{p\times n})$ are ***given*** "scaling matrices" defining the "structure" of the perturbation. Then the ***structured stability***

[1] The real stability radius is defined analogously. In spite of its prime importance it is not studied here, since even in the time-invariant setup only rudimentary results are available.

radius is

$$r_{\mathbb{C}}(A;B,C) = \inf\{\|D(\cdot)\|_{L_\infty} | D \in PC_b(\mathbb{R}_+, \mathbb{C}^{m\times p}) \text{ and (1.6) is not exponentially stable }\} \quad (1.7)$$

In the unstructured case $r_{\mathbb{C}}(A)$ is simply the distance of (1.1) from the set of not exponentially stable systems with respect to the L_∞-norm.

Remark 1.1 The following properties are easily obtained:

(a) $r_{\mathbb{C}}(A) = 0 \iff$ (1.1) is not exponentially stable

(b) $r_{\mathbb{C}}(\alpha A) = \alpha r_{\mathbb{C}}(A)$ for all $\alpha \geq 0$

(c) $A(\cdot) \mapsto r_{\mathbb{C}}(A)$ is continuous on $PC_b(\mathbb{R}_+, \mathbb{C}^{n\times n})$

2 Bohl exponent and Bohl transformation

For the stability behaviour of (1.1) the number

$$\begin{aligned} k_B(A) &:= inf\,\{-w \in \mathbb{R} | \exists M_w > 0 : \\ t \geq s \geq 0 &\Longrightarrow \|\phi(t,s)\| \leq M_w e^{-w(t-s)}\} \end{aligned} \quad (2.1)$$

introduced by Bohl [1] is useful. We call $k_B(A)$ the *Bohl exponent* of (1.1). It is possible that $k_B(A) = \pm\infty$. The following properties are easily seen.

Proposition 2.1 Let $A(\cdot) \in PC(\mathbb{R}_+, \mathbb{C}^{n\times n})$. Then

(a) $k_B(A) < 0 \iff$ (1.1) is exponentially stable

(b) If $A(\cdot) \equiv A \in \mathbb{C}^{n\times n}$ then

$$k_B(A) = \max_{i\in \underline{n}} Re\lambda_i(A), \quad \text{where } \lambda_i(A) \text{ are the eigenvalues of } A.$$

(c) In the scalar case, i.e. $n = 1$, we have

$$r_{\mathbb{C}}(A) = -k_B(A)$$

(d) For the matrix case only an inequality is valid:

$$r_{\mathbb{C}}(A) \leq -k_B(A)$$

Remark 2.2 We want to emphasize that $k_B(A)$ may be a bad indicator for the robustness margin of (1.1). Consider

$$A_k = -\begin{bmatrix} k & k^3 \\ 0 & k \end{bmatrix}, \quad D_k = k^{-1}\begin{bmatrix} 1 & 0 \\ -1 & 1 \end{bmatrix}, \quad \text{for } k \in \mathbb{N}$$

Then $\lim_{k\to\infty} k_B(A_k) = -\infty$. However, $\sigma(A_k+D_k) = \{\frac{1}{k}, \frac{1}{k}-2k\}$ although $\lim_{k\to\infty} \|D_k\| = 0$. Thus $\lim_{k\to\infty} r_{\mathbb{C}}(A_k) = 0$.

The following properties of the Bohl exponent can be found in [3].

Proposition 2.3 Let $A(\cdot) \in PC(\mathbb{R}_+, \mathbb{C}^{n\times n})$. Then

(a) $k_B(A)$ is finite if $A(\cdot)$ is *bounded.*

(b) $k_B(A)$ is finite iff $\sup_{0\le|t-s|\le 1} \|\phi(t,s)\| < \infty$.

(c) If $k_B(A) < \infty$ then

$$k_B(A) = \limsup_{s,t-s\to\infty} \frac{\log\|\phi(t,s)\|}{t-s}.$$

We now analyse the effect of time-varying linear coordinate transformations

$$z(t) = T(t)^{-1}x(t), \quad T(\cdot) \in PC^1(\mathbb{R}_+, GL_n(\mathbb{C})) \tag{2.2}$$

on the system (1.1) which yields

$$\dot z(t) = \hat A(t)z(t), \qquad \text{where } \hat A = T^{-1}AT - T^{-1}\dot T \tag{2.3}$$

These transformations will not, in general, preserve exponential stability. Therefore we introduce the set of *Bohl transformations* $\mathcal{B}_n$, i.e. the set of all $T(\cdot) \in PC^1(\mathbb{R}_+, GL_n(\mathbb{C}))$ such that

$$\inf\{\varepsilon \in \mathbb{R} | \exists M_\varepsilon > 0 : \forall t, s \ge 0 \Rightarrow \|T(t)^{-1}\| \cdot \|T(s)\| \le M_\varepsilon e^{\varepsilon|t-s|}\} = 0 \tag{2.4}$$

Remark 2.4 It is obvious that

(a) the set $\mathcal{B}_n$ forms a group with respect to (pointwise) multiplication

(b) $\mathcal{B}_n$ contains the group of *Lyapunov transformations*, i.e. all $T(\cdot) \in PC^1(\mathbb{R}_+, GL_n(\mathbb{C}))$ so that $T(\cdot), T(\cdot)^{-1}, \dot T(\cdot)$ are bounded,

(c) $k_B(A) = k_B(T^{-1}AT - T^{-1}\dot T)$ for all $T \in \mathcal{B}_n$

The following proposition shows that even in the time-invariant case a similarity transformation may drastically change the stability radius.

Proposition 2.5 [7] If $A \in \mathbb{C}^{n\times n}$ with $\sigma(A) \subset \mathbb{C}_-$ then $\{r_{\mathbb{C}}(T^{-1}AT);\ T \in GL_n(\mathbb{C})\}$ is equal to the interval $(0, -\max_{i\in \underline{n}} Re\lambda_i(A)]$ with possible exception of the right extremum.

In the scalar case we can prove

Proposition 2.5 If

$$\dot{x}(t) = a(t)x(t), \quad a(\cdot) \in PC(\mathbb{R}_+, \mathbb{C}) \tag{2.5}$$

has a *strict* Bohl exponent, i.e.

$$k_B(a) = \lim_{s,t-s\to\infty} \frac{\log ||\phi(t,s)||}{t-s}$$

then there exists $\Theta \in \mathcal{B}_1$ so that $z(t) = \Theta(t)^{-1}x(t)$ converts (2.5) into

$$\dot{z}(t) = k_B(a)z(t)$$

3 The perturbation operator

In the time-invariant setup, where $(A, B, C) \in \mathbb{C}^{n\times n} \times \mathbb{C}^{n\times m} \times \mathbb{C}^{p\times n}$, the structured stability radius can be characterized via the convolution operator

$$\begin{aligned} L_0 : L_2(0,\infty;\mathbb{C}^m) &\quad\to\quad L_2(0,\infty;\mathbb{C}^p) \\ u(\cdot) &\quad\mapsto\quad (t \mapsto \int_0^t Ce^{A(t-s)}Bu(s)ds) \end{aligned} \tag{3.1}$$

as follows

Proposition 3.1 [8] If $\sigma(A) \subset \mathbb{C}_-$ and $G(s) := C(sI_n - A)^{-1}B$ then

$$r_{\mathbb{C}}(A,B,C) = \begin{cases} ||L_0||^{-1} = \left[\max_{w\in\mathbb{R}} ||G(iw)||\right]^{-1} & \text{if } G \neq 0 \\ \infty & \text{if } G = 0 \end{cases}$$

In order to explore the possibility of obtaining similar results for time-varying systems, we consider the parametrized family of *perturbation operators* $(L_{t_0}^{\Sigma})_{t_0\in\mathbb{R}_+}$ defined by

$$L^{\Sigma}_{t_o} : L_2(t_0,\infty;\mathbb{C}^m) \rightarrow L_2(t_0,\infty;\mathbb{C}^p), \quad t_0 \geq 0$$
$$u(\cdot) \mapsto (t \mapsto \int_{t_0}^{t} C(t)\phi(t,s)B(s)u(s)ds) \tag{3.2}$$

associated with

$$\Sigma = (A,B,C) \in PC(\mathbb{R}_+,\mathbb{C}^{n\times n}) \times PC_b(\mathbb{R}_+,\mathbb{C}^{n\times m})\times$$
$$\times PC_b(\mathbb{R}_+,\mathbb{C}^{p\times n}),\ k_B(A) < 0 \tag{3.3}$$

Basic properties of $L^{\Sigma}_{t_0}$ are summarized in the following

Proposition 3.2 [6]

(a) $L^{\Sigma}_{t_0}$ is a bounded operator
(b) $t_0 \mapsto \|L^{\Sigma}_{t_0}\|$ is monotonically decreasing on $\mathbb{R}_+$
(c) $\|L^{\Sigma}_{t_0}\| = \|L^{\Sigma}_{t_1}\|$ for all $t_0, t_1 \in \mathbb{R}_+$ if A, B, C are periodic with a common period
(d) $\|L^{\Sigma}_{t_0}\|^{-1} \leq r_{\mathbb{C}}(A;B,C)$
(e) For the unstructured case, i.e. $B(\cdot) = C(\cdot) = I_n$, if $M, w > 0$ satisfy (1.2) then

$$\frac{w}{M} \leq \|L^{\Sigma}_{t_0}\|^{-1} \leq \lim_{t_0\to\infty} \|L^{\Sigma}_{t_0}\|^{-1} \leq r_{\mathbb{C}}(A)$$

As opposed to the time-invariant case $\|L^{\Sigma}_{t_0}\|^{-1}$ or $\lim\limits_{t_0\to\infty} \|L^{\Sigma}_{t_0}\|^{-1}$ do not necessarily coincide with $r_{\mathbb{C}}(A;B,C)$. Even in the simple case when $a(\cdot) \in PC(\mathbb{R}_+,\mathbb{R})$ is periodic and $b = c = 1$, we have worked out an example in [6] for which

$$\|L^{\Sigma}_{t_0}\|^{-1} = \|L^{\Sigma}_{t_1}\|^{-1} < r_{\mathbb{C}}(a) \quad \text{for all} \quad t_0, t_1 \in \mathbb{R}_+$$

However note that scalar Bohl transformations $\Theta \in \mathcal{B}_1$ do not change the stability radius but will change the norm of the perturbation operator. Let

$$\Sigma_\Theta := (A - \tfrac{\dot{\Theta}}{\Theta} I_n;\ B, C)$$

By using Proposition 2.5 and 3.1 one can show

Proposition 3.3 Suppose $a(\cdot) \in PC(\mathbb{R}_+,\mathbb{C})$ has a strict Bohl exponent $k_B(a) < 0$ and $b, c \in \mathbb{C}$. Then there exists a $\Theta \in \mathcal{B}_1$ such that

$$r_{\mathbb{C}}(a;b,c) = \|L^{\Sigma_\Theta}_{t_0}\|^{-1} \quad \text{for all} \quad t_0 \geq 0.$$

For the matrix case we have the following

Conjecture 3.4 Suppose $\Sigma = (A, B, C)$ satisfies (3.3), then

$$r_{\mathbb{C}}(A; B, C) = \sup_{\Theta \in \mathcal{B}_1} \left\{ \lim_{t_0 \to \infty} \| L_{t_0}^{\Sigma_\Theta} \|^{-1} \right\}$$

4 The associated parametrized differential Riccati equation

In the time-invariant setup another useful characterization of $r_{\mathbb{C}}(A; B, C)$ is possible via the parametrized *algebraic* Riccati equation, $\mathbf{ARE}_\rho$

$$A^* P + PA - \rho C^* C - PBB^* P = 0, \quad \rho \in \mathbb{R}$$

Proposition 4.1 [8] Suppose $(A, B, C) \in \mathbb{C}^{n \times n} \times \mathbb{C}^{n \times m} \times \mathbb{C}^{p \times m}$ and $\sigma(A) \subset \mathbb{C}_-$:

(a) If $-\infty < \rho < r_{\mathbb{C}}^2(A; B, C)$ then there exists a unique stabilizing Hermitian solution P_ρ of $\mathbf{ARE}_\rho$, that is a solution $P_\rho = P_\rho^*$ which satisfies
$\sigma(A - BB^* P_\rho) \subset \mathbb{C}_-$.
If $\rho = r_{\mathbb{C}}^2$ then there exists a unique Hermitian solution $P_{r_{\mathbb{C}}^2}$ of $\mathbf{ARE}_{r_{\mathbb{C}}^2}$ having the property $\sigma(A - BB^* P_{r_{\mathbb{C}}^2}) \subset \overline{\mathbb{C}_-}$.

(b) If there exists a Hermitian solution P_ρ of $\mathbf{ARE}_\rho$ then necessarily $\rho \leq r_{\mathbb{C}}^2(A; B, C)$.

Guided by this result we study, in the time-varying setting, the parametrized *differential* Riccati equation, $\mathbf{DRE}_\rho$

$$\dot{P}(t) + A^*(t)P(t) + P(t)A(t) - \rho C^*(t)C(t) - P(t)B(t)B^*(t)P(t) = 0, \quad t \geq t_0$$

associated with the system

$$\left. \begin{aligned} \dot{x}(t) &= A(t)x(t) + B(t)u(t), \quad x(t_0) = x_0 \in \mathbb{C}^n \\ y(t) &= C(t)x(t), \qquad\qquad\qquad t \geq t_0 \geq 0 \end{aligned} \right\} \tag{4.1}$$

Throughout this section we assume $\Sigma = (A, B, C)$ satisfies (3.3).

Kalman [9] and *Reghis and Megan* [11] among others, have studied differential Riccati equations, however their results cannot be applied to $\mathbf{DRE}_\rho$ if $\rho > 0$.

Just as in the time-invariant case we consider the parametrized optimal control problem $\mathbf{OCP}_\rho$:

Minimize over $u \in L_2(t_0, t_1; \mathbb{C}^m)$

$$J_\rho(x_0, (t_0, t_1), u(\cdot)) = \int_{t_0}^{t_1} [\|u(s)\|^2 - \rho\|y(s)\|^2] ds$$

where $y(\cdot)$ is defined via (4.1) and $\rho \in \mathbb{R}$. If $\rho < 0$ this is the usual linear quadratic regulator problem **LQR**, whereas in our situation $\rho > 0$, so that the state penalty is negative. To consider a cost functional with negative state penalty is quite natural in this context since we are concerned with a minimum norm *destabilization* problem while the classical **LQR** problem is concerned with stabilization.

The analysis of the $\mathbf{DRE}_\rho$ and its relation to the $\mathbf{OCP}_\rho$ is quite involved, details may be found in [6]. Here we only state the main results.

Proposition 4.2 (finite time) If $\rho < \|L_{t_0}^{\Sigma}\|^{-2}$, $0 \le t_0 < t_1 < \infty$, then

(a) there exists a unique Hermitian solution $P^{t_1}(\cdot)$ of $\mathbf{DRE}_\rho$ on $[t_0, t_1]$ with $P^{t_1}(t_1) = 0$

(b) $P^{t_1}(t) \le 0$ (resp. ≥ 0) for all $t \in [t_1, t_0]$ if $\rho \ge 0$ (resp. $\rho \le 0$).

(c) the minimal cost of $\mathbf{OCP}_\rho$ is

$$\inf_{u \in L_2(t_0, t_1; \mathbb{C}^m)} J_\rho(x_0, (t_0, t_1), u(\cdot)) \ = \ < x_0, P^{t_1}(t_0)x_0 >$$

(d) the optimal control is given by

$$u(t) \ = \ -B^*(t)P^{t_1}(t)x(t)$$

where $x(\cdot)$ solves $\dot{x}(t) = [A - BB^* P^{t_1}](t)x(t)$, $x(t_0) = x_0$.

The next proposition is obtained by studying what happens if $t_1 \to \infty$.

Proposition 4.3 (infinite time) If $\rho < \|L_{t_0}^{\Sigma}\|^{-2}$, $t_0 \ge 0$, then

(a) $P^+(t) = \lim_{t_1 \to \infty} P^{t_1}(t)$ exists for all $t \ge t_0$ and yields a bounded Hermitian solution of $\mathbf{DRE}_\rho$;

(b) $P^+(\cdot)$ is the only solution so that $k_B(A - BB^* P^+) < 0$;

(c) for any other bounded Hermitian solution $Q(\cdot) \in C^1(t_0', \infty; \mathbb{C}^{n \times n})$, $t_0' \ge t_0$ of $\mathbf{DRE}_\rho$ we have $Q(t) \le P^+(t)$ for all $t \ge t_0'$;

(d) the minimal cost is

$$\inf_{u \in L^2(t_0, \infty; \mathbb{C}^m)} J_\rho(x_0, (t_0, \infty), u(\cdot)) \ = \ < x_0, P^+(t_0)x_0 >;$$

(e) the optimal control is

$$u(t) \ = \ -B^*(t)P^+(t)x(t), \quad t \ge t_0$$

where $x(\cdot)$ solves

$$\dot{x}(t) = [A - BB^* P^+](t)x(t), \quad x(t_0) = x_0, \ t \ge t_0$$

As a partial converse of Proposition 4.3 we have

Proposition 4.4 If $Q(\cdot) \in C^1(t_0, \infty; \mathbb{C}^{n\times n})$ is a bounded Hermitian solution of $\mathbf{DRE}_\rho$ on (t_0, ∞) then necessarily $\rho \leq \|L_{t_0}^{\Sigma}\|^{-2}$.

While the previous two propositions yield a complete characterization of the norm $\|L_{t_0}^{\Sigma}\|$ in terms of the associated parametrized differential Riccati equation, they do not provide a full generalization of Proposition 4.1 to the time-varying case. To find a complete characterization of the stability radius $r_{\mathbb{C}}(A; B, C)$ for time-varying systems is an open problem.

5 Robust Lyapunov functions and nonlinear perturbations

The following proposition shows how solutions of the parametrized differential Riccati equation $\mathbf{DRE}_\rho$ can be used to construct *robust* Lyapunov functions for the system (1.1).

Proposition 5.1 Suppose $0 < \rho < \|L_{t_0}^{\Sigma}\|^{-2}$. If $P_\rho(\cdot)$ solves $\mathbf{DRE}_\rho$ then

$$V(t,x) \;:=\; -< x, P_\rho(t)x >, \quad t \geq t_0, \; x \in \mathbb{C}^n$$

is a common Lyapunov function for all perturbed systems

$$\dot{x}(t) = [A + BDC](t)x(t), \quad t \geq t_0, \; x(t_0) = x_0$$

with $\|D(\cdot)\|_{L_\infty}^2 < \rho$.

If (A, B, C) are constant matrices and $\sigma(A) \subset \mathbb{C}_-$ a similar result to Proposition 5.1 holds true for $\|D(\cdot)\|_{L_\infty} < r_{\mathbb{C}}(A; B, C)$, see [8].

Using the above Lyapunov function it is possible to extend our robustness analysis to nonlinear perturbations of the form $\Delta(t) = B(t)N(C(t)x, t)$ so that the perturbed system is

$$\dot{x}(t) = A(t)x(t) + B(t)N(C(t)x(t), t), \quad x(t_0) = x_0, \; t \geq t_0 \tag{5.1}$$

where (A, B, C) satisfy (3.3) and $N : \mathbb{R}^p \times \mathbb{R}_+ \to \mathbb{R}^m$ is continuously differentiable. We assume $N(0, t) = 0$ so that 0 is an equilibrium state of (5.1). The following result shows that no nonlinear perturbation with global gain smaller than $\|L_{t_0}^{\Sigma}\|^{-1}$ can destabilize the system.

Proposition 5.2 Suppose $\gamma < \|L_{t_0}^{\Sigma}\|^{-1}$ and

$$\|N(y,t)\| \leq \gamma \|y\| \quad \text{for all} \quad t \geq t_0, \; y \in \mathbb{C}^p$$

Then the origin is globally exponentially stable for the system (5.1).

It is not clear whether an analogous statement holds (over a suitable time interval (t_0, ∞)) if the gain of the nonlinear perturbation is strictly less than $r_{\mathbb{C}}(A; B, C)$.

References

[1] Bohl, P. (1913) *Ueber Differentialgleichungen*; J. Reine Angew. Math. 144, 284-313

[2] Coppel, W.A. (1978) *Dichotomies in Stability Theory*; Lect. Notes Math. 629, Springer-Verlag, Berlin et al.

[3] Daleckii, Ju.L. and Krein, M.G. (1974) *Stability of Solutions of Differential Equations in Banach Spaces*; AMS, Providence, Rhode Island

[4] Francis, B. and Zames, G. (1983) *Feedback minimax sensitivity and optimal robustness*; IEEE Trans. Autom. Control 28, 585-601

[5] Hahn, W. (1969) *Stability of Motion*; Die Grundlehren der math. Wissenschaften Bd. 138, Springer-Verlag, New York

[6] Hinrichsen, D., Ilchmann, A. and Pritchard, A.J. (1989) *Robustness of stability of time-varying systems*; J. Diff. Eqns Vol 82. 219-250.

[7] Hinrichsen, D. and Pritchard, A.J. (1986) *Stability radii of linear systems*; Syst. Control Lett. 7, 1-10

[8] Hinrichsen, D. and Pritchard, A.J. (1986) *Stability radius for structured perturbations and the algebraic Riccati equation*; Syst. Control Lett. 8, 105-113

[9] Kalman, R.E. (1960) *Contributions to the theory of optimal control*; Bull. Soc. Math. Mexico 5, 102-119

[10] Perron, O. (1930) *Die Stabilitätsfrage der Differentialgleichungen*; Math. Z. 32, 703-728

[11] Reghis, M. and Megan, M. (1977) *Riccati equation, exact controllability and stability*; Sem. Ecuat. Funct., Univ. Timisoara 42

[12] Zames, G. (1981) *Feedback and optimal sensitivity: model reference transformation, multiplicative seminorms and approximate inverses*; IEEE Trans. Autom. Control 16, 301-320

Parameterized stability domains and the robustness of integrodifferential systems

Stuart Townley

1 Introduction

One of the major themes in systems theory is the robustness aspect of controller design. Many different problems are analyzed :– design of robust H_∞ controllers (see [5]), computation of robust Liapunov functionals (see [8]) and of interest to us here, characterization of robustness measures for the system eigenvalues lying in a prescribed stability domain (see [9]). Quite often, for systems described by partial differential equations, it is an analysis of the spectrum which permits feedback design. (see e.g [2], [4]). Moreover various robustness questions can be answered via a spectrum analysis. For example the effect of time delays in the boundary stabilization of wave equations (see e.g [4]). It is the purpose of this paper to analyze robustness problems for general stability domains in the complex plane, in a functional analytic framework. A typical problem we have in mind is the following: Let Δ be the Laplacian defined on some open domain $\Omega \subset \mathbf{R}^n$ with sufficiently smooth boundary $\partial\Omega$. Consider the following partial differential equation defined on Ω,

$$\left.\begin{array}{ll} x_t(t,\xi) = \Delta x(t,\xi) & t \in \Omega \\ x(t,\xi) = 0 & t \in \partial\Omega \end{array}\right\} \tag{1}$$

Depending upon the boundary $\partial\Omega$ this results in a spectrum $\{\lambda_n\} \subset \mathbf{C}$, with corresponding eigenfunctions $\{\varphi_n\}$ forming a basis for $L^2(\Omega)$. Equation (1) defines a nominal system. If we consider the possibility of neglected transport terms or else perturbations to the boundary data then we must consider stability of the perturbed system

$$x_t(t,\xi) = \Delta x(t,\xi) + BDCx(t,\xi) \tag{2}$$

where the operators B and C are structured, unbounded operators abstracted from the problem and D (arbitrary within a given class of bounded operators) measures the size of the perturbation. Now the spectrum, $\{\lambda_n\}$, of the nominal system (1) will be located typically inside the set $\Sigma = \{\lambda \in \mathbf{C} | Re\lambda < -|Im\lambda|^2\}$. It is of interest then to characterize the range of $||D||$ for which this spectrum location is preserved. This is the problem which we address and moreover we characterize the minimum-norm destabilizing D such that the spectrum of (2) is not contained in Σ.

To motivate the ensuing analysis we consider firstly the case of A, B and C finite dimensional. In Hinrichsen and Pritchard [9] the following generalization of the robustness problem given in Hinrichsen and Pritchard [8] is analyzed. Let $(A, B, C) \in \mathbf{C}^{n\times n} \times \mathbf{C}^{n\times m} \times \mathbf{C}^{p\times n}$ and $\mathbf{C}_g \cup \mathbf{C}_b$ be a partition of the complex plane into a good set $\mathbf{C}_g$ and a bad set $\mathbf{C}_b$. The matrix A defines a nominal system

$$\dot{x}(t) = Ax(t), \quad x(0) \in \mathbf{C}^n \tag{3}$$

and we assume that $\sigma(A) = \{\lambda \in \mathbf{C} | det(\lambda I - A) = 0\} \subset \mathbf{C}_g$. The operators B and C determine a structured perturbation of A to $A + BDC$ corresponding to the system

$$\dot{x}(t) = (A + BDC)x(t), \quad x(0) \in \mathbf{C}^n \tag{4}$$

with $D \in \mathbf{C}^{m\times p}$. If $\mathbf{C}_g$ is an open set in $\mathbf{C}$ then the following robustness measure is well-defined.

Definition 1.1

$$r_{\mathbf{C}}(A; B, C; \mathbf{C}_b) = \inf\{||D|| \mid \sigma(A + BDC) \cap \mathbf{C}_b \neq \emptyset\}.$$

Using a simple spectrum analysis $r_{\mathbf{C}}(A; B, C; \mathbf{C}_b)$ is characterized as follows.

Proposition 1.2 (Hinrichsen, Pritchard)

$$r_{\mathbf{C}}(A; B, C; \mathbf{C}_b) = \inf_{s \in \partial \mathbf{C}_b} \frac{1}{||C(sI - A)^{-1}B||}$$

where $\partial \mathbf{C}_b$ denotes the boundary of $\mathbf{C}_b$. □

Remark
In [10], various specializations of this formula to the case $B \in \mathbf{C}^{n\times 1}$ and $\mathbf{C}_g = \mathbf{C}_- = \{\lambda \in \mid Re\lambda < 0\}$ or $\mathbf{C}_g = D_1 = \{\lambda \in \mathbf{C} \mid |\lambda| < 1\}$ are considered.

Unfortunately, arguments based purely on an analysis of the spectrum are not valid in an infinite dimensional setting. In [14], a robustness analysis for exponential stability of infinite dimensional systems is carried out via solutions of integral equations over L^2. It is this approach which we adopt here. Then for specific problem we recover a *spectrum type* analysis.

To continue with the motivation we assume that $\partial \mathbf{C}_b$ is parameterized by a function $f : \mathbf{C} \to \mathbf{C}$ in the following sense :–

i) f is rational, f has no poles in $\overline{\mathbf{C}_+}$,
ii) $f(\overline{\mathbf{C}_+}) = \mathbf{C}_b$, $f(i\mathbf{R}) = \partial\mathbf{C}_b$,
iii) $f(\infty) = \infty$.
Then

$$r_{\mathbf{C}}(A; B, C; \mathbf{C}_b) = \inf_{\omega \in \mathbf{R}} \frac{1}{\|C(f(i\omega)I - A)^{-1}B\|}.$$

For computational (see Hinrichsen, Kelb, Linnemann [7]) and theoretical purposes it is of interest to seek realizations of the *transfer function* $C(f(s)I - A)^{-1}B$. However, arbitrary minimal realization would lose the underlying role of the given structure operators B, C and as we see in the sequel, for some cases, other methods are more appropriate. Using properties i) and iii) of f we are guaranteed $g : \mathbf{C} \to \mathbf{C}$ such that

$$f(s) = sg^{-1}(s) + \lambda_0$$

Then

$$r_{\mathbf{C}}(A; B, C; \mathbf{C}_b) = \inf_{\omega \in \mathbf{R}} \frac{1}{\|g(i\omega)C(i\omega I - g(i\omega)(A - \lambda_0 I))^{-1}B\|}.$$

If g is analytic (no poles) in the closed right half plane then as a result $r_{\mathbf{C}}(A; B, C; \mathbf{C}_b)$ can be calculated as $\|L\|^{-1}$ where $L \in \mathcal{L}(L^2(0,\infty;\mathbf{C}^m), L^2(0,\infty;\mathbf{C}^p))$ is given by

$$(Lu)(t) = y(t) \begin{cases} y(t) = Cx(t) \\ \dot{x}(t) = \int_0^t \tilde{g}(t-s)((A - \lambda_0 I) + Bu(s))ds, \quad x(0) = 0 \end{cases}$$

and g is the Laplace transform of $\tilde{g}$.

Example 1.3
If $\mathbf{C}_g = \{\lambda \in \mathbf{C} | Re\lambda < -(Im\lambda)^2\}$ then $f(s) = s^2 + s$, $g(s) = (s+1)^{-1}$ and $\tilde{g}(t) = e^{-t}$. □

Hence robustness of the property of $\sigma(A + BDC) \subset \mathbf{C}_g$ is converted into robustness of exponential stability for the perturbed system

$$\dot{x}(t) = \int_0^t \tilde{g}(t-s)(A + BDC)x(s)ds \quad x(0) = x_0.$$

This perturbed integral equation can be embedded into a wider class of perturbation problems

$$\dot{x}(t) = (\tilde{A} + \tilde{B}\tilde{D}\tilde{C})x(t) + \int_0^t (K(t-s) + BH(t-s)C)x(s)ds \tag{5}$$

studied in [14] for the case of $\tilde{A}, \tilde{B}, \tilde{C}, K, B, C$ unbounded operators on appropriate complex Banach spaces. In the sequel we will use this framework for a robustness analysis of infinite dimensional systems with respect to parameterized stability domains. Firstly we summarize the well-posedness results which allow us a high degree of unboundedness.

Let B be a Banach space. Throughout the paper $L^2(0,\infty;B)$ denotes the Banach space of square integrable functions from $[0,\infty)$ into B, $< .,. >$ denotes the inner product on a Hilbert space.

2 Well-posedness of perturbed integro-differential systems

Let X be a complex Banach space and consider the following equation defined on X

$$\dot{x}(t) = \tilde{A}x(t) + \int_0^t K(t-s)x(s)ds, \quad x(0) = x_0 \tag{6}$$

Here $\tilde{A}$ and $K(\cdot)$ are closed, linear operators defined on X with $D(A) \subset D(\tilde{A}) \subset D(K(t))$ for some densely defined, closed linear operator A. (6) is said to be well-posed if there exists a strongly continuous family of maps $R(t) \in \mathcal{L}(X)$ satisfying

i) $R(0) = I_X$

ii) for all $x_0 \in D(A)$, $R(t)$ is continuously differentiable and

$$\begin{aligned} \dot{R}(t)x_0 &= \tilde{A}R(t)x_0 + \int_0^t K(t-s)R(s)x_0 ds \\ &= R(t)\tilde{A}x_0 + \int_0^t R(t-s)K(s)x_0 ds. \end{aligned}$$

Such a family $R(t)$ is called a resolvent operator for (6).

Remark
Many well-posedness results exist for such systems (see [1], [6] and [11]) with the most sophisticated being the analogue of the Hille-Yosida theorem for semi-groups due to Grimmer-Pruss[6].

Example 2.1
Let $\{\varphi_n,\ n = 1,2\cdots\}$ be a complete orthonormal basis for a Hilbert space X,

$\{\lambda_n,\ n = 1, 2 \cdots\} \subset \Sigma$. Set $\tilde{A} = 0$ and $K(t) = e^{-t}A$ where A is the generator of the semi-group

$$S(t)x = \Sigma e^{\lambda_n t} < x, \varphi_n > \varphi_n.$$

It is possible to show that (6) is well-posed in this case and $R(t) \in \mathcal{L}(X)$ is given by

$$R(t)x = \Sigma \left(\left(\nu_n e^{\mu_n t} - \mu_n e^{\nu_n t} \right) / \nu_n - \mu_n \right) < x, \varphi_n >$$

where ν_n and μ_n are the roots of $\lambda^2 + \lambda - \lambda_n = 0,\ n = 1, 2 \cdots$. This is the nominal system and corresponds to the case $\mathbf{C}_g = \Sigma$ and $\partial \mathbf{C}_b = \{s(s+1) | s \in i\mathbf{R}\}$.

In Pritchard-Townley [14] a general theory is presented for the perturbation of $\tilde{A}$ to $\tilde{A} + \tilde{B}\tilde{D}\tilde{C}$ and $K(\cdot)$ to $K(\cdot) + BH(\cdot)C$. $\tilde{B}$, $\tilde{C}, B, C$ play the role of both structured and unbounded operators. The main results concern characterizations of various robustness measures. We will apply their methods to our very special problem.

Set $\tilde{A} = \tilde{B} = \tilde{C} = 0$ and $K(t) = \tilde{g}(t)A$ where A is the generator of an exponentially stable semigroup $S(t),\ t \geq 0$ (possibly a spectral operator as in example 2.1). We assume uncertainty in A of the form $A + BDC$ with $B \in \mathcal{L}(U, X)$ and $C \in \mathcal{L}(\underline{X}, Y)$ fixed and $D \in \mathcal{L}(Y, U)$ arbitrary corresponding to the perturbed integrodifferential equation

$$\dot{x}(t) = \int_0^t \tilde{g}(t-s)(A + BDC)x(s)ds \quad x(0) = x_0. \tag{7}$$

As usual we interpret (7) in the mild form

$$x(t) = R(t)x_0 + \int_0^t R(t-s)B \int_0^s \tilde{g}(s-p)DCx(\rho)d\rho ds \tag{8}$$

Due to the specific nature of $R(t) \in \mathcal{L}(X)$ we make the following assumptions which guarantee well posedness of (7), (8):

(A1) $\underline{X} \subset X$ with continuous dense injections, $S(t)$ restricts to a semigroup on $\underline{X}$ and $R(t)$ restricts to a resolvent operator on $\underline{X}$ for the system

$$\dot{x}(t) = \int_0^t \tilde{g}(t-s)(\underline{A} + BDC)x(s)ds \quad x(0) = x_0.$$

Here $\underline{A}$ is the generator of $S(t)$ restricted to $\underline{X}$.

(A2) There exists $k > 0$ such that

$$||C\int_0^{\cdot} \tilde{g}_\varepsilon(\cdot - s)Rx_0ds||_{L^2(0,\infty;Y)} \leq ||x_0||_X \text{ for all } x_0 \in \underline{X},$$

and all sufficiently small $\varepsilon \geq 0$, where $\tilde{g}_\varepsilon(t) = e^{\varepsilon t}\tilde{g}(t)$.

(A3) There exist constants $M, \alpha > 0$ such that

$$||R(t)x_0|| \leq Me^{-\alpha t}||x_0||$$

(A4) The domain of A, $D_X(A) \subset \underline{X}$ with continuous dense injection.

Assumptions A2, A3 guarantee that $L : L^2(0,\infty;U) \rightarrow L^2(0,\infty;Y)$ defined by

$$(Lu)(t) = C\int_0^t \tilde{g}(t-s)\int_0^s R(s-\rho)Bu(\rho)d\rho ds$$

satisfies $L \in \mathcal{L}(L^2(0,\infty;U),\ L^2(0,\infty;Y))$.

Remark
In [13] there is a certain symmetry between the assumptions placed upon C and B (See also Pritchard and Salamon [12]). The appearance here of the term $\int_0^t \tilde{g}(t-s)R(s)x_0ds$ pushes the assumptions onto either B or C. This integral brings in a smoothing of $(A)^{-\frac{1}{2}}$ for example 2.1, but no smoothing of the resolvent operator because the ν_n, μ_n are asymptotic to a vertical line in $\mathbf{C}_-$. The analysis in the sequel indicates why we have chosen C in this setting.

Proposition 2.2
Assume (A1)–(A4) *hold and* $||D|| < 1/||L||$. *Then there exists a unique resolvent operator,* $R_D(t)$, *for the integrodifferential equation*

$$\dot{x}(t) = \int_0^t \tilde{g}(t-s)(A+BDC)x(s)ds, \quad x(0) = x_0.$$

Moreover, for all $x_0 \in X$ *and* $h(\cdot) \in L^2(0,\infty;X)$, $x_h(\cdot) \in L^2(0,\infty;X)$ *where*

$$\begin{aligned} x_h(t) &= R(t)x_0 + \int_0^t R(t-s)\int_0^s \tilde{g}(s-\rho)BDCx_h(\rho)d\rho ds + \\ &\quad + \int_0^t R(t-s)h(s)ds \end{aligned} \qquad (9)$$

Proof
See theorem 3.7, [14] for the essential details. □

In the general theory presented in [14], it is possible via solution of the inhomogeneous equation (9) to pose robustness problems for the general system (5) under much weaker assumptions than those presented here. In the special case here proposition 2.2 provides the basis of our robustness analysis.

3 Robustness analysis for parameterized stability domains

Let $\mathbf{C}_g \subset \mathbf{C}$ be such that assumptions i)–iii) hold for f, and g has no poles in $\overline{\mathbf{C}_+}$. We assume without loss of generality $\lambda_0 = 0$. Using the well-posedness result of theorem 2.3 we make the following.

Definition 3.1

$$r_{\mathbf{C}}(A;B,C;\mathbf{C}_b) = \sup\{r \mid \|D\| < r \text{ implies that there exists}$$

$$\text{an exponentially stable resolvent operator for (7)}\}$$

Lemma 3.2
Assume (A4) *holds, then there exists* $\delta < 0$ *such that for all* $\lambda \geq \delta$, $R(\lambda) \in \mathcal{L}(X)$, *where*

$$R(\lambda)x = \int_0^\infty e^{-\lambda t} R(t)x dt,$$

and

$$R(\lambda)x = (\lambda I - g(\lambda)A)^{-1}x.$$

□

Theorem 3.3
Assume that (A1)–(A4) *for* $\tilde{g}$ *parameterizing* $\mathbf{C}_b$, U *and* Y *are Hilbert spaces and*

$$\lim_{\varepsilon \downarrow 0} \sup_{\omega \in \mathbf{R}} \|C(f(i\omega - \varepsilon)I - A)^{-1}B\| = \sup_{\omega \in \mathbf{R}} \|C(f(i\omega)I - A)^{-1}B\| < \infty$$

then

$$r_{\mathbf{C}}(A;B,C;\mathbf{C}_b) = 1/\|L\| = 1/\left(\sup_{\omega \in \mathbf{R}} \|C(f(i\omega)I - A)^{-1}B\|\right)$$

$$= \inf_{s \in \partial \mathbf{C}_b} 1/\|C(sI-A)^{-1}B\|$$

Proof
The details follows analogously to those in [14]. The main difficulty occurs in establishing exactness of the formula above. To see this suppose

$$r_{\mathbf{C}}(A;B,C;\mathbf{C}_b) > \inf_{s \in \partial \mathbf{C}_b} 1/\|C(sI-A)^{-1}B\|.$$

Choose $\omega_0 \in \mathbf{R}$ and $u \in U$ such that

$$r_{\mathbf{C}}(A;B,C;\mathbf{C}_b) > 1/\left(\|C(f(i\omega_0)I-A)^{-1}Bu\|\right) > 1/\|L\|.$$

Define

$$\begin{array}{l} x_0 = (i\omega_0 I - g(i\omega_0)A)^{-1}Bu,\ y = g(i\omega_0)Cx_0,\ D \in \mathcal{L}(Y,U) \\ D\hat{y} = u < y, \hat{y} > /\|y\|^2,\ \|D\| = 1/\|y\| \text{ and} \\ h(t) = e^{i\omega_0 t}\int_t^\infty e^{i\omega_0 s} g(s)(A+BDC)x_0 ds \end{array}$$

It is then easy to show that $x(t) = e^{i\omega_0 t}$ satisfies

$$\dot{x}(t) = \int_0^t g(t-s)(A+BDC)x(s)ds + h(t),\quad x(0) = x_0 \in D(A).$$

But $h(\cdot) \in L^2(0,\infty;X)$ and $x(\cdot) \notin L^2(0,\infty;X)$ which together contradict the assumed exponential stability of $R_D(t)$, the resolvent operator for (7).

Example 3.4
Let $\{\varphi_n\}$, X and $S(t)$ be as in example 2.1. Define a subspace $\underline{X} \subset X$ as follows :–

$$\underline{X} = \{\Sigma x_n \varphi_n \mid \Sigma x_n^2 \gamma_n^2 < \infty\},\quad \varphi_n \neq 0,\ n = 1,2,$$

with the obvious norm

$$\|x\|_{\underline{X}}^2 = \Sigma x_n^2 \gamma_n^2.$$

Suppose that there is uncertainty in the operator A of example 2.1 of the form $(A+BDC)$ with

$$B_x = \Sigma b_n < x, \varphi_n >$$

$$C_x = \Sigma c_n < x, \varphi_n > .$$

It is easy to show that $B \in \mathcal{L}(X)$ and $C \in \mathcal{L}(\underline{X}, X)$ if $|b_n| \leq k$, $|c_n| \leq k|\gamma_n|$. Moreover (A1)–(A4) hold for $|c_n| = |\gamma_n|$ provided that $|c_n|^2 \leq k|\gamma_n|$ and $\inf\{|\lambda - \lambda_n|/|c_n| \mid n = 1, 2, \cdots, \lambda \in \mathbf{C}_b\} \geq \beta > 0$. Applying the formula for $r_{\mathbf{C}}(A; B, C; \mathbf{C}_b)$ to the case of $\mathbf{C}_g = \Sigma$ then

$$r_{\mathbf{C}}(A; B, C; \mathbf{C}_b) = \inf_{s \in \mathbf{C}_b} \inf_{n} \frac{|s - \lambda_n|}{|b_n c_n|} \tag{10}$$

If however $\mathbf{C}_g = \mathbf{C}_- = \{\lambda \in \mathbf{C} | Re\lambda < 0\}$ then $g = 1$ and

$$r_{\mathbf{C}}(A; B, C; \overline{\mathbf{C}}_+) = \inf_{s \in i\mathbf{R}} \inf_{n} \frac{|s - \lambda_n|}{|b_n c_n|}$$

In this case $R_D(t) \in \mathcal{L}(X)$, defines an exponentially stable semigroup (see [14]) and for all $x_0 \in D(A)$, $R_D(t)x_0$ is continuously differentiable and

$$\begin{aligned} R_D(t)x_0 &= (A + BDC)R_D(t)x_0 \\ &= R_D(t)(A + BDC)x_0. \end{aligned}$$

Corollary 3.5
Assume the conditions of theorem 3.3 hold. If $\|D\| < r_{\mathbf{C}}(A; B, C; \mathbf{C}_b)$ *then* $R_D(\lambda) \in \mathcal{L}(X)$ *for all* $\lambda \in \overline{\mathbf{C}}_+$ *where* $R_D(\lambda)x = \int_0^\infty e^{-\lambda t} R_D(t)x dt$.
In addition

$$\begin{aligned} R_D(\lambda)x &= R(\lambda)x + R(\lambda)BDy(\lambda) \\ y(\lambda) &= CR(\lambda)x + CR(\lambda)BDy(\lambda) \text{ for all } x \in X. \end{aligned} \tag{11}$$

If $(A + BDC)$ *is closable then* $(\lambda I - A_D)^{-1} \in \mathcal{L}(X)$ *for all* $\lambda \in \mathbf{C}_b$, *where* A_D *is the closure of* $(A + BDC)$.

Proof
Set $S_D(\lambda) = g(\lambda)R_D(\lambda)$. The pair of equations (11) show that $S_D(\lambda) \in D(A)$ for all $x \in X$ and $\lambda \in \mathbf{C}_+$. Let $x \in X$ and choose $\{x_n\} \subset D(A)$ with $x_n \to x$. It is easy to show that

$$(f(\lambda)I - A - BDC)S_D(\lambda)x_n = x_n.$$

But $x_n \to x$ and $S_D(\lambda)x_n \to S_D(\lambda)x$. Therefore $(f(\lambda)I - A_D)R_D(\lambda)x = x$ and $(f(\lambda)I - A_D)$ is onto X. Also $(f(\lambda)I - A_D)$ is one-one and therefore $(f(\lambda)I - A_D)^{-1}$ exists for all $\lambda \in \overline{\mathbf{C}}_+$. Moreover $\|(f(\lambda)I - A_D)^{-1}x\| = \|S_D(\lambda)x\| \leq k\|x\|$ for all $x \in D(A)$ and therefore $(f(\lambda)I - A_D)^{-1} \in \mathcal{L}(X)$. □

Remark
If $\mathbf{C}_g \subset \mathbf{C}_-$ and $\|D\| < r_{\mathbf{C}}(A; B, C; \mathbf{C}_b)$ then there exists a semigroup $S_D(t) \in \mathcal{L}(X)$ with generator $A_D : D(A) \to X$ satisfying $A_D x = (A + BDC)x$ for all $x \in D(A)$. Hence A_D is the closure of $(A + BDC)$ and by corollary 3.5, $(\lambda I - A_D)^{-1} \in \mathcal{L}(X)$ for all $\lambda \in \mathbf{C}_b$. This answers the problem of robustness measures with respect to parameterized stability domains.

Example 3.6
Let $\Omega = (0, 1)$ and consider the system

$$\begin{cases} x_t(t,\xi) = x_{\xi\xi}(t,\xi),\ t > 0,\ \xi \in (0,1) \\ x(t,0) = x(t,1) = 0,\ t \geq 0 \end{cases}$$

Define $Ax = \Delta x$, $A : D(A) = H^2(0,1) \cap H_0^1(0,1) \to L^2(0,1)$. A generates an exponentially stable semigroup $S(t)$ on $L^2(0,1)$ and the spectrum of $A = \{-n^2\pi^2, n = 1, 2, \cdots\}$. Moreover $\{\varphi_n\}$ with $\varphi_n = \sin \pi\xi$, $n = 1, 2, \cdots$ defines a complete orthonormal basis for $L^2(0,1)$. Consider the perturbed system

$$\begin{cases} x_t(t,\xi) = x_{\xi\xi}(t,\xi) + \alpha x_\xi(t,\xi),\ t > 0,\ \xi \in (0,1),\ \alpha \in \mathbf{C}, \\ x(t,0) = x(t,1) = 0,\ t \geq 0. \end{cases} \tag{12}$$

Then the perturbation term can be represented within our structure BDC by choosing $B = I$, $D \in \mathcal{L}(L^2(0,1))$ and $C \in \mathcal{L}(H^1(0,1), L^2(0,1))$ with $Cx = x_\xi$. In order to compute stability radii we choose a second basis $\{\psi_n\}$, $\psi_n = \cos n\pi_\xi$, for $L^2(0,1)$. Then

$$C(\lambda I - A)^{-1}x = \Sigma n\pi\psi_n < x, \varphi_n > 1/(\lambda - \lambda_n).$$

Suppose we consider $\mathbf{C}_g = \Sigma$. Assumptions (A1)–(A4) hold as in example 3.4 and

$$\begin{aligned} r_{\mathbf{C}}(A; B, C; \mathbf{C}_b) &= \inf_{s \in \mathbf{C}_b} \inf_{n} \frac{|s - \lambda_n|}{|b_n c_n|} \\ &= \sqrt{1 - 1/4\pi^2}. \end{aligned}$$

If instead we consider $\mathbf{C}_g = \mathbf{C}_-$ then

$$r_{\mathbf{C}}(A; B, C; \mathbf{C}_b) = \pi.$$

Hence if $\|D\| < \sqrt{1 - 1/4\pi^2}$ then there exists a semigroup $S_D(t)$ on $L^2(0,1)$ with generator A_D satisfying
i) $A_D x = (A + BDC)x$ for all $x \in D(A)$

ii) $(\lambda I - A_D)^{-1} \in \mathcal{L}(X)$ for all $\lambda \in \mathbf{C}_b$. Therefore the spectrum of $A_D \subset \mathbf{C}_b$ provided that $\|D\| < \sqrt{1 - 1/4\pi^2}$. Applying the abstract result to the specific case of system (12), the perturbed spectrum is contained in $\mathbf{C}_g$ provided that $|\alpha|^2 < 1 - 1/4\pi^2$. However a direct analysis of the spectrum shows that $|\alpha|^2 < \sqrt{4\pi^2 - 1}$. Note that the general theory allows arbitrary perturbations $D \in \mathcal{L}(L^2(0,1;\mathbf{C}))$ which accounts for this discrepency.

Remark

If we differentiate (7) for the system of example 3.5 then we obtain the system

$$\begin{pmatrix} x_t \\ x_{tt} \end{pmatrix} = \begin{bmatrix} 0 & I \\ \Delta & -I \end{bmatrix} + \begin{bmatrix} 0 \\ I \end{bmatrix} D(\partial_\xi, 0) \begin{pmatrix} x \\ x_t \end{pmatrix}. \tag{13}$$

Notice that the unbounded perturbation DC for the first order system becomes bounded for the second order system. This illustrates well the interplay between structure and unboundedness in the perturbation term when we have no smoothing due to the semigroup for (13). This compares with the role played by $\tilde{g}$ (in the resolvent operator case) in bringing in extra integration. System (13) gives another possibility for realizing the robustness problem with respect to Σ.

4 Conclusion

In this paper we have applied many of the general methods given in [14], for a robustness analysis of infinite dimensional systems, to the particular situation of parameterized stability domains. We use a functional analytic rather than spectrum based approach, but when the underlying systems are more regular we recover some information about the spectrum of the perturbed system. When applied to a simple diffusion problem, the robustness of the system is shown to be equivalent to that of exponential stability for a diffusion problem with memory or a wave equation with additional structure in the perturbation term.

References

[1] Chen. G, Grimmer. R, *Integral equations as evolution equations*. J. Diff. Eqns., Vol **45**, 1982.

[2] Curtain, R. F, *Stability of infinite dimensional systems by spectrum analysis: The spectrum determined growth assumption*. Systems and Control Letters, Vol **2**, 1982.

[3] Datko. R, *Uniform asymptotic stability of evolutionary processes in a Banach space*. SIAM J. Math. Analysis, Vol **3**, 1972.

[4] Datko. R, Lagnese. J, Polis. M. P, *An example of the effect of time delays in the boundary feedback stabilization of the wave equation.* SIAM J. control, Vol **24**, 1986.

[5] Francis. F, *A course in H^∞-Control Theory.* Lecture Notes in Control and Information Sciences, Vol **88**, Springer Verlag, 1987.

[6] Grimmer. R, Pruss. J, *On linear Volterra equations in Banach spaces.* Comput. Math. Appl., Vol 11, 1985.

[7] Hinrichsen. D, Kelb. B, Linnemann. A, *An algorithm for the computation of the structured stability radius with applications.* Automatica, 25 p771-775, 1989.

[8] Hinrichsen. D, Pritchard. A. J, *Stability radius for structured perturbations and the algebraic Riccati equation.* Systems and Control, Vol **8**, 1986.

[9] Hinrichsen. D, Pritchard. A. J, *New robustness results for linear systems under real perturbations.* In Proc. 27th IEEE Conference on Decision and Control, p1375-1379, Austin, 1988.

[10] Hinrichsen. D, Pritchard. A. J, *On robustness of root locations of polynomials under complex and real perturbations.* In Proc. 27th IEEE Conference on Decision and Control, Austin, 1988.

[11] Miller. R. K, Wheeler. R. L, *Well-posedness and stability of linear integrodifferential equations in abstract spaces.* Funkcialaj Ekvacioj, Vol **21**, 1978.

[12] Pritchard. A. J, Salamon. D, *The linear quadratic control problem for infinite dimensional systems with unbounded input and output operators.* SIAM J. Control, Vol **25**, 1987.

[13] Pritchard. A. J, Townley. S, *A stability radius for infinite dimensional systems.* Infinite Dimensional Systems, Proc. Vorau, 1986 Ed F. Kappel, Springer Verlag.

[14] Pritchard. A. J, Townley. S, *Robustness of linear systems*, J. Diff. Eqns Vol 77, p254-286, 1989.

On high-gain adaptive stabilization of infinite-dimensional systems with nonlinearities

Hartmut Logemann David H. Owens

Abstract

This paper develops an input-output theory of high-gain adaptive stabilization of certain infinite-dimensional processes with actuator and sensor nonlinearities. It is shown that there is a wide range of gain adaption rules achieving stability for the class of processes under consideration if proportional output feedback is used. The paper shows that the scope of applicibility of universal adaptive stabilization ideas extends far beyond finite-dimensional linear systems.

0 Nomenclature

$\mathbf{R}_+$:= set of non-negative real numbers.
$\mathbf{C}_\gamma := \{s \in \mathbf{C} | Re(s) > \gamma\}$, $\gamma \in \mathbf{R}$.
Let $J \subset \mathbf{R}$ be an interval, then $C(J, \mathbf{R}^n)$:= vector space of $\mathbf{R}^n$-valued continuous functions on J.
$C(J) := C(J, \mathbf{R})$.
$L^P(J)$:= vector space of real-valued p-integrable functions on J.
$LL^P(J)$:= vector space of real valued locally p-integrable functions on J.
$L^P_+ := L^P(\mathbf{R}_+)$, $LL^P_+ := LL^P(\mathbf{R}_+)$, $LL^P := LL^P(\mathbf{R})$.
$L^{P,\gamma}_+ := \{f : \mathbf{R}_+ \to \mathbf{R} \mid f\exp(\gamma\cdot) \in L^P_+\}$, $\gamma \in \mathbf{R}$.*
$BV(J, \mathbf{R}^{n\times n})$:= vector space of $\mathbf{R}^{n\times n}$-valued functions of bounded variation on J.
$H^\infty(\mathbf{C}_\gamma)$:= algebra of bounded holomorphic functions on $\mathbf{C}_\gamma$.
$H^\infty := H^\infty_0$.
$id_{\mathbf{R}}$:= the map $\mathbf{R} \to \mathbf{R}$, $x \mapsto x$.
Let $f \in LL^P_+$, then for all $t \geq 0$

$$(P_t f)(\tau) := \begin{cases} f(\tau), & 0 \leq \tau \leq t \\ 0, & \tau > t. \end{cases}$$

If $f \in C([0, \alpha))$, then for all $0 \leq t < \alpha$

$$(P_t f)(\tau) := \begin{cases} f(\tau), & 0 \leq \tau \leq t \\ 0, & \tau > t. \end{cases}$$

---

*Provided with the norm $||f||_{p,\gamma} := ||f\exp(\gamma\cdot)||_p$ $L^{P,\gamma}_+$ becomes a Banach space.

$S_+(\alpha,\beta)$ $(\beta \geq \alpha > 0)$ denotes the set of all Borel functions
$f : \mathbf{R}_+ \times \mathbf{R} \to \mathbf{R}$ such that $f(t,0) = 0\ \forall t \in \mathbf{R}_+$ and
$\beta x^2 \geq x f(t,x) \geq \alpha x^2\ \forall (t,x) \in \mathbf{R}_+ \times \mathbf{R}$.

$S_-(\alpha,\beta)$ $(\beta \geq \alpha > 0)$ denotes the set of all Borel functions
$f : \mathbf{R}_+ \times \mathbf{R} \to \mathbf{R}$ such that $f(t,0) = 0\ \forall t \in \mathbf{R}_+$ and
$(-\alpha)x^2 \geq x f(t,x) \geq (-\beta)x^2\ \forall (t,x) \in \mathbf{R}_+ \times \mathbf{R}$.

$S(\alpha,\beta) := S_+(\alpha,\beta) \cup S_-(\alpha,\beta)$.

$S := \bigcup_{\beta>\alpha>0} (S(\alpha,\beta)$.

Given $f \in S$, there exist number $\beta \geq \alpha > 0$ such that $f \in S(\alpha,\beta)$. We define :

> $sign(f) := +1$ if $f \in S_+(\alpha,\beta)$ and $sign(f) := -1$ if $S_-(\alpha,\beta)$ (notice that $sign(f) = sign(f(t,x))\ \forall (t,x) \in \mathbf{R}_+ \times (0,\infty)$.

For $f \in S$ and $x : \mathbf{R}_+ \to \mathbf{R}$ the symbol $f(\cdot, x(\cdot))$ denotes the function $t \mapsto f(t, x(t))$.

1 Introduction

The initial steps taken in the area of ***high-gain*** adaptive stabilization included the result by Willems and Byrnes [19] that linear, single-input single-output, minimum-phase systems with finite-dimensional state-space realizations (A, B, C) are stablized by the proportional output feedback law

$$u(t) = -sign(CB)k(t)y(t)$$

where the gain $k(t)$ is obtained from the differential equation

$$\dot{k}(t) = (y(t))^2.$$

These basic results appeared to depend upon the assumption of linearity, finite-dimensionality, the specific form of the control law and the gain evolution as a quadratic in the output. In particular, a state-space representation seemed to be necessary and exact implementation of $u(t)$ based upon exact measurements $y(t)$ appeared to be essential. For practical applications and robustness considerations an input-output theory would have advantages and could enable the removal of the finite-dimensionality assumption. Also, the inclusion of imperfect measurement and actuation in the form of nonlinear sensor and actuator characteristics would be of great benefit. These ideas and their implications are addressed in this paper. It is remarkable that, in essence, the results of previous studies (cf. the references below) carry over to these more complex situations with little change other than the need for a new ***scaling-invariance*** property in the controller gain characterization.

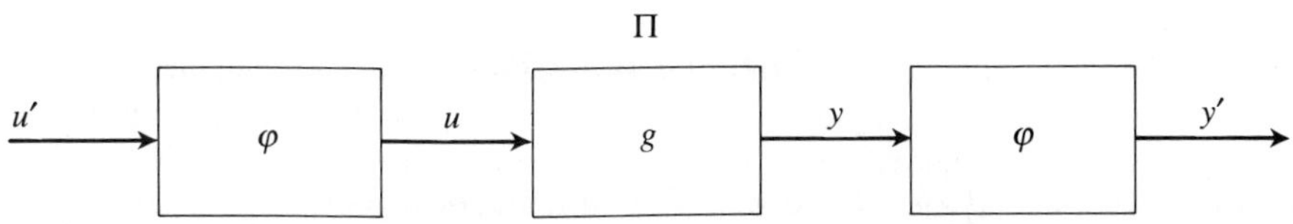

Figure 1.

However the proofs of the results in the linear finite-dimensional case do not carry over to the nonlinear infinite-dimensional situation and therefore the generalizations are far from being trivial.

Robust (non-adaptive) high-gain control of infinite-dimensional systems has been studied in some detail by the authors in Logemann and Owens [10]. In the present paper we shall develop an input-output approach to adaptive high-gain stabilization of certain single-input single-output nonlinear infinite-dimensional processes. The class of processes Π to be considered is shown in Figure 1. We suppose that φ and ψ are memoryless actuator and sensor nonlinearities and that g is the transfer function of a linear time-invariant (non necessarily finite-dimensional) system. Moreover we assume:

1. The functions φ and ψ are unbiased time-varying nonlinearities lying either in a positive or negative sector (i.e. φ and ψ are elements of g as defined in the list of notation).

2. The transfer function g is minimum phase (i.e. g has no zeros in the closed right-half plane) and there exists a real number $a \neq 0$ such that

$$sg(s) - a = O(s^{-1}) \text{as } |s| \to \infty \text{ in } \mathbf{C}_+.$$

   The above equation generalizes the *relative-degree one* condition for finite-dimensional systems when a is just CB in a state space realization (A, B, C)(cf. section 2).

Our adaptive stabilization results do not require parameter identification algorithms and can hence be regarded as being in the same spirit as the papers by Nussbaum [16], Willems and Byrnes [19], Dahleh and Hopkins [1], Mårtensson [13],

Owens *et al* [17] and Kobayashi [7]. The above references have in common that they are based on state-space methods. The articles by Nussbaum [16] and Willems and Byrnes [19] and the thesis by Mårtensson [13] deal with linear finite-dimensional systems, while Owens *et al* [17] consider finite-dimensional systems with certain non-linearities in the state. Dahleh and Hopkins [1] extend the main result of Willems and Byrnes [19] to a class of linear differential-delay systems. Generalizations to linear distributed parameter systems described by holomorphic semigroups are given by Kobayashi [7]. The theory developed in the present paper contains the result of Dahleh and Hopkins [1] and Kobayashi [7] and includes a considerably larger class of systems, feedback laws and gain adaption rules. In particular our results apply to retarded systems and Volterra integrodifferential systems. It is emphasized that the new theory can deal with systems subject to unbiased time-varying sensor and actuator nonlinearities lying either in a positive or negative sector, an important problem which has not been considered in the above references. Although input-output descriptions are used throughout the paper non-zero initial conditions are taken into account by using *initial-condition terms* (cf. Section 2).

In Section 2 we establish a number of preliminary results and define precisely the adaptive stabilization problem to be considered. In order to solve this problem it is useful to employ so-called *switching functions*, a concept which has its origin in the papers by Nussbaum [16] and Willems and Byrnes [19]. In particular it will turn out that the notion of a *scaling invariant* switching function introduced in Section 3 plays an essential role in the problem of dealing with the nonlinearities φ and ψ. A switching function is a locally (essentially) bounded function (taking on both positive and negative values) which enables the controller to *learn* the sign of the process, in other words it provides the controller gain with the *right* sign. A scaling invariant switching function has this property despite the presence of arbitrary unbiased time-varying actuator and sensor nonlinearities lying either in a positive or negative sector. A precise definition of these concepts is given in Section 3. As Owens *et al* [17] we allow switching as a function of both current and past gain and input data. this leads to a wide class of stabilizing adaptive high-gain controllers with convergence of the switching mechanism being independent of the gain adaption rules for generating $k(t)$. More precisely, we investigate the dynamics of the process Π under the control law

$$u(t) = \varphi[t, N(\xi(t))k(t)\psi(t, y(t))]$$

$$\dot{\xi}(t) = k(t)(\psi(t, y(t)))^2, \quad \xi(0) = \xi_0 \in \mathbf{R}$$

where $k(t)$ is a strictly positive time-varying gain and N is a scaling invariant switching function. It turns out that the above controller stabilizes the process Π in the sense that $y \in L^2_+ \cap L^\infty_+$ and $\lim_{t\to\infty} \xi(t)$ exists and is finite. Moreover if k is bounded it follows that $\lim_{t\to\infty} y(t) = 0$. In view of this result it is natural to consider the gain k as the image of a causal map operating on $\xi(\cdot)$ and $\psi(\cdot, y(\cdot))$. A large class of gain adaption rules is given ensuring closed-loop stability in the sense that

$y \in L^2_+ \cap L^\infty_+$, $\lim_{t\to\infty} y(t) = 0$, the limits $\lim_{t\to\infty} \xi(t)$ and $\lim_{t\to\infty} k(t)$ exist and are finite. These gain adaption laws include $\dot{k}(t) = \psi^2(t, y(t))$ and therefore fully generalize previous work. Section 4 discusses the problem of *exponential* stabilization. A sufficient condition for exponential stabilizability of the process Π is given and it is sketched out that an exponentially weighted version of the above control law achieve exponential stability.

2 Preliminaries

In the following sections a certain functional differential equation will play an important role. It is therefore useful to give a precise statement of existence and uniqueness of solutions. Consider the initial value problem

$$\dot{x}(t) = (Tx)(t) + f_1(t, x(t)) + f_2(t), \quad t \geq \alpha \tag{2.1a}$$

$$x|_{[0,\alpha]} = x_0 \in C([0,\alpha], \mathbf{R}^n), \quad \alpha \geq 0 \tag{2.1b}$$

where

1. $T : (LL^2_+)^n \to (LL^2_+)^n$. We assume that $T(0) = 0$ and that there exists $\mu > 0$ such that $\|P_t(Tx - Tx')\| \leq \mu \|P_t(x - x')\|$ $\forall x, x' \in (LL^2_+)^n, \forall t \geq 0$, i.e. T is unbounded, causal and of finite incremental gain.

2. $f_1 : \mathbf{R}_+ \times \mathbf{R}^n \to \mathbf{R}^n$ is a continuous and locally Lipschitzian function.

3. f_2 is a function in $(LL^1_+)^n$.

Of course, if $\alpha = 0$ in (2.1b), then $C([0,\alpha], \mathbf{R}^n) = \mathbf{R}^n$. In order to define what we mean by a solution of (2.1) on $[0,\beta)$ ($\alpha < \beta \leq \infty$) we have to give a meaning to Tx if $x \in C([0,\beta], \mathbf{R}^n)$ (remember that T operates on functions which are defined on $\mathbf{R}_+$ and are elements in $(LL^2_+)^n$). We set $(Tx)(t) = (TP_\tau x)(t)$ for $0 \leq t \leq \tau < \beta$. Since T is causal, this definition does not depend on the choice of τ.

2.1 Definition: A solution of (2.1) on $[0,\beta)$ is a function $x \in C([0,\beta], \mathbf{R}^n)$ ($\alpha < \beta \leq \infty$) such that
(i) x is absolutely continuous on $[\alpha,\beta)$ and satisfies (2.1a) a.e. on $[\alpha,\beta)$.
(ii) (2.1b) is satisfied by x.

2.2 Theorem: *The initial value problem* (2.1) *has a unique solution x on $[0,\beta)$, where $\alpha < \beta \leq \infty$. If $\beta < \infty$ and β cannot be increased, then there exist a sequence $\alpha < t_1 < t_2 < \cdots \to \beta$ such that* $\lim_{i\to\infty} |x(t_i)| = \infty$.

Although we hardly believe the above theorem to be unknown, we were unable

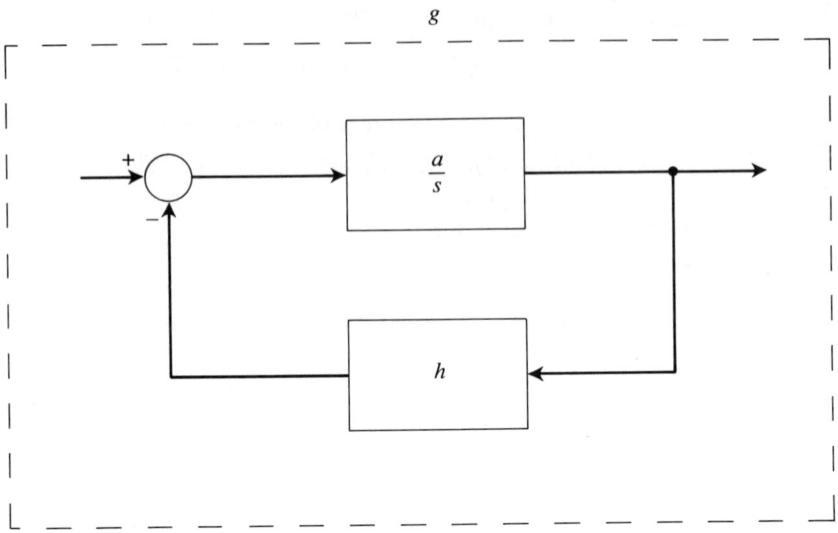

Figure 2.

to find it in the literature (e.g. the *classical* paper by Driver [2] and the book by Hale [5]). A proof is given in Logemann and Owens [11].

Let us consider the linear part of the nonlinear process Π shown in Figure 1. It is not assumed that the transfer function g is rational. Suppose that g is meromorphic on $\mathbf{C}_0$ and that

$$g^{-1}(s) = a^{-1}s + h(s), \tag{2.2}$$

where $a \in \mathbf{R}\backslash\{0\}$ and $h \in H^\infty$.

Of course, (2.2) is equivalent to

$$g(s) = (1 + \frac{a}{s}h(s))^{-1}\frac{a}{s}, \tag{2.3}$$

i.e. g is the feedback interconnection of the integrator $\frac{a}{s}$ and the transfer function h (cf. Figure 2).

2.3 **Remark:** (i) Suppose that g is meromorphic in some open set containing $\bar{\mathbf{C}}_0$. Then (2.2) holds iff g has no zeros in $\bar{\mathbf{C}}_0$ and $sg(s) - a = O(s^{-1})$ if $|s| \to \infty$ in $\mathbf{C}_0$. (ii) If g is the transfer function of a finite-dimensional system $\sigma : \dot{x} = Ax + bu$, $y = c^T x$, then (2.2) is satisfied if σ is minimum-phase and $c^T b \neq 0$.

In the following we shall assign an operator $H : L^2_+ \to L^2_+$ to the H^∞-function h. The operator H is given by $H = L^{-1}M_hL$, where L denotes the Laplace transform and M_h denotes the multiplication by h on the Hardy space H^2 (of the right-half plane). The operator H is linear bounded and shift invariant (in the sense of Vidyasagar [18]). As a consequence H is causal (cf. Vidyasagar [18]) and therefore it has a unique causal extension to LL^2_+. This extension will also be denoted by H. It should be mentioned that the converse is also true, i.e. given a linear bounded shift invariant operator $H : L^2_+ \to L^2_+$ there exists a function $h \in H^\infty$ such that $H = L^{-1}M_hL$(cf. Harris and Valenca [6] or Logemann [8]).

The function g can be thought of as being the transfer function of

$$\begin{aligned} \dot{y} &= a(u - (Hy + w)), \\ y(0) &= y_0, \end{aligned} \tag{2.4}$$

where $w \in L^2_+$ is due to a non-zero initial condition in the system with the transfer function h. The initial value problem (2.4) is of the form (2.1) and therefore has a unique solution for each $u \in LL^1_+$.

The *problem* is to find an adaptive control law $u'(t) = F(y'(\tau),\ 0 \le \tau \le t)$ (i.e. $u(t) = \varphi[t, F(\psi(\tau, y(t)),\ 0 \le \tau \le t)]$) which *stabilizes* the process Π shown in Figure 1 $\forall y_0 \in \mathbf{R}$ and $\forall w \in L^2_+$ under the assumptions that $a \neq 0$, $H : L^2_+ \to L^2_+$ is a linear bounded shift invariant operator and $\varphi, \psi \in S$. By *stability* we mean that $y,\ y' \in L^2_+ \cap L^\infty_+$ and $\lim_{t\to\infty} y(t) = \lim_{t\to\infty} y'(t) = 0$.

3 Adaptive high-gain stabilization

It will turn out that it is useful to introduce the concept of a *switching function* which has its origin in the paper by Nussbaum [16].

3.1 **Definition:** A function $N \in LL^\infty$ is called
(i) a switching function if for some $x_0 \in \mathbf{R}$

$$\sup_{x > x_0} \frac{1}{x - x_0} \int_{x_0}^{x} N(\lambda) d\lambda = +\infty \tag{3.1a}$$

and

$$\inf_{x > x_0} \frac{1}{x - x_0} \int_{x_0}^{x} N(\lambda) d\lambda = -\infty \tag{3.1b}$$

(ii) a *Nussbaum gain* if the function $id_{\mathbf{R}}N$ is a switching function.

A switching function N is called *scaling invariant* if the function $(\Gamma^\beta_\alpha o N)N$ is a switching function for all $\alpha,\ \beta > 0$, where the function Γ^β_α is given by

$$\Gamma_\alpha^\beta(\lambda) = \begin{cases} \alpha, & \lambda > 0 \\ 0, & \lambda = 0 \\ \beta, & \lambda < 0, \end{cases}$$

A Nussbaum gain N is called ***scaling invariant*** if $(\Gamma_\alpha^\beta oN)N$ is a Nussbaum gain for all $\alpha,\ \beta > 0$.

It is easily seen that, if conditions (3.1) are satisfied for some $x_0 \in \mathbf{R}$, they are satisfied for all $x_0 \in \mathbf{R}$. Moreover it is clear that a Nussbaum gain N is scaling invariant iff the switching function $id_{\mathbf{R}}N$ is .

We give an example of a scaling invariant switching function and a scaling invariant Nussbaum gain.

3.2 Example: Consider the function $N(\lambda) = \cos(\frac{1}{2}\pi\lambda)\exp(\lambda^2)$. In the paper by Nussbaum [16] it is shown that for all $x_0 \in \mathbf{R}$

$$\sup_{x>x_0} \int_{x_0}^{x} N(\lambda)d\lambda = +\infty$$

and

$$\inf_{x>x_0} \int_{x_0}^{x} N(\lambda)d\lambda = -\infty.$$

Using the same ideas it is possible to show that N is a scaling invariant switching function and a scaling invariant Nussbaum gain.

It is easy to find Nussbaum gain which are not scaling invariant. The Nussbaum gain

$$N(\lambda) = \begin{cases} 1, & n^2 \le |\lambda| < (n+1)^2,\ n \text{ even} \\ -1, & n^2 \le |\lambda| < (n+1)^2,\ n \text{ odd} \end{cases}$$

has been considered by Willems and Byrens [19]. It can be shown that $(\Gamma_\alpha^\beta oN)N$ is not a Nussbaum gain if $\alpha \neq \beta$ $(\alpha, \beta > 0)$. However it is not difficult to show that there exist ***bounded*** scaling invariant Nussbaum gains. An example is given by

$$N(\lambda) = \begin{cases} 1, & 0 \le |\lambda| < \lambda_0 \\ 1, & \lambda_n \le |\lambda| < \lambda_{n+1},\ n \text{ even} \\ -1, & \lambda_n \le |\lambda| < \lambda_{n+1},\ n \text{ odd}, \end{cases}$$

where $\lambda_{n+1} = (\lambda_n)^2$ and $\lambda_0 > 1$.

Let us consider the same process as in Section 2 (cf. Figure 1). We shall study the behaviour of the process Π if the following control law is applied

$$u(t) = \varphi[t, N(\xi(t))k(t)\psi(t, y(t))] \tag{3.2a}$$

$$\dot{\xi}(t) = k(t)(\psi(t, y(t)))^2, \ \xi(0) = \xi_0 \in \mathbf{R}, \tag{3.2b}$$

where N is a scaling invariant switching function and $k : \mathbf{R}_+ \to \mathbf{R}$ is a strictly positive function, i.e. $k(t) \geq \varepsilon > 0 \ \forall t \in \mathbf{R}_+$.

3.3 Lemma: *Let φ and ψ be in S and suppose that the feedback system given by* (2.4) *and* (3.2) *has a unique absolutely continuous solution which can be continued uniquely to the right as long as it remains bounded. Under these conditions the following is true:*
(i) $y, y' \in L^2_+ \cap L^\infty_+$.
(ii) $\lim_{t\to\infty} \xi(t)$ *exists and is finite.*
Moreover, if k is bounded, we have
(iii) $\lim_{t\to\infty} y(t) = \lim_{t\to\infty} y'(t) = 0$.

3.4 Remark: Suppose that k is continuous, N is continuously differentiable and φ and ψ are continuous with continuous partial derivatives $D_2\varphi$ and $D_2\psi$. Then it follows from Theorem 2.2 that the feedback system given by (2.4) and (3.2) has a unique solution which can be continued uniquely to the right as long as it remains bounded.

For the proof of Lemma 3.3 we need the following

3.5 Proposition: *Let $\varphi \in S(\delta_\varphi, \Delta_\varphi)$ and $\psi \in S(\delta_\psi, \Delta_\psi)$ and suppose that $sign(\varphi)sign(\psi) = +1$. Then the inequalities*

$$x\varphi(t, \lambda r\psi(t, x)) \leq \Gamma(\lambda)\lambda r(\psi(t, x))^2$$

and

$$x\varphi(t, \lambda r\psi(t, x)) \geq \Gamma'(\lambda)\lambda r(\psi(t, x))^2$$

hold $\forall (t, r, \lambda, x) \in \mathbf{R}^2_+ \times \mathbf{R}^2$, *where*

$$\Gamma(\lambda) := \begin{cases} \Delta_\varphi/\delta_\psi, & \lambda > 0 \\ 0, & \lambda = 0 \\ \delta_\varphi/\Delta_\psi, & \lambda < 0 \end{cases} \tag{3.3a}$$

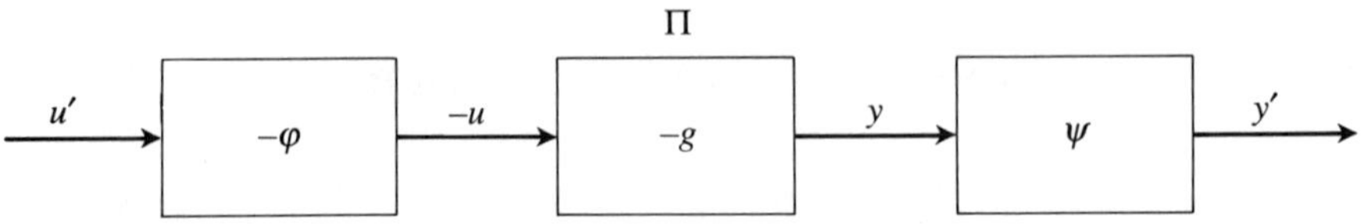

Figure 3.

and

$$\Gamma'(\lambda) := \Gamma(-\lambda). \tag{3.3b}$$

See [11] *for a proof.*

Proof of Lemma 3.3:

It follows from the equivalence of the diagrams in Figure 1 and Figure 3 that we can assume without loss of generality that $sign(\varphi)sign(\psi) = +1$. Furthermore let $\delta_\varphi, \Delta_\varphi, \delta_\psi$, and Δ_ψ be positive numbers such that $\varphi \in S(\delta_\varphi, \Delta_\varphi)$ and $\psi \in S(\delta_\psi, \Delta_\psi)$. Consider the equation

$$y\dot{y} = -a(yHy + yw) + ay\varphi[\cdot, (No\,)k\psi(\cdot, y(\cdot))] \tag{3.4}$$

Integrating (3.4) from 0 to t, using (3.3) and applying Proposition 3.5 yields

$$\frac{1}{2}(y^2(t) - y_0^2) \leq |a|(\|H\|\int_0^t y^2(\tau)d\tau + \|w\|_2(\int_0^t y^2(\tau)d\tau)^{\frac{1}{2}})$$

$$+ |a| \begin{cases} \int_0^t (\Gamma o N)(\xi(\tau))N(\xi(\tau))k(\tau)\psi^2(\tau, y(\tau))d\tau, & \text{if } a > 0 \\ -\int_0^t (\Gamma' o N)(\xi(\tau))N(\xi(\tau))k(\tau)\psi^2(\tau, y(\tau))d\tau, & \text{if } a < 0 \end{cases} \tag{3.5}$$

Using (3.2b) and the change of variables formula for Lebesgue integrable functions (cf. e.g. McShane and Botts [14], p. 195) it follows

$$\frac{1}{2}(y^2(t)-y_0^2) \le |a|(\|H\|\int_0^t y^2(\tau)d\tau + \|w\|_2(\int_0^t y^2(\tau)d\tau)^{\frac{1}{2}})$$

$$+|a|\begin{cases} \int_{\xi_0}^{\xi(t)} (\Gamma o N)(\lambda)N(\lambda)d\lambda, \text{ if } a>0 \\ -\int_{\xi_0}^{\xi(t)} (\Gamma' o N)(\lambda)N(\lambda)d\lambda, \text{ if } a<0 \end{cases} \tag{3.6}$$

Now realise that by (3.2b) and the properties of k

$$\begin{aligned} \int_0^t y^2(\tau)d\tau &\le \frac{1}{\varepsilon}\int_0^t k(\tau)y^2(\tau)d\tau \\ &\le \frac{1}{\varepsilon}\frac{1}{(\delta_\psi)^2}\int_0^t k(\tau)\psi^2(\tau,y(\tau))d\tau \\ &= \frac{1}{\varepsilon}\frac{1}{(\delta_\psi)^2}(\xi(t)-\xi_0) \end{aligned} \tag{3.7}$$

Setting $K_1 := |a|\frac{1}{\varepsilon}\frac{1}{(\delta_\psi)^2}\|H\|$ and $K_2 := \|w\|_2\frac{1}{\sqrt{\varepsilon}}\frac{1}{\delta_\psi}$ and using (3.6) and (3.7) we obtain

$$\frac{1}{2}y^2(t) \le \frac{1}{2}y_0^2 + (\xi(t)-\xi_0)\Bigg[K_1 + \frac{K_2}{\sqrt{\xi(t)-\xi_0}}$$

$$+|a|\frac{1}{\xi(t)-\xi_0}\left\{\begin{array}{l} \int_{\xi_0}^{\xi(t)} (\Gamma o N)(\lambda)N(\lambda)d\lambda, \text{ if } a>0 \\ -\int_{\xi_0}^{\xi(t)} (\Gamma' o N)(\lambda)N(\lambda)d\lambda, \text{ if } a<0 \end{array}\right\} \tag{3.8}$$

The equations (3.7) and (3.8) hold on each interval of the form $[0,\alpha)$ where the solution $(y(t),\xi(t))$ of the feedback system given by (2.4) and (3.2) exists. Since the r.h.s. of (3.8) has to be non-negative, it follows from the properties of the function N, that $\xi(t)$ and hence (by (3.8)) $y(t)$ remain bounded. As a consequence the solution $(y(t),\xi(t))$ exists on $[0,\infty)$ and it follows that $\lim_{t\to\infty}\xi(t)$ exists and is finite, which is (ii). Moreover, by (3.7) and (3.8), $y \in L_+^2 \cap L_+^\infty$ and hence $y' \in L_+^2 \cap L_+^\infty$, which is (i). Finally if k is bounded it follows from (2.4) and (3.2) via (i) and (ii) that $\dot{y} \in L_+^2$. Therefore we have $\lim_{t\to\infty} y(t) = 0$ and since $\psi \in S$ we obtain that

$\lim_{t\to\infty} y'(t) = 0$ which proves (iii).

3.6 **Remark:** In the linear case, i.e. $\varphi(t,x) = \psi(t,x) = x$ for all $(t,x) \in \mathbf{R}_+ \times \mathbf{R}$, Lemma 3.3 remains true for all (i.e. not necessarily scaling invariant) switching functions N. It should be regarded as a generalization of the result of Willems and Byrnes [19], which is just the special case given by:

- g is a *rational* transfer function
- $\varphi(t,x) = \psi(t,x) = x \ \forall (t,x) \in \mathbf{R}_+ \times \mathbf{R}$
- $k(t) = 1 \ (t \geq 0)$
- $N = N_0 id_{\mathbf{R}}$, where N_0 is a Nissbaum gain.

It is natural to regard the gain k as the image of a causal map Z operating on ξ and $\psi(\cdot, y(\cdot))$. Let us consider the following gain adaption rule

$$k = Z(\xi, \psi(\cdot, y(\cdot))) \tag{3.9}$$

where $Z : D_Z \to LL^1_+$ is a causal map whose domain D_Z contains the set $C(\mathbf{R}_+) \times \bigcup_{\psi\in S} \{\psi(\cdot, f(\cdot)) \mid f \in LL^\infty_+\}$. Moreover we assume that Z satisfies
(A1) $Z[D_Z \cap (L^\infty_+ \times (L^\infty_+ \cap L^2_+))] \subset L^\infty_+$.
(A2) $Z(f)$ is non-decreasing for all $f \in D_Z$.
(A3) There exists $\varepsilon > 0$ such that $\inf_{t\geq 0}\{Z(f)(t)\} \geq \varepsilon \ \forall f \in D_Z$.
Consider the following example:

$$\begin{aligned} Z: \quad & C(\mathbf{R}_+) \times LL^\infty_+ \to LL^1_+ \\ & (f_1, f_2) \mapsto k_0 + \int_0^{\cdot} f_2^2(\tau) P(|f_1(\tau)|, |f_2(\tau)|) d\tau \end{aligned} \tag{3.10}$$

where $k_0 > 0$ and P is a polynomial in two variables with positive coefficients. It is trivial that (A1)–(A3) are satisfied. The gain adaption induced by (3.10) can be written in the form of an ordinary differential equation

$$\begin{aligned} \dot{k}(t) &= \psi^2(t, y(t)) P(|\xi(t)|, |\psi(t, y(t))|), \\ k(0) &= k_0. \end{aligned} \tag{3.11}$$

The following theorem is a simple consequence of Lemma 3.3

3.7 **Theorem:** *Let φ and ψ be in S and suppose that* (A1)–(A3) *are satisfied. Moreover assume that the feedback system given by* (2.4), (3.2) *and* (3.9) *has a unique absolutely continuous solution* (y, ξ) *which can be continued uniquely to the*

right as long as it remains bounded. Under these conditions the following is true:
(i) $y, y' \in L^2_+ \cap L^\infty_+$.
(ii) $\lim_{t\to\infty} \xi(t)$ *exists and is finite.*
(iii) $\lim_{t\to\infty} k(t)$ *exists and is finite.*
(iv) $\lim_{t\to\infty} y(t) = \lim_{t\to\infty} y'(t) = 0$.

3.8 **Remark:** (i) Consider the special case that Z is given by (3.10). Suppose that N is continuously differentiable and that φ and ψ are continuous and have continuous partial derivatives $D_2\varphi$ and $D_2\psi$. Then it follows from Theorem 2.2 that there exists a unique solution (y, ξ, k) of the feedback system given by (2.4), (3.2) and (3.11) which can be continued to the right as long as (y, ξ) remains bounded.

(ii) Suppose that $\varphi(t, x) = x \ \forall (t, x) \in \mathbf{R}_+ \times \mathbf{R}$ and apply a L^2_+-input signal v to the closed-loop system given by (2.4), (3.2) and (3.9), i.e. (3.2a) becomes

$$u(t) = v(t) + N(\xi(t))k(t)\psi(t, y(t)).$$

It is easy to show that the conclusions of Theorem 3.7 remain true in this case. In particular it follows that the closed-loop operators $v \mapsto y$ and $v \mapsto y'$ are L^2-stable.

Using results of [3]–[5], [9] and [15] the input-output theory of this section can be applied to retarded systems and Volterrs integrodifferential systems which are minimum-phase and satisfy a *relative-degree one* – condition (see [11] for details). Moreover it can be shown that a certain class of spectral systems (containing the class of distributed parameter systems considered in [7]) fits into the framework introduced in Sections 2 and 3 (cf. [12]). It should be mentioned that in all three cases the adaptive control law given by (3.2) and (3.9) stablizes not only the output of the process but the internal variable too (cf. [11] and [12]).

4 Exponential adaptive stabilization

Throughout this section we assume that there exists $\mu > 0$ such that

$$h \in H^\infty(\mathbf{C}_{-\mu}) \tag{4.1}$$

and

$$w \in L^{2,\mu}_+, \tag{4.2}$$

where h is the function occuring in (2.2) and (2.3), and w is the *initial condition* – term in (2.4). It should be mentioned that (4.2) is a very reasonable assumption if (4.1) is satisfied. Indeed one should expect that the unforced motion of a linear system with ***stability margin*** μ is an element in $L_+^{2,\mu}$. Moreover it is clear that Remark 2.3(i) remains true if we replace $\mathbf{C}_0$ by $\mathbf{C}_{-\mu}$.

In order to achieve exponential stability we employ the following weighted version of (3.2b)

$$\dot{\xi}(t) = \exp(2\mu t)k(t)(\psi(t, y(t)))^2, \quad \xi(0) = \xi_0 \in \mathbf{R}. \tag{4.3}$$

Then it can be shown that the closed-loop system given by (2.4), (3.2a), (4.3) and (3.9) (where Z is supposed to satisfy (A1)–(A3)) is exponentially stable in the sense that there exist positive constants C and C' such that

$$|y(t)| \leq C\exp(-\mu t) \text{ and } |y'(t)| \leq C'\exp(-\mu t) \;\; \forall t \geq 0.$$

The proof is similar to that of Lemma 3.3. The reader should just notice that $H = L^{-1}M_h L$ (cf. (2.4)) is a causal linear bounded operator from $L_+^{2,\mu}$ into itself and he should consider the term $y\dot{y}\exp(2\mu\cdot) + \mu y^2\exp(2\mu\cdot)$ instead of $y\dot{y}$ in (3.4).

Acknowledgements

This work was suported by SERC under grant no. GR/D/45710.

References

[1] Dahleh, M. and W. E. Hopkins *Adaptive stabilization of single-input single-output delay systems*, IEEE Trans. Automatic Control, **AC-31**, 577-579, 1986.

[2] Driver, R. D. *Existence and stability of solutions of a delay-differential system*, Arch. Rational Mech. Anal. **10**, 401-426, 1962.

[3] Grossman, S. I. and R. K. Miller *Perturbation theory for Volterra integrodifferential systems*, J. Differential Equations, **8**, 457-474, 1970.

[4] Grossman, S. I. and R. K. Miller *Nonlinear Volterra integrodifferential systems with L^1-kernels*, J. Differential Equations, **13**, 551-566, 1973.

[5] Hale, J. K. *Theory of functional differential equations*, Springer-Verlag, New York, 1977.

[6] Harris, C. J. and J. M. E. Valenca *The stability of input-output dynamical systems*, Academic Press, London, 1983.

[7] Kobayashi, T. *Global adaptive stabilization of infinite-dimensional systems*, Systems & Control Letters, **9**, 215-223, 1987.

[8] Logemann, H. *Funktionentheoretische Methoden in der Regelungstheorie unendlichdimensionaler Systeme*, Doctoral Thesis, Institut für Dynamische Systeme der Universität Bremen, 1986.

[9] Logemann, H. *On the transfer matrix of a neutral system: characterizations of exponential stability in input-output terms*, System & Control Letters, **9**, 393-400,1987.

[10] Logemann, H. and D. H. Owens *Robust high-gain feedback control of infinite-dimensional minimum-phase systems*, IMA J. Math. Control & Information, **4**, 195-220, 1987.

[11] Logemann, H. and D. H. Owens *Input-output theory of high-gain adaptive stabilization of infinite-dimensional systems with nonlinearities*, to appear in Int. J. of Adaptive Control and Signal Processing.

[12] Logemann, H. and H. J. Zwart *Robust adaptive stablization of spectral systems*, in preparation.

[13] Mårtensson, B. *Adaptive stabilization*, Doctoral Thesis, Dept. of Automatic Control, Lund Institute of Technology, 1986.

[14] McShane, E. J. and T. A. Botts *Real analysis*, D. Van Nostrand Company, Princeton, New Jersey, 1959.

[15] Miller, R. K. *Asymptotic stability properties of linear Volterra integrodifferential equations*, J. Differential Equations, **10**, 485-506, 1971.

[16] Nussbaum, R. D. *Some remarks on a conjecture in parameter adaptive control*, Systems & Control Letters, **3**, 243-246, 1983.

[17] Owens, D. H., Prätzel-Wolters and A. Ilchmann *Positive-real structure and high-gain adaptive stabilization*, IMA J. Math. Control & Information, **4**, 167-181, 1987.

[18] Vidyasagar, M. *A note on time-invariance and causality*, IEEE Trans. Automatic Control, **AC-28**, 929-931, 1983.

[19] Willems, J. C. and C. I. Byrnes *Global adaptive stabilization in the absence of information on the sign of the high frequency gain*, Proc. of the INRIA Conf. on Analysis and Optimization of Systems (= Lecture Notes in Control and Information Sciences No. 62, Springer Verlag), pp 49-57, 1984.

Perturbation properties of a class of infinite-dimensional systems with unbounded control and observation

Jan Bontsema   Ruth F. Curtain

Abstract

We define a class of infinite-dimensonal systems with unbounded input and putput operators which are closed under state feedback, output injection and output feedback perturbations. We give an explicit critrion for spectral systems to belong to this class. As an application we extend the finite dimensional result on the equivalence of input-output stability and exponential stability to thisclass. The theory is illustrated by a p.d.e. model of a beam with viscous damping.

1 Introduction

The motivation for this study stems from our work on designing robust compensators for some partial differential models of flexible structures in Bontsema, Curtain and Schumacher [1]. We used the frequency domain approach from Curtain and Glover, [6], which results in a closed loop system which is only input-output stable. The question is whether one can also conclude exponential stability of the closed loop system. The finite-dimensional result states that one has equivalence if the state space realisation is detectable and stabilizable. In [7], [8] Jacobson showed that this result also holds for infinite-dimensional systems with bounded input and output operators. While this result may sound surprizing, the result in infinite-dimensions is not as nice as it sounds, because exponential stabilizability is a very strong concept in infinite dimensions; it implies that A has finitely many unstable eigenvalues seperated from the rest of the spectrum. While this does not include all systems, it does include a very large class, including our examples in [1], except that there the input and output operators are unbounded. This motivated the extension of Jacobson's results by Curtain, [4], to the class of systems with unbounded input and output operators studied by Pritchard and Salamon in [10]. This class includes the examples studied in [1], except that one can only obtain exponential stability on spaces related to, but different from, the original state space. Moreover the class did not allow for "co-located" actuators and sensors, $B = C^*$, an interesting case often used in applications.

The action of controlling a system (A, B, C) in some feedback configuration results in a closed loop system which is a combination of $(A + BF)-$, $(A + GC)-$ and $(A+BKC)-$type perturbations. It is therefore important to know if (A, B, C)

being in a certain class implies that feedback perturbations remain in the same class. Salamon has developed various theories in [10], [11] and [12] for the well posedness of (A, B, C) for unbounded B and C, which are appropriate for posing these questions. As mentioned above, the class defined in [10] is too small for our purpose while the class defined in [12] is too large to have nice perturbations properties; in particular, the $(A + BKC)$−perturbations may not be well posed. Here we identify a new class (A, B, C), which lies somewhere in between the two introduced in [10] and [12] and it uses the same mathematical formulation originally introduced by Salamon in [10]. Essentially it assumes some smoothing property of the semigroup and we give a nice sufficient condition for spectral systems to belong to this class and we show that the beam examples mentioned above are included.

The paper is organized in five sections as follows

1. This introduction.
2. Perturbation results for our class (A, B, C).
3. A characterization of spectral systems belonging to our class and an analysis of the p.d.e. model of the beam from [1].
4. An application of the perturbation results in section 2 to the problem of the equivalence of input-output stability and exponential stability. We illustrate this with the p.d.e. model of the beam from [1].
5. Conclusions.

2 Perturbation results

We begin by introducing our class of systems. Let W, H and V be real seperable Hilbert spaces such that

(A1) $\quad W \hookrightarrow H \hookrightarrow V$

where $\hookrightarrow$ means continuous, dense injection

(A2) Let A be the infinitesimal generator of the C_0−semigroup $S(t)$ on H .
(A3) Let Y be a real, seperable Hilbert space and suppose $C \in \mathcal{L}(W, Y)$ and the local observability map $\mathcal{C}(T) = \mathcal{L}(H, L_2(0, T; Y))$ where

$$\mathcal{C}(T)h := CS(t)h \text{ on } [0, T] \tag{2.1}$$

(A4) Let U be a real, seperable Hilbert space and suppose $B \in \mathcal{L}(U, V)$ and the local reachability map $\mathcal{B}(T) \in \mathcal{L}(L_2(0, T; U), H)$, where

$$\mathcal{B}(T)u := \int_0^T S(T - s)Bu(s)ds \tag{2.2}$$

for any $u \in L_2(0,T;U)$.
(A5) $D_V(A) \hookrightarrow W$ and for $Re\ \lambda > \omega$, the growth constant of $S(t)$, there exists an $M > 0$ such that

$$\int_0^\infty e^{-Re\lambda\sigma} \|S(\sigma)x\|_W d\sigma \leq M\|x\|_V \tag{2.3}$$

and the constant M is uniform in $Re\lambda \geq \omega + \varepsilon$, $\varepsilon > 0$.
(A6) For some results we shall need the related condition

$$\int_0^\infty e^{-Re\lambda\sigma} \|S(\sigma)B\|_{\mathcal{L}(U,W)} d\sigma \leq M_3 \tag{2.4}$$

(A6) follows from (A5) if U is finite-dimensional and both (A5) and (A6) are implied by the stronger condition

$$\int_0^\infty e^{-Re\lambda\sigma} \|S(\sigma)\|_{\mathcal{L}(V,W)} d\sigma \leq M \tag{2.4'}$$

We remark that (A1) and (A2) are standard assumptions in the formulation of Salamon, [11], and (A3) and (A4) are consistent with the well posedness definitions in Salamon [12]. In view of the recent results by Weiss in [14], they are in fact necessary for a well posed state formulation. The condition $D_V(A) \hookrightarrow W$ also appears in Pritchard and Salamon [10], and has the consequence that $S(t)$ generates as C_0−semigroup on all three spaces W, H and V and

$$\|S(t)x\|_X \leq M_2 e^{\omega t} \|x\|_X \tag{2.5}$$

for $x \in X = W$, H or V and real constants ω and $M_2 > 0$. Our new assumption is (A5) which implies some smoothing of the semigroup. In section 3 we see that spectral systems typically satisfy a (2.4') condition with $Re\lambda > \omega$, the growth constant of $S(t)$. We obtain nice perturbation results with (2.4') replacing (A5) and (A6), but although it would be nice to replace τ by ω, we were unable to prove that this property would be retained under feedback perturbations. So in general $\tau > \omega$. For the class of spectral systems assumptions (A1)–(A6) are weaker than the conditions imposed in Pritchard and Salamon, [10], but we still obtain nicer perturbation properties for our class for $(A + BF)$−, $(A + GC)$− and $(A + BKC)$−type perturbations. While parts of Lemma 2.1 have been proved elsewhere, for completeness we give explicit proofs here.

Lemma 2.1.

Suppose (A, B, C) satisfies (A1)–(A6) and $F \in \mathcal{L}(H, U)$, then there exists a

C_0–semigroup $S_F(t)$ on H which is the unique solution of

$$S_F(t)x = S(t)x + \int_0^t S(t-\sigma)BFS_F(\sigma)xd\sigma, \quad x \in H \tag{2.6}$$

If A_F is the infinitesimal generator of $S_F(t)$, then (A_F, B, C) satisfies (A1)–(A4) and $S_F(t)$ is well defined on $X = W$, H or V with

$$\|S_F(t)x\|_X \leq M_4 e^{\omega_2 t}\|x\|_X \tag{2.7}$$

for certain constants M_4 and ω_2.
If $F \in \mathcal{L}(V, U)$, then (A_F, B, C) satisfies (A5) also.

Proof

(a) To establish the existence and uniqueness of solutions of (2.6), we follow the technique used in [2], Theorem 2.31 and define the following sequence of operators.

$$T^0(t) := S(t)$$

$$T^n(t)x := \int_0^t S(t-\sigma)BFT^{n-1}(\sigma)xd\sigma, \quad x \in H$$

Then for $x \in H$, (A4) implies that $T^n(t)x$ is well defined and

$$\|T^n(t)x\|_H \leq \beta\|F\|_{\mathcal{L}(H,U)}\|T^{n-1}(\cdot)x\|_{L_2(0,t;H)}$$

where β denotes the norm of $\mathcal{B}(T)$. Using an induction argument and (2.3), it follows that

$$\|T^n(t)x\|_H \leq (\beta\|F\|)^n M_2 \max\{1, e^{\omega t}\}\left[\frac{t^n}{n!}\right]^{1/2}\|x\|_H$$

and so $T(t)x := \sum_{n=1}^{\infty} T^n(t)x$ is well defined and the convergence is uniform on $[0,T]$ with respect to the $\mathcal{L}(H)$ norm. $T(t)$ is a strongly continuous solution of (2.6) on $[0,T]$ and it extends to a C_0–semigroup on H using standard arguments. (c.f. [2], Theorem 2.31) since

$$D_V(A_F) = \{x \in V : (A+BF)x \in V\} = D_V(A) \hookrightarrow W$$

it follows that $S_F(t)$ is a well defined semigroup on all three spaces W, H and V and (2.7) holds for suitable constants M_4 and ω_2.

(b) To show that (A_F, B, C) satisfies (A4), we first estimate for $n = 1$

$$
\begin{aligned}
\|\int_0^t T^1(t-s)Bu(s)ds\|_H &\leq \|\int_0^t\int_0^{t-s} S(t-s-\sigma)BFS(\sigma)d\sigma Bu(s)ds\|_H \\
&= \|\int_0^t S(t-\sigma)B\int_0^\sigma FS(\sigma-s)Bu(s)dsd\sigma\|_H \\
&\leq \beta(\int_0^t \|\int_0^\sigma FS(\sigma-s)Bu(s)ds\|^2 d\sigma)^{1/2} \text{ using (A4)} \\
&\leq \beta^2\|F\|\sqrt{t}\|u\|_{L_2(0,t;U)}
\end{aligned}
$$

and using an induction argument, we obtain

$$
\|\int_0^t T^n(t-s)Bu(s)ds\|_H \leq \beta(\beta\|F\|)^n \left[\frac{t^n}{n!}\right]^{1/2} \|u\|_{L_2(0,t;U)}
$$

and since $S_F(t)x = \sum_{n=1}^{\infty} T^n(t)x$ it follows that

$$
\|\int_0^t S_F(t-s)Bu(s)ds\|_H \leq \beta(t)\|u\|_{L_2(0,t;U)}
$$

where $\beta(t) \leq \beta(1+\beta\|F\|\sqrt{t})e^{\beta^2\|F\|^2 t}$ and (A_F, B, C) satisfies (A4)

(c) To prove that (A_F, B, C) satisfies (A3), we need the stronger assumption (A6). Let us estimate the second term in (2.6)

$$
\|C\int_0^t S(t-\sigma)BFS_F(\sigma)xd\sigma\|_{L_2(0,T;Y)}
$$

$$
\leq \|C\|_{\mathcal{L}(W,Y)}\|\int_0^t \|S(t-\sigma)B\|_{\mathcal{L}(U,W)}\|F\|\;\|S_F(\sigma)x\|_H d\sigma\|L_2(0,T;Y)
$$

$$
\leq \|C\|e^{\tau T}\int_0^\infty e^{-\tau\sigma}\|S(\sigma)B\|_{\mathcal{L}(U,W)}d\sigma\|F\|\;\|e^{-\tau t}S_F(t)x\|_{L_2(0,T;H)}
$$

where $\tau > \omega$ the growth constant of $S(t)$ and using the property of convolution integrals,

$$
\leq \gamma(T)\|x\|_H
$$

using (2.4) and the usual properties of a C_0–semigroup on H. $\gamma(T)$ is a constant depending on T and so this estimate together with (2.6) and the fact that (A, B, C)

satisfies an (A3) condition shows that (A_F, B, C) does too.

(d) Under the stronger assumption $F \in \mathcal{L}(V, U)$, we see that considering $S(t)$ on the state space V, $A + BF$ is a bounded perturbation and by Corollary 9.4, [2], $S_F(t)$ is also the unique solution of the following integral equation on V

$$S_F(t)x = S(t)x + \int_0^t S_F(t-\sigma)BFS(\sigma)xd\sigma \tag{2.8}$$

Considering the second term, we estimate for $\tau > \omega$

$$\int_0^\infty e^{-\tau t}\|\int_0^t S_F(t-\sigma)BFS(\sigma)xd\sigma\|_W dt$$

$$= \int_0^\infty \|\int_0^t e^{-\tau(t-\sigma)}S_F(t-\sigma)BFe^{-\tau\sigma}S(\sigma)xd\sigma\|_W dt$$

$$\leq (\int_0^\infty \|e^{-\tau\sigma}S_F(\sigma)\|_{\mathcal{L}(V)}d\sigma)\|B\|\;\|F\|\int_0^\infty e^{-\tau\sigma}\|S(\sigma)x\|_W d\sigma$$

using the property of convolution integrals

$$\leq M_4\|B\|\;\|F\|(2(\tau-\omega_2))^{-1}M\|x\|_V$$

for $\tau > \max(\omega_2, \omega)$ by (2.7) and (2.3). We remark that the norm on F used above is in $\mathcal{L}(W, U)$ which is finite by the continuous injection in (A1). So combining this estimate with one for the first term in (2.8) we obtain an (A5) condition on the perturbed system (A_F, B, C) for $Re\lambda > \max(\omega_2, \omega) + \varepsilon$, $\varepsilon > 0$ sufficiently large.

In the case that $\omega_2 < \omega$, we can extend this to $s = Re\lambda > \omega_2$ the growth constant of S_F in the following way. Choose a fixed $\tau > \omega > s > \omega_2$, then

$$\frac{1}{\tau - s}\int_0^\infty e^{-s\tau}\|S_F(\sigma)x\|_W d\sigma = \left[\int_0^\infty e^{(s-\tau)\alpha}d\alpha\right]\int_0^\infty e^{-s\sigma}S_F(\sigma)x\|_W d\sigma$$

$$\int_0^\infty \int_\sigma^\infty e^{(s-\tau)\alpha}e^{-s\sigma}\|S_F(\sigma)x\|_W d\alpha d\sigma + \int_0^\infty \int_0^\sigma e^{(s-\tau)\alpha}e^{-s\sigma}\|S_F(\sigma)x\|_W d\alpha d\sigma$$

The first integral equals $\frac{1}{\tau - s}\int_0^\infty e^{-\tau\sigma}\|S_F(\sigma)x\|_W d\sigma$ and for the second integral we have

$$\int_0^\infty \int_0^\sigma e^{(s-\tau)\alpha} e^{-s\sigma} \|S_F(\sigma)x\|_W \, d\alpha d\sigma$$

$$= \int_0^\infty \int_\alpha^\infty e^{(s-\tau)\alpha} e^{-s\sigma} \|S_F(\sigma)x\|_W \, d\sigma d\alpha$$

$$= \int_0^\infty \int_0^\infty e^{-s\beta} e^{-\tau\alpha} \|S_F(\alpha+\beta)x\|_W \, d\alpha d\beta$$

(by a change of variables $(\sigma, \alpha) \to (\beta = \sigma - \alpha, \alpha)$)

$$\leq \int_0^\infty \int_0^\infty e^{-s\beta} \|S_F(\beta)\|_{\mathcal{L}(W)} e^{-\alpha\tau} \|S_F(\alpha)x\|_W \, d\alpha d\beta$$

$$= \frac{M_4}{s-\omega_2} \int_0^\infty e^{-\alpha\tau} \|S_F(\alpha)x\|_W \, d\alpha \text{ using (2.7)}$$

So

$$\int_0^\infty e^{-s\sigma} \|S_F(\sigma)x\|_W \, d\sigma \leq \left[1 + M_4 \frac{\tau - s}{s - \omega_2}\right] M\|x\|_V \leq \left[1 + M_4 \frac{\tau - \omega_2}{\varepsilon}\right] M\|x\|_V$$

and any fixed $\tau > \omega$ will do and so we have shown that (A5) holds for $Re\lambda \geq \omega_2 + \varepsilon$.

We remark that if $F \in \mathcal{L}(V, U)$, then the perturbed system (A_F, B, C) will inherit the (A1)–(A4) properties from (A, B, C) without the smoothing assumption (A5). However $F \in \mathcal{L}(H, U)$ alone is too weak to transmit the (A3) property to an $(A + BF)$–type perturbation.

As is well known (see Salamon [11]), the (A3) and (A4) conditions are dual to each other and so we can immediately deduce results about $(A + GC)$–type perturbations from Lemma 2.1 with the proviso that the perturbed semigroup $S_G(t)$ is the unique solution on H of

$$S_G(t)x = S(t)x + \int_0^t S_G(t-\sigma)GCS(\sigma)x d\sigma \tag{2.9}$$

Only in the bounded perturbation case ($G \in \mathcal{L}(Y, W)$) do we obtain a C_0–semigroup $S_G(t)$ which is the unique solution of the more usual integral equation.

$$S_G(t)x = S(t)x + \int_0^t S(t-\sigma)GCS_G(\sigma)x d\sigma \tag{2.10}$$

and the standard bounded argument from [2], Corollary 9.4 shows that the solutions of (2.9) and (2.10) are identical if $G \in \mathcal{L}(Y, W)$.

For completeness we state the precise results.

Corollary 2.2

Suppose (A, B, C) satisfies (A1)–(A6) and $G \in \mathcal{L}(Y, H)$, then there exists a C_0–semigroup $S_G(t)$ on H which is the unique solution of (2.9). If A_G is its infinitesimal generator, then (A_G, B, C) satisfies (A1)–(A4) and $S_G(t)$ is a well defined semigroup on $X = W$, H or V such that

$$\|S_G(t)x\|_X \leq M_5 e^{\omega_3 t}\|x\|_X \tag{2.11}$$

If $G \in \mathcal{L}(Y, W)$, then $S_G(t)$ is also the unique solution of (2.9) on H and (A_G, B, C) satisfies (A5) also.

Proof

(a) We appeal to the duality of the (A3) and (A4) conditions to obtain the existence and uniqueness of the solution $S_G(t)$ to (2.9) on H. That (A_G, B, C) satisfies an (A3) and (A4) condition also follows by duality and similarly that $S_G(t)$ is well defined on W, H and V and (2.11) holds.

(b) Let $G \in \mathcal{L}(Y, W)$. Considering $S_G(t)$ is the unique solution of (2.9) we estimate for the second term for $\tau > \omega$

$$\int_0^\infty e^{-\tau t}\|\int_0^t S_G(t-\sigma)GCS(\sigma)x d\sigma\|_W dt$$

$$\int_0^\infty \|\int_0^t e^{-\tau(t-\sigma)}S_G(t-\sigma)GCe^{-\tau\sigma}S(\sigma)x d\sigma\|_W dt$$

$$\leq (\int_0^\infty \|e^{-\tau\sigma}S_G(\sigma)GC\|_{\mathcal{L}(W)}d\sigma)\int_0^\infty e^{-\tau\sigma}\|S(\sigma)x\|_W d\sigma$$

by the property of convolution integrals

$$\leq M_5\|G\|\,\|C\|(\int_0^\infty e^{-\sigma(\tau-\omega_3)}d\sigma)M\|x\|_V$$

for sufficiently large τ, by (2.11) and (2.3) where the norm of G is in $\mathcal{L}(Y, W)$. So combining this with (2.9) and (A5) condition for (A, B, C), we see that for sufficiently large τ, (A_F, B, C) will satisfy an (A5) condition. This can be extended to $Re\lambda > \omega_3$ in (2.11) as in a) of the proof of Lemma 2.1.

For certain applications we need continuity properties concerning the input-output map.

Lemma 2.3

Suppose (A, B, C) satisfies (A1)–(a6)
(a) for $u \in L_2(0, T; U)$

$$x(t) = S(t)x_0 + \int_0^t S(t-s)Bu(s)ds \tag{2.12}$$

is a continuous function on $[0, T]$ with values in H. And

$$\begin{aligned} y(t) &= CS(t)x_0 + C\int_0^t S(t-s)Bu(s)ds \\ &= CS(t)x_0 + \int_0^t CS(t-s)Bu(s)ds \end{aligned} \tag{2.13}$$

is in $L_2(0, T; Y)$.

(b) For $Re\lambda > \omega$, the transfer function is defined by

$$G(\lambda) = C(\lambda - A)^{-1}B \tag{2.14}$$

and G is analytic and bounded in $Re\lambda \geq \omega - \varepsilon$, for $\varepsilon > 0$.

Proof

a) That $x(t)$ is continuous is a consequence of (A2) and (A4). (A1)–(A4) are insufficient to ensure that (2.13) is well defined; we need to invoke (A5) and (A6) to show that

$$\int_0^t S(t-s)Bu(s)ds \in L_2(0, T; W)$$

$$(\int_0^T \| \int_0^t S(t-s)Bu(s)ds\|_W^2 dt)^{1/2}$$

$$\leq e^{\tau T}(\| \int_0^T \| \int_0^t e^{-\tau(t-s)}S(t-s)Be^{-\tau s}u(s)ds\|_W^2 dt)^{1/2} \qquad \text{where } \tau > \omega$$

$$\leq e^{\tau T}(\int_0^T e^{-\tau t}\|S(t)B\|_{\mathcal{L}(U,W)}dt(\int_0^T e^{-2\tau s}\|u(s)\|^2 ds)^{1/2}$$

$\leq const\ \|u\|_{L_2(0,T;U)}$ by (2.4)

Then (A3) shows that $y(t)$ defined by (2.13) is in $L_2(0, T; Y)$. Similarly one can show that $\int_0^t CS(t-s)Bu(s)ds \in L_2(0,T;Y)$ directly using (2.4).

(b) Recall that for $Re\lambda > \omega$, the resolvent of A is defined by

$$(\lambda - A)^{-1}x = \int_0^\infty e^{-\lambda t}S(t)x dt$$

and so (2.3) implies that $(\lambda - A)^{-1} \in \mathcal{L}(V,W)$ for $Re\lambda > \omega$. This together with (A3) and (A4) shows that $C(\lambda - A)^{-1}B \in \mathcal{L}(V,W)$ for $Re\lambda > \omega$ and is analytic and uniformly bounded in $Re\lambda \geq \omega - \varepsilon$ for any $\varepsilon > 0$.

It is interesting to note that we used the stronger condition (A6) to establish the time domain input-output map, not the frequency domain map.

We now consider the type of perturbations which arise in direct output feedback where $u = Ky$ for some $K \in \mathcal{L}(Y,U)$ and $y = Cx$ is the observation. This is not covered by Lemma 2.1 since the feedback $F = KC \in \mathcal{L}(W,U)$ is too unbounded. Although this type of feedback is a common one in applications, it is difficult to fit into the general semigroup set-up proposed by Salamon in [12]. One needs to impose some extra conditions to ensure that the closed loop system results in a well posed semigroup; here we show that the extra assumptions (A5) and (A6) are sufficient.

Theorem 2.4

Suppose (A,B,C) satisfies conditions (A1)–(A6) then there exists a C_0-semigroup on H, $S_K(t)$ which is the unique solution on H of

$$S_K(t)x = S(t)x + \int_0^t S(t-s)BKCS_K(s)xds \tag{2.15}$$

If A_K is the infinitesimal generator of $S_K(t)$, then (A_K,B,C) also satisfies (A1)–(A6).

Proof

(a) Using an idea due to Townley and Pritchard [13] we consider the pair of equations for $K \in \mathcal{L}(Y,U)$

$$x(t) = S(t)x_0 + \int_0^t S(t-\sigma)BKCx(\sigma)d\sigma \tag{2.16}$$

$$y(t) = CS(t)x_0 + \int_0^t S(t-\sigma)BKy(\sigma)d\sigma. \tag{2.17}$$

Let us introduce the map $J_R : L_2(0,\infty;Y) \to L_2(0,\infty;Y)$ defined by

$$(J_R y)(t) = C\int_0^t e^{-R(t-\sigma)}S(t-\sigma)BKy(\sigma)d\sigma. \tag{2.18}$$

Then $J_R \in \mathcal{L}(L_2(0,\infty;Y))$ for sufficiently large R since

$$\begin{aligned} \|J_R y\|_{L_2} &\le \|C\|\int_0^\infty \|e^{-R\sigma}S(\sigma)B\|_{\mathcal{L}(U,W)}\|K\|d\sigma\|y\|_{L_2} \\ &\le \|C\|M_3\|y\|_{L_2} \text{ for } R > \omega \text{ by (2.4)} \end{aligned}$$

Moreover

$$\begin{aligned} \int_0^\infty e^{-R\sigma}\|S(\sigma)B\|d\sigma &= \int_0^\infty e^{-(R-\tau)\sigma}e^{-\tau\sigma}\|S(\sigma)B\|d\sigma \text{ provided } \tau > \omega \\ &\le \int_0^\varepsilon e^{-\tau\sigma}\|S(\sigma)B\|d\sigma + e^{-(R-\tau)\varepsilon}\int_0^\infty e^{-\tau\sigma}\|S(\sigma)B\|d\sigma \end{aligned}$$

and so $\|J_R\|$ can be made less than one for R sufficiently large. Thus for sufficiently large R, $I - J_R$ has a bounded inverse. Rewriting (2.17) as

$$\begin{aligned} e^{-Rt}y(t) &= CS(t)e^{-Rt}x_0 + C\int_0^t e^{-R(t-\sigma)}S(t-\sigma)BKe^{-R\sigma}y(\sigma)d\sigma \\ &= CS(t)e^{-Rt}x_0 + J_R(e^{-R\cdot}y(\cdot))(t) \end{aligned}$$

we see that (2.17) has a unique solution in $L_2(0,T;Y)$ for any $x_0 \in H$:

$$y(t) = e^{Rt}(I - J_R)^{-1}(CS(\cdot)e^{-R\cdot})(t)x_0$$

and

$$\begin{aligned} \|y\|_{L_2(0,T;Y)} &\le \frac{e^{RT}}{1 - \|J_R\|}\|CS(\cdot)e^{-R\cdot}x_0\|_{L_2(0,\infty;Y)} \\ &\le \frac{\gamma M_2 e^{2RT}}{(1 - \|J_R\|)(e^{2TR} - e^{2T\omega})^{1/2}}\|x_0\|_H \end{aligned} \tag{2.19}$$

where γ is the norm of the observability operator from (A3) and ω, M_2 are the constants associated with $S(t)$ from (2.4). The last estimate comes from

$$\begin{aligned}\int_0^\infty \|CS(t)e^{-Rt}x_0\|^2 dt &= \sum_{k=0}^\infty \int_0^T \|CS(t)S(kT)e^{-Rt}e^{-RkT}x_0\|^2 dt \\ &\leq \sum_{k=0}^\infty e^{-2RkT} \int_0^\infty \|CS(t)S(kT)x_0\|^2 dt \\ &\leq \sum_{k=0}^\infty e^{-2RkT}\gamma^2 \|S(kT)x_0\|_H^2 \text{ by (A3)} \\ &\leq \gamma^2 \left[\sum_{k=0}^\infty e^{-2RkT}e^{2\omega kT}\right] M_2^2\|x_0\|_H^2 \text{ by (2.5)} \\ &= \frac{\gamma^2 M_2^2 \|x_0\|_H^2}{1-e^{-2T(R-\omega)}} \text{ provided } R > \omega\end{aligned}$$

Defining the $H-$valued function

$$\bar{x}(t) = S(t)x_0 + \int_0^t S(t-\sigma)BKy(\sigma)d\sigma \tag{2.20}$$

we see that for $x_0 \in H$

$$\|\bar{x}(t)\|_H \leq M_2 e^{\omega t}\|x_0\|_H + \beta\|K\|\|y\|_{L_2(0,T;Y)}$$

using (2.5) and (A4)

$$\leq \text{constant } \|x_0\|_H \text{ from (2.19).}$$

Now $\bar{x}(t)$ is a solution of (2.15) and there exists a $T(t) \in \mathcal{L}(H)$ for each $t \in [0,T]$ defined by

$$\bar{x}(t) = T(t)x_0 \tag{2.21}$$

and $T(t)$ satisfies (2.15) for $x \in H$.
$T(t)$ is strongly continuous in t since

$$\begin{aligned}\|T(t)x_0 - x_0\|_H &\leq \|S(t)x_0 - x_0\|_H + \|\int_0^t S(t-\sigma)BKy(\sigma)d\sigma\|H \\ &\leq \|S(t)x_0 - x_0\|_H + \beta\|K\|\ \|y\|_{L_2(0,t;Y)} \text{from (A4)}\end{aligned}$$

$\rightarrow 0$ as $t \rightarrow 0$ since $S(t)$ is a C_0-semigroup and

$$\|y\|_{L_2(0,t;Y)} \to 0 \text{ as } t \to 0$$

and the semigroup property and the extension to $[0,\infty)$ follows in the usual way. So there exists a C_0-semigroup $T(t)$ on H which is a solution of (2.15) and $CT(t)x_0$ is the *unique* solution of (2.17), which implies that $T(t)$ is the unique solution of (2.15).

(b) To show that (A_K, B, C) satisfies (A3) we recall that from a) we had that $CS_K(t)x_0 = y(t)$ and so (A3) follows from (2.19).

(c) We now show that (A_K, B, C) satisfies (A5). Notice that J_R defined by (2.18) is also a bounded map from $L_1(0,\infty;Y)$ to itself and

$$\begin{aligned}\|J_R\|_{L_1(0,\infty)} &\leq \int_0^\infty \|e^{-R\sigma}S(\sigma)B\|d\sigma\|K\|\|y\|_{L_1(0,\infty)} \\ &\leq \|C\|M_3\|K\|\|y\|_{L_1(0,\infty)}\end{aligned}$$

for $R > \omega$ in (2.4).

Analagously to the estimate (2.19) we obtain

$$\begin{aligned}\|E^{-R\cdot}y(\cdot)\|_{L_1(0,\infty;Y)} &\leq \frac{1}{1-\|J_R\|}\|CS(\cdot)e^{-R\cdot}x_0\|_{L_1(0,\infty;Y)} \\ &\leq \frac{\|C\|}{1-\|J_R\|}\int_0^\infty e^{-R\sigma}\|S(\sigma)x_0\|_W\,d\sigma \qquad (2.22) \\ &\leq \frac{M\|C\|}{1-\|J_R\|}\|x_0\|_V\end{aligned}$$

by (2.3) choosing $R > \omega$ sufficiently large. (Here $\|J_R\|$ denotes the norm on $L_1(0,\infty;Y)$. Now we estimate on the second term in (2.15), noting that $y(t) = CS_K(t)x_0$

$$\begin{aligned}&\int_0^\infty e^{-\tau t}\|\int_0^t S(t-\sigma)BKy(\sigma)d\sigma\|_W\,dt \\ &\leq \int_0^\infty\int_0^t e^{-\tau(t-\sigma)}\|S(t-\sigma)BKe^{-\tau\sigma}y(\sigma)\|_W\,d\sigma dt \\ &\leq \int_0^\infty e^{-\tau\sigma}\|S(\sigma)B\|d\sigma\|K\|\int_0^\infty e^{-\tau\sigma}\|y(\cdot)\|_Y\,d\sigma\end{aligned}$$

using the property of convolution integrals

$$\leq M_3 \frac{\|K\|M\|C\|}{1-\|J_R\|}\|x_0\|_V \text{ from (2.4) and (2.22).}$$

This combined with (2.15) shows that (A_K, B, C) inherits the (A5) property from (A, B, C). (The extension to $Re\lambda >$ the growth constant of S_K can be done as Lemma 2.1 (d)).

(d) Finally we show that (A_K, B, C) satisfies an (A4) assumption. Consider the following estimate on the second term in (2.15)

$$\|\int_0^T \int_0^{T-s} S(T-s-\sigma)BKCS_K(\sigma)d\sigma Bu(s)ds\|_H$$

$$\leq \int_0^T \|\int_0^{T-\sigma} S(t-s-\sigma)BKCS_K(\sigma)Bu(s)ds\|_H d\sigma$$

interchanging the order of integration

$$\leq \int_0^T \beta(\int_0^{T-\sigma} \|KCS_K(\sigma)Bu(s)\|^2 ds)^{1/2} d\sigma \text{ by (A4) on } (A, B, C)$$

$$\leq \int_0^T \beta\|K\|\|C\|\|S_K(\sigma)B\|_{\mathcal{L}(U,W)} d\sigma \|u\|_{L_2}$$

$$\leq \text{const } \|u\|_{L_2} \text{ since } (A_K, B, C) \text{ satisfies (A6).}$$

3 Spectral systems

We now show the conditions (A1)–(A6) introduced in section 2 can be readily checked for the class of spectral systems discussed in [3] and used in [1], [4], [5]; special classes were also considered in [10].

Let A be a discrete spectral operator with countably many eigenvalues $\{\lambda_n\}_{n=1}^{\infty}$ and a Riesz basis of corresponding eigenvectors $\{\phi_n\}_{n=1}^{\infty}$, [3], where for simplicity it is assumed that all eigenvalues have multiplicity 1. If A^* has the corresponding eigenvalues and eigenvectors $\{\bar{\lambda}_n, \varphi_n\}_{n=1}$, by suitable normalization $\{\phi_n, \varphi_n\}$ forms a biorthogonal sequence for H, i.e.

$$< \phi_n, \varphi_m >= \delta_{mn} \tag{3.1}$$

and A and $S(t)$ have the decompositions

$$Ax = \sum_{n=1}^{\infty} \lambda_n < x, \varphi_n >_H \phi_n \tag{3.2(i)}$$

for

$$x \in D(Z) = \{x \in H : \sum_{n=1}^{\infty} |\lambda_n|^2| < x, \varphi_n > |^2 < \infty\} \tag{3.2(ii)}$$

$$S(t)x = \sum_{n=1}^{\infty} e^{\lambda_n t} < x, \varphi_n >_H \phi_n \tag{3.3}$$

Furthermore $x \in H$ also has the decomposition

$$x = \sum_{n=1}^{\infty} < x, \varphi_n > \phi_n \tag{3.4}$$

and without loss of generality we may take the norm of x on H to be

$$||x||^2 = \sum_{n=1}^{\infty} | < x, \varphi_n > |^2 \tag{3.5}$$

This is because $\{\phi_n\}$ is a Riesz basis for H ([15], p. 31, 32). As in [2], Example 2.40, it follows readily from (3.3) that A generates a C_0-semigroup if

$$\sup_n Re\lambda_n < \infty \tag{3.6}$$

One way of obtaining V and W spaces to satisfy (A1) is to introduce the following weighted spaces as in [10].

$$H_\alpha = \{x = \sum_{n=1}^{\infty} x_n \phi_n : \sum_{n=1}^{\infty} \alpha_n |x_n|^2 < \infty\} \tag{3.7}$$

for the positive sequence $\{\alpha_n\}$; $n = 1, \cdots, \infty$, with inner product

$$< x, h >_\alpha = \sum_{n=1}^{\infty} \alpha_n x_n h_n; \quad x, h \in H_\alpha \tag{3.8}$$

Suppose now that for each $n = 1, 2, \cdots$ the following are well defined

$$c_n = C\phi_n \in Y \tag{3.9(i)}$$

$$b_n = B^* \varphi_n \in U \tag{3.9(ii)}$$

and consider the maps $C_\gamma : H_\gamma \to Y$ and $B_\beta : U \to H_\beta$ defined by

$$C_\gamma x = \sum_{n=1}^{\infty} x_n c_n \tag{3.10(i)}$$

$$B_\beta u = \sum_{n=1}^{\infty} < b_n, u >_U \phi_n \tag{3.10(ii)}$$

For simplicity we write B for B_β and C for C_γ.

Then it is clear that $C \in \mathcal{L}(H_\gamma, Y)$ (resp. $B \in \mathcal{L}(U, H_\beta)$) if (3.11)(i) (resp. (3.11)(ii) holds

$$\sum_{n=1}^{\infty} \frac{1}{\gamma_n} \|c_n\|_Y^2 < \infty \tag{3.11(i)}$$

$$\sum_{n=1}^{\infty} \beta_n \|b_n\|_U^2 < \infty \tag{3.11(ii)}$$

An easy calculation shows that (A, B, C) satisfies (A3) if (3.11)(i) and (3.12)(i) and (3.12)(ii) hold and (A4) if (3.11)(ii) and (3.12)(iii) and (3.12)(iv) hold.

$$\sum_Q \|c_n\|_Y^2 < \infty \tag{3.12(i)}$$

$$\sum_P \frac{\|c_n\|_Y^2}{|Re\lambda_n|} < \infty \tag{3.12(ii)}$$

$$\sum_Q \|b_n\|_U^2 < \infty \tag{3.12(iii)}$$

$$\sum_P \frac{\|b_n\|_U^2}{|Re\lambda_n|} < \infty \tag{3.12(iv)}$$

where

$$Q = \{n \in \mathbf{N} | Re\lambda_n > -1\} \quad \text{and}$$

$$P = \{n \in \mathbf{N} | Re\lambda_n \leq -1\}.$$

We now give a simple condition to check when such a spectral system satisfies (A1)–(A6).

Lemma 3.1

Suppose A is a discrete spectral operator which satisfies (3.1)–(3.6) and its eigenvectors form a Riesz-basis for the Hilbert space H. Then we can find Hilbert spaces $V = H_\beta$ and $W = H_\gamma$ defined by (3.7), (3.8) for suitable weighting sequences $\{\beta_n\}$ and $\{\gamma_n\}$ which satisfy (A1) and (A, B, C) satisfies (A2)–(A6) if the following conditions hold

$$\sum_Q \|b_n\|_U^2 < \infty \tag{3.13(i)}$$

$$\sum_P \frac{\|b_n\|_U^2}{|Re\lambda_n|^\alpha} < \infty \tag{3.13(ii)}$$

$$\sum_Q \|c_n\|_Y^2 < \infty \tag{3.13(iii)}$$

$$\sum_P \frac{\|c_n\|_Y^2}{|Re\lambda_n|^\alpha} < \infty \tag{3.13(iv)}$$

for some real α, $0 \leq \alpha < 1$.

Proof

The assumptions on A ensure that (A2) holds. Now (3.13) implies (3.12) since $|RE\lambda_n| \geq |Re\lambda_n|^\alpha$, for $Re\lambda_n \leq -1$ and $0 \leq \alpha < 1$.

For W we choose H_γ with weights $\gamma_n = |Re\lambda_n|^\alpha$, $n \in P$ and $\gamma_n = 1$, $n \in Q$. For V we choose H_β with weights $\beta_n = |Re\lambda_n|^{-\alpha}$, $n \in P$ and $\beta_n = 1$, $n \in Q$. Then $W \hookrightarrow H \hookrightarrow V$, so (A1) holds. (3.13) implies (3.11) and (3.12) and so (A3) and (A4) hold. In addition

$$D_V(A) = \{x \in V : \sum |\lambda_n|^2 |x_n|^2 \beta_n < \infty\} \hookrightarrow W = \{x : \sum \gamma_n |x_n|^2 < \infty\}$$

provided that $|\lambda_n| \geq$ constant $|\frac{\gamma_n}{\beta_n}|$, $n \in P$, and this is satisfied since $\frac{\gamma_n}{\beta_n} = |Re\lambda_n|^{2\alpha}$; $0 \leq \alpha < 1$, $n \in P$.

Now for $x \in V$

$$\|S(t)x\|_W^2 = \sum_{n=1}^{\infty} e^{2Re\lambda_n t}\gamma_n |x_n|^2 =$$

$$\sum_Q e^{2Re\lambda_n t}\beta_n |x_n|^2 + \sum_P |Re\lambda_n|^{2\alpha} e^{2Re\lambda_n t}\beta_n |x_n|^2$$

$$\leq \sum_Q e^{2\omega t}\beta_n |x_n|^2 + \sum_P |Re\lambda_n|^{2\alpha} e^{2Re\lambda_n t}\beta_n |x_n|^2,$$

with ω the growth constant of $S(t)$. On $(-\infty, 0]$ the function $F(x) = |x|^{2\alpha}e^{2xt}$ has a maximum of $\frac{\alpha_1^2}{t^{2\alpha}}$ for $x = -\frac{\alpha}{t}$, where $\alpha_1 = \alpha^\alpha e^{-\alpha}$. Therefore

$$\|S(t)x\|_W^2 \leq \max\left[\frac{\alpha_1^2}{t^{2\alpha}}, e^{2\omega t}\right] \sum_{n=1}^{\infty} \beta_n |x_n|^2$$

and

$$\|S(t)\|_{\mathcal{L}(V,W)} \leq \max\left[\frac{\alpha_1}{t^\alpha}, e^{\omega t}\right],$$

which implies (2.4') holds for $Re\lambda > \omega$ since $0 \leq \alpha < 1$ and hence (A5) and (A6) are satisfied.

We remark that the conditions imposed in Salamon and Pritchard [10] require that

$$\sum_P \frac{\|b_n\|_U \|c_n\|_Y}{|Re\lambda_n|^\alpha} < \infty$$
$$\sum_Q \|b_n\|_U \|c_n\|_Y < \infty \qquad (3.13b)$$

hold for $0 \leq \alpha < \frac{1}{2}$, which is stronger than our condition (3.13). Notice also that (3.13b) with $\alpha = 1$ is sufficient for the system to have a nuclear Hankel operator if we make the extra assumption $\sup_n Re\lambda_n < -\varepsilon$. Now (3.13b) with $\alpha = 1$ is implied by (3.12). We remark that the assumption that A is discrete and spectral is not absolutely essential to the proof: all we needed is that A and $S(t)$ have decompositions (3.2) and (3.3) and the assumption (3.6) together with the assumption that $\{\phi_n\}$ is a Riesz basis for H.

An interesting example was studied in [1] and [4] which satisfies these assumptions, but has a finite accumulation point and its spectrum comprises two parts: one which has $Re\lambda_n \to \delta$ and one with $Re\lambda_n \to -\infty$. Lemma 3.1 applies to this example too.

Example Flexible beam.

Let us consider the p.d.e. model for a beam with viscous damping considered in [1].

$$\begin{aligned}
&w_{tt} + \alpha A w_t + Aw = \delta(0)u_1(t) - \delta'(0)u_2(t) \\
&w_{xx}(-1) = 0 = w_{xx}(1); \quad w_{xxx}(-1) = 0 = w_{xxx}(1) \\
&-1 \le x \le 1; \quad t \ge 0 \\
&y_1(t) = w(0,t); \quad y_2(t) = w_x(0,t)
\end{aligned} \tag{3.14}$$

where $w(x,t)$ represents the deflection of the beam at position x at time t and A is the positive, self adjoint operator on $L_2(-1,1)$ given by

$$\left\{ \begin{array}{l} A = \dfrac{d^4}{dx^4} \\ D(A) = \left\{ \begin{array}{l} h \in L_2(-1,1) | w_x, w_{xx}, w_{xxx}, w_{xxxx} \in L_2(-1,1) \\ w_{xx}(-1) = w_{xx}(1) = 0, \quad w_{xxx}(-1) = 0 = w_{xxx}(1) \end{array} \right\} \end{array} \right. \tag{3.15}$$

As state space we choose $H = D(A^{1/2}) \oplus L_2(-1,1)$ with the inner product

$$\begin{aligned} < \begin{bmatrix} v_1 \\ v_2 \end{bmatrix}, \begin{bmatrix} w_1 \\ w_2 \end{bmatrix} >_H &= < v_1, w_1 >_{L_2} + < A^{1/2}v_1, A^{1/2}w_1 >_{L_2} \\ &+ < v_2, w_2 >_{L_2} \end{aligned} \tag{3.16}$$

So that on H (3.14) can be formulated as the abstract system

$$\dot{z} = \mathcal{A}z + Bu; \quad y = Cz \tag{3.17}$$

where

$$\begin{aligned} \mathcal{A} &= \begin{bmatrix} 0 & 1 \\ -A & -\alpha A \end{bmatrix} \\ B &= \begin{bmatrix} 0 & 0 \\ \delta(0) & -\delta'(0) \end{bmatrix} \end{aligned} \tag{3.18}$$

$$C = \begin{bmatrix} v_1 \\ v_2 \end{bmatrix} = \begin{bmatrix} v_1(0) \\ \frac{dv_1}{dx}(0) \end{bmatrix}; \quad z = \begin{bmatrix} w \\ w_t \end{bmatrix} \tag{3.19}$$

$\mathcal{A}$ has the eigenvalues μ_n, μ_{-n}; $n = 0, 1, 2, \cdots$ given by

$$\mu_n, \mu_{-n} = \frac{-\alpha\lambda_n^4 \pm \sqrt{\alpha^2\lambda_n^8 - 4\lambda_n^4}}{2} \tag{3.20}$$

where λ_n are the roots of

$$\left.\begin{aligned} \sinh\lambda_n \cos\lambda_n + \cosh\lambda_n \sin\lambda_n = 0; \quad n = 0, 2, 4, \cdots \\ \sinh\lambda_n \cos\lambda_n - \cosh\lambda_n \sin\lambda_n = 0; \quad n = 1, 3, 5, \cdots \end{aligned}\right\} \tag{3.21}$$

The normalised eigenvectors are

$$\phi_n = \eta_n \begin{bmatrix} v_n \\ \mu_n v_n \end{bmatrix}, \quad \phi_{-n} = \eta_{-n} \begin{bmatrix} v_n \\ \mu_{-n} v_n \end{bmatrix}, \qquad \text{with } \eta_{\pm n} = (1 + \lambda_n^4 + |\mu_{\pm n}|^2)^{-\frac{1}{2}} \tag{3.22}$$

where for even n

$$v_n = \alpha_n \left\{ \cos\lambda_n x + \cos\lambda_n \frac{\cosh\lambda_n x - \cos\lambda_n x}{\cosh\lambda_n + \cos\lambda_n} \right\} \tag{3.23}$$

$$\alpha_n = \frac{\cosh\lambda_n + \cos\lambda_n}{\sqrt{\cosh^2\lambda_n + \cos^2\lambda_n}} \tag{3.24}$$

The appropriate expressions for odd n are obtained by replacing cos by sin and cosh by sinh in (2.23) and by replacing α_n by

$$\rho_n = \frac{\sinh\lambda_n + \sin\lambda_n}{\sqrt{\sinh^2\lambda_n - \sin^2\lambda_n}} \tag{3.25}$$

The eigenvectors of A^* are

$$\begin{aligned} f_n &= \begin{bmatrix} v_n \\ -\left[\dfrac{1+\lambda_n^4}{\lambda_n^4}\right]\mu_n v_n \end{bmatrix}; \\ f_{-n} &= \begin{bmatrix} v_n \\ -\left[\dfrac{1+\lambda_n^4}{\lambda_n^4}\right]\mu_{-n} v_n \end{bmatrix} \end{aligned} \tag{3.26}$$

and $\{\phi_n, \varphi_n\}$ from a biorthogonal sequence in H where

$$\varphi_n = \frac{1}{< f_n, \varphi_n >_H} f_n \tag{3.27}$$

We need to know that B^* is given by

$$B^* \begin{bmatrix} w_1 \\ w_2 \end{bmatrix} = \begin{bmatrix} w_2(0) \\ \frac{d}{dx} w_2(0) \end{bmatrix} \tag{3.28}$$

and then we can systematically check that all the conditions of Lemma 3.1 hold. First we estimate

$$\begin{aligned} C\varphi_n &= \eta_n \begin{bmatrix} v_n(0) \\ \frac{d}{dx} v_n(0) \end{bmatrix} = \begin{bmatrix} \eta_n \alpha_n \\ 0 \end{bmatrix} \text{for } |n| \text{ even} \\ &= \begin{bmatrix} 0 \\ \eta_n \lambda_n \rho_n \end{bmatrix} \text{for } |n| \text{ odd} \qquad (3.29) \\ B^* \varphi_n &= -\frac{1}{< f_n, \varphi_n >_H} \left[\frac{1 + \lambda_n^4}{\lambda_n^4} \right] \frac{\mu_n}{\eta_n} C\varphi_n \end{aligned}$$

and so for large n, using the fact that $\lambda_n = O(n)$, and $\eta_n < f_n, \varphi_n >_H = O(1)$, we obtain

$$\|B^* \varphi_n\| \sim \mu_n \|C\varphi_n\| = \left\{ \begin{array}{l} O(\frac{1}{n^2}) \ n \text{ positive, even} \\ O(\frac{1}{n}) \ n \text{ positive, odd} \\ O(1) \ n \text{ negative, even} \\ O(n) \ n \text{ negative, odd} \end{array} \right\} \tag{3.30}$$

$$|Re\mu_n| \sim \left\{ \begin{array}{l} O(1) \ n \text{ positive} \\ O(n^4) \ n \text{ negative} \end{array} \right\} \tag{3.31}$$

and so (3.13) holds for $\alpha : \frac{3}{4} < \alpha < 1$. Condition (3.13b) holds for $0 < \alpha < \frac{1}{2}$. This was verified in [4], but there the formulation of Pritchard and Salamon [10] was used which led to an artificial choice of V and W spaces which were not contained in and did not contain the natural state space H. So there we obtained well posedness in two new spaces, but not on the original state space, which was not entirely satisfactory. Moreover the treatment there could not handle the interesting case of co-located sensors, namely B as above, but the new sensor operator $C_2 = B^*$. For this we obtain the estimates

$$\|C_2\varphi_n\| = \begin{cases} O(\frac{1}{n^2}) & n \text{ positive, even} \\ O(\frac{1}{n}) & n \text{ positive, odd} \\ O(1) & n \text{ negative, even} \\ O(n) & n \text{ negative, odd} \end{cases} \tag{3.32}$$

and again (3.13) holds for $\frac{3}{4} < \alpha < 1$, but (3.13b) does not hold.

So both of these examples are well-posed on the original space H and they satisfy (A1)–(A6) conditions. Of course the theory of Salamon in [12] would give well posedness (A1)–(A4) conditions as well, but no information about $(A+BKC)$−type perturbations.

As a consequence of the (A1)–(A6) conditions we know that any feedback configuration of (A, B, C) with any other system satisfying (A1)–(A6) conditions will remain in the same class and hence be well posed.

4 Application to the equivalence of input-output stability and exponential stability

In [4] the equivalence of input-output stability and exponential stability was shown to hold for a class of systems which, particularising to spectral systems, corresponds to a (3.13b) condition with $\alpha < 1/2$. While the results were adequate for delay systems, for spectral systems we can obtain better results by exploiting the perturbation results from section 2. First we need to recall some basic definitions from [4]. Throughout this section we suppose that (A, B, C) satisfies (A1)–(A6).

Definition 4.1 μ-Exponential stabilizability and detectability

$S(t)$ is μ-exponentially stabilizable if there exists $F \in \mathcal{L}(H, U)$ such that the $(A+BF)$−perturbed semigroup $S_F(t)$ is exponentially stable on H with decay rate exceeding μ. (We say that $S_F(t)$ is μ-exponentially stable on H).

$S(t)$ is μ-exponentially detectable if there exists $G \in \mathcal{L}(Y, H)$ such that the $(A + GC)$-perturbed semigroup $S_G(t)$ is μ-exponentially stable on H .

Definition 4.2 Spectrum decomposition assumption

A satisfies the spectrum decomposition assumption at δ if $\sigma_\delta^+(A) = \sigma(A) \cap \{\lambda \in \mathbf{C} : Re\lambda \geq \delta\}$ is bounded and seperated from $\sigma_\delta^-(A) = \sigma(A) \cap \{\lambda \in \mathbf{C} : Re\lambda < \delta\}$ such that a rectifiable, simple, closed curve can be drawn so as to enclose an open set containing $\sigma_\delta^+(A)$ in its interior and $\sigma_\delta^-(A)$ in its exterior. The spectral projection, P_δ, corresponding to $\sigma_\delta^+(A)$ induces the following decompositions on H

$$H = H^{+} \oplus H^{-} \tag{4.1}$$

$$A = \begin{bmatrix} A^{+} & 0 \\ 0 & A^{-} \end{bmatrix};$$

$$S(t) = \begin{bmatrix} S^{+}(t) & 0 \\ 0 & S^{-}(t) \end{bmatrix} \tag{4.2}$$

where A^{+} is bounded on H^{+} and A^{-} generates the C_0–semigroup $S^{-}(t)$ on H^{-}. Under assumptions (A3) and (A4) then the following decompositions with respect to P_δ are well defined

$$B = \begin{bmatrix} B^{+} \\ B^{-} \end{bmatrix}, \quad C = (C^{+}, C^{-}) \tag{4.3}$$

in the sense that $B^{+} \in \mathcal{L}(U, H^{+})$, $C^{+} \in \mathcal{L}(H^{+}, Y)$ and $(A^{-}, B^{-}C^{-})$ satisfies (A1)–(A6) assumptions with respect to V^{-} and W^{-}, where we have the induced spectral projections

$$V = V^{+} \oplus V^{-}; \quad W = W^{+} \oplus W^{-} \tag{4.4}$$

Consequently $G(\lambda) = C(\lambda - A)^{-1}B$ has the decomposition

$$G(\lambda) = C^{+}(\lambda - A^{+})^{-1}B^{+} + C^{-}(\lambda - A^{-})^{-1}B^{-} \tag{4.5}$$

But this is only valid for $Re\lambda > \tau$. Lemma 2.3 shows that for $Re\lambda > \omega$ the growth constant of $S(t)$, we have

$$\begin{aligned} G(\lambda) \;&=\; C^{+}(\lambda - A^{+})^{-1}B^{+} + C^{-}(\mu - A^{-})^{-1}B^{-} + \\ &\quad + (\mu - \lambda)C^{-}(\mu - A^{-})^{-1}(\lambda - A^{-})^{-1}B^{-} \end{aligned} \tag{4.5a}$$

$C^{+}(\lambda - A^{+})^{-1}B^{+}$ is well defined since this is a finite-dimensional system and (4.5a) can be extended to $Re\lambda > \omega'$ the growth constant of $S^{-}(t)$.

The main result of [4] concerns the consequences of exponential stabilizability.

Theorem 4.3

Suppose (A, B) is a system satisfying (A1), (A2), (A4) and U *is finite dimensional.* Then (A, B) is μ-exponentially stabilizable on H if and only if A satisfies the spectrum decomposition assumption at $-\mu^{-}$ for some $\mu^{-} > \mu$ and the following holds

(i) H^+ is finite dimensional
(ii) (A^+, B^+) is controllable
(iii) $S^-(t)$ is μ-exponentially stable on H

where we refer to the spectral decomposition of (A, B) with respect to $-\mu^-$ on H in (4.1)–(4.3), and μ^- may be taken arbitrarily close to μ. (We have assumed B to be one-one, of course).

A dual result holds for μ-exponential detectability. Finally we need an appropriate definition of input-output stability, which is a much weaker concept that the one used in [4].

Definition 4.4

The $p \times m$ matrix transfer function $G(\cdot)$ is μ-input-output stable if it is analytic and bounded in $Res > \mu - \varepsilon$, for some $\varepsilon > 0$.

We now obtain a nice generalization of the well known finite-dimensional result concerning the equivalence of input-output and exponentially stability.

Theorem 4.5

Suppose (A, B, C) is a system satisfying (A1)–(A5) and suppose that U and Y are finite-dimensional. Then A is μ-exponential stable if and only if (A, B) is μ-exponentially stabilizable, (C, A) is μ-exponentially detectable and its transfer function is μ-input-output stable.

Proof

(a) The necessity follows from definitions 4.1 and 4.2 and Lemma 2.3(b): If A is μ-exponentially stable, then its growth constant is $-\mu - \varepsilon$ for some $\varepsilon > 0$ and its transfer function given by (2.14) is analytic and bounded in $Re\lambda > -\mu - \varepsilon$.

(b) Sufficiency: from Theorem 4.3 A satisfies a spectrum decomposition at $-\mu-\varepsilon$ for some arbitrarily small $\varepsilon > 0$ and so the transfer function has a decomposition (4.5a) for $Re\lambda > \omega'$ the growth constant of A^-. Theorem 4.3(iii) shows that $\omega' = \mu - \varepsilon'$ for some $\varepsilon' > 0$. From Lemma 2.3(b) we may conclude that the transfer function of (C^-, A^-, B^-) is analytic and bounded in $Re\lambda > \mu - \varepsilon'$ (see (a)) and by assumption the transfer function of (A, B, C) is analytic and bounded in $Re\lambda > \mu - \varepsilon_1$ for another $\varepsilon_1 > 0$. Let us choose $\varepsilon < \varepsilon_2 = \min(\varepsilon_1, \varepsilon')$, which is always possible by Theorem 4.3. Then the transfer function of (C^+, A^+, B^+) is analytic and bounded in $\{Re\lambda > -\mu - \varepsilon_2\} \supset \{Re\lambda \geq -\mu - \varepsilon\}$. But theorem 4.3 and its dual show that (C^+, A^+, B^+) is a finite-dimensional, controllable and observable system with its spectrum in $\{Re\lambda \geq -\mu - \varepsilon\}$. Hence $A^+ = 0$ and $S(t) = S^-(t)$ which is μ-exponentially stable.

Theorem 4.5 has important implications for our example considered in section

3. In [1] finite-dimensional controllers were designed for this example using the frequency domain methods from [6]; one obtains input-output stability of the closed loop system. For the case of displacement measurements (3.29) we showed in [4] that this was equivalent to exponential stability of the closed loop system on certain Hilbert spaces. We could not establish exponential stability on the original state space (3.16) and the theory did not apply to the case of velocity measurements, the co-located case. In section 3 we showed that both types of measurements satisfy (A1)–(A6) conditions with respect to the original state space and so Theorem 4.5 is applicable. In particular, if we stabilize the system in the input-output sense by finite-dimensional compensators (which always satisfy (A1)–(A6)), the resulting closed loop system will have an exponentially stable system operator provided that it is exponentially stabilizable and detectable. In [4] it was shown that the closed loop system was detectable and stabilizable if and only if the realisations of the original system and the compensator were. We can always construct a minimal realisation of the finite-dimensional compensator and the detectability and stabilizability of spectral systems is easy to check (Lemma 5.1, [4]). Suppose that A satisfies the spectrum decomposition assumption at $Re\lambda = -\mu - \varepsilon$ and that it has finitely many eigenvalues in $Re\lambda \geq -\mu - \varepsilon$. Denoting this set by $Q(\mu)$, we have that (C, A) is μ-exponentially detectable if and only if

$$C\phi_n \neq 0 \text{ for all } n \text{ such that } \lambda_n \in Q(\mu) .$$

and (A, B) is μ-exponentially stabilizable if and only if

$$B^*\varphi_n \neq 0 \text{ for all } n \text{ such that } \lambda_n \in Q(\mu) .$$

Since all the coefficients $C\phi_n, C_2\phi_n$ and $B^*\varphi_n$ in our example are nonzero, we have μ-exponential detectability and stabilizability provided $\mu > -\frac{1}{\alpha}$, where we have an accumulation point. So for this example, with both displacement and velocity measurements, we can conclude that μ-input-output stability implies μ-exponential stability of the closed loop system for all $\mu > -\frac{1}{\alpha}$.

5 Conclusions

We have identified a class of infinite-dimensional systems which is closed under perturbations of the type $(A+BF)$, $(A+GC)$ and $(A+BKC)$, where B and C may be unbounded input and output operators. The property of perturbations remaining in the same class has important applications in control design where feedback control results in a perturbed system operator of the above type. Here we have applied the results to the problem of the equivalence of input-output stability, a frequency domain concept, and exponential stability of the system operator on its original state space. We have given sufficient conditions for spectral systems to belong to this class and have illustrated the above results with a p.d.e. model of a flexible beam with viscous damping.

Acknowledgements

We would like to thank Stuart Townley, Tony Pritchard, and Dietmar Salamon of the University of Warwick for very useful discussions on this topic.

References

[1] J. Bontsema, R. F. Curtain and J. M. Schumacher, *Robust control of flexible structures: a case study*, Automatica, **24**, pp.177-186, 1988.

[2] R. F. Curtain and A. J. Pritchard, *Infinite dimensional linear systems theory*, LNCS **8**, Springer Verlag, Berlin, 1978.

[3] R. F. Curtain, *Spectral Systems*, Int. J. Control, **39**, pp. 657-666, 1984.

[4] R. F. Curtain, *Equivalence of input-output stability and exponential stability for infinite-dimensional systems*, Technical Report TW-176, 1987, Mathematics Institute, University of Groningen.

[5] R. F. Curtain, K. Glover and J. Lam, *Reduced order models for distributed systems based on Hankel-norm approximations*, Ed. L. Meirovitch, Proc. 5th VPI and US/AIAA Symposium on Dynamics and Control of Large Structures, June 12-14, 1985, Blacksburg, Virginia.

[6] R. F. Curtain and K. Glover, *Robust stabilization of infinite dimensional systems by finite dimensional controllers*, Systems and Control Letters **7** (1986) pp. 41-47.

[7] C. A. Jacobson, *The structure of exponential stabilization of a class of linear distributed parameter systems*, Ph.D. Thesis, ECSE Dept. Rensselaer Polytechnic Institute, Troy, New York, 1986.

[8] C. A. Jacobson and C. N. Nett, *Linear state space systems in infinite-dimensional space: the role and characterization of joint stabilizability/detectability*, Proc. 1987 American Control Conference, June 1987.

[9] A. Pazy, *Semigroups of linear operators and applications to partial differential equations*, Springer Verlag, 1983.

[10] A. J. Pritchard and D. Salamon, *The linear quadratic optimal control problem for infinite dimensional systems with unbounded input and output operators*, SIAM J. Control and Optimiz., 1987, **25**, pp. 121-144.

[11] D. Salamon, *Control and observation of neutral systems*, RNM 91, Pitman, London, 1984.

[12] D. Salamon, *Infinite-dimensional linear systems with unbounded control and observation: a functional analytic approach*, Trans. Amer. Math. Soc. 1987, **300**, pp.383-431.

[13] S. Townley and A. J. Pritchard, Private communication.

[14] G. Weiss, *Admissibility of unbounded control operators*, (submitted to SIAM J. Control and Optimization).

[15] R. M. Young, *An Introduction to Nonharmonic Fourier Series*, Academic Press, London, 1980.

The L_2 approximation of infinite-dimensional systems

Jonathan R. Partington

1 Introduction

Given a stable continuous-time linear time-invariant system (which may be infinite-dimensional) with impulse response $h(t) \in L_2(0,\infty)$, and transfer function $G(s)$, its Laplace transform, the problem of *model reduction* – approximating the system by a system of lower degree – is one that has been much studied (see, for example, Adamjan, Arov and Krein [1], Glover [5], Glover, Curtain and Partington [7], and Glover and Partington [12].) Recall that we have

$$\|h\|_2 = \left(\int_0^\infty |h(t)|^2\,dt\right)^{1/2} = \left(\frac{1}{2\pi}\int_{-\infty}^\infty |G(i\omega)|^2\,d\omega\right)^{1/2} \tag{1.1}$$

In this paper we restrict ourselves to Single Input-Single Output systems: however many of the results go through to the matrix-valued case, with very little alteration.

Rational L_2 approximation has been found useful in system identification problems (Wahlberg and Ljung [21]), but can give multiple local minima as well as numerical difficulties (Baratchart [2] and Braess [4]).

In many approximation schemes, the achievable error is known to be related to properties of the Hankel operator $\Gamma : L_2(0,\infty) \to L_2(0,\infty)$, defined by

$$(\Gamma u)(t) = \int_0^\infty h(t+\tau)u(\tau)\,d\tau. \tag{1.2}$$

(See, for example, [17] for basic properties of such operators.)

Recently a theoretical lower bound for L_2 approximation was given by Glover [6] as follows: Let $\Theta : L_2(0,\infty) \to L_2(0,\infty)$ be the scaled Hankel operator given by

$$(\Theta u)(t) = \frac{1}{\sqrt{\pi}}\int_0^\infty t^{-1/4}h(t+\tau)\tau^{-1/4}u(\tau)d\tau \tag{1.3}$$

Then $\|\Theta\|_{HS}$, the Hilbert-Schmidt norm of Θ, is equal to the L_2 norm of h, and the rank of Θ is equal to the degree of the system with impulse response h. It follows that the L_2 error in approximating h by the impulse response h_0 of a system of degree k is bounded below as follows:

$$\|h - h_0\|_2 \geq \left(\sigma_{k+1}^2 + \sigma_{k+2}^2 + \ldots\right)^{1/2}, \tag{1.4}$$

where $\sigma_1 \geq \sigma_2 \geq \cdots \geq 0$ are the singular values of Θ.

For a large class of infinite-dimensional systems it is possible to calculate an asymptotic expression for the singular values of Θ, and thus to determine an asymptotic expression for the lower bound in L_2 approximation. One such class of systems is delay systems, which have been widely studied in the context of model reduction (see, for example, Glover, Lam and Partington [8].) In addition to deriving a calculable lower bound for L_2 errors, it is possible to employ concrete algorithms which produce a convergence rate that is asymptotically the same as that given by Glover's theoretical lower bound.

In this paper, we provide in Section 2 various estimates of the singular values of a scaled Hankel operator under various conditions on the system's impulse response and transfer function. Section 3 gives some methods which can be used for L_2 model reduction with optimal rates of convergence. Finally in Section 4 we discuss the use of modal expansions of delay systems in the L_2 problem (the analogous problem for H_∞ was discussed in [22] and [18].) Here the convergence rates can be determined but are not optimal in general.

2 Asymptotic results on scaled singular values

In this section we consider the scaled Hankel operator Θ given by (1.3), whose Hilbert-Schmidt norm equals the L_2 norm of h. In view of the remarks in the introduction, we are concerned with establishing results on the asymptotic behaviour of its singular values $\sigma_i(\Theta) = \alpha_i$, say. It is convenient to begin with functions h which are in L_1 and of compact support $[0, A]$, say. Our first result is simple, but useful in that it enables us to employ existing results on the singular values of Hankel operators, notably those from [8] and [9].

Proposition 2.1. If $h(t) = 0$ for $t \geq A$, then the singular values of Θ satisfy $\alpha_i \geq A^{-1/2}\pi^{-1/2}\sigma_i(\Gamma)$, where Γ is the Hankel operator $(\Gamma u)(t) = \int_0^\infty h(t+\tau)u(\tau)d\tau$.

Proof. Let M be the operator on L_2 which consists of multiplication by the function $t^{1/4}\chi_{[0,A]}$; thus $\|M\| = A^{1/4}$. Then $\Gamma = \sqrt{\pi}M\Theta M$ so that

$$\begin{aligned} \sigma_i(\Gamma) &\leq \sqrt{\pi}\|M\|\alpha_i\|M\| \text{ (see, e.g. [17])} \\ &= \pi^{1/2}A^{1/2}\alpha_i, \end{aligned} \tag{2.1}$$

as required.

One particular function which is of great importance, not only for its application to delay systems, but also in its role as a fundamental discontinuous function, is the step function

$$h(t) = \begin{cases} 1, & \text{if } t < T \\ 0, & \text{if } t > T, \end{cases} \tag{2.2}$$

the impulse response corresponding to the transfer function

$$G(s) = (1 - e^{-sT})/s. \tag{2.3}$$

For this particular example, it is possible to obtain an asymptotic formula for α_i, which shows that in fact the inequality given in Proposition 2.1 is realistic.

Theorem 2.2 Let $h(t) = \chi_{[0,T]}(t)$ and Θ the scaled Hankel operator as in (1.3). Then the singular values of Θ satisfy

$$i\alpha_i \to C\pi^{-3/2}T^{1/2} \quad \text{as} \quad i \to \infty \tag{2.4}$$

where

$$C = \Gamma(\frac{3}{4})^2/\Gamma(\frac{3}{2}) = 1.69\ldots \tag{2.5}$$

Proof Since the integral kernel $t^{1/4}h(t+\tau)\tau^{1/4}\pi^{-1/2}$ is real and symmetric, it follows that Θ is Hermitian (it is also compact) and it is necessary to determine its eigenvalues. Suppose that $\lambda u(t) = \frac{1}{\sqrt{\pi}}\int_0^{T-t} u(\tau)t^{-1/4}\tau^{-1/4}d\tau$. Writing $u(t) = t^{-1/4}v(t)$, we obtain

$$\lambda v(t) = \frac{1}{\sqrt{\pi}}\int_0^{T-t} v(\tau)\tau^{-1/2}d\tau. \tag{2.6}$$

This gives

$$\lambda \dot{v}(t) = -\frac{1}{\sqrt{\pi}}\frac{v(T-t)}{\sqrt{T-t}} \tag{2.7}$$

Thus

$$\ddot{v}(t) - \frac{\dot{v}(t)}{2(T-t)} + \frac{1}{\pi\lambda^2}\,\frac{v(t)}{\sqrt{t}\sqrt{T-t}} = 0 \tag{2.8}$$

i.e.

$$\frac{d}{dt}\left(\dot{v}(t)\sqrt{T-t}\right)+\frac{v(t)}{\pi\lambda^2\sqrt{t}}=0 \tag{2.9}$$

and

$$v(T)=0=\dot{v}(0). \tag{2.10}$$

Such Sturm-Liouville-type equations may be analysed by established techniques (see, e.g. [15], pp.274-6, and [16], pp.214-7). Briefly, the substitutions

$$z=\frac{1}{\pi}\left(\frac{T-t}{t}\right)^{1/8}v \tag{2.11}$$

$$x=\int_0^t \frac{1}{\sqrt{\pi}}\,\frac{1}{t^{1/4}(T-t)^{1/4}}dt \tag{2.12}$$

reduce the equation to the form

$$z''-f(x)z+\frac{z}{\lambda^2}=0, \tag{2.13}$$

over the interval

$$0\le x\le \int_0^T \frac{1}{\sqrt{\pi}}t^{-1/4}(T-t)^{-1/4}dt=2\sqrt{\frac{T}{\pi}}\int_0^{\pi/2}\sqrt{\sin\theta\,\cos\theta}\,d\theta. \tag{2.14}$$

This last integral equals $T^{1/2}C\pi^{-1/2}$, where

$$C=B(\frac{3}{4},\frac{3}{4})=\Gamma(\frac{3}{4})^2/\Gamma(\frac{3}{2})=1.69\ldots \tag{2.15}$$

The eigenvalues of (2.13) are asymptotic to those of the equation $z''+z/\lambda^2=0$ on the same interval, with the same boundary conditions, that is

$$\frac{1}{\lambda_i^2}\sim\frac{(i+1/2)^2\pi^2}{TC^2/\pi}\quad,\ \text{or}\ \ i\lambda_i\to C\pi^{-3/2}T^{1/2} \tag{2.16}$$

as required. □

The singular values of the Hankel operator induced by the function $h(t)$ above satisfy $i\sigma_i\to T/\pi$, [8], [9], and so in this case the inequality given in Proposition 2.1 is conservative to a factor of 1.69.

Although the numbers (α_i) are primarily used to supply a lower bound for L_2 approximation, it is convenient to be able to estimate their size from above as well as below. Our next result should be compared with Theorem 4.2 of [9], where a similar inequality is given for the singular values of a Hankel operator.

Theorem 2.3. If $h(t) = 0, t \geq A$ and $h(t) = -\int_t^A \dot{h}(\tau)d\tau$, with $\dot{h} \in L_2$, then

$$\alpha_{2i}(\Theta_h) \leq \frac{A}{\pi i}(1 + o(1))\ \alpha_i(\Theta_{\dot{h}}) \qquad i = 2, 3, \ldots \tag{2.17}$$

Proof. Let $J : L_2[0, A] \to L_2[0, A]$ denote the integral operator

$$(Jf)(\tau) = \int_t^A f(\tau)d\tau\ . \tag{2.18}$$

Now

$$\begin{aligned} \Theta_h u(t) &= \pi^{-1/2}\int_0^\infty h(t+\tau)u(\tau)t^{-1/4}\tau^{-1/4}d\tau \\ &= -\pi^{-1/2}\int_0^\infty\int_t^\infty \dot{h}(r+\tau)u(\tau)t^{-1/4}\tau^{-1/4}drd\tau \qquad (2.19) \\ &= -\int_t^\infty t^{-1/4}r^{1/4}(\Theta_{\dot{h}}u)(r)dr = -M^{-1}JM\Theta_{\dot{h}}u(t), \end{aligned}$$

for, by Fubini's theorem, we may interchange the order of integration on a dense set of u and deduce that $\Theta_h = -M^{-1}JM\Theta_{\dot{h}}$ everywhere since all operators are bounded.

It may be verified that $M^{-1}JM$ is a bounded operator whose singular values satisfy $i\sigma_i \to A/\pi$ as $i \to \infty$ (see [11, Lemma 3.3]), and hence, since for any operators P and Q it is true that the singular values satisfy

$$\sigma_{m+n+1}(PQ) \leq \sigma_{m+1}(P)\sigma_{n+1}(Q) \tag{2.20}$$

(see, for instance, Gohberg and Krein [13] or [17]) for all $m, n \geq 0$, the result follows on writing $m = i, n = i - 1, P = J$ and $Q = \Theta_{\dot{h}}$. □

Using Theorem 2.3 repeatedly it is straightforward to deduce that if $h(t)$ is sufficiently smooth, then its scaled singular values tend to zero rapidly. Note that, for $h \in L_2(0, A)$,

$$n\alpha_{2n}^2 \leq \alpha_{n+1}^2 + \ldots + \alpha_{2n}^2 \to 0, \tag{2.21}$$

and hence the scaled singular values are $o(n^{-1/2})$.

Theorem 2.4 Suppose that $0 < T < A$, that $h(t)$ is absolutely continuous on $[0,T)$ and $(T,A]$ and that $\dot{h}(t)$ is square integrable on $[0,T)$ and $(T,A]$, that $h(t)=0$ for $t \geq A$, and that $L = \lim_{t\to T-} h(t) - \lim_{t\to T+} h(t)$ exists and is finite. Then for the scaled Hankel operator given by (1.3), the singular values satisfy

$$i\alpha_i \to |L|C\pi^{-3/2}T^{1/2}, \quad \text{as} \quad i \to \infty, \quad \text{where } C = \Gamma(\frac{3}{4})^2/\Gamma(\frac{3}{2}), \tag{2.22}$$

and hence

$$\left\{\inf\|h-\tilde{h}\|_2 : \text{ degree } (\tilde{h}) \leq k\right\} \geq |L|C\pi^{-3/2}T^{1/2}k^{-1/2} + O(k^{-1}) \tag{2.23}$$

as $k \to \infty$.

Proof. Writing $h(t) = h_1(t) + h_2(t)$, where $h_1(t) = L\chi_{[0,T]}$ and $\dot{h}_2 \in L_2[0,A]$,

we have that $i\alpha_i(h_1) \to |L|C\pi^{-3/2}T^{1/2}$ by Theorem 2.2

and that $\alpha_i(h_2) = O(i^{-3/2})$ by Theorem 2.3.

Now, by the Ky Fan theorem (see, e.g. Gohberg and Krein [13]) that for any operators S_1, S_2, if $i^r\sigma_i(S_1) \to a$ and $i^r\sigma_i(S_2) \to 0$ for some $r > 0$, then $i^r\sigma_i(S_1 + S_2) \to a$, we obtain (2.22) and (2.23) follows from (1.4). □

As in [8], it is possible to make deductions about the singular values of Θ even when h has non-compact support, but decays to 0 rapidly at infinity. We give one such result.

Corollary 2.5. Let $h, \ldots, h^{(r+1)}$ be as above with $h^{(r+1)} \in L_2$, and suppose that $h(t), \ldots, h^{(r+1)}(t)$ decay to zero faster than any power of t, as $t \to \infty$. Then

$$\alpha_n(\Theta_h) = O(n^{-(r+3/2)+\epsilon}) \qquad \text{for any } \epsilon > 0. \tag{2.24}$$

Proof. Write $h = h_1 + h_2$, where h_1 is smooth with compact support and $\|h_2\|_2$ is sufficiently small (depending on n). The remainder of the proof is similar to the proof of the analogous result for Hankel operators in [8]. □

In practice, it is common to be able to specify delay systems by their transfer functions rather than their impulse responses and indeed L_2 lower bounds may be obtained using transfer functions. Techniques similar to those used in [9] may be used to calculate the scaled singular values of such systems.

Let

$$h_1(s) = \sum_0^{n_1} p_i(s)e^{-\gamma_i s}, \tag{2.25}$$

where $0 = \gamma_o < \gamma_1 < \ldots < \gamma_{n_1}$, the p_i are polynomials of degree δ_i, and $\delta_i < \delta_o$ for $i \neq 0$. Suppose also that

$$h_2(s) = \sum_0^{n_2} q_i(s)e^{-\beta_i s}, \tag{2.26}$$

with $0 \leq \beta_o < \ldots < \beta_{n_2}$, where the q_i are polynomials of degree $d_i < \delta_o$ for each i. Transfer functions of the form $G(s) = h_2(s)/h_1(s)$ with h_1, h_2 coprime were considered in [22], [18] and, more recently, in [9] and [10]. Using results of Bellman and Cooke [3], it was shown that such $G(s)$ have only finitely many poles in any half plane $\{\text{Re s} > \gamma\}$ and asymptotic expressions for their Hankel singular values were given in [9]. By using the previous results of this section, it is possible to give asymptotic expressions for the scaled singular values (α_i) and hence the lower bound in L_2 approximation can be extended to these systems also. We illustrate the method by a simple example.

Example 2.6. Let $G(s) = \dfrac{e^{-Ts}}{s + e^{-s}}$ with $T \geq 0$.

One technique for calculating an asymptotic form for the singular values is to perform an expansion of the following form:

If $T > 0$ write $G(s) = G_1(s) + G_2(s)$, namely

$$\frac{e^{-Ts}}{s + e^{-s}} = \frac{e^{-Ts}}{s+1} + \frac{e^{-Ts}(1 - e^{-s})}{(s+1)(s+e^{-s})}. \tag{2.27}$$

As in Theorem 2.2 it can be shown that

$$i\alpha_i(G_1) \to C\pi^{-3/2}T^{1/2} \tag{2.28}$$

as $i \to \infty$, and by taking the decomposition further and using Corollary 2.5, that

$$\alpha_i(G_2) = O(i^{-2}). \tag{2.29}$$

It follows that $\alpha_i(G)$ is asymptotic to $\alpha_i(G_1)$, which gives a lower bound for the L_2 error of an n-state approximant of order $n^{-1/2}$.

For $T = 0$ write

$$\begin{aligned} \frac{1}{s + e^{-s}} &= \frac{1}{s+1} + \frac{1 - e^{-s}}{(s+1)(s+e^{-s})} \\ &= \frac{1}{s+1} - \frac{e^{-s}}{(s+1)^2} + \frac{e^{-s} - 1}{(s+1)^2(s+e^{-s})} \end{aligned} \tag{2.30}$$

The second term dominates the others, implying that the scaled singular values are asymptotically of order n^{-2} and the L_2 lower bound is of order $n^{-3/2}$.

3 Approximation errors

Several techniques are available for L_2 model reduction, and we begin by enumerating a few here.

(i) Truncated Balanced realizations. These are discussed in [7], where a theoretical error bound is derived. In numerical examples these tend to perform better than this L_2 error bound would suggest.

(ii) Modal expansions. These are discussed in Section 4. In general these do not converge at the optimal rate.

(iii) Padé techniques. These are considered by Lam [14], as well as in [11], and the following characterisation of the frequency response error in approximating e^{-sT} can be used to prove optimal convergence rates for delay systems with no imaginary axis poles except perhaps at the origin.

Proposition 3.1 (Lam) The frequency response error of e^{-sT} approximated by its $[n/n]$ Padé approximant denoted by $R_n(sT)$ satisfies (for $\psi = 2(\sqrt{2}/e)^{1/2}$)

$$|e^{-i\omega T} - R_n(i\omega T)| \leq \begin{cases} 2(\frac{\omega T}{\psi n})^{2n+1} & \omega T \leq \psi n \\ 2 & \omega T > \psi n. \end{cases} \tag{3.1}$$

In the remainder of this section we consider a further technique, which can be used in combination with some of the others to obtain optimal convergence rates.

(iv) Orthogonal functions. The following result is standard.

Proposition 3.2. Let $L_n(t) = \frac{e^t}{n!}\frac{d^n}{dt^n}(t^n e^{-t})$, $n = 0, 1, 2\ldots$; then, writing $\phi_n(t) = e^{-t/2}L_n(t)$, the functions $(\phi_n)_{n=0}^{\infty}$ form a complete orthonormal set in $L_2(0,\infty)$. Further, each L_n is a polynomial of degree n, satisfying the differential equation

$$t\ddot{y}(t) + (1-t)\dot{y}(t) + ny(t) = 0. \tag{3.2}$$

Moreover, if $\alpha > 0$, the functions $e^{-\alpha t}L_n(2\alpha t)\sqrt{2\alpha}$ also form a complete orthonormal set in $L_2(0,\infty)$.

The Laguerre functions above can be used directly to good effect provided that a suitable value of α is used, as our next result illustrates.

Theorem 3.3. Let

$$h(t) = \left\{ \begin{array}{ll} 1, & \text{if } 0 \leq t < 1 \\ 0, & \text{if } t > 1 \end{array} \right\}$$

be the pulse, and let $N > 0, n \geq 2N$, and

$$E_n = \inf_{a_o, \ldots, a_{n-1}} \|h - e^{-2Nt}(a_o + a_1 t + \ldots + a_{n-1}t^{n-1})\|_2 \tag{3.3}$$

then $E_n = O(N^{-1/4}n^{-1/4})$, so that for $n = 2N$, $E_n = O(n^{-1/2})$, in accordance with the theoretical lower bound obtained using Theorem 2.2.

Proof. $E_n = \frac{1}{(4N)^{1/2}} E'_n$, where

$$E'_n = \inf_{b_o, \ldots, b_{n-1}} \|g(x) - e^{-x/2}(b_o + \ldots + b_{n-1}x^{n-1})\|_2 \tag{3.4}$$

and $g(x) = \chi_{[0,4N]}(x)$ (writing $x = 4Nt$). Let

$$C_n = \int_1^{4N} e^{-x/2} L_n(x) dx. \tag{3.5}$$

Szegö [20, (8.22.9)] gives the following asymptotic formula:

$$\begin{aligned} e^{-x/2} L_n(x) = (-1)^n (\pi \sin \phi)^{-1/2} x^{-1/4} n^{-1/4} \\ \{\sin[(n + 1/2)(\sin 2\phi - 2\phi) + \tfrac{3\pi}{4}] + O(1) n^{-1/2} x^{-1/2}\}, \end{aligned} \tag{3.6}$$

where $x = (4n + 2)\cos^2 \phi$, $\epsilon \leq \phi \leq \frac{\pi}{2} - \epsilon n^{-1/2}$, uniformly for a fixed $\epsilon > 0$. Using this once can show that

$$C_n = O(N^{1/4} n^{-3/4}). \tag{3.7}$$

Moreover,

$$\int_0^1 e^{-x/2} L_n(x) dx = O(n^{-3/4}), \tag{3.8}$$

so that since the functions $(e^{-x/2} L_n(x))$ form a complete orthonormal set

$$E'_n = \left(\sum_{k>n} \left(\int_0^{4N} e^{-x/2} L_k(x) dx \right)^2 \right)^{1/2} = O(N^{1/4} n^{-1/4}), \tag{3.9}$$

so that $E_n = O(N^{-1/4} n^{-1/4})$, as required. □

Note that, in the frequency domain, we are approximating the Laplace transform of h by a rational function whose poles are all at one point – in this case at $\alpha = -2N$. To attain the optimal order of convergence, it is necessary to let α depend on n, and indeed $\alpha = n$ is appropriate.

With this choice, one may estimate the constants involved to obtain

$$E_n \le 2^{1/4} \pi n^{-1/2} (1 + o(1)). \tag{3.10}$$

The orthogonal polynomial technique is more widely applicable for general functions with several derivatives with fast decay at infinity. The following result has several applications.

Lemma 3.4. Let $\alpha > 0$, let $f \in L_2(0, \infty)$ and

$$C_n^\alpha(f) = \int_0^\infty f(t) e^{-\alpha t} L_n(2\alpha t) \sqrt{2\alpha}\, dt.$$

If f is such that $\ddot{f}(t)$ exists, and $\dot{f}$ is the integral of $\ddot{f}$ (hence absolutely continuous), and $tf, \dot{f}, t\ddot{f} \in L_2(0, \infty)$, then

$$C_n^\alpha(f) = \frac{-1}{n} C_n^\alpha \left(t\ddot{f}/2\alpha + \dot{f}/2\alpha + f/2 - \frac{\alpha t f}{2} \right). \tag{3.11}$$

Proof. Writing $x = 2\alpha t$, we have

$$C_n^\alpha(f) = \int_0^\infty f(x/2\alpha) L_n(x)\ dx/\sqrt{2\alpha}.$$

Now $(xe^{-x} L'_n)' + ne^{-x} L_n = 0$, by (3.2), so integrating by parts,

$$C_n^\alpha(f) = \int_0^\infty (f(x/2\alpha)e^{x/2})' \frac{xe^{-x}L_n'}{n\sqrt{2\alpha}}\, dx$$
$$= -\int_0^\infty ((f(\tfrac{x}{2\alpha})e^{x/2})'xe^{-x})' L_n dx/n\sqrt{2}\alpha$$
$$= -\tfrac{1}{n}\int_0^\infty \left(\tfrac{1}{4\alpha^2}f''(\tfrac{x}{2\alpha})x + \tfrac{1}{2\alpha}f'(\tfrac{x}{2\alpha}) + \tfrac{1}{2}f(\tfrac{x}{2\alpha}) - \tfrac{1}{4}xf(\tfrac{x}{2\alpha})\right)$$
$$e^{-x/2}L_n(x)\, dx/\sqrt{2\alpha} \tag{3.12}$$

which yields (3.11) on substituting $t = x/2\alpha$ again.

Corollary 3.5. Suppose that $t^p f^{(q)}(t) \in L_2(0,\infty)$ for $0 \le p \le m$, $0 \le q \le 2m$, and that $f, \dot{f}, \ldots, f^{(2m-1)}$ are absolutely continuous. Let $C_n(f) = \int_0^\infty f(t)e^{-t/2}L_n(t)dt$. Then $C_n(f) = o(n^{-m})$, and hence

$$\inf \{\|f - \tilde{f}\|_2 : \text{rank } (\tilde{f}) \le k\} = o(k^{1/2-m}). \tag{3.13}$$

Proof. This follows by induction on m, using Lemma 3.4. If $m = 0$, $f \in L_2(o,\infty)$ and $C_n(f)$ is square summable, and hence $o(1)$. The induction step is given by (3.11).
Finally

$$\inf \{\|f - \tilde{f}\|_2 : \text{rank } (\tilde{f}) \le k\} \le \left(\sum_{n>k} C_n(f)^2\right)^{1/2} = o(k^{1/2-m}). \tag{3.14}$$

□

Using the decomposition technique outlined in Section 2, combined with the orthogonal polynomial or Padé techniques, it is possible to obtain approximants showing optimal L_2 convergence rates. For example, in approximating the functions given in example 2.6, the order of convergence is the same as that of the function $e^{-Ts}/(s+1)$ if $T > 0$, or the function $-e^{-s}/(s+1)^2$ if $T = 0$. Fuller details will be given in [11].

4 L_2 modal expansions for delay systems

In this section we consider functions $G(s) = h_2(s)/h_1(s)$ as in sections 2 and 3. Modal expansions for such functions were considered in [22] and [18], where conditions were given for the convergence of a partial fraction expansion

$$G(s)=\sum_{j=1}^{m}\sum_{i=1}^{n_j}\frac{c_{ij}}{(s-z_j)^i}+\sum_{j=m+1}^{\infty}\frac{\mathrm{Res}(G,z_j)}{s-z_j} \tag{4.1}$$

where $z_1,\ldots,z_m$ are the multiple poles of G (finitely many) and the remaining poles $z_j=x_j+iy_j$ are simple.

The results of [22] and [18] require the notion of the ***distribution diagram*** of $h_1(s)=\sum_{i=0}^{n_1}p_i(s)e^{-\gamma_i s}$. As defined by Bellman and Cooke [3], this is the polygonal line K which is the upper boundary part of the convex hull of the points $(0,0),(\gamma_i,\delta_i)$ $(i=0,\ldots,n_1)$, and $(\gamma_{n_1},0)$. We make the standing assumption that no segment K_j of K contains more than two points (γ_i,δ_i).

The parameter τ is given by

$$\tau=\min_{0\le i\le n_2} f_i \tag{4.2}$$

where f_i is such that (β_i,d_i+f_i) lies on K. Provided that $0\le\alpha_i\le\gamma_{n_1}$ and (α_i,d_i) lies below K for each i, then $\tau>0$ and $\mathrm{Res}(G,z_j)=O(|z_j|^{-\tau})$.

In [18] it was shown that the L_∞ error obtained by truncating the modal expansion to n terms is $O(\frac{\log n}{n^\tau})$ as $n\to\infty$. One thus has uniform convergence over the right half plane.

The following is an analogous estimate for the L_2 case.

Theorem 4.1

Let $G(s)=h_2(s)/h_1(s)$ and τ be as above. Then, provided that $\tau>\frac{1}{2}$, the modal expansion converges in L_2 and the truncation error satisfies

$$\left\|\sum_{j=n+1}^{\infty}\frac{\mathrm{Res}(G,z_j)}{s-z_j}\right\|=O\left(\frac{\log n}{n^{\tau-\frac{1}{2}}}\right) \tag{4.3}$$

Proof. As in [18] it is enough to consider the value of

$$S=\sum_{y_j\ge n}\left|\frac{\mathrm{Res}(G,z_j)}{s-z_j}\right| \tag{4.4}$$

at a point $s=i\omega$. The following estimates are taken from [18].

(i) If $\omega\le n-1$, then $S=O(\frac{\log n}{n^\tau})$.

(ii) If $\omega\ge n-1$, then there are three contributions to S:

(a) $y_j>\omega+1$: $O(\frac{\log\omega}{\omega^\tau})$; (b) $|y_j-\omega|<1$: $O(\frac{1}{\omega^\tau\log\omega})$; (c) $n\le y_j\le\omega+1$: $O(\frac{\log n}{n^\tau})$ if $\omega\le 2n$ and

$$\frac{K}{(\omega-2)^\tau 1}+\frac{K}{(\omega-3)^\tau 2}+\ldots\frac{K}{(\omega-k)^\tau(k-1)}, \tag{4.5}$$

if $\omega \geq 2n$, where $n-1 < \omega - k \leq n$, which is seen to be $O(\frac{\log \omega}{\omega^\tau}) + O(\frac{n^{1-\tau}}{\omega})$, by splitting the sum into its first $[\omega/2]$ terms followed by the rest.

Thus the error S satisfies $S = O(\frac{\log n}{n^\tau})$ if $\omega \leq 2n$ and $S = O(\max(\log \omega/\omega^\tau, \frac{n^{1-\tau}}{\omega}))$ if $\omega \geq 2n$. If we now square and integrate we obtain a contribution of $O((\log n)^2 n^{1-2\tau})$ from each part of the range, which establishes the result.

Thus for $G(s) = e^{-\alpha s}/(s + e^{-s})$, we have $\tau = 1 - \alpha$ and the truncation error is $O(\log n/n^{1/2-\alpha})$, so that the modal expansion converges in L_2 if $\alpha < 1/2$.

References

[1] Adamjan, V.M., Arov, D.Z. and Krein, M.G., *Analytic Properties of Schmidt Pairs for a Hankel Operator and the Generalised Schur-Takagi Problem,* Math. USSR Sbornik, Vol.15 No.1, 31-73, 1971.

[2] Baratchart, L., Recent and new results in rational L^2-approximation, in "*Modelling, Robustness and Sensitivity Reduction in Control Systems,*" Ed. R.F. Curtain, Springer, 119-126, 1987.

[3] Bellman, R. and Cooke, K.L., "*Differential-Difference Equations,*" Academic Press, 1963.

[4] Braess, D., "*Nonlinear approximation theory,*" Springer-Verlag, 1986.

[5] Glover, K., All Optimal Hankel-Norm Approximations of Linear Multivariable Systems and their L_∞ Error Bounds, *Int. J. Contr.*, Vol.39 No.6, 1115-1193, 1984.

[6] Glover, K., Model Reduction: A Tutorial on Hankel-Norm Methods and Lower Bounds on L_2 Errors, *IFAC World Congress*, Vol.X., Pergamon Press, 288-293, 1987.

[7] Glover, K., Curtain, R.F. and Partington, J.R., Realisation and Approximation of Linear Infinite Dimensional Systems with Error Bounds, *SIAM J. Control and Optimization*, 26, 863-898, 1988.

[8] Glover, K., Lam, J. and Partington, J.R., Balanced Realisation and Hankel-Norm Approximation of Systems Involving Delays, *Proc. I.E.E.E. Conference on Decision and Control, Athens*, 1810-1815, 1986.

[9] Glover, K., Lam, J. and Partington, J.R., Rational approximation of a class of infinite-dimensional systems I: Singular values of Hankel operators, *Math. Control Sig. Sys.*, 3, 325-344, 1990.

[10] Glover, K., Lam, J. and Partington, J.R., Rational approximation of a class of infinite-dimensional systems II: Optimal convergence rates of L_∞ approximants, *Math. Control Sig. Sys.*, to appear.

[11] Glover, K., Lam, J. and Partington, J.R., Rational approximation of a class of infinite-dimensional systems III: The L_2 case, *J. Approx. Theory*, to appear.

[12] Glover, K. and Partington, J.R., Bounds on the achievable accuracy in model reduction, in "*Modelling, Robustness and Sensitivity Reduction in Control Systems*," Ed. R.F. Curtain, Springer, 95-118, 1987.

[13] Gohberg, I.C. and Krein, M.G., "*Introduction to the theory of linear nonselfadjoint operators*," Amer. Math. Soc., 1969.

[14] Lam, J., "*Model Reduction of Delay Systems*," Ph.D. Thesis, University of Cambridge, 1988.

[15] Margenau, H. and Murphy, G.M., "*The mathematics of Physics and Chemistry*," Van Nostrand, 1964.

[16] Olver, F.W.J., "*Asymptotics and special functions*," Academic Press, 1974.

[17] Partington, J.R., "*An introduction to Hankel Operators*," Cambridge University Press, 1988.

[18] Partington, J.R., Glover, K., Zwart, H.J. and Curtain, R.F., L_∞ Approximation and Nuclearity of Delay Systems, *Sys. & Contr. Lett.*, 10, 59-65, 1988.

[19] Reade, J.B., Eigenvalues of smooth kernels, *Math. Proc. Cambridge Philos. Soc.*, 95, 135-140, 1984.

[20] Szegö, G., "*Orthogonal polynomials*," Amer. Math. Soc., 1939.

[21] Wahlberg, B. and Ljung, L., Design variables for bias distribution in transfer function estimation, *IEE Trans. Auto. Cont.*, AC-31, 134-144, 1986.

[22] Zwart, H.J., Curtain, R.F., Partington, J.R. and Glover, K., Partial fraction expansions for delay systems," *Sys. & Contr. Lett.*, 10, 235-244, 1988.

Cost functionals for non-commensurate delay systems

J. E. Marshall K. Walton A. Barnes

Abstract

This paper is concerned with extending results obtained for systems where delays are related integers (commensurate delays) to cases where the delays are 'general'. The principal topic is that of the *calculation* of cost functionals for systems with non-commensurate delays.

1 Introduction

The evaluation of cost functionals of integral of squared error (I.S.E.) and related functions for the case of *delay-free systems* is a classical one. By the use of Parseval's theorem and complex variable techniques it is possible to find expressions for the values of such integrals in terms of parameters of an appropriate Laplace transform. Such integrals are listed for low order cases [1, 2] and iterative methods are given for higher order cases [3].

Even for systems with one delay the corresponding I.S.E. are more difficult to obtain in closed form. The difficulty stems in part from the nature of the roots of expressions such as

$$F(s,h) = A_0(s) + A_1(s)\exp(-sh) = 0$$

The roots are infinite in number and methods based on finding residues and summing the resulting infinite series are technically difficult, even when $A_0(s)$ and $A_1(s)$ are of low order. It is usual to consider cases for which the order, n, of $A_0(s)$ is of higher order than that of $A_1(s)$, which is usually the case in practice.

The use of the Parseval identity, and a technique described as 'direct' enables the integrals to be found in the single delay case, which involves the roots of the finite polynomial [4].

$$W(-s^2) = A(s).A(-s) - C(s)C(-s) = 0.$$

The integral is found by summing the finite number of residues over the roots of this polynomial. This polynomial is of the order n in s^2 [5].

This polynomial is also related to a 'direct' method of stability for time-delay systems [4, 6].

This method for the evaluation of cost functionals for a single delay extends to cases where the delays are related by integrals, for example where $E(s)$, the Laplace transform of the errors is of the form, [7],

$$E(s) = \frac{B_0(s) + B_1(s)\exp(-sh) + B_2(s)\exp(-2sh) + \cdots}{A_0(s) + A_1(s)\exp(-sh) + A_2(s)\exp(-2sh) + \cdots}$$

The polynomial corresponding to W (above) is now of order $n \times m$ in s^2, where n is the order of $A_0(s)$, assumed to be of higher order than the other A terms, and mh is the maximum delay.

2 Independent delays

Consider control systems with two delays. There are many examples of this, for instance, a single delay system with a predictor controller using a delay model. The model delay may well be different from the system delay, either through ignorance of the exact system delay, or may be deliberately chosen to be different as mismatch of the system delay and its model improves performance in some cases, [12].

The exact methods referred to earlier are applicable only to cases where the delays are related by integers, which is unlikely to be so in general. Further, in the case of the delays being related by integers, when these are not small, the resulting polynomial W will be of high order. Though in principle progress can be made, the large number of roots of W means the evaluation of many residues, and the roots will be known numerically, not algebraically. The integers will obviously be large if the delay system and model delays are unequal but close in value.

Due to the large algebraic and computational burden when attempting to find exact cost functionals via commensurate methods and the implication of this when delays are irrationally related, it is essential to seek good approximate methods for which algebraic expressions are possible. The case of two delays close in value suggests an alternative approach, one which combines the features of a low order commensurate method with that of the Padé approximant, which is known to be useful in the context of modelling delays.

Recall that the diagonal entries of a Padé table for $\exp(x)$ may be written as $P_N(-x)/P_N(x)$, where $P_N(x)$ is a polynomial of order N. This approximant agrees with the Taylor expansion of the exponential as far as the term in x^{2N}. When x is purely imaginary the approximant has unity modulus.

We shall replace the transfer function $\exp(-s\Delta)$ of a delay Δ, by its Padé approximant. The gain factor is unity at all frequencies, and the phase term is good for small Δ, approaching $-N\pi$ at large frequencies.

It is a convenient notation to write $\exp(-sh) = Z$, $\exp(-sh_0) = Z_0$.

As an example we choose a prediction scheme with matched sub-plant for which the output $y(t)$, and input $x(t)$ are related by the Laplace transform expression

$$Y(s) = \frac{CGZ}{1 + CG + CG(Z - Z_0)}.X(s),$$

where C is the controller transfer function, G the sub-plant transfer function, $Z = \exp(-sh)$ the plant-delay, and $Z_0 = \exp(-sh_0)$ the delay model transfer functions respectively.

Defining errors as $x(t) - y(t+h)$, a natural definition in this context, we obtain the following expression for the Laplace transfer of error:

$$E(s) = \frac{1 + CG(Z - Z_0)}{1 + CG + CG(Z - Z_0)}.X(s)$$

If h_0 differs from h such that $h_0 = h + \Delta$, a case of overestimation, $Z_0 = Z.\exp(-s\Delta)$.
$\approx Z.\dfrac{P_N(-s)}{P_N(+s)}$.
Then

$$E(s) = X(s).\frac{P_N(s) + CGZ\,(P_N(s) - P_N(-s))}{P_N(s)(1 + CG) + CGZ(P_N(s) - P_N(-s))}.$$

This reduces, after the various Laplace transforms are made explicit to the form

$$E(s, \Delta) = \frac{B_0(s, \Delta) + B_1(s, \Delta)Z}{A_0(s, \Delta) + A_1(s, \Delta)Z}$$

for which the direct method of evaluation may be used to obtain $\int_0^\infty e^2(t)\;dt$ in terms of h, and Δ; $J(h, h_0) = J(h, h + \Delta)$.

For the case of 'underestimation' where Δ would take negative values, implying an unstable Padé model, we write $h = h_0 + \Delta$, and produce a similar expression for $E(h_0, \Delta)$ and hence J in terms of h_0, and Δ, for which $J(h, \Delta)$ may be found simply.

There is an extension of this method to the mismatch of a commensurate case, where for integers p, q, $h = pH$, $h_0 = qH + \Delta$, which with the notation $Z = \exp(-sH)$, and modelling Δ with the diagonal Padé model $\exp(-s\Delta) \approx P_-/P_+$, with dependence on N, and s understood, leads to

$$E(s) = \frac{P_+ + CG\,(Z^p P_+ - Z^q P_-)}{P_+(1 + CG) + CG(Z^p P_+ - Z^q P_-)}X(s)$$

which reduces to a standard commensurate form for $E(s)$. This results in an expression for J in terms of H, and Δ.

It is known that error estimates are difficult to obtain in the context of the Padé approximant; but we may take advantage of knowing J exactly at commensurate values of h_0. For example, $J(h, h)$ in a *strict* Smith prediction control scheme is known from delay free calculations, $J(h, \frac{h}{2})$ may be found from a commensurate calculation.

$J(h, 0)$ is also calculable as it is a single delay calculation.

3 Evaluation of $J_p(h, h_0)$

In the example which we explore in some detail we consider the dependence of J on h and h_0 for a prediction scheme for which $C = K$, constant, $G = \frac{1}{s}$ ($= G_0$, matched), and a unit step input. It is known from simulation, and other studies, [12], that for small h (to be quantified) an underestimation of h, ie $h < h_0$ leads to better performance (lower J) than in the Smith case $h_0 = h$.

We give the calculation of $J_p(h, h_0)$, (using a Padé approximant for h_0), and compare, the graphs with those of $J_p(h, h)$ and $J_p(h, \frac{h}{2})$.

Exact agreement should be obtained when $h_0 = 0$ of $J_p(h, 0)$ with $J(h, 0)$, for which an expression is known.

We are free to choose any value of N in the Padé approximant. In this example we take the simplest choice of $N = 1$ for which $P_1(s) = 1 + \frac{sh_0}{2}$. It is algebraically helpful to write $\frac{h_0}{2} = \alpha$, for now. The expression for $E(s)$ is given by

$$E(s) = \frac{1}{s} \cdot \frac{\alpha s^2 + s(1 + \alpha K) - K + K(1 + \alpha s)Z}{s(\alpha s + 1 + 2\alpha K) + K(1 + \alpha s)Z}$$

equivalent to the single delay case

$$E(s) = \frac{1}{s} \cdot \frac{B(s) + D(s)Z}{A(s) + C(s)Z}.$$

It will be helpful to note that $D(s) = C(s) = K(1 + \alpha s)$, and $A(s) = B(s) + C(s)$.

Suppress the obvious s dependence, and use the relation $A(-s) = \bar{A}$. By Parseval's theorem we have

$$J_p(h, 2\alpha) = \frac{1}{2\pi i} \int_{-i\infty}^{+i\infty} -\frac{1}{s^2} \cdot \frac{B + DZ}{A + CZ} \cdot \frac{\bar{B} + \bar{D}/Z}{\bar{A} + \bar{C}/Z} \cdot ds$$

Now

$$\frac{B + DZ}{A + CZ} = \frac{CZ + A - (A - B)}{CZ + A} = 1 - \frac{C}{CZ + A},$$

and it is found convenient to rewrite the integral as

$$\frac{1}{2\pi i} \int_{-i\alpha}^{+i\alpha} -\frac{1}{s^2} \cdot \left(1 - \frac{C}{CZ + A}\right) \cdot \left(1 - \frac{\bar{C}}{\bar{C}/Z + \bar{A}}\right) ds.$$

Each bracketted term is zero at $s = 0$, so that the contour may be indented to the right. As the calculation for an integral on the right infinite semi-circle is 0 we may close the contour to the right (from which the origin is excluded) so then the integral becomes

$$\frac{1}{2\pi i}\int_{\Gamma R(Clockwise)}\left(+\frac{1}{s^2}\cdot\frac{\bar{C}}{\bar{C}/Z+\bar{A}}-\frac{1}{s^2}\cdot\frac{C\bar{C}}{(CZ+A)(\frac{\bar{C}}{Z}+\bar{A})}\right)ds.$$

The poles within this contour are roots of $\bar{C}+\bar{A}Z=0$; the roots of $CZ+A=0$ lie in the left half plane by the supposition of stability, which is easily verified for some, h, α, K [4]. At these RH roots $Z=\frac{-\bar{C}}{\bar{A}}$ and the second expression becomes

$$-\frac{1}{s^2}\cdot\frac{C\bar{C}}{(A-\frac{C\bar{C}}{\bar{A}})(\bar{C}/Z+\bar{A})}=-\frac{1}{s^2}\frac{(C\bar{C}).\bar{A}}{(A\bar{A}-C\bar{C}).(\bar{C}/Z+\bar{A})}.$$

This introduces new poles, the roots of $A\bar{A}-C\bar{C}=0$ which we denote by W, and is a finite polynomial in s^2. After further indention of the contour, and integrating in the LH semicircle, as we may do [4,7], we find that the contribution from the s terms (which now lie in the contour) is $-h$, and the W terms are given by

$$-\sum_{W=0}^{res}\cdot\frac{1}{s^2}\cdot\frac{C\bar{C}.\bar{A}}{(\bar{A}+\bar{C}/Z).W(s)}$$

where

$$\begin{aligned}W(s) &= \alpha^2s^4-s^2(1+\alpha K)^2-K^2(1-\alpha)^2=0\\ &= \alpha^2(s-i\omega_c)(s+i\omega_c).(s-\frac{K}{\alpha\omega_c}).(s+\frac{K}{\alpha\omega_c});\end{aligned}$$

as the infinite roots of $\bar{A}+\bar{C}/Z$ now lie outside the contour we are left with a summation of a finite number of residues which by obvious symmetry of roots leads (after some algebraic effort) to the expression for J in terms of α, k, and h.

$$J(k,h,\alpha)=-h+$$

$$\frac{K^2(1+\alpha^2\omega^2)}{2\alpha^2\omega^2(\omega^2+\beta^2)}\cdot\frac{(1+2K\alpha+\alpha^2\omega^2)\cos\omega h+2K\alpha^2\omega\sin\omega h}{K(1+\alpha^2\omega^2)+2K\omega^2\cos\omega h-\omega(1+2K\alpha a+\alpha^2\omega^2)\sin\omega h}+$$

$$\frac{K^2(1-\alpha^2\beta^2)}{2\alpha^2\beta^2(\beta^2+\omega^2)}\cdot\frac{(1+2K\alpha-\alpha^2\beta^2)\cos h(\beta h)-2K\alpha^2\beta\sin h(\beta h)}{K(1-\beta^2\omega^2)-2K\alpha^2\beta^2\cos h(\beta h)+\beta(1+2K\alpha-\alpha^2\beta^2)\sin h(\beta h)}$$

where $-\omega^2$, β^2 are the real roots of

$$\alpha^2x^2-x(1+K\alpha)(1+3K\alpha)-K^2=0.$$

This expression is exact if a Padé model is used in the controller. Otherwise it

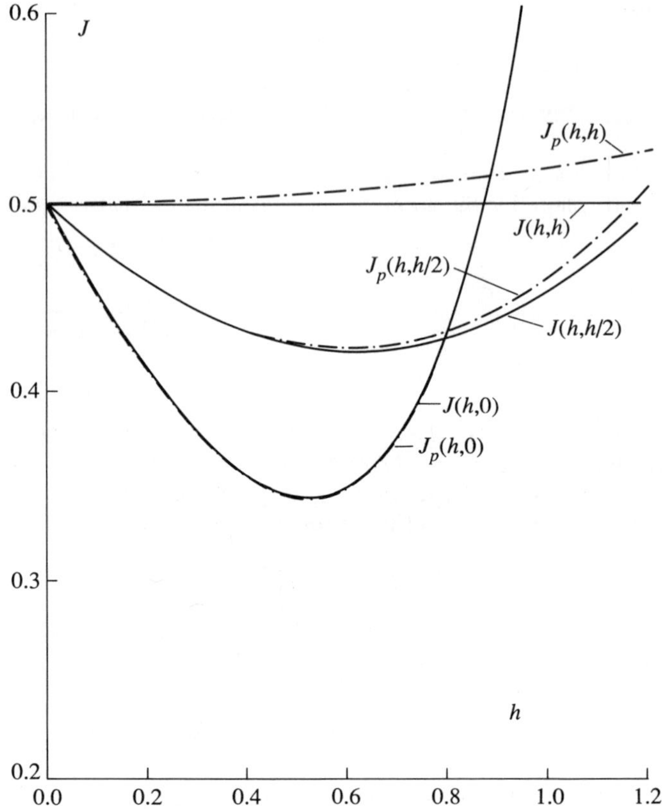

Figure 3.1. Comparison of exact and approximate values of Cost Functionals.

is an approximation to the case where $Z_0 = \exp(-s2\alpha)$. To check the degree of agreement we may use exact expressions for the case where $h_0 (= 2\alpha) = h$, where $h_0 = 0$, and where $h_0 = h/2$.

When $\alpha = 0$ there is exact agreement, which is to be expected. At $h_0 = h$, the strict Smith prediction $J(K, h, h) = \frac{1}{2K}$, constant.

At $h_0 = h/2$, a commensurate calculation [7], $J(K, h, h/2)$ is given by

$$\frac{1}{K}\left\{1 - \frac{hK}{2} + \frac{h^2K^2}{4} - \frac{1}{2}\frac{\cos\frac{(Kh)}{2}}{1 + \sin\frac{(Kh)}{2}}\right\}$$

Cases corresponding to these formulae, when $K = 1$, are given in Figure 3.1.

When we replace *both* delays by Padé expressions, we find the approximation,

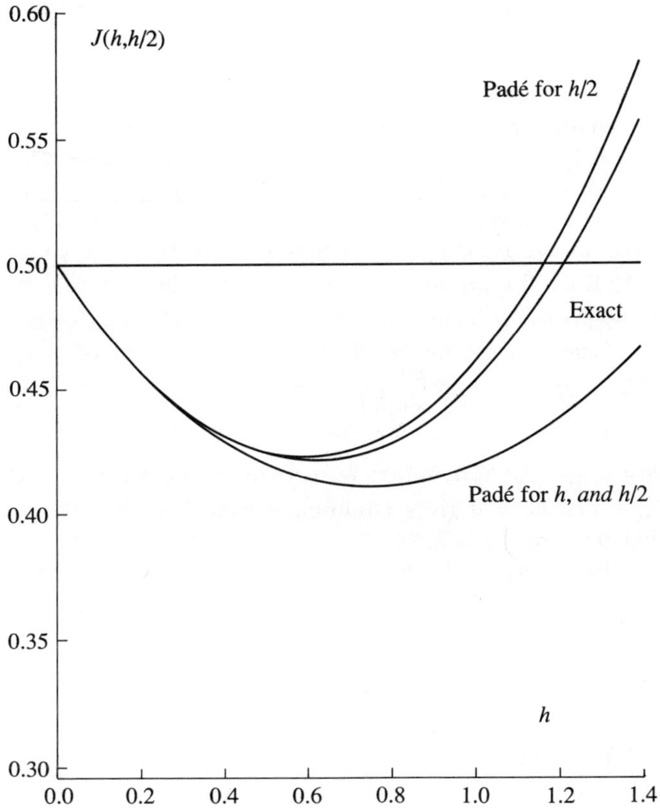

Figure 3.2. Effects of using Padé models for one, and both delay terms in $E(s, \exp(-sh), \exp(-sh_0))$.

at $K = 1$, by delay free methods

$$J_{p,p}(h, h/2) = \frac{24 - 6h + 7h^2 + h^3}{2(24 + 6h + h^2)}$$

Figure 3.2, shows that this gives a much poorer approximation, and is in error at its minimum value by 14%, and the value of $h_{\min}$ is incorrect by 28%.

4 Comments

We note that the Padé approximant applied to the approximate calculation of the cost functional in the example consists of three terms. The first term h would seem to suggest a quicker route to its evaluation. The obvious route where $E(s)$ is *not* expressed as a difference would be expected to yield this term more quickly. Surprisingly, this appears not to be the case.

The second term ("the trig. term"), and the third term ("the hyperbolic term") are obviously related, due to the nature of the roots of $W(s^2)$.

The final term is of the order in α^4, which for a mismatch problem would be small. The difference in the value of J when this term is omitted is negligible for small $\alpha < 1$ in this example.

An expansion of the second term in powers of α using the computer algebra system **REDUCE** [13, 14] produces a series, of which the term in α is $\frac{\partial}{\partial h_0} J_p(K, h, h_0)_{h_0=0}$. This coefficient is identical to that of the first derivative w.r.t. h_0 in the exact case calculated independently. The term independent of α, i.e. $J(K, h, o)$ is given by $J(K, h, o) = \frac{\cos Kh}{2K(1 - \sin Kh)} - h$, which agrees (as it should) with the direct single delay calculation, giving the minimum value of J at $Kh = \frac{\pi}{6}$, when $h_0 = 0$. Equating $\frac{\partial}{\partial h_0}(J_p(K, h, h_0))$ to zero results in a condition for the maximum value of h, for which $h_0 = 0$ gives a minimum value less than the SP values, the condition being that $\cos Kh = Kh$.

REDUCE also confirmed the expected dependence on α of the hyperbolic term . Note that β is inversely proportional to α. Use of the Padé approximant, introduces new roots (of the order of $(\frac{1}{\alpha})$) into the $W(\cdot)$ polynomial, as well as perturbing the others.

5 Conclusions

It is possible to obtain good approximation for perturbations of a delay parameter in a two, or more, delay system, typically a prediction control example.

The extension to systems with mismatched sub-plant, and sub-plants of a higher order is possible, in principle, as well as expansions using higher order (diagonal) Padé approximants.

The increase in algebraic burden is evident, but it is expected that **REDUCE**, used skillfullly, will enable these higher order extensions to be achieved, as well as giving further insights (via Taylor series methods) to a more critical evaluation of principle error terms when Padé approximants are used in this context.

References

[1] Newton, G. C., Gould, L. A., Kaiser, J. F., 1957. *Analytical design of linear feedback controls*, Wiley.

[2] James, H. M., Nichols, N. B., Philips, R. S., 1947. *Theory of Servomechanisms*. M.I.T.

[3] Astrom, K. J., 1970. *Introduction to Stochastic Theory*, Academic Press.

[4] Walton, K., Marshall, J. E., 1987. *Direct method for TDS stability analysis.* Proc. IEE **134**(D), 2, 101-107.

[5] Walton, K., Marshall, J. E., 1985. *Evaluation of time-delay cost functionals,* 4th IMA Int. Conf. on Control Theory. Ed. Cook, P A.

[6] Cooke, K., van den Driesche, P., 1986. *On the zeroes of some transcendental functions.* Funkcialaj Ekvacioj, **29**, 77-90.

[7] Walton, K., Marshall, J. E., 1987. *Evaluation of cost functionals for systems with commensurate delays.* Proc. IEE **134**(D), 2, 108-114.

[8] MacDonald, N., 1987. *An interference effect of independent delays.* Proc. IEE **134**(D), 1, 38-42.

[9] Walton, K., Marshall, J. E., 1987. *Systems with more than one time delay.* IEE Colloquim Digit No.: 1987/115.

[10] Gorecki, H., Korytowski, A., Marshall, J. E., Walton, K. *Integral performance criteria for time delay systems,* Ellis Horwood. (In preparation).

[11] Hydon, P. E., Marshall, J. E., Walton, K. *All-pass feed-back networks.* (This conference).

[12] Hocken, R. D., Salehi, S. V., Marshall, J. E., 1983. *Time-delay mismatch and the performance of prediction control schemes.* Int. J. Control, **38**, (2), 433-447.

[13] Hearn, A. C., 1987. **REDUCE-3** *Users Manual,* version 3.3. Rand Corporation Publications CP78 (7/78).

[14] Davenport, J. H., Siret, Y., Tournier, E., 1988. *Computer Algebra Systems and Algorithms for Algebraic Computation.* Academic Press.

All-pass feed-back systems

P. E. Hydon J. E. Marshall K. Walton

Abstract

Systems with transfer function $F(s)$, such that $F(s)F(-s) = 1$ have the all-pass property that input sinusoids of any frequency are transmitted, in the steady state, unchanged in amplitude.

Stability and a result related to cost-functional evaluation are discussed for systems in which the feed-back transfer function is 'all-pass'.

Results derived in the paper enable comparisons to be made between systems with pure-delay feedback and the corresponding system where the delay is replaced by a Padé model, and also find application in the analysis of time-delay control schemes of the predicator type [9, 10].

The paper explores limits to which finite methods may be used in time-delay system problems. The principal tools used are related to recent direct methods for the analysis of time-delay systems [22, 24].

1 Introduction

The work described in this paper had its origins in the context of the control of time-delay systems. As in other branches of system theory there is a rich interplay between the topics of stability and the evaluation of cost functionals. For example, it is evident that cost functional values, integral of squared error say, will not be defined if the system under investigation in unstable. Recent work in the evaluation of cost functionals for time-delay systems has shown that if there is only one delay, or if delays are related by integers, then closed-form formulae are possible for cost values [21, 23, 25]. However in the independent (non-commensurate) case closed-form formulae are not possible in general and it is necessary to use approximate methods, but with the need to quantify the accuracy [5, 6, 8, 9, 10].

The Padé approximant has been an important tool in the investigation of time-delay systems [4, 8, 9, 20].

We shall be concerned only with Padé approximants to the exponential functions, and in particular the diagonal entries of the Padé table for exp (-sh) [2-4, 14-17].

These diagonal entries are ratios of polynomials of equal order, n, and when expanded agree with the Taylor expansion of exp (-sh) as far as (and including) the term in s^{2n}.

Diagonal entries are chosen so that when s is replaced by $i\omega \ : \ 0 \leq \omega < +\infty$ the modulus of the approximant is unity.

Clearly this all-pass property is shared by the pure delay h, with transfer function exp (-sh). The difference between the delay and approximant is that of phase, which is $\arg \dfrac{P_N(-i\omega)}{P_N(+i\omega)}$ for the Padé approximant $\dfrac{P_N(-s)}{P_N(s)}$ and is of course $-\omega h$ for the pure delay.

The second section of the paper investigates the use of finite methods for the solution of stability problems of time delay systems [7-11]. It is shown that a direct method for time-delay system stability may be employed in the delay-free case, because of the common all-pass property. It is important to show to what extent delay-free methods may be used here, in view of the known difficulties [9]. Examples are given which draw attention to what qualitative features of stability behaviour may be found by Padé methods.

The third section concerns the difference between two cost functionals which may be defined for systems with feedback elements [12]. A formula for this difference which applies to general all-pass feedback systems is applied to pure delay feedback and to the Padé model feedback. Again the results offer insights into the use of finite methods for time-delay system analysis.

2 Stability of time-delay systems

In this section the direct method of Walton and Marshall [25] is described briefly. This recent method is applicable to commensurate as well as single delay systems. In the derivation of the method the property that $\exp(-sh).\exp(+sh) = 1$ is exploited, from which it follows that the method is applicable to all systems, where $F(s).F(-s) = 1$, whether they be time-delay systems or not. It follows that the stability analysis of time-delay problems and systems with Padé all-pass transfer function elements have much in common. With the common properties identified we are in a good position to explain what differences remain, and hence to quantify to what extent Padé models may be used as replacements for delay in stability studies, and in the wider context of control.

2.1 The direct method [25]

Consider the system with characteristic equation

$$A(s) + C(s)\exp(-sh) = 0 \tag{2.1}$$

$A(s)$, $C(s)$ are polynomials with real coefficients of order n, m respectively, where $n > m$. If this equation admits of purely imaginary roots, for some h, they will be in conjugate pairs, so that at these values of ω we have

$$A(-s) + C(-s)\exp(+sh) = 0 \text{ also.} \tag{2.2}$$

Hence the common roots of (2.1) and (2.2) on the imaginary axis satisfy the real polynomial in ω^2

$$W(\omega^2) = A(i\omega)\cdot A(-i\omega) - C(i\omega)\cdot C(-i\omega) = 0. \tag{2.3}$$

The values of ω^2 satisfying (2.3) are found; any negative values, corresponding to complex ω are rejected.

Those values for which ω^2 is positive are used in (2.1) to find corresponding values of h. As h increases through these values the roots of (2.1) pass (in pairs) from, or into, the RH plane. Such values of ω and h are spoken of as stabilising, or destabilising values, respectively.

These values are easily distinguished by the relation [25],

$$\operatorname{sgn}.\frac{\partial Res}{\partial h} = \operatorname{sgn}\frac{dW}{d(\omega^2)} \tag{2.4}$$

evaluated at the corresponding values of ω. Due to the continuity of $W(\omega^2)$ such values of W alternate (unless there be repeated roots, easily checked) so that it is not usually necessary to *calculate* $\frac{dW}{d\omega^2}$.

This information together with the behaviour at $h = 0$, checked by Routh-Hurwitz say, enables the full stability to be assessed.

2.2 Stability windows

The method of 2.1 is now applied to an example for which $C(s) = \sqrt{2}$, and $A(s) = s^2 + s + 2$, i.e., for what values of h do the roots of

$$s^2 + s + 2 + \sqrt{2}\exp(-sh) = 0$$

lie in the LH plane. It follows that $W(\omega^2) = \omega^4 - 3\omega^2 + 2 = (\omega^2 - 2)(\omega^2 - 1)$. Solving for h : $-\omega^2 + i\omega + 2 = -\sqrt{2}(\cos(\omega h) - i\sin(\omega h))$

$$\text{at } \omega = 1 = \omega_1 \text{ say } \omega_1 h_{1,q} = \frac{3\pi}{4} + 2q\pi \tag{2.5}$$

$$\text{at } \omega = \sqrt{2} = w_2 \text{ say } w_2 h_{2,q} = \frac{\pi}{2} + 2q\pi. \tag{2.6}$$

where q is any integer such that the corresponding values of h, $h_{1,q} = \pi(2q + \frac{3}{4})$, $h_{2,q} = \frac{\pi}{2\sqrt{2}}(1 + 4q)$, are positive. There are no RH roots at $h = 0$, by inspection.

As $W(0)$ is positive, it follows that $h_{1,q}$ are stabilising, $h_{2,q}$ destabilising.

It is convenient to display the values of $h_{1,q}$ $h_{2,q}$ in ascending order of magnitude in a simple table as shown below. It is evident that failure to alternate indicates an upper value of delay above which there are always RH roots. The ranges of h for which there is stability are called ***stability windows***. In this example there are three stability windows $0 < h < .35\pi$, $0.75\pi < h < 1.76\pi$ and $2.75\pi < h < 3.18\pi$ only.

$h_{1,q}$	STAB	0		$.75\pi$		2.75π			4.75π
$h_{2,q}$	DESTAB		$.35\pi$		1.76π		3.18π	4.56π	

The numbers of windows has no simple relationship to the degree of $A(s)$, or $C(s)$.

For the example $s^2 + s + b^2 + b\exp(-sh) = 0$, it may be shown that for large b the number of windows is approximately $2b^2 - \frac{b}{\pi}$. (In the example above $b = \sqrt{2}$, and there are 3 windows; when $b = 20$ there are 793 windows!).

These results are surprising at first sight, and emphasise the need for caution in any intuitive approach to time-delay control.

2.3 The corresponding results for the Padé case

Replacing the $\exp(-sh)$ of the previous worked example by $F(s) = (1 - \frac{sh}{2})/(1 + \frac{sh}{2})$, we have

$$A(s) + C(s).F(s) = 0, \text{ and} \tag{2.7}$$

$$A(-s) + C(-s).F(-s) = 0. \tag{2.8}$$

Now

$$F(s).F(-s) = 1 \tag{2.9}$$

so that we have the same polynomial

$$W(\omega^2) = A(i\omega)A(-i\omega) - C(i\omega)C(-i\omega) = 0, \text{ so that}$$

the values of ω^2 are *identical* to those of the first example $\omega_1 = 1, \omega_2 = \sqrt{2}$.

Solving $s^2 + s + 2 + \sqrt{2}(1 - \frac{sh}{2})/(1 - \frac{sh}{2}) = 0$, which is linear in h, we find that $h_1 = 2\sqrt{2} + 1$, $h_2 = \sqrt{2}$. Again no RH root at $h = 0$.

The result, Equn. (2.4) may be shown to be applicable here also.

So we have 2 stability windows, and stability for large h!

$$0 < h < \sqrt{2}, \quad 2\sqrt{2} + 1 < h < +\infty.$$

2.4 Comparisons

From this and similar examples it is seen that stability behaviour can be erroneous at large values of h, which given the origin of the Padé approximation as a Taylor series is not surprising. The number of stability windows when a finite order Padé expansion is used can not exceed the order of the polynomial $A(s)$ (usually assumed in practice to be greater than the order of $C(s)$). The number of windows can only be increased by increasing the order of the Padé approximant, but then only modestly.

In the example we see that the destabilising delay value of the first window is larger in the Padé case. That this will always be so, at least for first order Padé, may be seen by invoking the following: –

As the ω_1 values are the same, the phase shift due to the all-pass network must be the same in each case, as the contribution from $A(s)$ and $C(s)$ is the same, and the total phase-shift is the same (Bode). The pure delay phase shift is $-\omega_1 h$, that for the Padé is $-2\tan^{-1}(\frac{\omega_1 h_p}{2})$. For these to be equal

$$\frac{\omega_1 h}{2} = \tan^{-1}(\frac{\omega_1 h_p}{2}) < \frac{\omega_1 h_p}{2} \tag{2.10}$$

$$\text{so, } h < h_p. \tag{2.11}$$

This result has extended to Padé approximants of higher order.

Also it has been shown that as the order of the Padé approximant increases, the value of h_p approaches that at h, always from above. The convergence is rapid for the first, and arguably the most important stability window.

Results on convergence to the dominant roots of $A(s)+(C(s)\text{esp}(-sh)) = 0$ for a given h have been obtained by numerical studies, and formally [17, 18]. Convergence to the dominant roots is good [19].

3 Cost functionals [1, 12, 13, 20-23]

Consider the feedback system of Figure 3.1, where G has an explicit $\frac{1}{s}$ term, and $F(s).F(-s) = 1$ Define cost functionals : $J_B = \int_0^\infty (x-y)^2\, dt$, and $J_F = \int_0^\infty (x - z)^2\, dt$, both of which find use in parametric optimal control. It has been found that calculation of J_B is usually easier than that of J_F, so that if the value of $J_B - J_F$ may be found then each may be found simply from the other.

It is instructive to consider the case where $F = \exp(-sh)$, for $x(t)$ a step input H(t)

$$J_F = \int_0^\infty (x-z)^2 dt = \int_0^h (x-z)^2 dt + \int_h^\infty (x-z)^2 dt \tag{3.1}$$

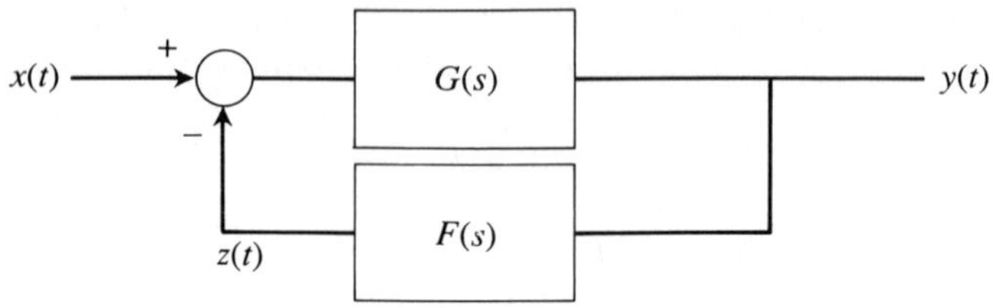

Figure 1.

But $z(t) = 0$ for $t < h$, and $x(t) = x(t-h)$ for $t > h$

$$J_F = \int_0^h x^2 dt + \int_h^\infty (x(t-h) - y(t-h))^2 dt \tag{3.2}$$

Let

$$t - h = \tau;$$

$$\begin{aligned} J_F &= \int_0^h x^2 dt + \int_0^\infty (x(\tau) - y(\tau))^2 d\tau \\ &= h + J_B \end{aligned} \tag{3.3}$$

$$J_B - J_F = -h \tag{3.4}$$

3.1 $J_F - J_B$ for the all-pass case $F(s).F(-s) = 1$

The proof in the delay case depended critically on time delay behaviour ie. that $z(t) = 0$ for $t < h$. For the general $F(s)$ this will not be true.

Again we assume a step input and give the Laplace transforms of the errors. It is convenient to denote by E_B, E_F, X, Y, Z the Laplace transforms of the corresponding quantities, and to denote $G(-s)$ by $\bar{G}$, and $F(-s)$ by $\bar{F}$, etc. Now

$$E_F = X(\frac{1-(1-F)G}{1+FG}),\ E_B = X(\frac{1}{1+FG}) \tag{3.5}$$

By Parseval's theorem

$$J_F = \frac{1}{2\pi i}\int_{-i\infty}^{+i\infty} E_F\bar{E}_F.ds \tag{3.6}$$

$$J_B = \int_{-i\infty}^{+i\infty} E_B.\bar{E}_B.ds \tag{3.7}$$

So that $J_B - J_F = \frac{1}{2\pi i}\int_{-i\infty}^{+i\infty}(E_B\bar{E}_B - E_F\bar{E}_F)ds$, which after some manipulation *and using* $F\bar{F} = 1$ we write as

$$J_F - J_B = -\frac{1}{2\pi i}\int_{-i\infty}^{+i\infty}\frac{1}{s^2}\cdot(\frac{G(1-F)}{1+FG} + \frac{\bar{G}(1-\bar{F})}{1+\bar{F}\bar{G}})ds \tag{3.8}$$

Recall $F(0) = 1$ so that $1 - F(0) = 0$, hence no s^2 contribution. Indent the line from $-i\infty$ to $+i\infty$ to the left of the origin and close the contour to the right for the first product and to the left for the second product to obtain

$$J_F - J_B = -\frac{1}{2\pi i}(\int_{\Gamma_R}\frac{1}{s^2}\cdot\frac{G(1-F)}{1+FG}ds + \int_{\Gamma_R}\frac{1}{s^2}\frac{\bar{G}(1-\bar{F})}{1+\bar{F}\bar{G}}ds). \tag{3.9}$$

where Γ_R, Γ_L are indented to L at origin and are closed by R or L infinite semicircles respectively. The second term has all its poles in RH plane if $1 + FG = 0$ has no RH roots, ie. if stability occurs.

Similarly the LH roots of $1 + FG$ are not included in the first contour so that we have

$$J_F - J_B = \underset{s=0}{\text{res}}\ \frac{1}{s^2}\cdot\frac{G(1-F)}{1+FG} \tag{3.10}$$

i.e.

$$\frac{d}{ds}((1-F)\cdot\frac{1}{\frac{1}{G}+F})_{s=0} = -F'(0)\cdot\frac{1}{F+\frac{1}{G}}|_{s=0} + (1-F(0))\frac{d}{ds}(\frac{1}{F+\frac{1}{G}})$$

$$= -F'(0) \text{ as } F(0) = 1 \text{ and } \lim_{s\to 0}\frac{1}{G} = 0.$$

Therefore

$$J_B - J_F = F'(0) \tag{3.11}$$

Check: when $F(s) = \exp(-sh)$ $F'(0) = -h$, which agrees.

3.2 $J_B - J_F$ when $F(s)$ is Padé all-pass

In this case

$$F(s) = \frac{P_-(sh)}{P_+(sh)} \approx \exp(-sh).$$

Now the nth order diagonal Padé approximant agrees with the Taylor expansion as far as the term is $(sh)^{2n}$
i.e.

$$F(s) = 1 - sh + \frac{(sh)^2}{2!} \cdots$$

so that for all the diagonal Padé approximants to $\exp(-sh)$

$$F'(0) = -h \text{ so that}$$

$$J_B - J_F = -h \tag{3.12}$$

for all Padé cases also.

4 Conclusions

Strengths and weaknesses of the use of the Padé (diagonal entry) approximants have been demonstrated in the context of stability. The minimum value of delay for which a system (which is stable at $h = 0$) exhibits instability may be found by using a Padé approximant. The accuracy of this critical value of h improves rapidly from above, as the Padé order increases. However it is clear that Padé approximants of sensible order can not capture the behaviour for increasing h, even for systems of modest order.

A relationship, helpful in the context of finding cost functional expressions, has been derived which applies to linear systems which have an all-pass feed-back element. Though obvious for the pure-delay case this result also applies to all diagonal Padé models of delay, and to other non-Padé models possessing the property $F(s) \cdot F(-s) = 1$.

Applications of these results are to be found in the context of mismatch of predictor type controllers for time-delay systems.

References

[1] Astrom, K., (1970), *Introduction to stochastic control theory*, Academic Press.

[2] Baker, G. A., Jr. (1975), *Essentials of Padé approximants*, Academic Press.

[3] Baker, G. A., Jr. and Graves-Morris, P. R., *Padé Approximants (Vol I)*, Adison-Wesley.

[4] Fifer, S., (1961), *Analogue Computation (Vol IV)*, McGraw-Hill.

[5] Hocken, R. D. and Marshall, J. E., (1982), *Mismatch and the Optimal Control of Linear Systems with Time-Delays*, Opt. Cont. Applns. & Methods, **3**, 211-218.

[6] Hocken, R. D. and Marshall, J. E., (1983). *The Effects of Mismatch on an Optimal Control Scheme for Linear Systems with Control Time-Delays*, Opt. Cont. Applns. & Methods, **4**, 47-69.

[7] Hocken, R. D., Salehi, S. V. and Marshall, J. E., (1983), *Time-Delay Mismatch and the Performance of Prediction Control Schemes*, Int. J. Control **38**, (2) 433-447.

[8] Landers, P. M., (1983), *Correspondence re Improvement of System Performance by the use of Time-Delay Elements*, IEE Proc., **130**, Pt.D, No. 2, 92.

[9] Marshall, J. E.,(1979), *The Control of Time-Delay systems*, IEE Control Engineering Series, Vol. 10, Peter Peregrinus.

[10] Marshall , J. E., (1987), *Smith Predictors*, in Systems and Control Encyclopedia, Pergamon.

[11] Marshall, J. E. and Salehi, S. V., (1982), *Improvement of System Performance by the use of Time-Delay Elements*, IEE Pt.D, **127**, No 5, 177-181.

[12] Marshall, J. E. and Walton, K., (1985), *A Review of Analytical Methods for control of Time-Delay Systems*, in Fourth IMA international Conference on Control theory, Cambridge.

[13] Newton, G. C., Gould, L. A. and Kaiser, J. F., (1957), *Analytical Design of Linear Feedback Controls*, Wiley.

[14] Padé, H., (1892), *Sur la representation approchee d'une fonction par des fractions rationelles*, Ann. de L'Ecole Normale Sup. 3-ieme Serie, **9**, Suppl., 3-93.

[15] Ralston, A. and Wilf, H. S., (1960), *Mathematical Methods for Digital Computers*, Wiley.

[16] Saff, E. B. and Varga, R. S., (1975), *On the zeros and poles of Padé approximants to e^z*, Num. math., **25**, 1-14.

[17] Saff, E. B. and Varga, R. S., (1977), *On the zeros and poles of Padé approximants to e^z II*, in Saff, E. B. and Varga, R. S. (eds.), *Padé and rational approximation*, Academic Press.

[18] Saff, E. B., Varga, R. S. and Ni, W.-C., (1976), *Geometric convergence of rational approximations to e^z in infinite sectors*, Num. Math., **26**, 211-225.

[19] Takahashi, S., Yamanaka, K. and Yamada, M., (1987), *Detection of dominant poles of systems with time-delay by using Padé approximation*, Int. J. Control, **45**, 251-154.

[20] Walton, K. and Gorecki, H., (1984), *On the evaluation of cost functionals with particular emphasis on time-delay systems*, IMA J. of Mathl. Control and Infn., **I**, (3), 283-306.

[21] Walton, K., Ireland, B. and Marshall, J. E., (1986), *Evaluation of weighted quadratic functionals for time-delay systems*, Int. J. Control, **44**, No. 6, 1491-1498.

[22] Walton, K. and Marshall, J. E., (1984a), *Closed-form solution for time-delay systems cost functionals*, Int.J. Control, **39**, No. 5, 1063-1071.

[23] Walton, K. and Marshall J. E., (1984b), *Mismatch in a predictor control scheme: some closed-form solutions*, Int. J. Control, **40**, 403-419.

[24] Walton, K. and Marshall, J. E., (1987a), *A direct method for TDS stability analysis*, Proc. IEE (D), **134**, (2), 101-107.

[25] Walton, K. and Marshall, J. E., (1987b), *Evaluation of cost functionals for systems with commensurate delays by a finite polynomial method*, Proc. IEE (D), **134**, (2), 108-114.

[26] Walton, K., Marshall, J. E., Gorecki, H. and Korytowski, *Integral Performance Criteria for time-delay systems*, Ellis Horwood (in preparation).

Multivariable controller design for an industrial distillation column

F. Allgöwer J. Raisch

Abstract

In this paper, we report our experience with the design of linear multivariable controolers for a staged binary distillation column with 40 trays. Controller design methods include H_∞-minimization, DNA-design, CL-design and H_2minimization. Results obtained at the real plant are shown.

1 Introduction

Distillation is one of the most important processes in the chemical industries. In distillation columns, a mixture of chemical substances is seperated into products that are more valuable. Control of distillation columns is essential to meet product specifications, to reduce investment and energy costs and to limit environmental impact. These goals can be achieved by maintaining product purities within given bounds.

The column considered in this paper is a staged binary column with 40 trays, which satisfies industrial standards.

In the sequel, a number of linear multivariable controllers and a conventional proportional-integral controller are designed for this column. For multivariable design, we use H_∞- and H_2-minimization method [2] [3] [11], the Characteristic Locus method [4] and the Direct Nyquist Array method [9]. We are especially interested in assessing the effort needed to get a 'good' controller: That is, of course, closely related to the question of how easily engineering specifications can be included in the design process.

This paper is restricted to binary separation. The same methods, however, can be used to control multi-component distillation columns.

2 The plant

2.1 The distillation column

The high-purity distillation plant to be controlled is a staged column for the separation of a binary mixture. It operates continuously and consists of 40 trays, a condenser and a reboiler. The column is in pilot plant scale and satisfies industrial standards (Table 2.1.1).

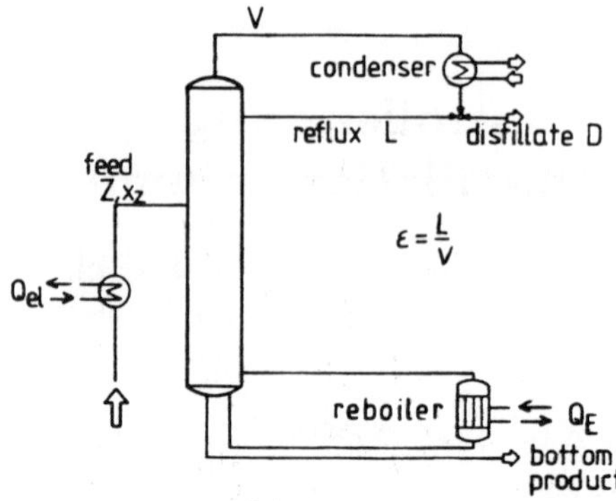

Figure 2.1.1. Schematic representation of the distillation column

The column exhibits most of the problems that appear in industrial binary distillation and is therefore well suited to test controllers under realistic conditions.

As an example, we consider a mixture of two alcohols, methanol and propanol, that is to be separated. Propanol is the high-boiling residue and therefore removed at the base, methanol is the light-boiling distillate and is withdrawn at the column head.

material	glass
trays	40 tubble cap trays, 10cm diameter
height	10cm
reboiler	electricaly heated 0 to 12KW, level controlled
condenser	water cooled, level controlled
pressure	atmospheric
feed	on tray 12 (numbered from the top), electricaly preheated
preheater	with temperature control
heatjacket	12 sectors independently controlled

Table 2.1.1: Technical data of the plant

The fractioning tower is heated from below with the energy flow Q_E, which is used as one manipulated variable. The second manipulated variable is the reflux-ratio ϵ, the ratio between reflux flow L and vapor flow V (Figure 2.1.1). There are temperature sensors on each tray and a process-chromatograph at the head to measure the concentration of one component there.

The column is divided into two parts, namely the stripping section at the base and the enrichment section above it. The mixture to be seperated is fed into the fractioning tower as a continuous stream, the feed Z, between stripping and enrichment section. Before entering the column, the feed is electrically preheated to a constant temperature (Figure 2.1.1).

The most important disturbances that occur in practice are changes in flow Z and feed composition x_z. These two disturbances are considered in this paper.

2.2 The nonlinear model

The dynamic behaviour of the distillation column can be described fairly accurately by a mixed set of differential and algebraic equations.

The nonlinear model is based on the following assumptions:

- neglected hold-up of the vapor phase
- ideal stages, i.e., the mixture is at equilibrium temperature on each tray (Murphree efficiency is introduced to correct errors resulting from this assumption)
- hydrodynamical effects are neglected
- the volume of the liquid on each stage is considered to be a constant

Material balances, energy balances and thermodynamical correlations for each stage then yield a set of 320 nonlinear differential and algebraic equations.

To give a flavour of the severe nonlinearities, the calculation of the boiling-point equilibrium is shown as an example:

$$y_{ik} = y_{i\,k+1} + E(y^*_{ik} - y_{i\,k+1}) \tag{2.2.1}$$

where

$$y^*_i = \frac{p_{is}(T)}{p}\gamma_i x_i, \quad i = 1, 2 \tag{2.2.2}$$

$$\sum_{i=1}^{2} y^*_i = 1 \tag{2.2.3}$$

The activity coefficients γ_i are due to Wilson

$$\ln \gamma_i = -\ln \sum_{j=1}^{2} x_j \Gamma_{ij} + 1 - \sum_{j=1}^{2} \frac{x_j \Gamma_{ji}}{\sum_{l=1}^{2} x_l \Gamma_{jl}}, \quad i = 1, 2 \tag{2.2.4}$$

and the saturation vapor pressure $p_{is}(T)$ is due to Antoine

$$p_{is}(T) = e^{A_i - \frac{B_i}{T + C_i}}, \quad i = 1, 2. \tag{2.2.5}$$

where:

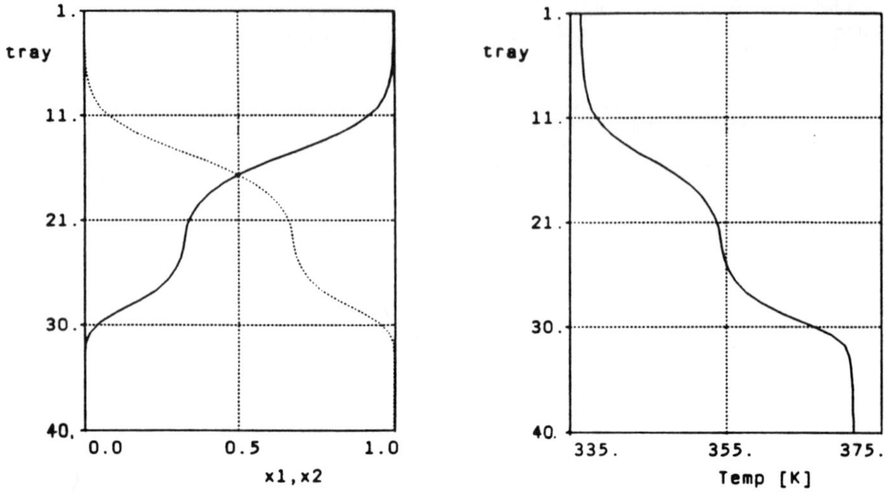

Figure 2.3.1. Concentration and temperature profiles of the stationary column

y_i ⋯ vapor concentration of the ith component
y_i^* ⋯ vapor concentration of the ith component at eqilibrium
p ⋯ total pressure
x_i ⋯ liquid concentration of the ith component
T ⋯ temperature in Kelvin
Γ_{nm} ⋯ Wilson-parameter
$\left.\begin{matrix} A_i \\ B_i \\ C_i \end{matrix}\right\}$ ⋯ Antoine constants

Models of the reboiler and the condenser consist of another 7, respectively 5, nonlinear equations. Thus a set of 332 nonlinear algebraic-differential equations describes the dynamic behaviour.

2.3 Nonlinear simulation

Based on the nonlinear dynamic equations referred to in Chapter 2.2, a rather accurate nonlinear simulation of the column is possible. The software package DIVA (**D**ynamicische **SI**mulation **V**erfahrenstechnischer **A**nlagen [5]) was used for simulation.

Dynamic behaviour of the uncontrolled plant can be understood best from concentration- and temperature-profiles along the vertical axis of the column. Figure 2.3.1 shows the profiles of the stationary column.

In binary separation, the temperature on each stage is 1-to-1 to the concentration on that stage. Therefore, considering only temperature profiles and responses is

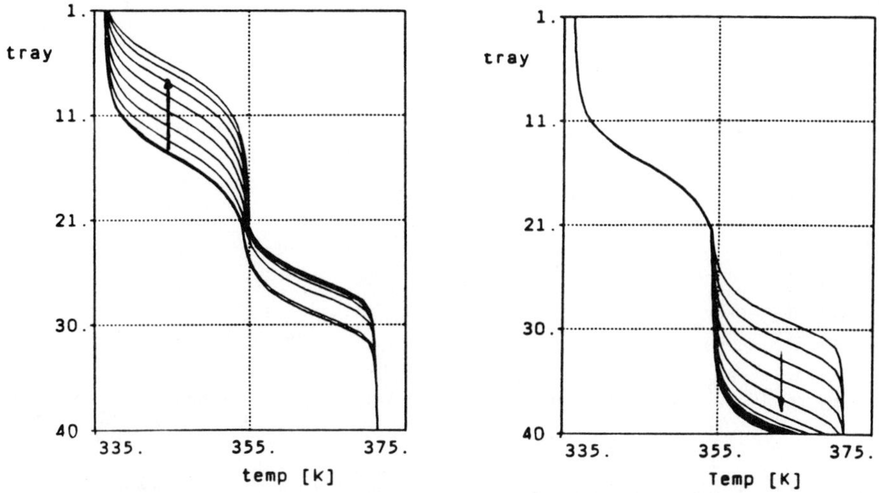

Figure 2.3.2. Responses to step changes of the reflux-ratio by -1% and of the feed flow by +20% (temperature-profiles)

sufficient. Qualitative behaviour of the column due to changes in the manipulated variable (Q_E, ϵ) and changes in the disturbances (feed flow and feed composition) can be described by a movement of the entire profile either towards the head or towards the bottom of the column (plus slight changes in the shape of the profile) [8].

As an example, the effects of a -1% step change of the reflux-ratio and a +20% step change of the feed flow are shown in Figure 2.3.2. A good controller should keep the profile nearly stationary in the presence of disturbances. It is obvious that the product specifications, namely high purity at the top and bottom of the column, are met this way.

If the profile is nearly stationary, we observe largest temperature variations on the 14th and 28th stage. Therefore, temperatures on these stages, which are approximately located at the points of inflection of the profile, are taken as measurements (T_{14}, T_{28}). Figure 2.3.3 shows temperature time responses on these stages corresponding to Figure 2.3.2. (Throughout the paper, temperature is measured in Kelvin K, energy flow Q_E in Kilowatt [KW]).

Additional temperature measurements on other stages improve information about process dynamics only slightly. This is especially true for temperature sensors at the top and bottom stages, which would give direct knowledge of distillate and bottom concentration theoretically. [6]

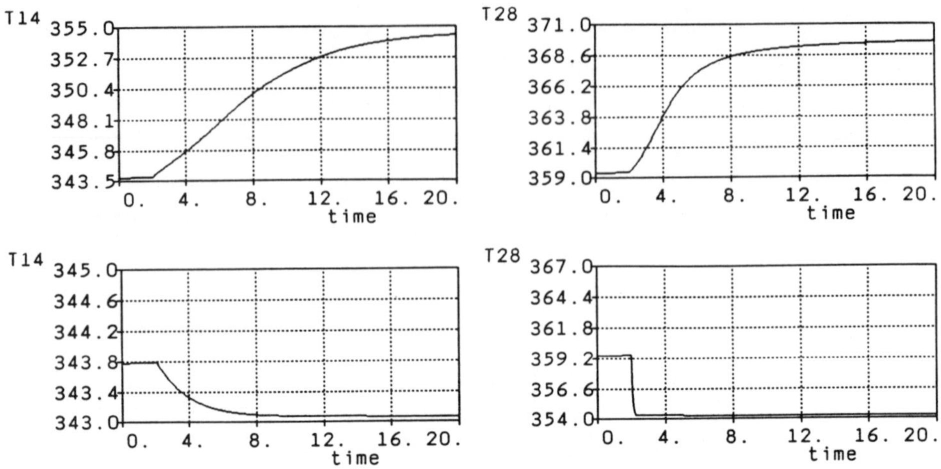

Figure 2.3.3. Temperature time responses on the 14th and 28th tray to a -1% step change of the reflux-ratio and a +20% step change of the feed flow

2.4 Plant specifications

By inspection of nonlinear simulation results, sufficient purity of bottom product and distillate is guaranteed, if temperature variations from the set points $T_{14}^{stat} = 343.8K$ and $T_{28}^{stat} = 359.4K$ are restricted to certain bounds $T_{14}^{\min}, T_{14}^{\max}$ and $T_{28}^{\min}$, $T_{28}^{\max}$. Set point changes are not of practical importance and therefore not considered here.

In simulation, the appropriate steady-state value of the energy flow Q_E is 2.48KW. Maximal and minimal admissible values are $Q_E^{\max} = 12$KW and $Q_E^{\min} = 2$KW. Additionally, changes in Q_E should not be too sudden, because the energy flow can be affected only indirectly (lag-element with a time constant of about 3 minutes).

Due to actuator saturation only reflux-ratios between $\epsilon^{\min} = 0.16$ and $\epsilon^{\max} = 0.65$ are acceptable. In simulation, the steady-state reflux-ratio is $\epsilon^{stat} = 0.5355$.

Usually, the feed into the distillation column is the output of a superposed process unit, e.g., a chemical reactor. Changes in its operating condition directly affect the column feed. It is assumed that disturbances in the feed flow are limited to $\pm 20\%$ of the steady-state value of 14.08l/h. Disturbances in the feed concentration are assumed to vary by a maximum of $\pm 15\%$ of the stationary value of $x_{methanol}^{stat} = 0.32$. Both disturbances may very well occur as steps.

2.5 Operation of the plant

Industrial distillation columns are complex chemical processes. For example, there is not only the reflux-ratio and the energy-flow to influence but also the amount of distillate taken off, the amount of bottom product taken off, the coolant flow for the condenser, the preheating of the feed and a lot more. To operate such a system is much more than just to control concentration purity. A complex process-operating-system, here a PLS80, is necessary.

The controllers considered in this paper were implemented on a MicroVAX II that is directly connected to the process-operating-system. All communication between plant, controller and operator is realized via the process-operating-system, which also supervises the 20 underlying control loops.

2.6 The linear model

Controller design based on the complex nonlinear model mentioned in Chapter 2.2 is hardly possible. One way to derive a controller is to use a reduced nonlinear model that exhibits the essential dynamics of the process [8] [6]. Often, these model-based design methods are only suited for a special kind of separation and require a lot of effort.

Alternatively, a controller based on a linear model can be designed. Because of the severe nonlinearities (for the same operating conditions, plant gain and dominant time constants are highly dependent on sign and size of step changes), the use of a high-order linear model yields hardly an improvement compared to a low-order model. A very simple model exhibits the essential plant dynamics well enough for control purposes.

The linear model was identified from nonlinear simulations, because we did not have access to the real distillation column until the controllers were implemented. Temperature responses on stages 14 and 28 to steps in the manipulated variable yield:

$$\underline{Y}(s) = \underbrace{\begin{pmatrix} -\frac{1220}{6.1s+1} & \frac{100}{(1.2s+1)(1.5s+1)} \\ -\frac{1500}{2.7s+1} & \frac{320}{1.6s+1} \end{pmatrix}}_{G(s)} \underline{U}(s) \tag{2.6.1}$$

where

$$Y(1) \quad \cdots \quad T_{14} - T_{14}^{stat} \text{ (in Kelvin)}$$

$$Y(2) \quad \cdots \quad T_{28} - T_{28}^{stat} \text{(in Kelvin)}$$

$$U(1) \quad \cdots \quad \epsilon - \epsilon^{stat}$$

$$U(2) \quad \cdots \quad Q_E - Q_E^{stat} \text{(in Kilowatt)}$$

Time is referenced in hours.

As gain and time constants of each transmission channel vary by up to 300% depending on the size and sign of the respective identification step input, roboustness should be a main goal of linear controller design.

3 Controller design

Several linear controller design methods were applied to the model $G(s)$ (eqn. 2.6.1). The resulting compensators have to meet the following design specifications:

- actuator limits (see Section 2.4)
- performance requirements (see Section 2.4)
- to achieve asymptotic rejection of disturbance steps, we require the largest singular value of the sensitivity matrix S to approach zero for small frequencies:

$$\bar{\sigma}[S] = \bar{\sigma}[(I + GK)^{-1}] \rightarrow 0 \text{ for } \omega \rightarrow 0 \tag{3.1}$$

- we require the closed loop bandwidth ω_B (frequency, where the maximum singular value of the complementary sensitivity matrix T crosses the -3dB line) to lie between 40 (hour)$^{-1}$ and 50 (hour)$^{-1}$:

$$\bar{\sigma}[T] = \bar{\sigma}[(I + GK)^{-1}GK] < -3\text{dB for } \omega > \omega_B \tag{3.2}$$

  Reasons for this are:

  - Suppression of measurement noise.
  - Process dynamics with frequencies faster than ω_B is completely neglected in our model; multiplicative modelling errors tend to infinity for $\omega > \omega_B$. Thus, in a conservative design like ours, $\bar{\sigma}[T]$ has to drop for frequencies above ω_B to ensure closed loop stability [1].

As usual, the process of controller design consists of four steps:

- linear design
- singular value analysis to check requirements (3.1) and (3.2)
- nonlinear simulation to check performance specifications and non-violation of actuator limits
- implementation of the controller at the real plant and – if necessary – on-line tuning.

For the first two steps we used the design packages STABLE-II [7] and KEDDC [10]. Nonlinear simulation was performed with the 42nd order model described in Section 2.3 (software package DIVA [5]).

3.1 H_∞-Design

The cost function to be minimized over the set of all finite-dimensional, proper and stabilizing compensator matrices $K(s)$ is given by

$$||H(s)||_\infty = \max_\omega \bar{\sigma}[H(j\omega)] \tag{3.1.1}$$

where H is chosen as

$$H = \begin{pmatrix} W_1(s)S(s)W_2(s) \\ W_3(s)T(s)W_2(s) \\ W_4(s)K(s)S(s)W_2(s) \end{pmatrix} \tag{3.1.2}$$

and the weighting matrices W_2, W_4 are:

$$W_2(s) = diag\{0,1\} \tag{3.1.3}$$

$$W_4(s) = diag\{0.01\} \tag{3.1.4}$$

$W_1(s)$ and $W_3(s)$ are diagonal matrices with minimum-phase entries depicted in Figure 3.1.1. This choice of weighting matrices reflects requirements (3.1) and (3.2). The entries of W_4 were chosen very small, because we did not encounter any difficulties with actuator limits either in linear or nonlinear simulation. The matrices W_i, $i = 1, \cdots, 4$ are a reasonable guess; thus, we did not need many iterations to find these particular weights. The resulting compensator is of order 8 and has three poles near $-\infty$. This very fast part of the compensator was approximated by nondynamic elements. Thus, we get a 5th order controller with two stable poles close to the origin, a pair of poles at $(-0.245 \pm j0.383)$ $(\text{hour})^{-1}$ and a pole at -0.617 $(\text{hour})^{-1}$. Minimum and maximum singular values of K, S and T are shown in Figures 3.1.2 and 3.1.3.

Nonlinear simulation of closed loop time responses to step changes in disturbances was done. Feed flow was varied by $\pm 20\%$, feed concentration of component 1 by $\pm 15\%$. Because of lack of space, only two responses can be presented here (Figure 3.1.4 and Figure 3.1.5)). As a step change in feed flow by $+20\%$ proved to be hardest to control, only responses of the controlled plant to this disturbance are given here (Figure 3.1.6). The controlled plant exhibited a steady-state temperature offset on the 28th tray of approximately 0.4 Kelvin. This is due to a steady decline in the feed concentration of the light-boiling component (ramp disturbance). It occurs, because in our experiments, distillate and bottom product are mixed and reused as feed. It is not anticipated that this will be a problem in industrial use. Thus, redesign was not deemed to be necessary.

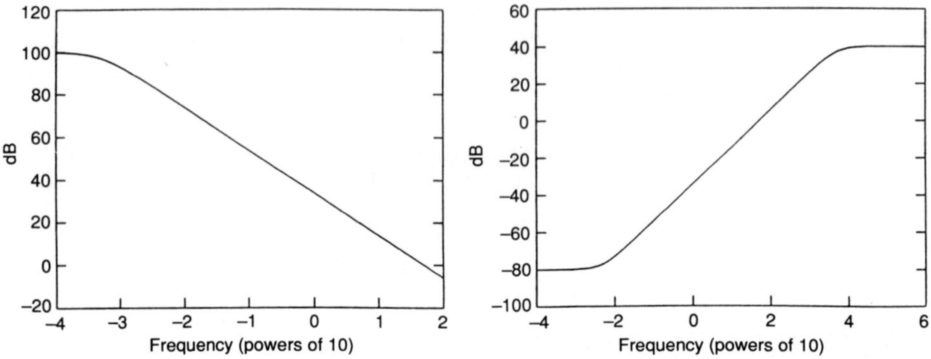

Figure 3.1.1. Elements of weighting matrices W_1 and W_3

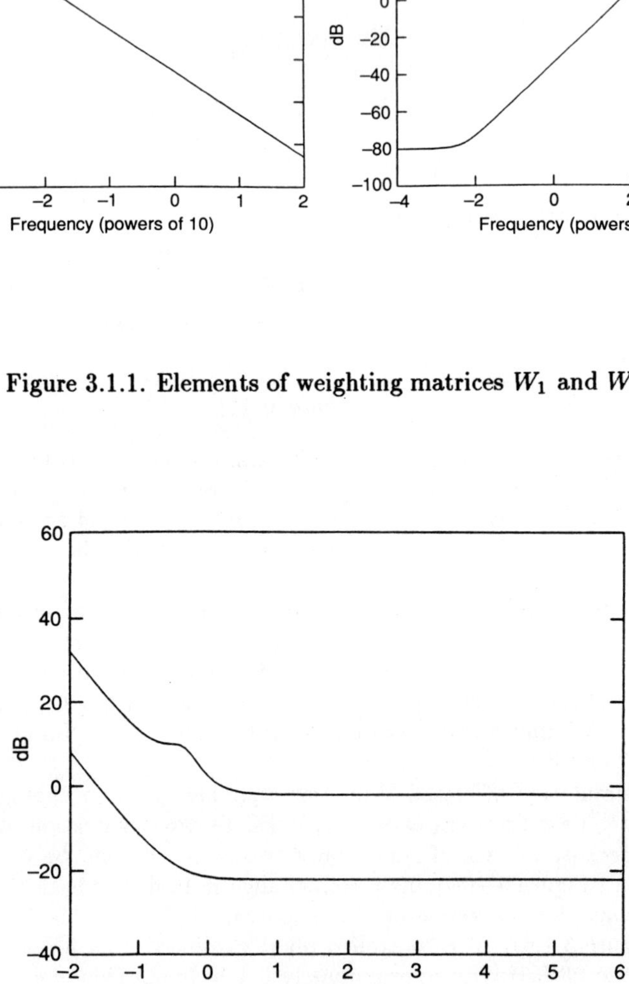

Figure 3.1.2. Minimum and maximum singular value of K

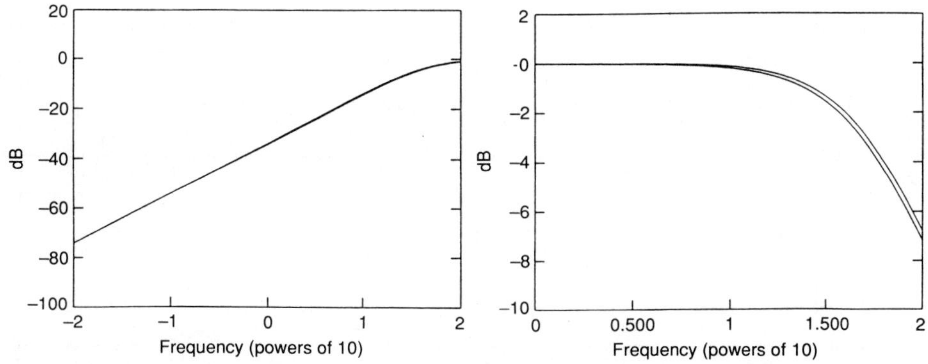

Figure 3.1.3. Minimum and maximum singular values of S and T

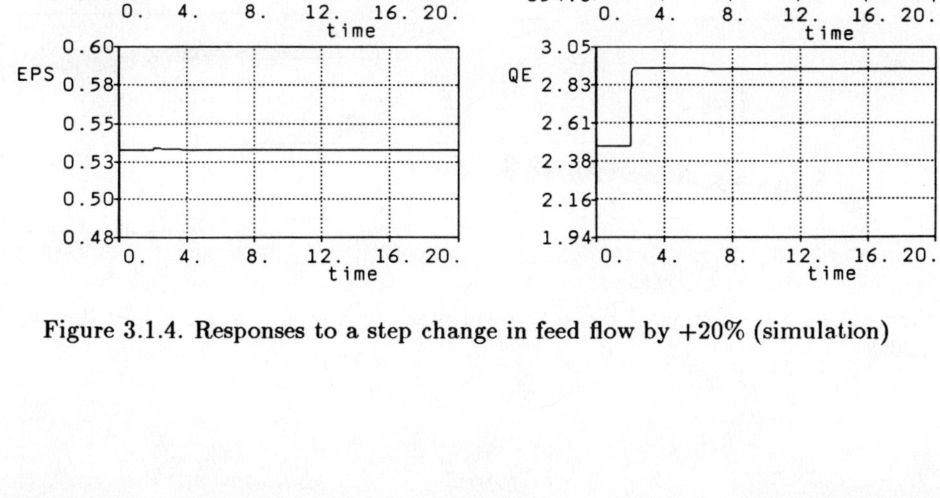

Figure 3.1.4. Responses to a step change in feed flow by +20% (simulation)

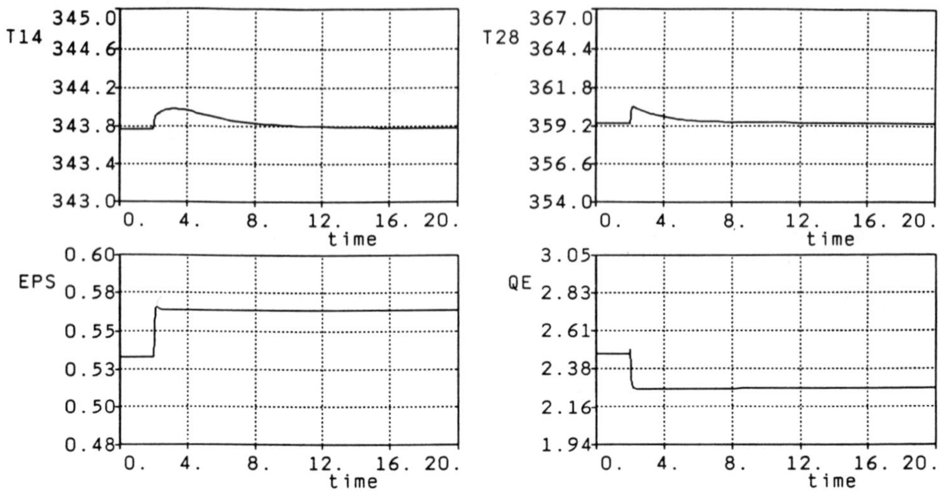

Figure 3.1.5. Responses to a step change in the feed concentration of component 1 by - 15% (simulation)

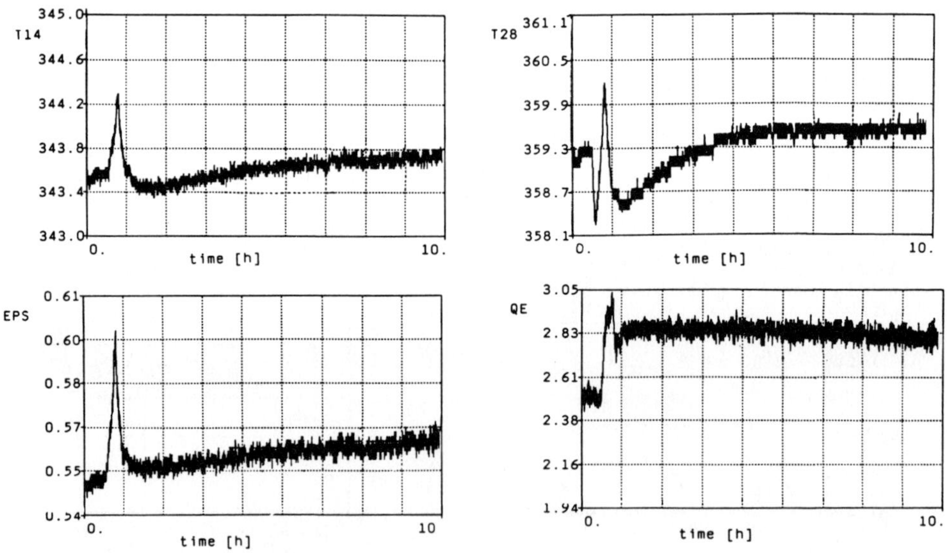

Figure 3.1.6. Responses of the controlled plant to a step change in feed flow by +20%

3.2 Characteristic Locus design method [4]

To achieve satisfactory closed loop performance, the characteristic locus design method relies on the manipulation of the eigenvalues (characteristic loci) and eigenvectors (characteristic directions) of the open loop frequency response matrix $G(j\omega)K(j\omega)$. The design proceeds in three stages:

- High-frequency alignment

  At high frequencies, the characteristic directions of GK should approach the standard basis direction set. This will guarantee small high-frequency interaction between (linear) control loops. Moreover, in this frequency range, the moduli of the closed loop charateristic loci will be good approximations for the singular values of the complementary sensitivity matrix T. Good alignment is achieved by postmultiplying G with

$$K_H = \begin{pmatrix} -0.137 & 0.0 \\ -0.376 & 0.4 \end{pmatrix} \tag{3.2.1}$$

  K_H is obtained as a real approximation of the frame matrix of $G^{-1}(j\omega_H)$, $\omega_H = 30(\text{hour})^{-1}$. K_H also contains appropriate scaling factors for the characteristic loci of G in this frequency range. Success of the alignment procedure can be judged from Figure 3.2.1, where the moduli of the characteristic loci and the misalignment angles between characteristic directions and standard basis vectors are shown.

- Medium-frequency adjustment

  As ω_B is well within the region of small alignment angles, closeness of open loop characteristic loci at this frequency is equivalent to closeness of $\underline{\sigma}[T(\omega_B)]$ and $\bar{\sigma}[T(\omega_B)]$. This is achieved fairly (but not very) well through an additional compensator

$$K_M(s) = A \begin{pmatrix} k_M(s) & 0.0 \\ 0.0 & 1.0 \end{pmatrix} A^{-1} \tag{3.2.2}$$

  where A is a real approximation to the eigenvector frame of $G(j\omega_M)K_H, \omega_M = 20\ (\text{hour})^{-1}$ and k_M is a first-order lag with corner frequency 10 $(\text{hour})^{-1}$ and attenuation 9.5dB. The result of this loci adjustment procedure can be checked from the Nyquist plots in Figure 3.2.2.

- Low-frequency compensation

  At this stage, we inject additional low-frequency gain into the loops and balance up the moduli of the characteristic loci in the low-frequency range. This is accomplished by an additional compensator

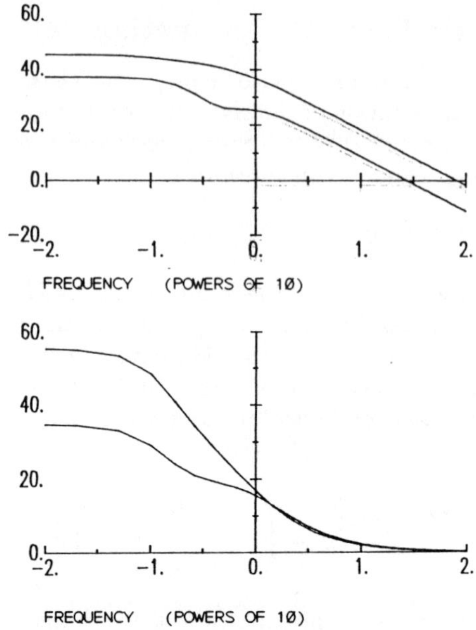

Figure 3.2.1. Moduli of characteristic loci and misalignment angles after high-frequency compensation

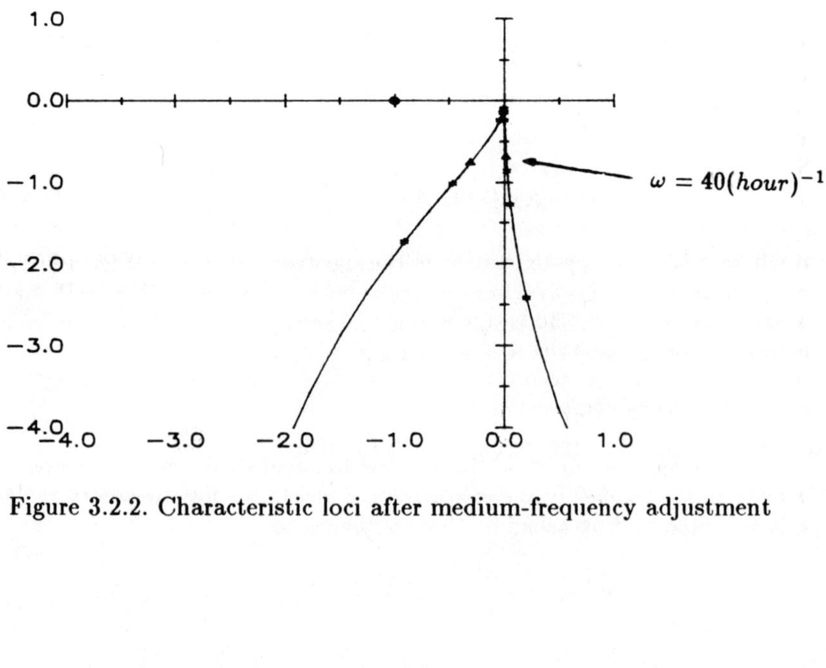

Figure 3.2.2. Characteristic loci after medium-frequency adjustment

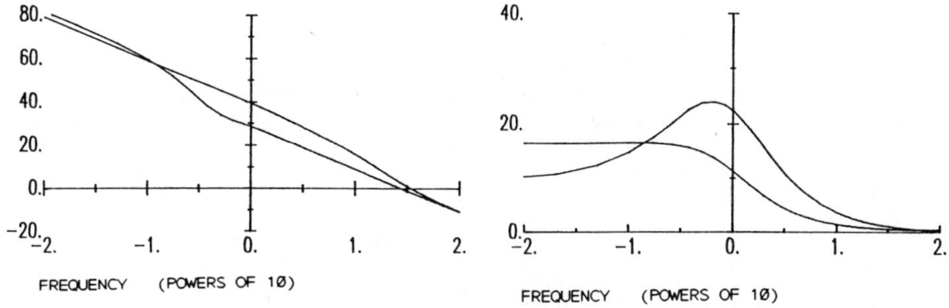

Figure 3.2.3. Moduli of characteristic loci and misalignment angles of GK

$$K_L(s) = \frac{1}{s} B \begin{pmatrix} 1.5 & 0.0 \\ 0.0 & 0.5 \end{pmatrix} B^{-1} + I \tag{3.2.3}$$

where B is a real approximation to the eigenvector frame of $GK_H K_M$ at $\omega_L = 0.1\ (\text{hour})^{-1}$.

Thus, the result of this design is a third order controller with two poles at the origin and one pole at $s = -10\ (\text{hour})^{-1}$. Moduli of the open loop characteristic loci and misalignment plots are shown in Figure 3.2.3. Minimum and maximum singular values of S and T are given in Figure 3.2.4.

Nonlinear simulation of closed loop time responses and real responses of the controlled plant to a step change in feed flow are presented in Figures 3.2.5. and 3.2.6. Due to a software error, data acquisition was stopped too early. The controlled plant exhibited a steady-state temperature offset on the 28th tray of approximately 0.5 K. For the same reasons as in Section 3.1, this is not considered to be a problem. Thus, additional integral action was not included in the controller. Steady states of the manipulated variables are different from those in Figure 3.1.6. This is also due to the decline in the feed concentration of the light-boiling component.

3.3 Direct Nyquist array design method [9]

In DNA design, the multivariable problem is reduced to a number of scalar problems by introducing a suitable 'nearly decoupling' precompensator K_P. Especially,

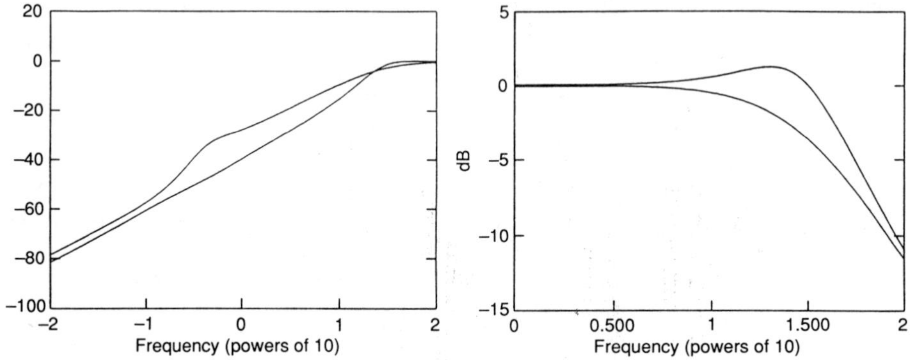

Figure 3.2.4. Minimum and maximum singular values of S and T

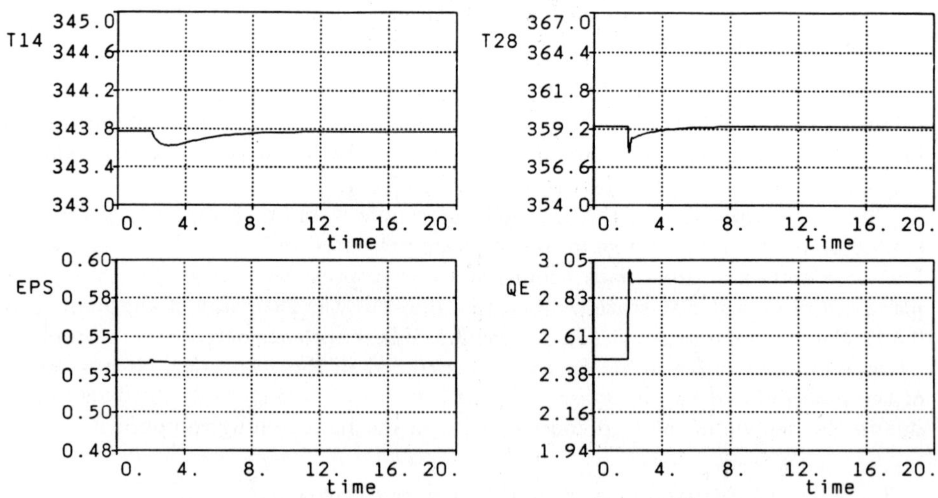

Figure 3.2.5. Responses to a step change in feed flow by +20% (simulation)

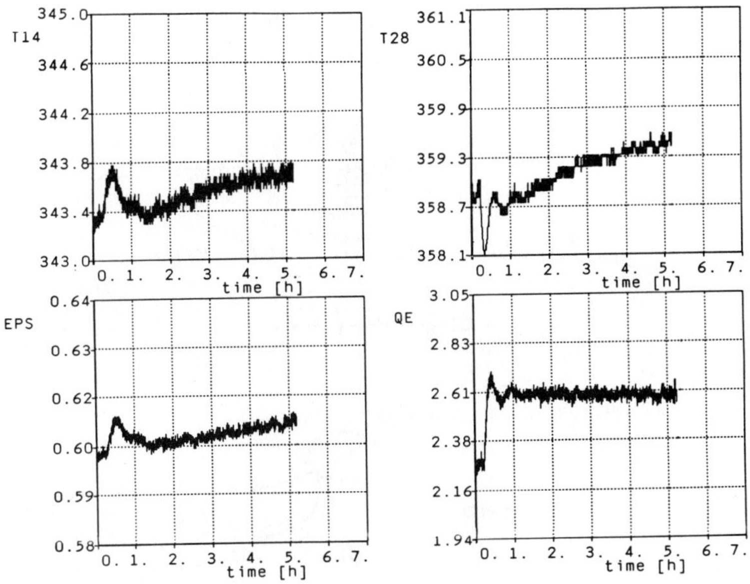

Figure 3.2.6. Responses of the controlled plant to a step change in feed flow by +20%

if $G(j\omega)K_P(j\omega)$ is column diagonal dominant, two scalar controllers $k_{C_1}(s)$ and k_{C_2} stabilizing their individual loops will guarantee that

$$K_C(s) = \begin{pmatrix} k_{C_1}(s) & 0.0 \\ 0.0 & k_{C_2} \end{pmatrix} \tag{3.3.1}$$

stabilizes the multivariable control system.

The precompensator matrix $K_P(s)$ was chosen as

$$K_P(s) = \begin{pmatrix} -\dfrac{0.135 + 0.365s}{0.625 + s} & 1 \\ -1 & 12.17 \end{pmatrix} \tag{3.3.2}$$

From Figure 3.3.1 it is clear that the uncompensated system is not column diagonal dominant, whereas for the compensated system diagonal dominance is achieved (Figure 3.3.2). Thus, scalar controller design for the two diagonal elements of $G(s)K_P(s)$ is justified. For both systems, we choose PI controllers

$$K_{C_1}(s) = 0.562\frac{1+s}{s} \tag{3.3.3}$$

$$K_{C_2}(s) = 0.013\frac{1+s}{s} \tag{3.3.4}$$

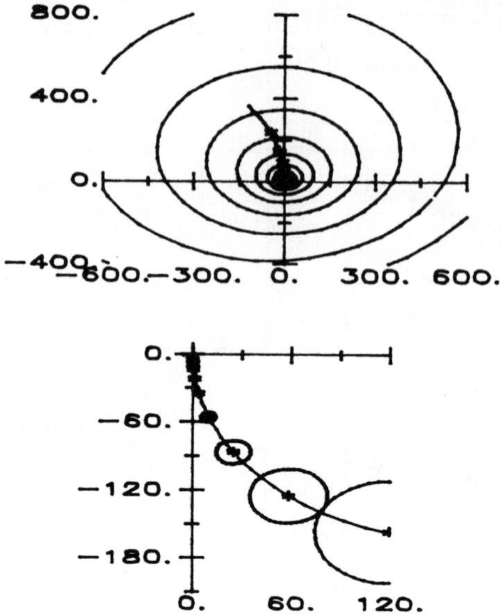

Figure 3.3.1. Gershgorin bands of uncompensated system

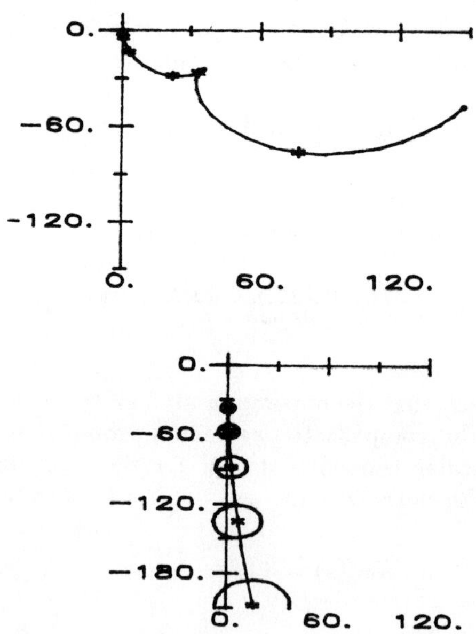

Figure 3.3.2. Gershgorin bands of compensated system

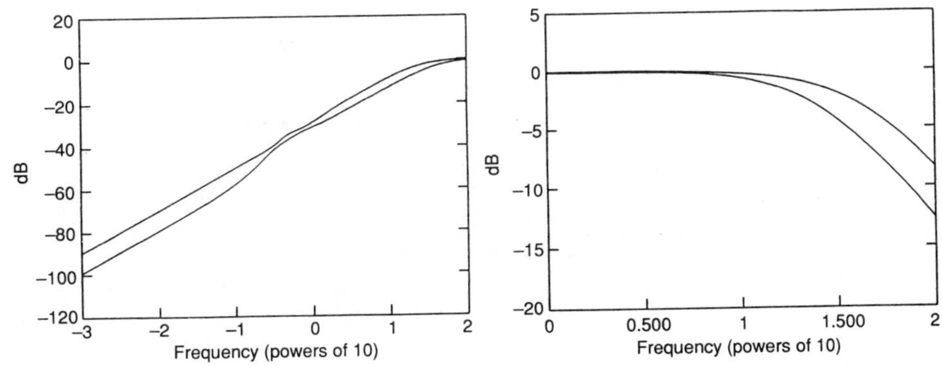

Figure 3.3.3. Minimum and maximum singular values of S and T

Apart from introducing integral action, they approximately balance up open loop gains.

Minimum and maximum singular values of S and T are given in Figure 3.3.3.

Nonlinear simulation of closed loop time responses and real responses of the controlled plant to a step change in feed flow are presented in Figures 3.3.4 and 3.3.5.

3.4 H_2-Minimization [11]

The cost function to be minimized is given by

$$\|H\|_2 = \sqrt{\frac{1}{2\pi}\int_{-\infty}^{+\infty} tr[H^*(j\omega)H(j\omega)]\, d\omega} \tag{3.4.1}$$

and H is the same as in Chapter 3.1. The resulting compensator is of order 8 and has two poles near $-\infty$. Those very fast modes were approximated by nondynamic elements. Thus, we end up with a 6th order controller with two poles close to the origin, a pair of poles at $(-0.245 \pm j0.383)(\text{hour})^{-1}$ and two poles at $-55.2\ (\text{hour})^{-1}$.

Maximum and minimum singular values of the H_2-controller matrix in Figure 3.4.1 should be compared to those of the H_∞-controller. Maximum and minimum singular values of S and T are given in Figure 3.4.2.

Unfortunately, we did not have the opportunity to implement this controller at the plant. Thus, only simulation results are given (Figure 3.4.3).

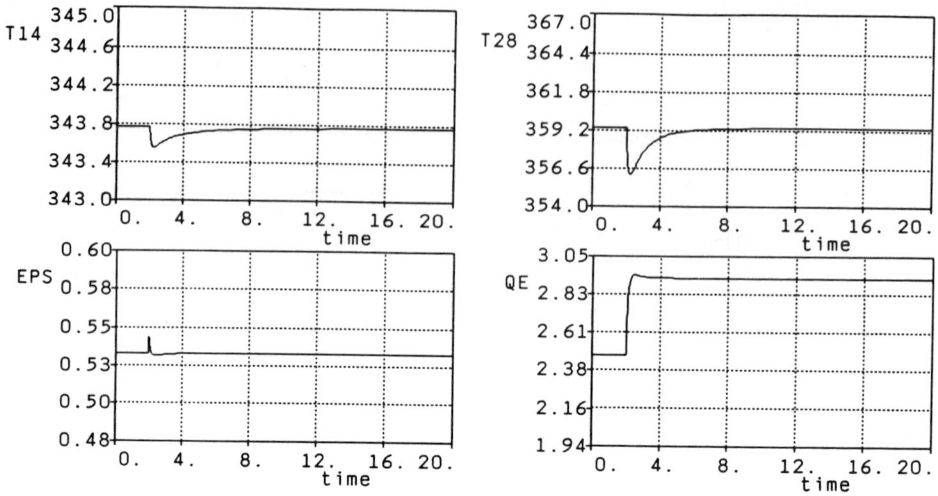

Figure 3.3.4. Responses to a step change in feed flow by +20% (simulation)

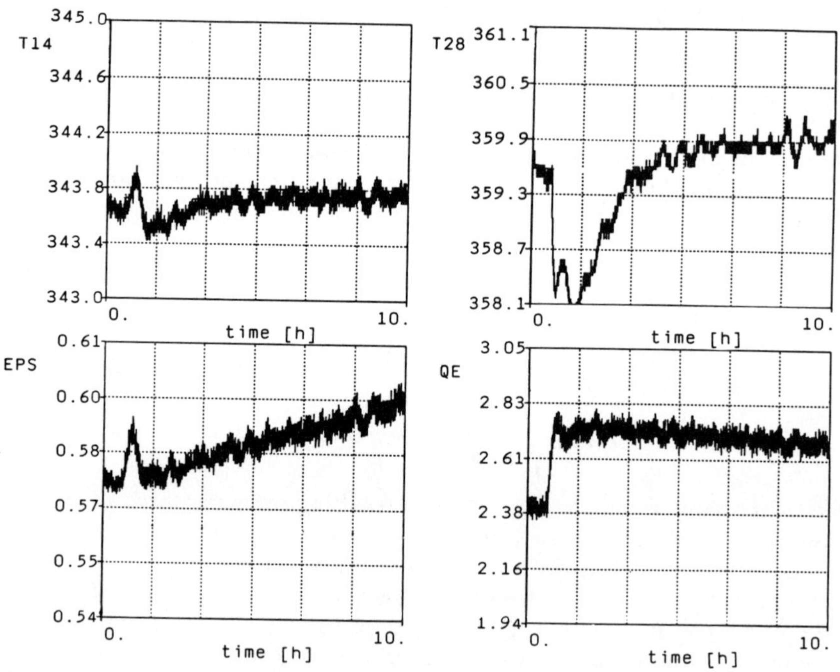

Figure 3.3.5. Responses of the controlled plant to a step change in feed flow by +20%

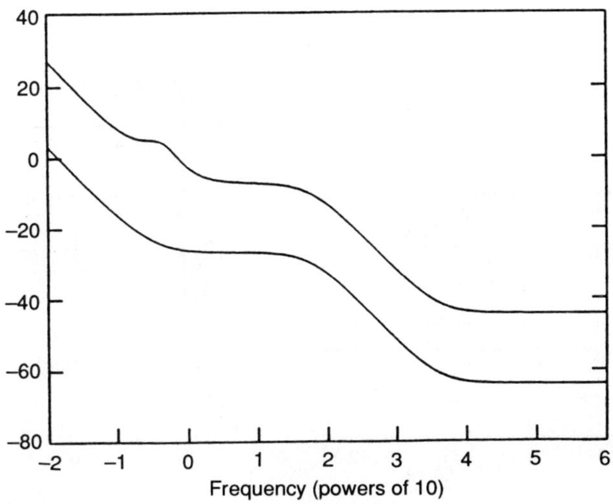

Figure 3.4.1. Minimum and maximum singular value of K

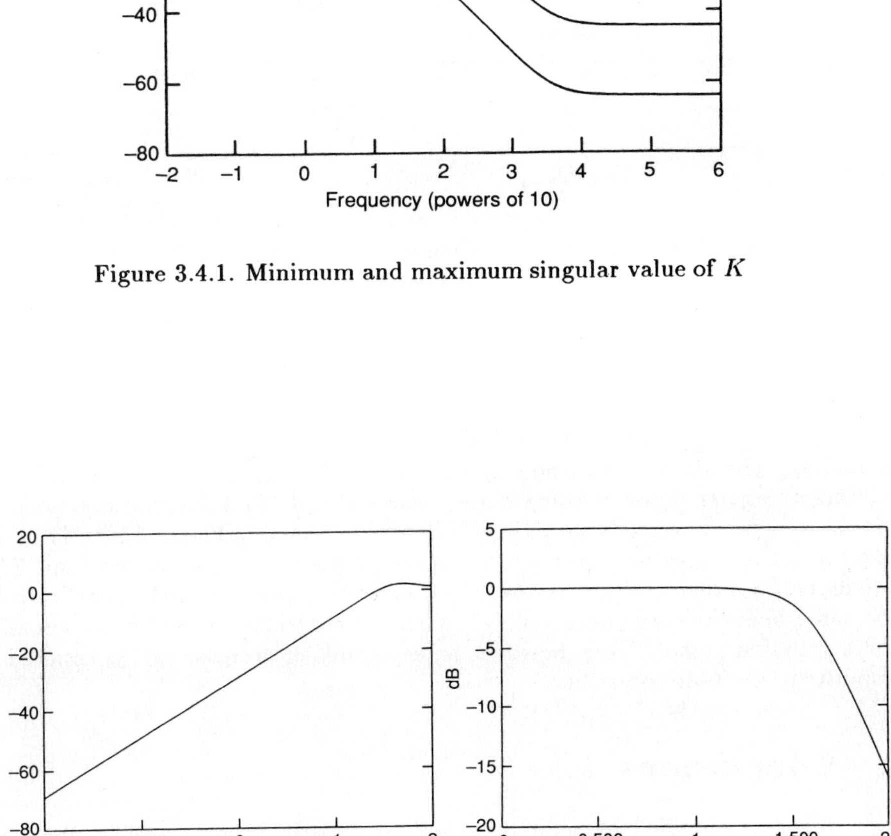

Figure 3.4.2. Minimum and maximum singular values of S and T

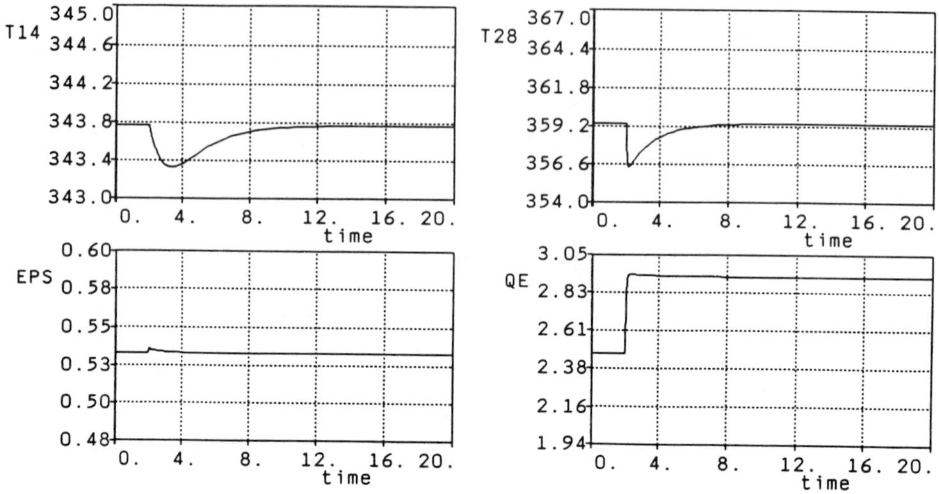

Figure 3.4.3. Responses to a step change in feed flow by +20% (simulation)

3.5 Conventional PI-design

For comparison with the multivariable techniques of the preceeding chapters, we have designed conventional PI-controllers k_1 and k_2 for the diagonal elements $g_1(s)$ and $g_2(s)$ of $G(s)$. They were chosen to approximately balance up the gains of g_1 and g_2. The gains of $g_1 k_1$ and $g_2 k_2$ are shown in Figure 3.5.1. Maximum and minimum singular values of S and T are given in Figure 3.5.2. Nonlinear simulation of time responses to a step change in feed flow is presented in Figure 3.5.3. The drop in temperature T_{28} is only stopped, when the profile moves into the bottom. This is reflected in a remarkable decrease of bottom purity (Figure 3.5.4). Qualitatively the same behaviour was observed, when the PI-controllers were implemented at the distillation plant. There, however, an even smaller decrease in T_{28} resulted in impurity of the bottom product.

4 Conclusions

All controllers obtained by multivariable design methods work well in nonlinear simulation and achieve satisfactory performance at the real plant. This is, however, not true for the conventional PI-controllers.

As our designs are conservative, there is considerable scope for on-line tuning of controller parameters. This has been done. As expected, it improves performance. It has to be stressed, however, that - with the exception of the decoupled PI-

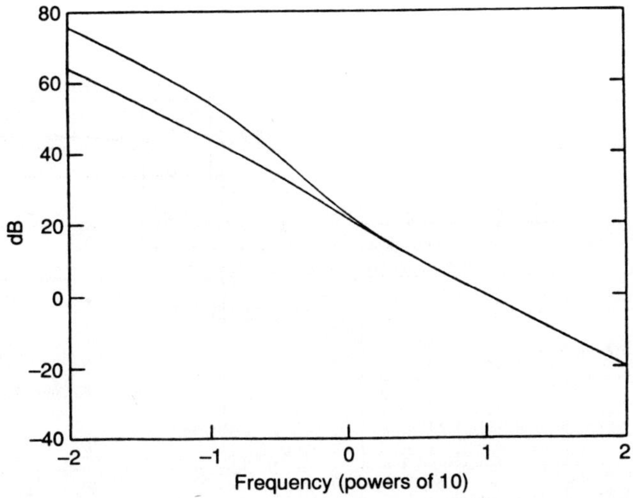

Figure 3.5.1. Gain of g_1k_1 and g_2k_2

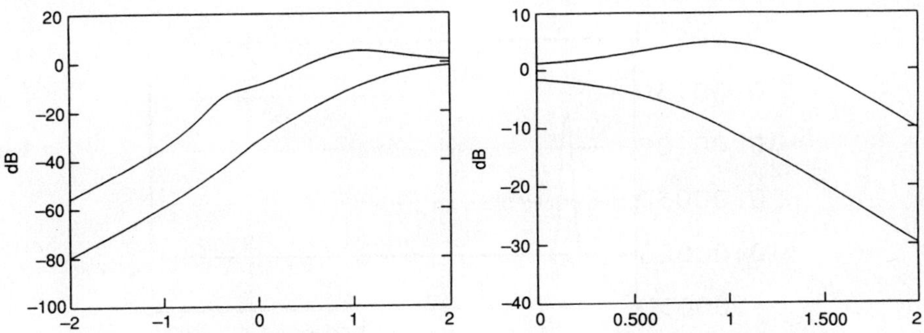

Figure 3.5.2. Minimum and maximum singular values of S and T

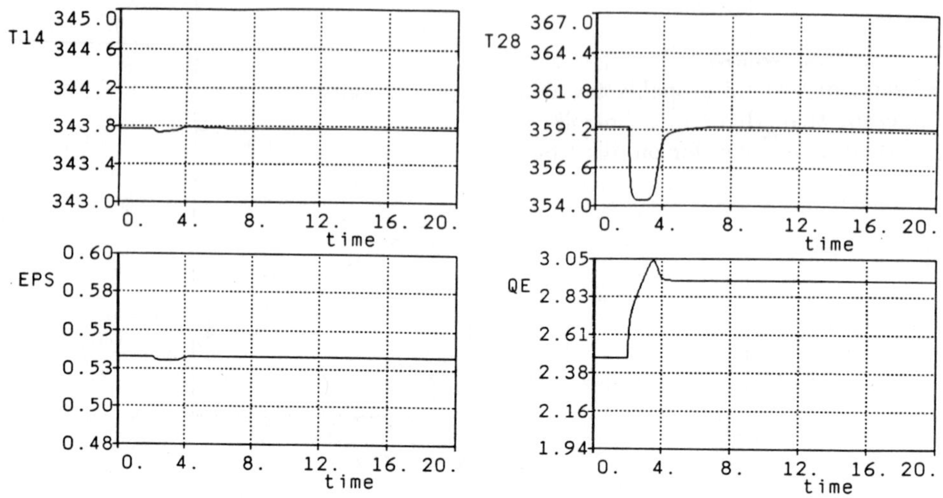

Figure 3.5.3. Responses to a step change in feed flow by +20% (simulation)

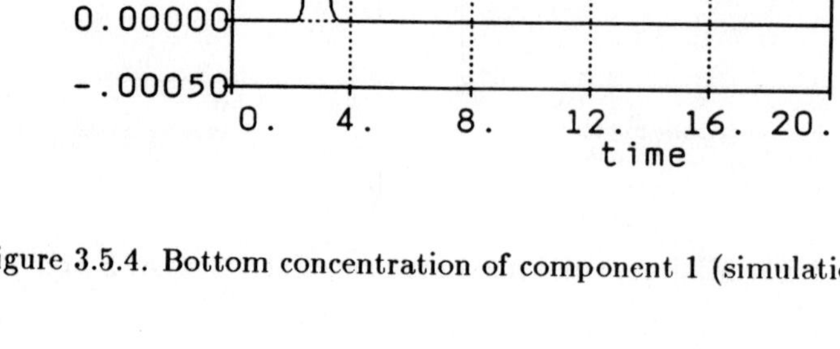

Figure 3.5.4. Bottom concentration of component 1 (simulation)

controllers - on-line tuning is not necessary to meet performance specifications. Thus, the results are not given here.

For the disturbances considered in this paper, it is possible to trade off increased design effort (multivariable design) and on-line tuning (conventional PI-design). If, however, a situation occurs, where both multivariable design and on-line tuning are necessary, one would probably favour DNA or CL controllers; they can be tuned more easily than H_2 or H_∞ controllers.

If there is no need for on-line tuning, it is difficult to tell which multivariable design method is easiest to handle. It seems to be pretty much a question of personal preference and experience. In our case (comparably low level of experience with any multivariable design), the amount of time necessary to achieve satisfactory results was about the same for each method.

Acknowledgements

Thanks are due to L. Lang, M. Wurst, A. Kröner, P. Holl, H. P. Kurrle and E. Hensel.

References

[1] John C. Doyle and Gunter Stein. Multivariable feedback design: concepts for a classical/modern synthesis. *IEEE Transactions on Automatic Control*, **26**:4-16, 1981.

[2] Bruce A. Francis. *A Course in H_∞ Control Theory*. Springer-Verlag, Berlin, 1987.

[3] Bruce A. Francis and John C. Doyle. Linear control theory with an H_∞ optimality criterion. *SIAM Journal on Control and Optimization*, **25**:815-844, 1987.

[4] A. G. J. MacFarlance and B. Kouvaritakis. A design technique for linear multivariable feedback systems. *International Journal of Control*, **25**:837-874, 1977.

[5] W. Marquardt, P. Holl, D. Butz, and E. D. Gilles. DIVA, a flow-sheet oriented dynamic process simulator. *Chemical Engineering & Technology*, **10**:164-173, 1987.

[6] Wolfgang Marquardt. *Nichtlineare wellenausbreitung - Ein Weg zu reduzierten dynamischen Modellen von Stofftrennprozessen*. VDI Verlag, Düsseldorf, 1988.

[7] Ian Postlethwaite, S. D. O'Young, and D.-W. Gu, *stabe-H User's Guide*. Technical Report, University of Oxford, 1987.

[8] B. Retzbach. *Mathematische Modelle von Destillationskolonnen zur Synthese von Regelungskonzepten*. VDI Verlag, Düsseldorf, 1986.

[9] H. H. Rosenbrock, *Computer-Aided Control Systems Design*. Academic Press, London, 1974.

[10] C. Schmid, *KEDDC Benutzerhandbuch*. Technical Report, Ruhr-Universität Bochum, 1987.

[11] G. Stein and M. Athans. The LQG/LTR procedure for multivariable feedback control design. *IEEE Transactions on Automatic Control*, **32**:105-114, 1987.

Fuel optimal control of an automotive CVT powertrain

R. P. Jones

Abstract

This paper considers the problem of computing a fuel optimal engine and transmission control strategy for an automotive powertrain incorporating a continuously variable transmission (CVT). The control study is based on an idealised state variable model, in which it is assumed that the engine torque can be modulated instantaneously. With this idealised model, the variable engine power is represented by inequality constraints, which define the maximum and minimum allowable engine speeds, and which bound the variable engine torque by a peak torque curve and peak overrun torque curve, both of which are dependent on engine speed. The approach invilves minimising a cost functional involving the fuel flow rate, which is a nonlinear function of engine torque and speed. The computational technique is based on an iterative gradient algorithm for state and inequality constrainted optimal control.

1 Introduction

Automotive vehicles, such as cars and heavy goods vehicles, are propelled by energy released through the combustion of petrol, or diesel, in an internal combustion engine. This results in mechanical power, in the form of speed of rotation and torque at the engine flywheel, which is delivered to the vehicle drive wheels via the transmission and other drivetrain components, which normally include the propshaft, final drive and half shafts. The transmission usually consists of a clutch, fluid coupling or torque converter, to provide a means of decoupling the engine from the drive wheels during gear changes or when the vehicle is at rest, together with a gearbox.

Automotive gearboxes enable the power produced by the engine to be matched to the varying power demands of the vehicle as its speed and acceleration requirements change. Conventionally, this is achieved in a car by the use of a manually operated gearbox, capable of providing 4 or 5 discrete ratios, or an automatic gearbox, providing 3 or 4 discrete ratios. An alternative approach, which is receiving considerable attention within the automotive industry worldwide, is to employ an advanced form of gearbox, viz. a continuously variable transmission (CVT), capable of providing a continuous variation in gear ratio within an upper and lower limit.

Existing automatic transmissions are mostly controlled by hydraulic systems which implement crude gear shift laws in isolation of any engine control. However, recent work in the area of automotive powertrain control indicates a trend toward increasing use of on-board electronic systems, capable of integrating the control of the engine and transmission. Integrated engine/transmission control provides the opportunity for engine parameters such as fuel consumption and exhaust emissions to be optimised throughout the operating range of a vehicle. To effect this, an optimal strategy for the combined control of the engine and transmission is required.

Previous work on the application of optimisation techniques to automotive power systems has concentrated on the isolated control of engines for maximising fuel economy [1]. Recently, a method of computing fuel optimal engine and transmission control strategies has been developed and applied to a powertrain incorporating a discrete ratio transmission [2]. The latter optimisation technique involves a sequential decision process based on dynamic programming principles.

This paper considers the problem of computing a fuel optimal engine and transmission control strategy for an automotive powertrain incorporating a CVT. The computational approach adopted is based on an iterative gradient algorithm for state and inequality constrained optimal control. The paper builds on expertise resulting from an earlier modelling study of an experimental vehicle incorporating a CVT powertrain [3].

2 Powertrain model

The powertrain of an automotive CVT vehicle can be modelled by the nonlinear dynamic equations

$$\begin{aligned}(\alpha + \beta x_2^2)\dot{x}_1 &= -x_2(\gamma + \delta x_2 x_1 + \epsilon x_2^2 x_1^2) + u_1 \\ &\quad -\beta x_2 x_1(-x_2 + u_2) - \mu x_2 u_3\end{aligned} \tag{2.1}$$

$$\dot{x}_2 = -x_2 + u_2 \tag{2.2}$$

in which x_1 represents the engine flywheel speed, x_2 the (continuous) gear ratio, u_1 the gross engine torque, u_2 the ratio demand, and u_3 the braking force. The parameters α, β and μ denote the engine flywheel inertia, vehicle inertia and effective wheel radius (including final drive), respectively, and the term $\gamma + \delta x_2 x_1 + \epsilon x_2^2 x_1^2$ represents the vehicle resistance to motion (road load), reflected at the propshaft. The corresponding vehicle speed is given by

$$y = \mu x_2 x_1 \tag{2.3}$$

The dynamic equations (2.1) and (2.2) represent the idealised case in which power losses in the CVT are ignored, and it is assumed that the engine torque u_1

can be modulated instantaneously. This model assumes that the response of the CVT to a demanded change in ratio can be approximated by a first order lag with unit gain and time constant. This model also ignores driveline compliance effects, which are considered negligible for the control problem considering in this paper.

The available engine power is represented by the inequality constraints

$$u_1^{\min}(x_1) \leq u_1 \leq u_1^{\max}(x_1) \tag{2.4}$$

and

$$x_1^{\min} \leq x_1 \leq x_1^{\max} \tag{2.5}$$

Constraint (2.5) defines the maximum and minimum allowable engine speed, and constraint (2.4) bounds the engine torque by a peak torque curve $u_1^{\max}$ and peak overrun torque curve $u_1^{\min}$, which are both dependent on engine speed.

The braking action available at the wheels is constrained by

$$0 \leq u_3 \leq u_3^{\max} \tag{2.6}$$

where $u_3^{\max}$ represents the maximum available braking force, and the upper and lower limits on the CVT ratio are represented by the inequality constraints

$$o \leq x_2^{\min} \leq x_2 \leq x_2^{\max} \tag{2.7}$$

Note that it is possible for the minimum gear ratio $x_2^{\min}$ of a CVT to be zero, an effect which is called geared neutral [3].

The engine power constraints (2.4) and (2.5) are derived as regression models from torque and speed data obtained on an engine test bed. Specifically, the peak torque curve $u_1^{\max}$ and peak overrun torque curve $u_1^{\min}$ are represented by the polynomial expressions

$$u_1^{\max}(x_1) = a_1 x_1^2 + a_2 x_1 + a_3 \tag{2.8}$$

and

$$u_1^{\min}(x_1) = b_1 x_1 + b_2 \tag{2.9}$$

which are described graphically in Figure 1. Finally, the remainder of the parameters defining the CVT powertrain model are based on data corresponding to an experimental vehicle.

The powertrain model consists of the nonlinear state variable equations (2.1)–(2.3) in combination with the control and state inequality constraints (2.4)–(2.7). The constraints on the control variable u_1 given in expression (2.4) are dependent on the state variable x_1, whereas the constraint on the control variable u_3 given in expression (2.6) are independent of the states.

3 Control problem

This paper is concerned with the computation of a schedule for engine torque and gear ratio which optimises the fuel consumption when a CVT vehicle experiences a load demand, expressed in terms of a desired speed profile $y_d(t)$ defined on a time interval $[0, T]$. The approach adopted is based on a nonlinear optimal control problem with state and control inequality constraints.

The optimal control problem involves the minimisation of a cost functional in the form

$$J = \nu[y_d(T) - y(T)]^2 + \frac{1}{T}\int_0^T \{\lambda[y_d(t) - y(t)]^2 + f(x_1(t), u_1(t))\}\, dt \qquad (3.1)$$

with respect to the triple $(u_1(\cdot), u_2(\cdot), u_3(\cdot))$, and subject to the nonlinear state variable model (2.1)–(2.3) and state and control inequality constraints (2.4)–(2.7). In this cost functional, $f(x_1, u_1)$ represents the fuel flow rate which is dependent on engine torque and speed, and the term $[y_d - y]^2$ represents the error between the desired and actual vehicle speed. The parameters λ and ν provide a means of trading off the requirement to minimise fuel usage against the requirement for the vehicle speed y to track the desired speed y_d.

The fuel flow rate characteristics are derived as a regression model based on fuel flow, torque and speed measurements obtained on an engine test bed. Specifically, the fuel flow rate is represented as a function of engine torque and speed by the expression

$$f(x_1, u_1) = c_1x_1^3 + c_2u_1^3 + c_3x_1^2u_1 + c_4x_1^2 + c_5x_1u_1 + c_6x_1 + c_7 \qquad (3.2)$$

which is described graphically in the contour plot presented in Figure 2.

The Hamiltonian for this optimal control problem takes the form

$$\begin{aligned} H(x,p,u) \;=\; & \lambda[y_d - \mu x_2x_1]^2 \\ & + \{c_1x_1^3 + c_2u_1^3 + c_3x_1^2u_1 + c_4x_1^2 + c_5x_1u_1 + c_6x_1 + c_7\} \\ & + p_1[\alpha + \beta x_2^2]^{-1}\{-x_2(\gamma + \delta x_2x_1 + \epsilon x_2^2x_1^2) + u_1 \\ & -\beta x_2x_1(-x_2 + u_2) - \mu x_2u_3\} + p_2(-x_2 + u_2) \end{aligned} \qquad (3.3)$$

and the corresponding costate equations are

$$\begin{aligned} -\dot{p}_1 \;=\; & -2\lambda\mu x_2[y_d - \mu x_2x_1] \\ & + \{3c_1x_1^2 + 2c_3x_1u_1 + 2c_4x_1 + c_5u_1 + c_6\} \\ & + p_1[\alpha + \beta x_2^2]^{-1}\{-x_2(\delta x_2 + 2\epsilon x_2^2x_1) - \beta x_2(-x_2 + u_2)\} \end{aligned} \qquad (3.4)$$

and

$$
\begin{aligned}
-\dot{p}_2 = {} & -2\lambda\mu x_1[y_d - \mu x_2 x_1] \\
& - 2\beta p_1 x_2[\alpha + \beta x_2^2]^{-2}\{-x_2(\gamma + \delta x_2 x_1 + \epsilon x_2^2 x_1^2) + u_1 \\
& - \beta x_2 x_1(-x_2 + u_2) - \mu x_2 u_3\} \\
& + p_1[\alpha + \beta x_2^2]^{-1}\{-(\gamma + 2\delta x_2 x_1 + 3\epsilon x_2^2 x_1^2) \\
& -\beta x_1(-2x_2 + u_2) - \mu u_3\} - p_2
\end{aligned} \tag{3.5}
$$

with associated final values

$$p_1(T) = -2\nu\mu x_2(T)[y_d(T) - \mu x_2(T)x_1(T)]$$

and

$$p_2(T) = -2\nu\mu x_1(T)[y_d(T) - \mu x_2(T)x_1(T)]$$

The gradient $\frac{\partial H}{\partial u}$ of the Hamiltonian is given by

$$\frac{\partial H}{\partial u_1} = \{3c_2 u_1^2 + c_3 x_1^2 + c_5 x_1\} + p_1[\alpha + \beta x_2^2]^{-1} \tag{3.6}$$

$$\frac{\partial H}{\partial u_2} = -\beta x_2 x_1 p_1[\alpha + \beta x_2^2]^{-1} + p_2 \tag{3.7}$$

$$\frac{\partial H}{\partial u_3} = -\mu x_2 p_1[\alpha + \beta x_2^2]^{-1} \tag{3.8}$$

which define a local descent direction for the optimal control problem.

This paper investigates the use of an iterative gradient algorithm for the solution of this state and inequality constrained optimal control problem. The computational method is based on the simple iteration

$$u^{(i+1)}(t) = u^{(i)}(t) - K^{(i)}\frac{\partial H}{\partial u}(u^i(t)) \tag{3.9}$$

which is guaranteed to produce a reduction in the value of the cost functional $J(u^i)$, providing that the positive parameter $K^{(i)}$ defining the improvement in $u^{(i)}$ at iteration i, is sufficiently small.

A pseudocode description of the algorithm is provide in table 1. Referring to the table, the *if* $\cdots$ *then* structure controlled by the logical test $J(u^{(i+1)}) > J(u^{(i)})$ ensures that a value of $K^{(i)}$ is used at each iteration which reduces the value of the cost functional $J(u^i)$. This is achieved by repeatedly replacing $K^{(i)}$ by $0.5K^{(i)}$ until $J(u^{(i+1)}) \leq J(u^{(i)})$. The choice for $u^{(0)}$ is made to ensure that the state and control inequality constraints (2.4)–(2.7) are initially satisfied. At subsequent iterations, a clipping-off procedure [4] is applied such that $u^{(i)}$ is clipped off at the upper and lower limits in the state independent constraints (2.6), whenever these constraints are violated. The *if* $\cdots$ *then* structure controlled by the logical test on the violation of inequality constraints (2.4), (2.5) and (2.7) ensures that these state dependent constraints are not violated at subsequent iterations, again by repeatedly replacing $K^{(i)}$ by $0.5K^{(i)}$ until the constraint are satisfied. The logical test $K^{(i)}\|\frac{\partial H}{\partial u}(u^{(i)})\| > tol\|u^{(i)}\|$ which controls the *while* $\cdots$ *do* structure implements the relative convergence criterion

$$\|u^{(i+1)} - u^{(i)}\| \leq tol\ \|u^{(i)}\| \tag{3.10}$$

for an appropriate tolerance *tol*, where the norm of the function $u(\cdot)$ is given by

$$\|u\|^2 = \int_0^T u'(t)u(t)\, dt$$

4 Computational results

A computational investigation of this nonlinear constrained optimal problem has been carried out using the model parameters presented in table 2. For this investigation, a typical vehicle speed time history demand, as presented in Figure 3, was utilised. This speed profile is made up from a phase consisting of a constant acceleration from rest to $13.33ms^{-1}(48kmh^{-1})$ over $15s$ followed by a steady $20s$ period; a second phase consisting of a further constant acceleration to $22.22ms^{-1}(80hmh^{-1})$ over $15s$ followed by a steady $15s$ period; and a final phase consisting of a constant deceleration to $17.778ms^{-1}(64kmh^{-1})$ over $20s$ followed by a steady $15s$ period.

Figure 4 presents the convergence characteristics of the optimisation algorithm, in the form of a plot of the value of the cost functional J against iteration count i, resulting from the particular choice of weighting parameters $\lambda = 30$ and $\nu = 10$. The associated nominal and converged values of engine torque u_1 and braking effort u_3 are displayed in Figures 5 and 6, and the corresponding nominal and converged values of engine speed x_1 and CVT ratio x_2 are given in Figures 7 and 8. Note that The CVT ratio demand u_2 is not displayed since it closely resembles x_2. Figure 4 indicates that the optimisation algorithm convergence slowly, requiring approximately 100 full iterations for a choice of relative convergence tolerance $tol = 1.0 \times 10^{-4}$. Each of these iterations involves integration of the state equations (2.1) and (2.2) and the costate equations (3.4) and (3.5). To ensure that $J_i < J_{i-1}$at each full iteration i a number of subiterations involving further integration of the state equations were needed. The maximum number of such subiterations was found to be 5, with an average of 2. Clipping of the controls was applied at every iteration to prevent the minimum braking effort u_3 becoming negative, but action was not required to prevent violation of the other state dependent constraints.

A comparison of the desired and converged vehicle speed time histories is shown in Figure 3, where it can be seen that the tracking accuracy has been influenced by the (conflicting) requirement to minimise the fuel usage. The total quantity of fuel used in executing the optimal manoeuvre was found to be $80.97g$ in this case.

5 Conclusion

This paper has considered the problem of computing a fuel optimal engine and transmission control strategy for an automotive powertrain incorporating a CVT. The control study has been based on a nonlinear state variable model of a powertrain where the variable engine power is represented by inequality constraints, which define the maximum and minimum allowable engine speeds, and which bound the available engine torque by a peak torque curve and peak overrun torque curve. The approach involved the minimisation of a cost functional involving a combination of fuel flow rate and error between the desired and actual vehicle speeds. The computational technique was based on a simple iterative gradient algorithm for state and inequality constrained optimal control.

A computational study of this simple iterative algorithm has been carried out for a typical demanded vehicle speed profile. This has demonstrated that convergence to an optimal solution can be achieved but that a large number of iterations is required for convergence. The study has further demonstrated that several subiterations are also required at each full iteration to ensure that the gradient method reduces the value of the cost functional.

The investigation reported in this paper indicates that numerical optimisation techniques have potential for use in the computation of automotive powertrain control strategies. However, the results obtained so far suggest that it would be worthwhile to consider a more sophisticated iterative gradient algorithm. In particular, the likely reduction in the number of iterations needed for convergence resulting from the use of a conjugate algorithm should be studied. Here little additional computing overhead would be expected to result from the requirement for a line search at each iteration, since several subiterations are currently needed in the simple iterative algorithm. This would almost certainly be the case if a bracking procedure employing quadratic approximation were used in the line search.

References

[1] Tennant, J. A., Cohen, A. I., Rao, H. S. and Powell, J. D., *Computer-aided procedures for optimization of engine controls*, in *Application of control theory in the automotive industry*, 1-12, Inderscience Enterprises, 1983.

[2] Kuzak, D. M., Shields, B. D., Freedman, R. J., Lee, J. J. and Rittmueller, D. L., *Powertrain control strategy determination for computer-controlled transmissions*, Int. J. Vehicle Design, **8**, 13-36, 1987.

[3] Jones, R. P., *et al.*, ***Modelling and simulation of an automotive powertrain incorporating a Perbury continuously variable transmission***, in ***Integrated Engine/Transmission Systems***, 41-49, IMechE, 1986.

[4] Quintana, V. H. and Davison, E. J., ***Clipping-off gradient algorithms to compute optimal controls with constrained magnitude***, Int. J. Control, **20**, 243-255, 1974.

define $u^{(0)}$ and $K^{(0)}$
set $i = 0$
integrate equations (2.1) and (2.2) for $x^{(0)}$
compute $J(u^{(0)})$ from equation (3.1)
integrate equations (3.4) and (3.5) for $p^{(0)}$
compute $\frac{\partial H}{\partial u}(u^{(0)})$ from equation (3.6)–(3.8)
while $K^{(i)}\|\frac{\partial H}{\partial u}(u^{(i)})\| > tol\|u^{(i)}\|$ do
    compute $u^{(i+1)}$ from equation (3.9)
    clip $u^{(i+1)}$ if inequality constraint (2.6) is violated
    integrate equation (2.1) and (2.2) for $x^{(i+1)}$
    compute $J(u^{(i+1)})$ from equation (3.1)
    if $J(u^{(i+1)}) > J(u^{(i)})$ them
        replace $K^{(i)}$ by $0.5K^{(i)}$
    else if inequality constraints (2.4), (2.5) and (2.7) are violated then
        replace $K^{(i)}$ by $0.5K^{(i)}$
    else
        integrate equations (3.4) and (3.5) for $p^{(i+1)}$
        compute $\frac{\partial H}{\partial u}(u^{(i+1)})$ from equations (3.6)–(3.8)
        set $K^{(i+1)} = 2K^{(i)}$
        replace i by $i + 1$
    end if
end while

Table 1. Iterative gradient algorithm

engine power constraints	a_1	-4.86×10^{-4}	fuel flow model	c_1	1.24×10^{-8}
	a_2	3.08×10^{-1}		c_2	3.59×10^{-7}
	a_3	8.66×10^{1}		c_3	7.87×10^{-8}
	b_1	-3.81×10^{-2}		c_4	-1.13×10^{-5}
	b_2	-9.55		c_5	1.73×10^{-5}
	x_1^{min}	7.85×10^{1}		c_6	5.15×10^{-3}
	x_1^{max}	5.76×10^{2}		c_7	-2.07×10^{-1}
powertrain dynamics	α	1.09×10^{-1}	CVT ratio constraints	x_2^{min}	0
	β	7.96		x_2^{max}	2
	γ	1.36×10^{-1}			
	δ	-4.29×10^{-2}			
	ϵ	3.53×10^{-2}	braking constraints	u_3^{max}	6.5×10^{3}
	μ	7.7×10^{-1}			

Table 2. Parameters used in computational study

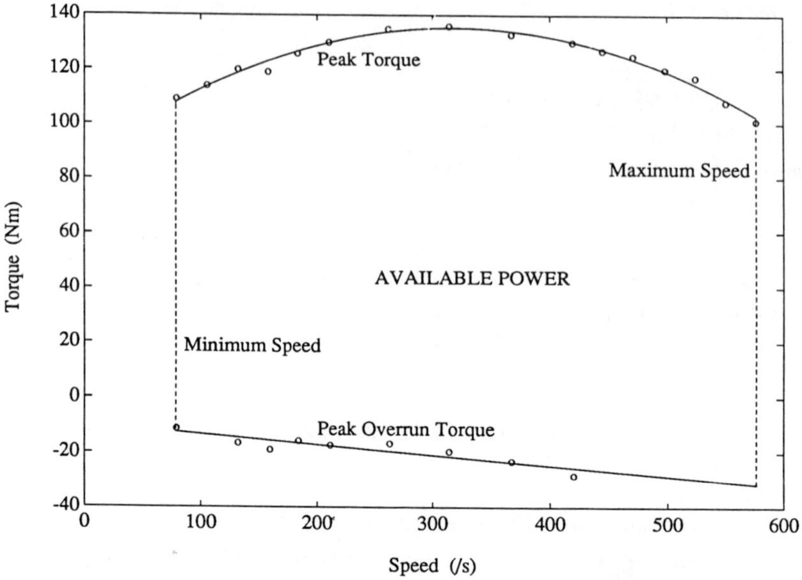

Figure 1. Engine power constraints.

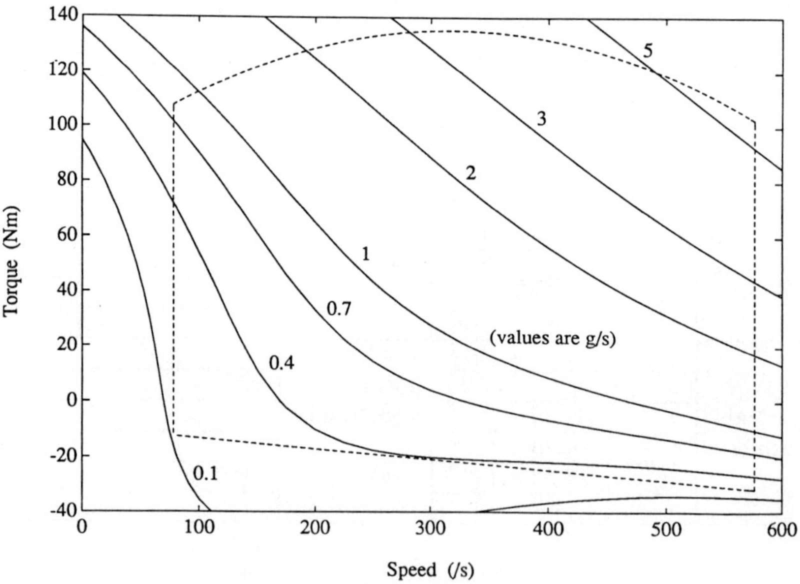

Figure 2. Fuel flow control plot.

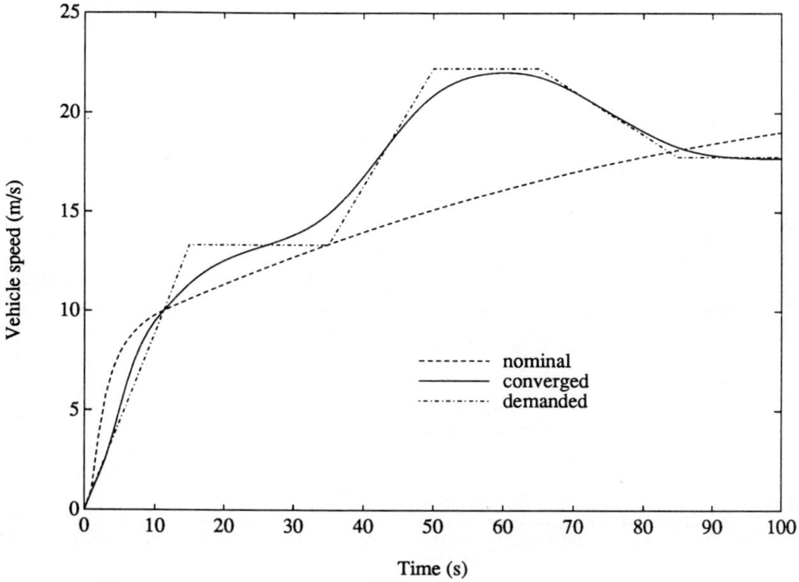

Figure 3. Tracking of vehicle speed.

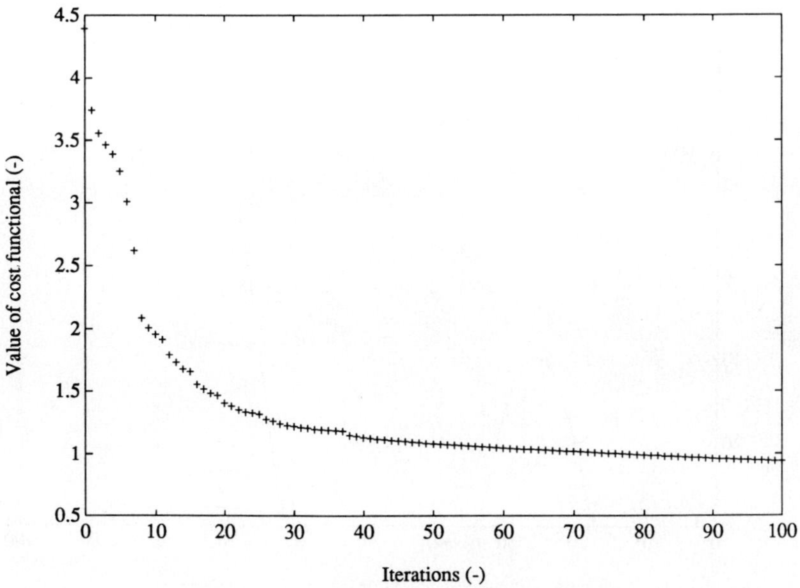

Figure 4. Convergence of algorithm.

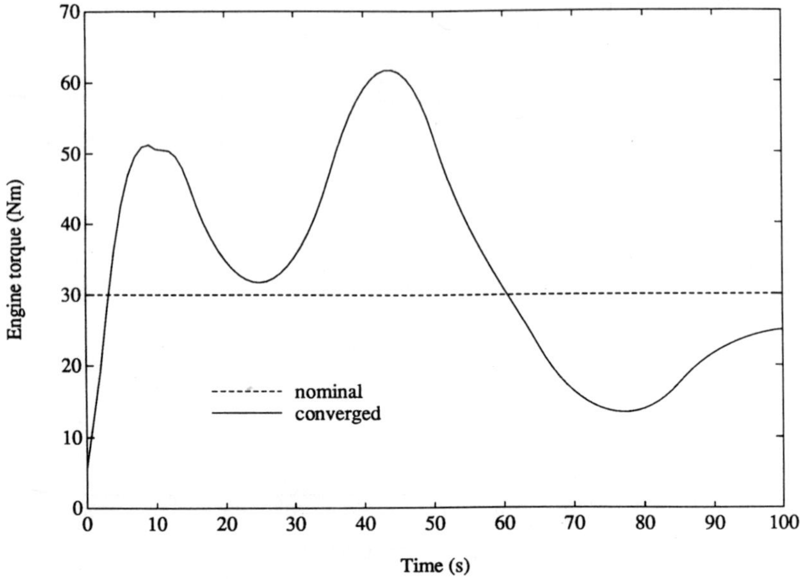

Figure 5. Engine torque time history.

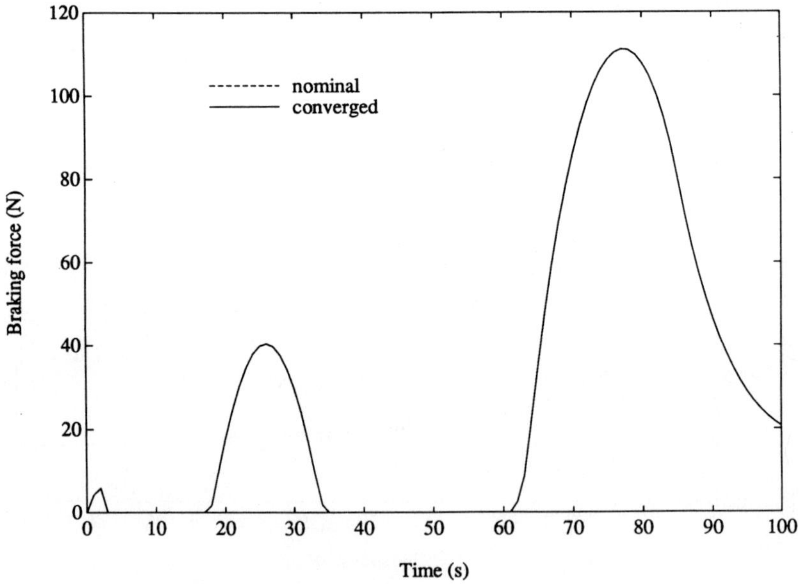

Figure 6. Braking force time history.

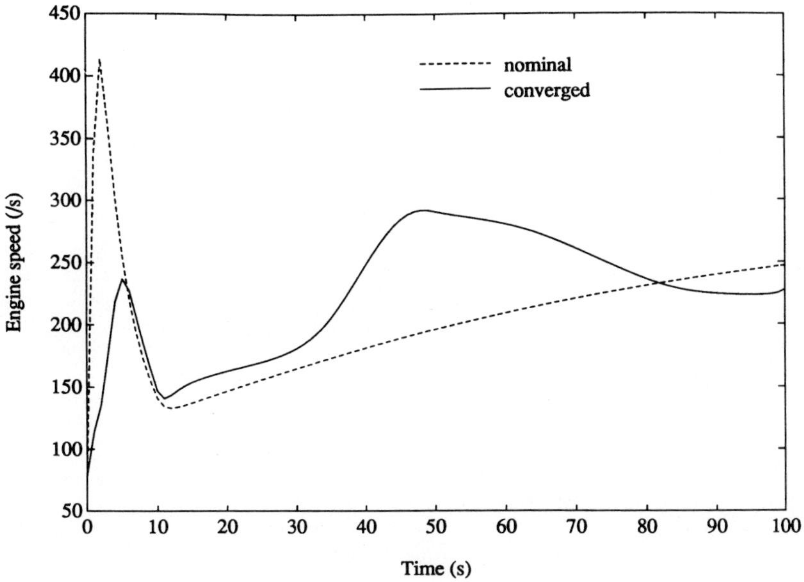

Figure 7. Engine speed time history.

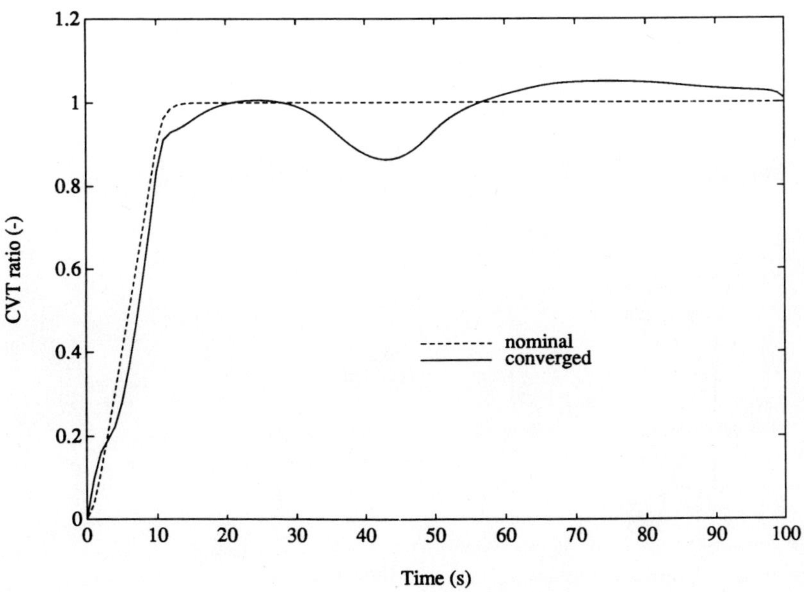

Figure 8. CVT ratio time history.

Optimal shape design for differential inequalities

R. Butt J. E. Rubio

Abstract

The purpose of this paper is to obtain an optimal shape design for a system described by a differential inequality. An electrochemical machining moving boundary problem is formulated, after a change of variable, as an elliptic variational inequality. The computations are done by a finite element method, the gradient of the criteria with respect to the coordinate of the moving nodes is computed and the performance criterion is then minimized by a gradient method.

1 Introduction

The optimal shape problem can be solved for systems described by differential equations (e.g., Pironneau [5]). Here we shall discuss this problem for a system described by a differential inequality. We take the example of an electrochemical machining problem. A metal part can be shaped by placing it as an anode in an electrolytic cell; applying a potential difference across the cathode and anode, between which lies an appropriate electrolyte, causes a chemical reaction on the anode surface resulting in the removal of anode metal. This process is called electrochemical machining (E. C. M.). One would like to predict for a given cathode shape the anode shape as a function of machining time. We note that this is a moving boundary problem as the unknown anode surface changes with time. A detailed account of the physical and practical details of the process and its industrial uses are given by McGeough [4]. Essentially, the anode is moved towards the cathode, or vice versa, at a constant velocity and the product of the erosion of the anode is swept away by the electrolyte which is pumped through the space between the electrodes. Several authors have proposed and studied a quasi-steady mathematical model in which the electrodes are assumed to be equipotential surfaces, and the electrolyte to be homogeneous and isotroptic with constant conductivity. The physical set-up is depicted schematically in Figure 1. The region inside the cathode surface C is denoted by Ω; A_t is the anode surface at time t, i.e., the moving boundary of the anode shrinking region $A(t)$, and the region occupied by the electrolyte is denoted by $D(t)$, so that $\Omega \equiv A(t) \cup D(t)$. Let $\psi(x,t)$ denote the potential field. Then the quasi-steady model of E.C.M. consists of Laplace's equations for ψ and an equation describing the rate of removal of anode metal derived from Faraday's and Ohm's laws. Let $V(t) > 0$ denote the potential difference between the electrodes at time

t. Then the mathematical problem is to find $\psi(x,t)$ and $l(x)$ such that

$$\nabla^2\psi(x,t) = 0 \quad x \in D(t) \quad t \in [0,T] \tag{1.1}$$

$$\psi(x,t) = 0 \quad x \in C \quad t \in [0,T] \tag{1.2}$$

$$\psi(x,t) = V(t) \quad x \in A_t \quad t \in [0,T] \tag{1.3}$$

$$\nabla\psi \cdot \nabla\ell = L, \tag{1.4}$$

where L is a positive constant.

The problem (1.1) is similar to a one-phase Stefan problem describing the freezing of some substance, and Elliott [2] formulated a variational inequality for this problem and drew attention to the close analogy with the Hele-Shaw injection problem. To this end we extend ψ to all of Ω by $\psi(x,t) = V(t)$ in $\Omega\backslash D(t)$ and consider the Baiocchi transformation

$$\varphi(x,t) = \begin{cases} \displaystyle\int_{l(x)}^{t} \{V(\gamma) - \psi(x,\gamma)\}d\gamma & x \in \Omega\backslash D(0); \ t > l(x) \\ \displaystyle\int_{0}^{t} \{V(\gamma) - \psi(x,\gamma)\}d\gamma & x \in D(0); \ t > 0 \end{cases} \tag{1.5}$$

Let $f(x) \in L^2(\Omega)$ be defined by

$$f(x) \equiv \begin{cases} -L & x \in A(0) \quad \text{and} \\ 0 & x \in D(0), \end{cases} \tag{1.6}$$

then $\varphi(x,t)$ satisfies in the sense of distributions the equation

$$\begin{aligned} -\nabla^2\,\varphi(x,t) &= f \quad x \in D(t) \\ \varphi(x,t) &= 0 \quad x \in A(t). \end{aligned} \tag{1.7}$$

$$\begin{aligned} \varphi(x,t) &= \int_0^t V(\gamma)d\gamma \ (\equiv W(t)) \quad x \in C \\ \varphi &= \varphi_n = 0 \qquad\qquad\qquad x \in A_t \end{aligned} \tag{1.8}$$

It can be shown (Elliott [2]) that in the interior of Ω, ψ satisfies the linear complementarity problem

$$-\nabla^2\varphi - f \geq 0 \qquad u \geq 0$$

$$(-\nabla^2\varphi - f)\varphi = 0.$$

For each fixed time t the unknown boundary problem consists now in finding $\{\varphi(x,t),\ D(t),\ A_t\}$ such that (1.7) and (1.8) hold. Thus the anode surface at any time t can be obtained by solving a single elliptic free boundary problem and the anode surface depends solely on the number $W(t)$, so that t enters only through the boundary conditions $\varphi = W(t)$. Define the inner products $(.,.)$ and $a(.,.)$ on $L^2(\Omega)$ and $H_0^1(\Omega)$ respectively, by

$$(f,g) = \int_\Omega fg\, dx \quad \forall f,g \in L^2(\Omega)$$

and

$$a(\varphi,\phi) = \int_\Omega \nabla\varphi . \nabla\phi\, dx \quad \forall\varphi,\phi \in H_0^1(\Omega),$$

with the associated norms being denoted by $|f|^2 = (f,f)$ and $||\varphi||^2 = a(\varphi,\varphi)$. For each $t \geq 0$ let $K(t)$ be a closed convex subset of H^1 defined by

$$K(t) = \{\phi \in H^1(\Omega) : \ \phi = W(t) \ \text{ on } C \text{ and } \phi \geq 0 \text{ on } \Omega\}. \tag{1.9}$$

If $\varphi(x,t)$ is a solution of (1.7) and (1.8) then $\varphi(t) \in K(t)$ and for every $\phi \in K(t)$ it satisfies

$$\begin{aligned} a(\varphi,\phi-\varphi) &= \int_{D(t)} \nabla\varphi . \nabla(\phi-\varphi)dx \\ &= \int_{D(t)} -\nabla^2\varphi(\phi-\varphi)dx + \int_{CUA_t} \varphi_n(\phi-\varphi)ds \\ &= \int_{D(t)} f(\phi-\varphi)dx \geq \int_\Omega f(\phi-\varphi)dx = (f,\phi-\varphi). \end{aligned}$$

Thus the problem for φ consists in finding for each $t \geq 0$ the solution $\varphi(t) \in K(t)$ to the variational inequality

$$a(\varphi(t),\phi-\varphi(t)) \geq (f,\phi-\varphi(t)) \quad \forall\phi \in K(t). \tag{1.10}$$

Elliott [2] proved some regularity results for $\varphi(t)$ and gave some numerical results obtained by a finite-element approximation to (1.10) with error estimates.
We note that the bilinear form $a(.,.)$ in (1.10) is elliptic, i.e.,

$$a(\phi,\phi) \geq \alpha||\phi||^2 \quad \alpha > 0,\ \forall\phi \in H_0^1(\Omega) \tag{1.11}$$

Note 1. The main idea of our problem is to change the anode surface to a given fixed shape so as to control the shape A_t at a fixed time t; the study of a physical system modelled by elliptic variational of form (1.10) leads to the problem of controlling the set $A(t)$, or its boundary A_t, by modifying the cathode surface. Thus, we must choose the boundary of Ω so as to keep set $A(t)$ as close as possible to a given (measurable) subset E of Ω, representing the shape we wish to reach. Thus a natural choice for the performance index is

$$\int_\Omega |\chi_\varphi(x) - \chi_E(x)|^2 dx \tag{1.12}$$

where χ_φ and χ_E are the characteristic functions of $A(t)$ and E, respectively. However, since the function $\varphi \to \int_\Omega |\chi_\varphi(x) - \chi_E(x)|^2 dx$ is not locally lipschitzian on $L^2(\Omega)$, we shall approximate the cost functional in a manner described by Saguez [6]; for $\lambda > 0$, put:

$$\int_\Omega |\lambda/(\lambda + \varphi) - \chi_E(x)|^2 dx \tag{1.13}$$

For small $\lambda > 0$, the value of this functional is close to that of (1.12), see Saguez [6].

Note 2. The optimal shape problem can be solved for systems described by differential equations. The simple examples to consider are the optimization of nozzles, and some examples of airfoil design (Angrand [1]). So to solve the problem for the system described by a differential inequality we shall introduce the penalized equation as follows,

$$A\varphi_\varepsilon + 1/\varepsilon\varphi_\varepsilon^- = f, \quad \varphi_\varepsilon \in H_0^1(\Omega), \tag{1.14}$$

(where $V^- = -\sup(-V, 0)$, and $A : V \to V'$ is a linear continuous and symmetric operator satisfying the coercivity condition, i.e., $(A\varphi, \phi) = a(\varphi, \phi)$, i.e., $A = -\nabla.\nabla$), whose solution φ_ε tends to the solution of (1.10) when $\varepsilon \to 0$; for the existence and uniqueness of a solution of this equation see Lions [3].

Before introducing an algorithm using this approximation we shall briefly review the method of finite elements.

2 Finite element approximation

To illustrate the method, let (1.14) be made discrete by triangulation elements of degree 1. In variational form (1.14) becomes

$$\int_\Omega (\nabla\varphi_\varepsilon . \nabla \omega + F(\varphi_\varepsilon)\omega - f\omega)\, dx = 0 \quad \omega \in H^1(\Omega), \varphi_\varepsilon \in H^1(\Omega); \tag{2.1}$$

where $F(\varphi_\varepsilon) = 1/\varepsilon\varphi_\varepsilon^-$.

Let τ_h be a triangulation of Ω, $\cup T_k = \Omega$. The parameter h is the size of the largest side or edge, and we assume that we have a family of triangulations of τ_h. Let P^m be the space of polynomials of degree m on Ω, and denote by

$$H_h^m(\Omega_h) = \{\omega_h \in C^0(\Omega_h) \ : \omega_h|_{T_k} \in P^m \ \ \forall T_k \in \tau_h\} \tag{2.2}$$

the space of continuous piecewise polynomial functions on Ω_h. We know that H_h^m is of finite dimension; (2.1) becomes

$$\int_{\Omega_h} (\nabla\varphi_{h,\varepsilon}.\,\nabla\,\omega_h + 1/\varepsilon\varphi_\varepsilon^-\omega_h - f\omega_h)\,dx = 0; \tag{2.3}$$

if $\{\omega^i\}_1^N$ is a basis for H_h^m, (2.3) is equivalent to

$$A\varphi = F \tag{2.4}$$

where $A_{ij} = \int_{\Omega_h} \nabla\,\omega^i.\ \nabla\,\omega^j + 1/\varepsilon\varphi_\varepsilon^-\omega^j)\,dx; \quad F_i = \int_{\Omega_h} f\omega^i\,dx$

$$\varphi_h = \sum_1^N \varphi_i\omega^i. \tag{2.5}$$

The $\{\omega^i\}$ are polynomials of degree $\leq m$ on T_k, so A_{ij} can be computed exactly. In the case $m = 2$, if $\{q^j\}$ denotes the middles of the sides of vertices $\{q^j, q^k\}$, ω^i is uniquely determined by

$$\omega^i(q^j) = \delta_{ij} \quad \forall i,j = \{1,\cdots,N'\} \cup (\{1,\cdots,N'\} \times \{1,\cdots,N'\}).$$

3 A first computational algorithm

Since we know that algorithms exist for solving optimal shape problems for systems described by partial differential equations (Pironneau [5]), as a first approach here we shall use the ε-approximation of our variational inequality with ε small introduced above in (1.14). The optimal shape is found by successive approximation starting from an initial guess Ω_h^0, the algorithm is then developed by means of a gradient method. We note that the problem has been discretized, so that the shape Ω_h is defined by the coordinates of the nodes; then, considering the expression

$$E_\varepsilon(\Omega_h) = \int_{\Omega_h} (\lambda/(\lambda + \varphi_{h,\varepsilon}) - \chi_E)^2\,dx \tag{3.1}$$

(where $\varphi_{h,\varepsilon}$ is solution of (1.14) with ε small) for the cost function E_ε, we must compute its gradient. The following theorem has been adapted from Pironneau [5] to compute the gradient G of the cost function E_ε.

Theorem A. *If E_ε is given by* (1.13), (2.3), *and* (2.2) *with $m = 1$, then*

$$\begin{aligned} G_l^k = & \int_{\Omega_h} \partial/\partial x_l\{\omega^k(\lambda/(\lambda+\varphi_{h,\varepsilon}) - \chi_E(x))^2\}\, dx + \\ & \int_{\Omega_h} \{(\nabla\varphi_{h,\varepsilon}\cdot\nabla\omega^k)\partial P_{h,\epsilon}/\partial x_l - (\nabla\varphi_{h,\varepsilon}\cdot\nabla P_{h,\epsilon})\partial\omega^k/\partial x_l\}\, dx + \\ & \int_{\Omega_h} \{\partial/\partial x_l(fP_{h,\epsilon}\omega^k) - (f\omega^k\partial P_{h,\epsilon}/\partial x_l)\}\, dx + \\ & \int_{\Omega_h} \{(F(\varphi_{h,\varepsilon})\nabla\omega^k\partial P_{h,\epsilon}/\partial x_l) - \partial/\partial x_l(\omega^k F(\varphi_{h,\varepsilon})P_{h,\epsilon})\}\, dx, \end{aligned} \tag{3.2}$$

$l = 1, 2$ and $k = 1, \cdots, n$, where $P_{h,\varepsilon}$ is solution of

$$\begin{aligned} & \int_{\Omega_h} (\nabla P_{h,\varepsilon}\cdot\nabla\omega_h + F'(\varphi_{h,\varepsilon})P_{h,\varepsilon}\omega_h)\, dx = \\ & -2\int_{\Omega_h} (\lambda/(\lambda+\varphi_{h,\varepsilon}) - \chi_E(x))\lambda/(\lambda+\varphi_{h,\varepsilon})\omega_h\, dx,\ \omega_h, P_{h,\epsilon} \in H_h^1, \end{aligned} \tag{3.3}$$

where $F'(\varphi_{h,\varepsilon}) = 1/\varepsilon d/d\varphi(\varphi_\varepsilon^-)$, and (3.3) *is equivalent to our second penalized equation,*

$$AP_{h,\varepsilon} + 1/\varepsilon(d/d\varphi\varphi_\varepsilon^-)P_{h,\varepsilon} = -2[\lambda/(\lambda+\varphi_{h,\varepsilon})^2](\lambda/\lambda+\varphi_{h,\varepsilon}) - \chi_E) \equiv f_1 \tag{3.4}$$

We note that the function $\varphi\ \varphi^-$ is not differentiable at $\varphi = 0$; we have *defined* $F'(0) = 0$. This choice turns out to be unimportant because $\varphi_{h,\varepsilon} > 0$ on Ω, with the exception of a (zero-measure) subset of Γ. In Reference [8] we have related these ideas to the theory of optimization of non-differentiable functions; the vector G with components G_l^k given by (3.2) and (3.3) can be identified with a local subgradient which defines a direction of descent (see Shor [9], especially Theorem 2.18).

Thus, we define a first algorithm to solve the optimal shape problem:

Algorithm 1

0. Choose Ω_h^0, i.e., $\{q^{k,0}\}$
1. Compute $\varphi_{h,\varepsilon}^m$ by (2.3) and (2.2) (with $m = 1$). This is an approximation to the solution for the domain Ω_h^0.
2. Compute $P_{h,\varepsilon}^{m'}$ by (3.3). This is a function needed to compute the gradient below.
3. Compute G_l^k by (3.2), $l = 1, 2$.

4. Let $q^{k,m'}(\rho) = q^{k,m'} + \rho G_l^k$. Compute $\rho^{m'}$, an approximation of

$$\arg \min_{0<\rho<\rho_{\max}} E_\varepsilon(\{q^{k,m'}(\rho)\}),$$

where E_ε is given by (1.13) and (2.3). This step involves a one-dimensional optimization in the direction of the gradient; here $\rho_{\max}$ is an appropriate value.

5. Set $q^{k,m'+1} = q^{k,m'}(\rho^{m'})$.

6. After some terminal check, if necessary go on with the same procedure with a new estimate of Ω defined by the nodes $q^{k,m'+1}$, i.e., go back to 1.

4 A second computational algorithm

We shall solve now the optimal shape problem for the differential inequality by taking limits, as ε tends to zero, of the corresponding expressions in the previous algorithm. As before, the optimal shape is found by successive approximations starting from an initial guess Ω_h^0; the algorithm is then developed by means of a gradient method. As we noted in Algorithm I the problem has been discretized, so that the shape Ω_h is defined by the coordinates of the nodes.

The first penalized equation (2.3) can be put in the form:

$$\int_{\Omega_h} F(\varphi_{h,\varepsilon})\omega_h \, dx = -\int_{\Omega_h} (\nabla\varphi_{h,\varepsilon} \cdot \nabla\omega_h - f\omega_h) \, dx \tag{4.1}$$

so, by using (4.1) in Theorem A, we obtain

$$\begin{aligned} G_l^k = &\int_{\Omega_h} (\lambda/(\lambda+\varphi_{h,\varepsilon}) - \chi(x))^2 \partial/\partial x_l(\omega^k) \, dx + \\ &\int_{\Omega_h} \{(\nabla\varphi_{h,\varepsilon} \cdot \nabla\omega^k)\partial/\partial x_l(P_{h,\varepsilon}) + (\nabla P_{h,\varepsilon} \cdot \nabla\omega^k)\partial/\partial x_l(\varphi_{h,\varepsilon})\} \, dx + \\ &\int_{\Omega_h} \{(\nabla\varphi_{h,\varepsilon} \cdot \nabla(\partial/\partial x_l(\omega^k)))P_{h,\varepsilon} + (\nabla P_{h,\varepsilon} \cdot \nabla(\partial/\partial x_l(\varphi_{h,\varepsilon})))\omega^k\} \, dx + \\ &\int_{\Omega_h} \{\partial/\partial x_l[f\omega^k P_{h,\varepsilon} - f\omega^k \partial/\partial x_l(P_{h,\varepsilon})] - fP_{h,\varepsilon}\partial/\partial x_l(\omega^k)\} dx, \end{aligned} \tag{4.2}$$

where $P_{h,\varepsilon}$ is solution of (3.4).

Now we come to the implementation of the main idea of our problem, i.e., we take the limits as $\varepsilon \to 0$, of the quantities and functions defined and computed in section 2. Since we know (Lions [3]) that

$$\begin{aligned} &\varphi_{h,\varepsilon} \to \varphi_h \quad \text{in } H_h^2(\Omega_h) \text{ weakly as } \varepsilon \to 0; \text{ and also,} \\ &\varphi_{h,\varepsilon} \to \varphi_h \quad \text{in } L^2(\Omega_h) \text{ strongly as } \varepsilon \to 0; \end{aligned} \tag{4.3}$$

so, (3.1) becomes:

$$E(\Omega_h) = \int_{\Omega_h} (\lambda/(\lambda + \varphi_h) - \chi_E)^2 \, dx \tag{4.4}$$

Now we shall compute the gradient G of the cost functional by using the expression in (4.2). Since we know that

$$\lim_{\varepsilon \to 0} \int_{\Omega_h} (\nabla P_{h,\varepsilon} \cdot \nabla \varphi_{h,\varepsilon}) \, dx = \int_{\Omega_h} (\nabla P_h \cdot \nabla \varphi_h) \, dx,$$

and

$$\partial/\partial x_l(\varphi_{h,\varepsilon}) \to \partial/\partial x_l(\varphi_h) \text{ in } L^2(\Omega_h) \text{ weakly as } \varepsilon \to 0,$$

so, by using these conditions and by taking the limit, then (4.2) becomes

$$\begin{aligned} G_l^k \;=\; & \int_{\Omega_h} (\lambda/(\lambda + \varphi_h) - \chi(x))^2 \partial/\partial x_l(\omega^k) \, dx + \\ & \int_{\Omega_h} \{(\nabla \varphi_h \cdot \nabla \omega^k)\partial/\partial x_l(P_h) + (\nabla P_h \cdot \nabla \omega^k)\partial/\partial x_l(\varphi_h)\} \, dx \\ & \int_{\Omega_h} \{(\nabla \varphi_h \cdot \nabla(\partial/\partial x_l(\omega^k)))P_h + (\nabla P_h \cdot \nabla(\partial/\partial x_l(\varphi_h)))\omega^k\} \, dx + \\ & \int_{\Omega_h} \{\partial/\partial x_l[f\omega^k P_h - f\omega^k \partial/\partial x_l(P_h)] - fP_h \partial/\partial x_l(\omega^k)\} dx, \end{aligned} \tag{4.5}$$

where φ_h is is solution of (1.10); P_h is defined in the following theorem:

Theorem B. *As $\varepsilon \to 0$, $P_{h,\varepsilon} \to P_h$ in H_h^2, P_h being the solution of the variational inequality for $\omega_h \in H_h^2$*

$$a(P_h, \omega_h - P_h) \geq (f_1, \omega_h - P_h) \tag{4.6}$$

where φ_h is solution of (1.10) *and f_1 is defined by* (3.4).

In this algorithm it has been found necessary to use a second-order approximation, that is, $m = 2$, to be able to compute some of the limits appearing above.

We define then a second algorithm to solve the optimal shape problem for systems described by an differential inequality:

Algorithm 2

0. Choose Ω_h^0, i.e., $\{q^{k,0}\}$
1. Compute $\varphi_h^{m'}$ by (1.10) and (2.2) (with $m = 2$)

2. Compute $P_h^{m'}$ by (4.6)
3. Compute G_l^k by (4.5), $i = 1, 2$.
4. Let $q^{k,m'}(\rho) = q^{k,m'} + \rho G_l^k$. Compute $\rho^{m'}$, an approximation of

$$\arg \min_{0<\rho<\rho_{\max}} E(\{q^{k,m'}(\rho)\}),$$

   where E is given by (1.13) and (1.10).
5. Se $q^{k,m'+1} = q^{k,m'}(\rho^{m'})$.
6. After some terminal check, if necessary go back to 1.

5 Description of the program and algorithms used

The implementation of the second algorithm ($m = 2$) will be described here. The optimum design program is composed of the following modules.

1. A module for solving the direct problem (or state problem). We give again the formulation of this problem:
Find $\varphi_h \in H_h^2$ such that

$$\int_{\Omega_h} (\nabla\varphi_h \cdot \nabla\omega_h - f\omega_h)\, dx \geq 0 \quad \forall\omega_h \in H_h^2 \tag{5.1}$$

or find $\varphi_h \in H_h^2$ such that

$$I(\varphi_h) \leq I(\omega_h) \quad \forall\omega_h \in H_h^2 \tag{5.2}$$

for an appropriate quadratic function $I(\varphi_h)$, which may be written $I(\varphi_1, \varphi_2, \cdot, \varphi_{N(h)})$ to emphasize the dependence of φ_h on the coefficients in (2.3). The problem (5.2) is solved by the relaxation method, which can be briefly described as follows: let

$$\varphi_h^0 = (\varphi_1^0, \cdots\cdots, \varphi_{N(h)}^0) \text{ given in } H_h^2$$

with φ_h^n known: then φ_h^{n+1} is determined coordinate by coordinate; further iterations in the algorithm are given by ($\varphi_h = (\varphi_1, \cdots, \varphi_{N(h)})$)

$$\varphi_i^{n+1} = \varphi_i^n + \omega(\varphi_i^{n+1/2} - \varphi_i^n);$$

here ω is the relaxation parameter. $0 < \omega < 2$.

The process described above, is stopped when

$$\sum_{i=1}^{N(h)} |\varphi_i^{n+1} - \varphi_i^n| / \sum_{i=1}^{N(h)} |\varphi_i^{n+1}| \leq \varepsilon_r$$

(In our computational experiments we took $\varepsilon_r = 10^{-4}$).

2. A module for solving the adjoint-state problem, whose solution is needed to compute the descent direction (the vector G). The adjoint state P_h is given by the solution of the following variational inequality:
Find $P_h \in H_h^2(\Omega)$ such that

$$\int_{\Omega_h} (\nabla P_h \cdot \nabla \omega_h - f_1 \omega_h)\, dx \geq 0 \quad \text{for all every } \omega_h \in H_h^2$$

where φ_h is given and is the solution of the state problem. For this problem, we use the same method as we used in the case of the state problem.

3. A module for the computation of the descent direction when we know the solution φ_h of the state problem and the solution P_h of the adjoint state problem. In the formula we must account for the variability of the domain. For each moving node we can compute the real vector or the projected vector if a displacement direction is imposed.

4. A module minimizing the criterion functional when we know a descent direction. we used the gradient method with optimal choice of step length ρ and eventually projection.

5. A drawing module for the plotting of the results related to a given geometry. This is convenient for quickly analyzing computational results.

The finite element method (on triangles, using second order polynomials) was used to solve the equations (1.10), (4.5), and (4.6) with $f = -L = 0$, $V(t) = 1$, $\lambda = 0.1$, $t = 1$. The triangulation is composed of 1071 nodes and 500 triangles. The main idea of our method is to change the anode surface so that A_t is close to a given fixed shape at any time t; in our example we took the fixed shape as a square as shown in Figure 2 and we can also see in Figure 2 the domain Ω_h where the criterion E

$$E(\Omega_h) = \int_{\Omega_h} (\lambda/(\lambda + \varphi_h - \chi_E)^2\, dx$$

is evaluated. The starting value of the criterion is $E(\Omega^0) = 2.52987$ with $\lambda = 0.1$ at iteration zero. We can see the new shapes of the anodes in Figures 3 and 4 at iterations 5 and 12 with criterion $E(\Omega^5) = 1.2872$ and $E(\Omega^{12}) = 0.56995$ respectively at time $t = 1$. The figures 5 and 6 show the final shape of the anode at the same time $t = 1$ with final criterion $E(\Omega^{20}) = 0.05996$ after 20 iterations, and the relation between the performance criterion E and number of iterations respectively. We have been successful, therefore, in developing a procedure for optimizing shapes for a system described by a differential inequality. Further developments of the method will be reported elsewhere.

Appendix

The main purpose of this appendix is to prove Theorem B. First, we prove two lemmas:

Lemma A. *Let $a(P_{h,\varepsilon}, \phi_{h,\varepsilon})$ be a bilinear, continuous form on $H_h^2 \times H_h^2$ such that*

$$a(P_{h,\varepsilon}, P_{h,\varepsilon}) \geq 0 \ \forall P_{h,\varepsilon} \in H_h^2$$

Then the function $\phi_{h,\varepsilon} \to a(\phi_{h,\varepsilon}, \phi_{h,\varepsilon})$ is lower semi-continuous with respect to the weak topology.

Proof. From the bilinearity, we have $\forall P_{h,\varepsilon} \in H_h^2, \phi_{h,\varepsilon} \in H_h^2$

$$\begin{aligned} a(\phi_{h,\varepsilon}, \phi_{h,\varepsilon}) &= a(P_{h,\varepsilon}, P_{h,\varepsilon}) + [a(P_{h,\varepsilon}, \phi_{h,\varepsilon} - P_{h,\varepsilon}) + \\ a(\phi_{h,\varepsilon} &- P_{h,\varepsilon}, P_{h,\varepsilon}) + a(P_{h,\varepsilon} - \phi_{h,\varepsilon}, P_{h,\varepsilon} - \phi_{h,\varepsilon})] \end{aligned} \tag{A.1}$$

since

$$a(P_{h,\varepsilon}, P_{h,\varepsilon}) \geq 0 \ \ \forall \ P_{h,\varepsilon} \in H_h^2,$$

then

$$\begin{aligned} a(\phi_{h,\varepsilon}, \phi_{h,\varepsilon}) \ \ &\geq \ \ a(P_{h,\varepsilon}, P_{h,\varepsilon}) + \\ & [a(P_{h,\varepsilon}, \phi_{h,\varepsilon} - P_{h,\varepsilon}) + a(\phi_{h,\varepsilon} - P_{h,\varepsilon}, P_{h,\varepsilon})] \end{aligned}$$

Now, let $\phi_{h,\varepsilon} \to P_h$ weakly. So from the continuity of "a", and the fact that

$$\liminf_{\phi_{h,\varepsilon} \to P_h} a(\phi_{h,\varepsilon}, \phi_{h,\varepsilon}) \geq a(P_h, P_h)$$

the map $\phi_{h,\varepsilon} \to a(\phi_{h,\varepsilon}, \phi_{h,\varepsilon})$ is lower-semicontinuous. □

Before proving Theorem B we state two theorems; for their proofs, see Stampacchia [7].

Theorem C. *Suppose that $\mathcal{A} : X \to X'$ is monotone and hemicontinuous $[D(\mathcal{A}) = X]$. Then for any bounded closed convex subset K of X, there exists a $u \in K$ such that*

$$(\mathcal{A}u, v - u) \geq 0 \ \ \forall \ v \in K. \tag{A.2}$$

Theorem D. *Let K be a bounded closed convex subset of X and suppose that $\mathcal{A} : D(\mathcal{A}) \to X'$ is monotone, $D(\mathcal{A}) = K$, and $\mathcal{A}$ is continuous on finite-dimensional subspaces of $D(\mathcal{A})$. Then there exists a $u \in K$ such that* (A.2) *holds.* □

Lemma *Let $\mathcal{A} : K \to X'$ be monotone and hemicontinuous, then u satisfies* (A.2) *if and only if*

$$(\mathcal{A}v, v - u) \geq 0 \quad \forall\, v \in K; \tag{A.3}$$

here, K is any bounded closed convex set.

Proof of Lemma

Let us suppose that u satisfies (A.2), i.e.,

$$(\mathcal{A}v, v - u) \geq 0 \quad \forall\, v \in K;$$

then by the monotonicity of $\mathcal{A}$

$$(\mathcal{A}v, v - u) \geq (\mathcal{A}u, v - u);$$

thu, (A.2) $\Rightarrow$ (A.3). To prove the converse, note that, for any $\omega \in K$,

$$v = t\omega + (1 - t)u = u + t(\omega - u) \text{ is in } K, \text{ if } 0 < t < 1.$$

So by using (A.3) and by taking limits st $t \to 0$, we obtained the (A.2). □

Theorem B *As $\varepsilon \to 0, P_{h,\varepsilon} \to P_h$ in H_h^2, P_h being the solution of the variational inequality*

$$a(P_h, \omega_h - P_h) \geq (f_1, \omega_h - P_h) \tag{A.4}$$

where φ_h is solution of (1.10) *and f_1 is defined by* (3.4).

Proof of Theorem B

Consider the second penalized equation (3.3) in variational form, with $\omega_h = P_{h,\varepsilon}$:

$$\begin{aligned} &\int_{\Omega_h} ((\nabla P_{h,\varepsilon} \cdot \nabla P_{h,\varepsilon}) + 1/\varepsilon((d/d\varphi\varphi_{h,\varepsilon}^-)P_{h,\varepsilon}, P_{h,\varepsilon}))\, dx \\ &= \int_{\Omega_h} (f_1, P_{h,\varepsilon})\, dx \end{aligned} \tag{A.5}$$

Since we know that $F'(\varphi_{h,\varepsilon}) = 1/\varepsilon(d/d\varphi\varphi_{h,\varepsilon}^-) \geq 0$, then

$$\int_{\Omega_h} (d/d\varphi\varphi_{h,\varepsilon}^-)P_{h,\varepsilon}^2\, dx \geq 0;$$

then (1.11) and (A.5) imply that

$$\begin{aligned} &\alpha\|P_{h,\varepsilon}\|^2 \le C^1\|P_{h,\varepsilon}\|, \quad (C^1 = \|f_1\|), \text{ and then} \\ &\|P_{h,\varepsilon}\|^2 \le C \qquad (C = C^1/\alpha \text{ is a constant, independent of } \varepsilon). \end{aligned} \tag{A.6}$$

A subsequence, also denoted by $P_{h,\varepsilon}$, can be extracted from the sequence $P_{h,\varepsilon}$, such that

$$P_{h,\varepsilon} \longrightarrow P_h \text{ weakly in } H_h^2.$$

From (A.5) and (1.11), we have

$$\alpha\|P_{h,\varepsilon}\|^2 + 1/\varepsilon|(d/d\varphi(\varphi_{h,\varepsilon}^-))|\ \|P_{h,\varepsilon}\|^2 \le C^1\|P_{h,\varepsilon}\|^2$$

or

$$1/\varepsilon|(d/d\varphi(\varphi_{h,\varepsilon}^-))|\ \|P_{h,\varepsilon}\|^2 \le C^1\|P_{h,\varepsilon}\|^2 - \alpha\|P_{h,\varepsilon}\|^2,$$

$$1/\varepsilon|(d/d\varphi(\varphi_{h,\varepsilon}^-))|\ \|P_{h,\varepsilon}\|^2 \le \|P_{h,\varepsilon}\|^2(C^1/\|P_{h,\varepsilon}\| - \alpha).$$

By letting $\varepsilon \longrightarrow 0$, in above equation, we obtain

$$(d/d\varphi\ \varphi_h^-) = 0 \ \ (\text{because } \|P_h\| \ne 0). \tag{A.7}$$

By using (3.3), we can write (A.5) in the following way:

$$\begin{aligned} &a(P_{h,\varepsilon}, \omega_h - P_{h,\varepsilon}) - (f_1, \omega_h - P_{h,\varepsilon}) = \\ &\quad 1/\varepsilon[((d/d\varphi\varphi_{h,\varepsilon}^-)P_{h,\varepsilon}, P_{h,\varepsilon} - \omega_h)], \\ &\quad \text{for all } \omega_h \in H_h^2(\Omega_h). \end{aligned} \tag{A.9}$$

The monotonicity of $F'(\varphi_h)$ implies that (putting $\mathcal{A} = (\varphi_h)$, $u = \omega_h$, $v = P_{h,\varepsilon}$)

$$(F'(\varphi_h)\omega_h, P_{h,\varepsilon} - \omega_h) \ge 0 \quad K = H_h^2, \tag{A.9a}$$

which is equivalent ot (A.2). By using (A.8), we can write (A.9) as:

$$\begin{aligned} a(P_{h,\varepsilon}, \omega_h - P_{h,\varepsilon}) - (f_1, \omega_h - P_{h,\varepsilon}) &= \\ (F'(\varphi_{h,\varepsilon})P_{h,\varepsilon} - F'(\varphi_h)\omega_h, P_{h,\varepsilon} - \omega_h), \end{aligned} \tag{A.9b}$$

(because $\varphi_{h,\varepsilon} \longrightarrow \varphi_h$ in $L^2(\Omega)$, see Lions [3]) for all $\omega_h \in H_h^2(\Omega_h)$; then (A.9a) gives rise to:

$$\begin{aligned} &a(P_{h,\varepsilon}, \omega_h - P_{h,\varepsilon}) - (f_1, \omega_h - P_{h,\varepsilon}) = \\ &(F'(\varphi_{h,\varepsilon})P_{h,\varepsilon}, P_{h,\varepsilon} - \omega_h) - (F'(\varphi_h)\omega_h, P_{h,\varepsilon} - \omega_h) \ge 0, \end{aligned} \tag{A.9c}$$

so that the lemma implies that

$$\begin{aligned} a(P_{h,\varepsilon}, \omega_h - P_{h,\varepsilon}) - (f_1, \omega_h - P_{h,\varepsilon}) = \\ (F'(\varphi_{h,\varepsilon})P_{h,\varepsilon}, P_{h,\varepsilon} - \omega_h) \geq 0, \end{aligned} \tag{A.9d}$$

which is (A.3). Thus,

$$a(P_{h,\varepsilon}, \omega_h) - (f_1, \omega_h - P_{h,\varepsilon}) \geq a(P_{h,\varepsilon}, P_{h,\varepsilon}), \quad \forall \omega_h \in H_h^2; \tag{A.10}$$

letting $\varepsilon \to 0$ in (A.10), we obtain

$$a(P_h, \omega_h) - (f_1, \omega_h - P_h) \geq \liminf_{\varepsilon \to 0} a(P_{h,\varepsilon}, P_{h,\varepsilon}). \tag{A.11}$$

By applying the lemma A, we obtain now

$$a(P_h, \omega_h) - (f_1, \omega_h - P_h) \geq a(P_h, P_h), \tag{A.12}$$

which implies that

$$a(P_h, \omega_h - P_h) \geq (f_1, \omega_h - P_h) \tag{A.13}$$

which shows that P_h is solution of variational inequality (4.6).

References

[1] Angrand, F., *Méthodes numerique pour la conception optimale en aerodynamique*, 3 cycle Thesis, University of Paris 6, 1980.

[2] Elliott, C. M., *On a variational inequality formulation of an electrochemical machining moving boundary problem and its approximation by the finite element methods.* Journal of the Institute of Mathematics and its Applications, **25**, 121-131, 1980.

[3] Lions, J. L., *Some topic on variational inequalities and applications*, in *New developments in differential equations*, W. Eckaus (ed.); North-Holland Publishing Company, 1976.

[4] McGeough, J. A., and Rasmussen, H., *On the derivation of the quasi-steady model in electrochemical machining*, Journal of the Institute of Mathematics and its Applications, **13**, 13-21, 1974.

[5] Pironneau, O., *Optimal shape design for elliptic system.* Springer-Verlag, New York, 1984.

[6] Saguez, C., *Contrôle d' un systéme gouverné par une inequation variationnelle parabolique; observation du domain de contact*; C. R. Acad. Sci. Paris, **287**, 957-959, 1978.

[7] Stampacchia, G., *Variational inequalities, theory and applications of monotone operators*, Proc. NATO Adv. study Inst., Oderisi-Gubbio, 101-192, 1969.

[8] Butt, R., *Optimal shape design for differential inequalities*, Thesis, Leeds University, England, 1988.

[9] Shor, N. Z., *Minimization methods for non-differentiable functions*. Springer-Verlag, New York, 1979.

$$\begin{aligned} A(t) &= \text{Anode region} \\ A_t &= \text{Boundary of the Anode region} \\ A_0 &= \text{Initial Anode} \\ C &= \text{Cathode} \end{aligned}$$

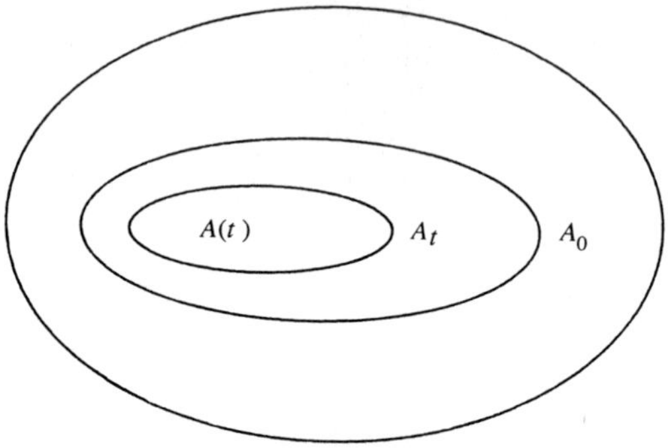

Figure 1. Optimal shape design

Total no. of nodes = 1071
Total no. of triangles = 500
Total no. of iterations = 0
$\lambda = 0.1,\quad E_0 = 2.52987,\quad E = 0.00000$

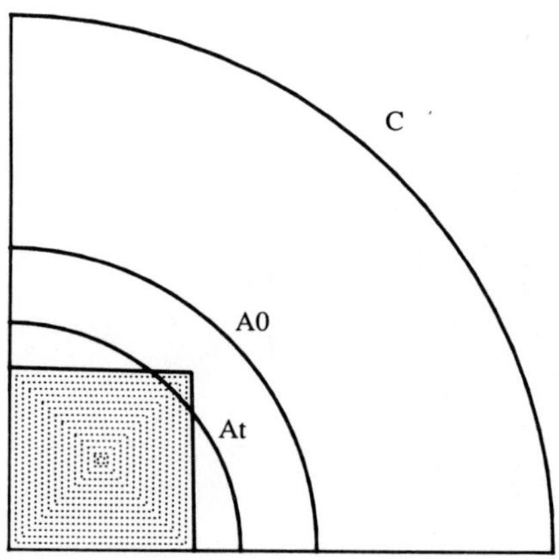

Figure 2. Optimal shape design (for quadratic case)

Total no. of nodes = 1071
Total no. of triangles = 500
Total no. of iterations = 5T
$\lambda = 0.1, \quad E_0 = 2.52987, \quad E = 1.28715$

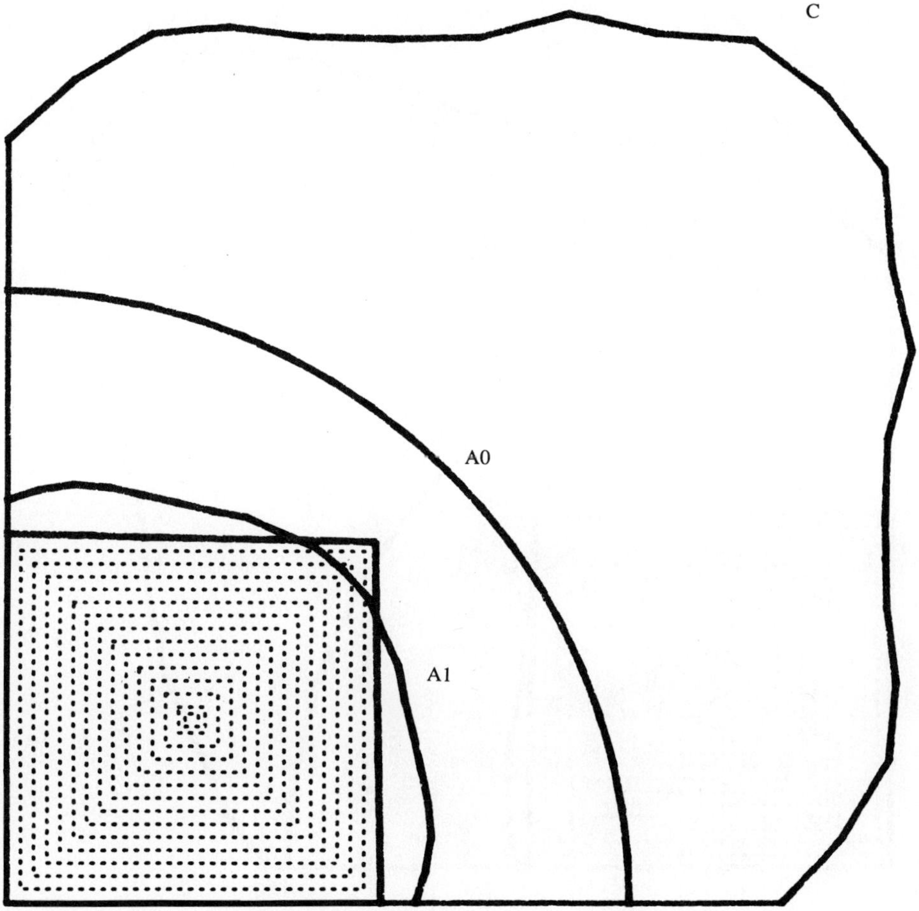

Figure 3. Optimal shape design (for quadratic case)

Total no. of nodes = 1071
Total no. of triangles = 500
Total no. of iterations = 12
$\lambda = 0.1, \quad E_0 = 2.52987, \quad E = 0.56995$

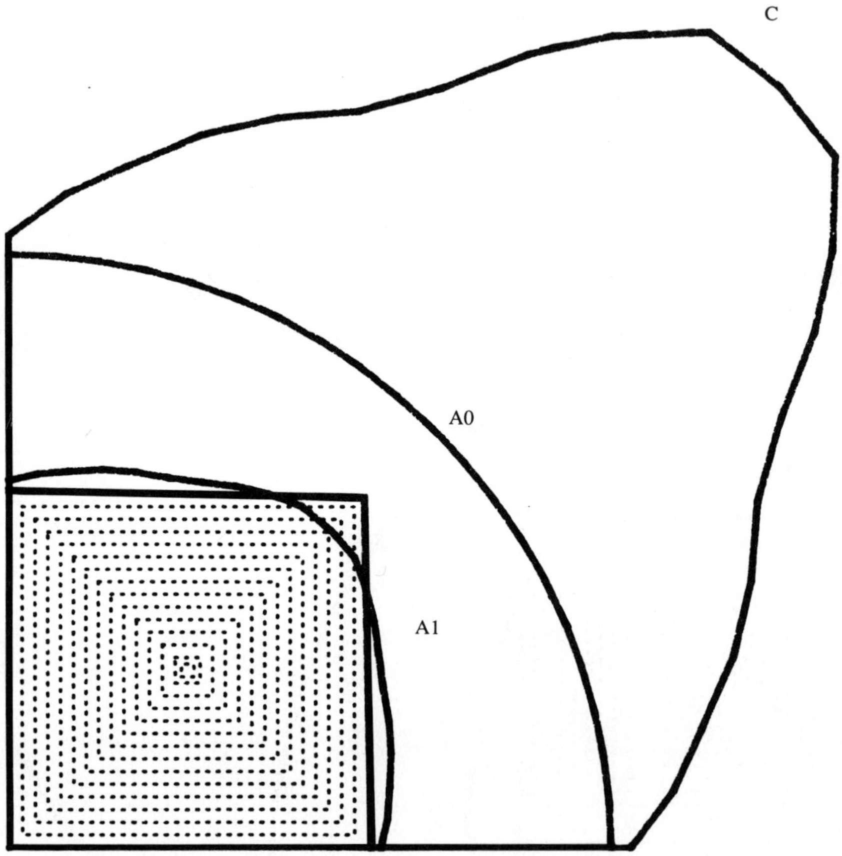

Figure 4. Optimal shape design (for quadratic case)

Total no. of nodes = 1071
Total no. of triangles = 500
Total no. of iterations = 20
$\lambda = 0.1$, $E_0 = 2.52987$, $E = 0.05999$

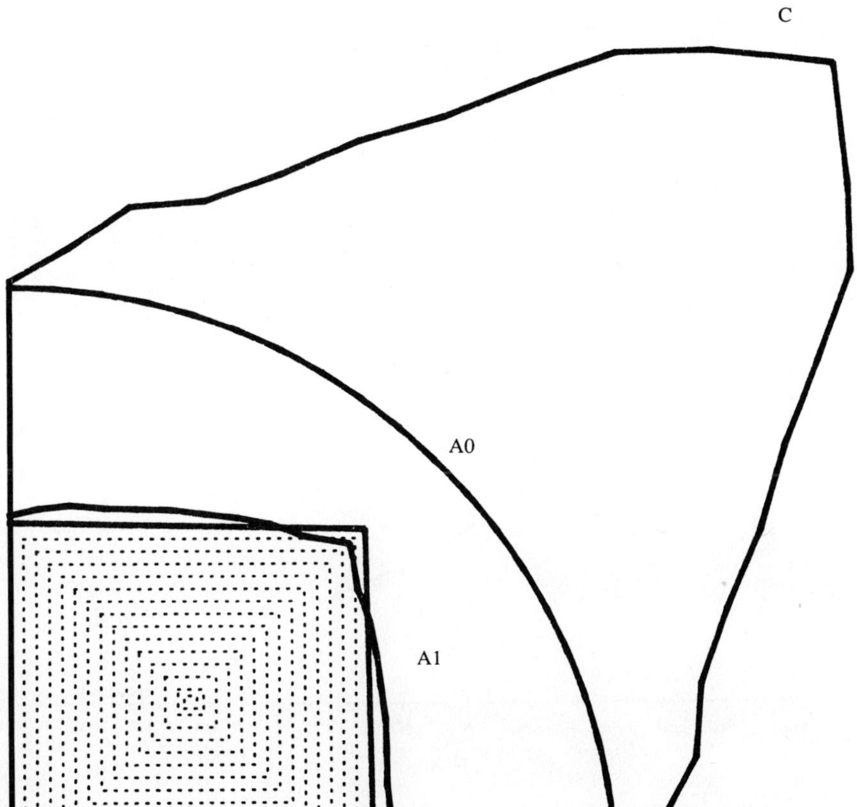

Figure 5. Optimal shape design (for quadratic case)

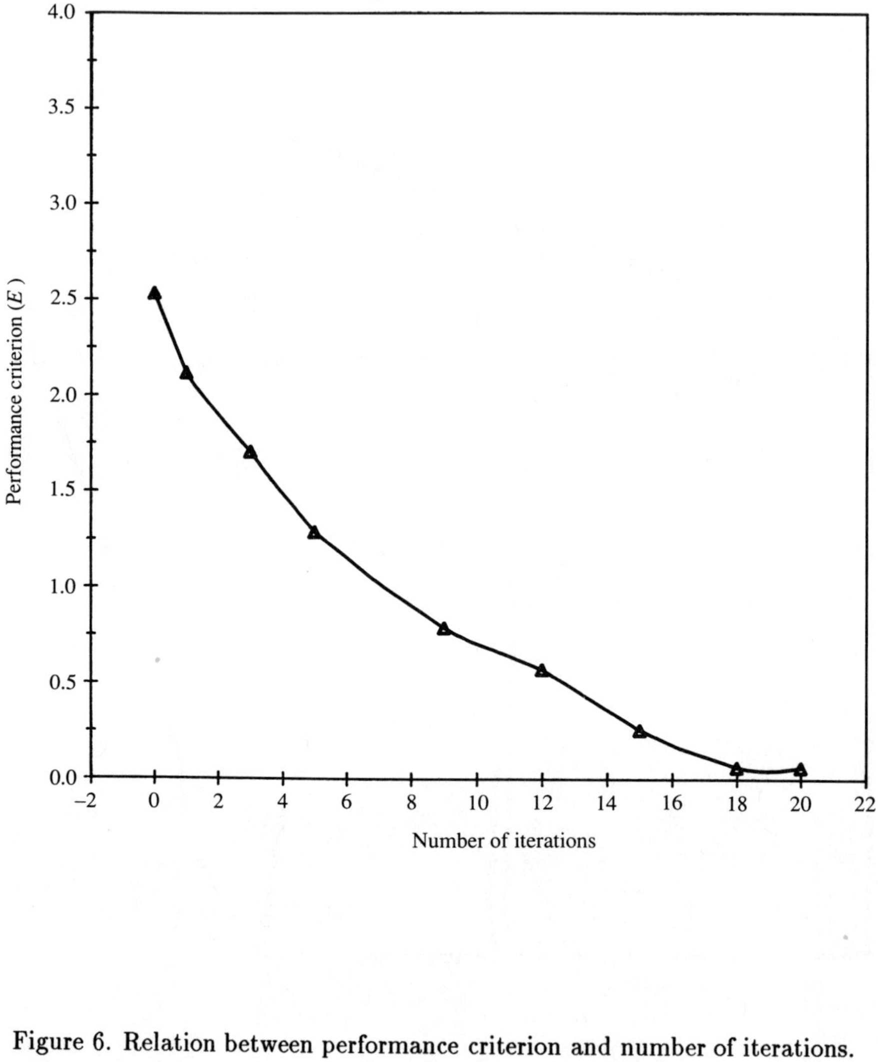

Figure 6. Relation between performance criterion and number of iterations.

Some Recent Theoretical Developments in Process Control and Their Relevance to Industry

P.V.S. Ponnapalli Johnson, M.A. Grimble, M.J.
Owens, D.H.

Abstract

The problem of system and controller structure selection is an important step in control system implementation. While various theoretical techniques have been proposed to aid the engineer to evaluate and select proper structures, these techniques did not receive widespread use in industry for various reasons. This paper highlighted some of the important new developments in theory and the perception of the industrial engineers of these developments. While considerable gaps still exists between theory and practice, many exciting and useful avenues are open which can bridge the gap and make theory more directly useful in practice.

1 Introduction

The gap between theoretical advances and industrial application in the area of control theory as applied to process control has been attracting considerable attention in recent times [7], [11]. It has also been seen that majority of the controllers employed by the industry are still PID-based. While a number of important issues concerning the controller design, robustness, integrity, and stability for example have been addressed by modern control theory, the important task of system and controller structure selection has been somewhat neglected until recently.

A typical process is a multivariable system. In such a situation, the selection of suitable variables that could be used as inputs and outputs for the design of a control system requires careful consideration. The associated problems are: the selection of suitable control strategy (for example multivariable or multi-loop), choice of appropriate input-output pairings and the use of auxiliary measurements, that is the selection of external structure of the plant and the controller (Figure 1). These aspects have been considered from different viewpoints in control system literature, for example interaction analysis, decoupling, input-out/loop pairing, controllability, observability and robustness. A number of useful tools for the analysis and synthesis have been developed. In this paper, a brief review of the developments based on the concepts of Relative Gain Array, Singular Value analysis and Internal Model Control is presented and an assessment is made on the gap between theoretical developments and the needs of the industry.

2 Relative gain array (RGA)–based measures

The RGA technique was proposed by Bristol [3] as a measure to determine the proper pairing of inputs and outputs in a multivariable system so that minimum interaction occurs among control loops. The relative gain between an input u_j and an output y_i is expressed as

$$\Phi_{ij} = \begin{cases} (\partial y_i/\partial u_j) \mid u_k = 0; k \neq j \\ \\ (\partial y_i/\partial u_j) \mid y_l = 0; l \neq i \end{cases} \tag{2.1}$$

where the numerator represents the steady–state gain between u_j and y_i when the system is open–loop and the denominator represents the steady–state gain between the same variables when all other outputs y_l are held at their normal value.

An array formed from $\Phi_{ij}, i = 1,..,n; j = 1,..,n$ is called the Relative Gain Array. Only the steady–state gains of a plant are required to compute the RGA. The interpretation of the RGA is as follows:

(1)	$\Phi_{ij} = 0$	$\Rightarrow$	u_j does not affecty_i
(2)	$\Phi_{ij} = 1$	$\Rightarrow$	$y_i$is affected only by u_j and not by any other inputs, i.e., this loop is completely decoupled
(3)	$0 < \Phi_{ij} < 1$	$\Rightarrow$	Interaction exists
(4)	$\Phi_{ij} < 0$	$\Rightarrow$	Severe interection, problems
or	$\Phi_{ij} > 1$	$\Rightarrow$	with stability

Thus the RGA is a simple measure to ascertain the extent of interactions in a multivariable system. Its computational simplicity, minimal data requirements and its empirically found reliability has led to its popularity in the process control industry [17].

The original RGA was restricted to systems with square transfer function matrices and also did not reflect the effects of dynamic interaction among process control loops. However, it was shown that the RGA can be extended to provide information on the dynamic closed–loop properties of a decoupled system [4], [17]. Other important properties which may be determined include controllability, stabilizability, failure tolerance and robustness [16].

A number of useful extensions have been made to the basic RGA method. These include the Relative Dynamic Array [24], the Average Relative Gain method [12] and the Dynamic Block Relative Gain (DBRG) Array methods [1], [22]. The DBRG was formulated both in terms of a transfer function matrix as well as in terms of Model Reference Control Schemes (MRS) [1]. It can be used for analysis as well as for the synthesis of decentralized control structures which may not necessarily be single–input single–output.

These techniques, however, are not complete in the sense that the control structures suggested by them may not always be the right choice (see [19] for example). The applicability of these techniques for large multivariable systems (*e.g.* those with

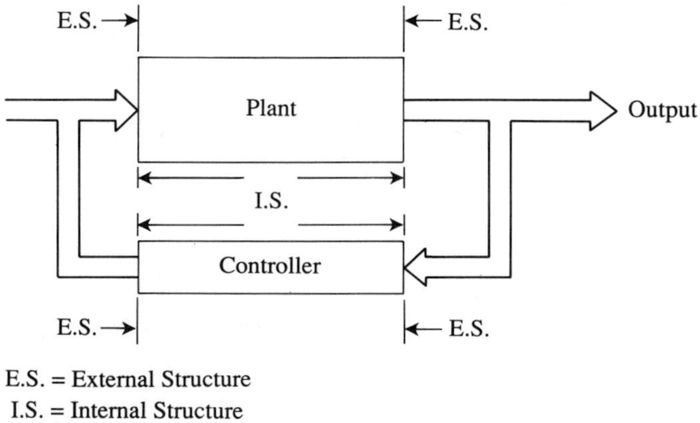

Figure 1. The structure of a typical feedback system

transfer function matrices of $min(r, m) > 4$) may require further simulation studies to confirm the input–output pairing derived. In such cases, the amount of computation time required for using the DBRG may be considerably higher than that using RGA.

3 Singular value–based methods

The Singular Value Decomposition (SVD) methods are widely used in stability and performance robustness theory [20]. The geometrical interpretation of the SVD of a plant transfer function matrix provides a useful insight into the sensitivity and interaction effects of a control system [9].

The SVD methods extend the use of condition numbers to obtain an indication of the ability of the system to tolerate modelling accuracies. The principal problem associated with this approach is the scale–dependency of the singular vectors which lead to changes in control structures with scaling.

The concept of Structured Singular Value (SSV) [6] has led to many useful extensions to analyse linear systems with structured uncertainties and to develop interaction measures for the performance and stability studies of systems under decentralized control [2], [18]. The SSV also overcomes the scaling problems of the ordinary singular value analysis. A robust H_2–type internal Model Controller using the SSV concept has also been proposed for the synthesis of robust controllers for multivariable open–loop stable or unstable systems [25]. The characterization and efficient computation of the SSV is also reported [10], [8].

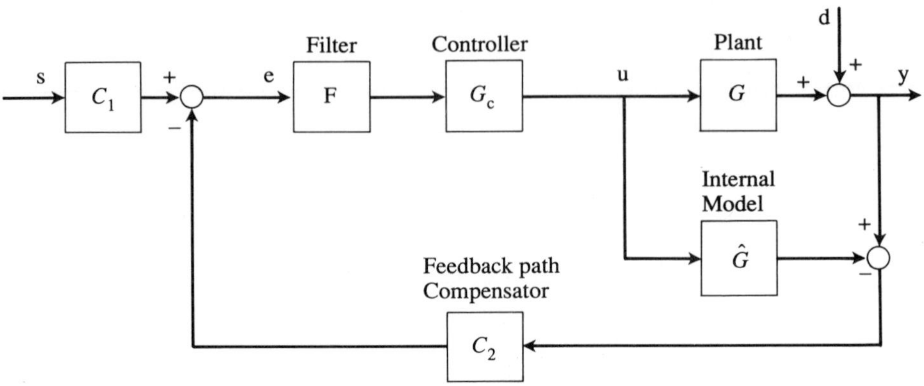

Figure 2. Structure of Internal Model Control Scheme

4 Internal model control (IMC) schemes

The IMC scheme was proposed by Garcia and Morari both for SISO and MIMO discrete systems [13]–[15]. Although the development was motivated by other existing control techniques such as inferential control [5], the IMC scheme has a conceptual simplicity which permits easy interpretation in terms of other popular control methods like the Smith Predictor or the Ogunnaike and Ray time–delay compensator [15].

The basic philosophy behind the IMC is to use a model $\hat{G}(z)$ and compare the performance of the model to that of the actual plant $G(z)$ (see Fig. 2). The error (which also consists of additional noise/disturbances) is fed back to be used for control calculation. A filter, F is introduced in the forward path to improve the robustness of the IMC loop in the presence of model inaccuracies. The filter can also be used to shape output response as well as to prevent excessive control excursions. Formulation of the control problem as a k–step ahead predictive problem [13] provides a set of 'tunable' parameters to achieve an acceptable compromise in terms of regulation, servo performance and stability considerations. Additional degrees of freedom can be generated using compensators C_1 and C_2.

The model $\hat{G}(z)$ will often have to be approximated as exact models are usually difficult to find. It may further involve factorization of $\hat{G}(z)$ such that only the 'invertable' part of $\hat{G}(z)$ (containing no nonminimum phase or time–delay characteristics) is used in the controller design. The effectiveness of IMC has been illustrated in a number of application areas such as Distillation Columns (*e.g.* [21]). The implications of IMC methods for PID controllers are discussed in [23].

5 Discussion

The RGA and derivatives, as discussed earlier, have already found some use in the industry for interaction analysis and input–output/loop pairing. Problems remain concerning their reliability, completeness (in terms of theoretical proof) and relative usefulness.

The singular value analysis has many useful properties and has been shown to be useful both for structure selection, analysis and synthesis. The structured singular value is mainly aimed at studying the robustness properties of plants with uncertainties. A number of analysis and synthesis techniques have been proposed using the SSV. However, certain issues concerning the completeness of proof and computational aspects are still undergoing extensive research. Work is required in clarifying the various properties of SSV and the relation between robustness and the structures of system and controller.

The IMC has proved to be a useful framework for the study of various design and analysis techniques using a common framework. A similar reasoning could form the basis for developing a tool to compare and analyse different possible system and controller structures.

6 The views and needs of the industry

As the ultimate goal of developments in control theory is to be useful in practice, it is necessary that the views and needs of the industry must be given adequate importance if this aim is to be achieved. In the following, a cross–section of such views obtained during direct discussions, via questionnaires and through published sources is presented and the implications for theoretical developments are discussed.

6.1 State–of–the–art

The process control industry is still largely traditional in its design philosophy, dominated by SISO/Multi–loop PID controllers. The design is predominantly based on classical design techniques such as Nyquist design, Bode Design, with experience, heuristics and rules–of–thumb playing a major role. Proven reliability and cost effectiveness are the watchwords as far as practical control system design is concerned. Apart from a few adaptive (self–tuning) and optimal controllers, very few of the new theoretical developments made an impression on the industry. Some of the main reasons for such a state–of–affairs seem to lie in (a) lack of awareness of new developments among the industrial world; (b) lack of simple tools; (c) lack of personnel with adequate theoretical background to follow the new developments, and (d) lack of clarity in the literature regarding the applicability, advantages and limitations of new techniques.

6.2 Industrial viewpoint

On the other hand, industrial engineers view many of the theoretical techniques as not being of much use for application. Some of the main reasons for such a view, as pointed out by them are:

- Simplistic assumptions on the models of plant and disturbances in theoretical studies;
- Lack of clarity regarding the implementation aspects (sensors/actuators needed, tuning problems, cost effectiveness *etc.*);
- Dearth of ready–to–use software for testing new techniques;
- Few simple diagnostic measures.

6.3 Industrial needs

The needs and requirements of the industry can therefore be briefly summarised as follows:

- Real problems should be addressed rather than selecting "academic" problems;
- Any proposed objective measure should be simple, relevant, understandable and physically meaningful;
- Guidance as to how many sensors/actuators are needed;
- Simple guides and explanations must be provided to demystify the process detailing why it is difficult to control a given process;
- Some projection of the estimated cost of implementation (or achievable performance) must be indicated;
- More collaboration between the academia and industry should be encouraged to keep the practising engineers abreast with new advances in theory.

6.4 Future directions

Based on the above considerations, certain broad directions can be identified which could help in making the new developments in theory more accessible to the industrial engineer as well as identifying the usefulness and limitations of some of these new developments.

- Comparative studies using benchmark problems should be performed to highlight the relative advantages and limitations of various techniques on a common framework.

- Simple diagnostic measures (like the RGA) must be developed to help the decision making process of an industrial engineer in plant and controller structure selection.
- Presentation of new techniques in algorithmic form must be encouraged so that they are easily programmable; easy to use software packages must be developed and made accessible to the industry for quick assessment of new techniques.
- Practical constraints of specific application areas must be considered while developing and evaluating new techniques.

7 Conclusions

The problem of system and controller structure selection is an important step in control system implementation. While various theoretical techniques have been proposed to aid the engineer to evaluate and select proper structures, these techniques did not receive widespread use in industry for various reasons. This paper highlighted some of the important new developments in theory and the perception of the industrial engineers of these developments. While considerable gap still exists between theory and practice, many exciting and useful avenues are open which can bridge the gap and make theory more directly useful in practice.

Acknowledgements

This work is carried out under SERC contract No. SO/517/86. The authors acknowledge the support and encouragement of Prof. J.S. Anderson and Prof. P. Hammond of the SERC Control and Instrumentation Subcommittee.

References

[1] Arkun, Y., Dynamic block relative gain and its connection with the performance and stability of decentralized control structures, Int. J. Control, 46, 4, 1187–1193, 1987.

[2] Arkun, Y. and Morgan, C.O. III., Robustness analysis of distillation column control schemes using structured singular values, Proc. ACC, 1772–1777,1987.

[3] Bristol, E.H., On a new measure of interaction for multivariable process control, IEEC Trans, on Aut. Control, 133–134, Jan. 1966.

[4] Bristol, E.H., Recent results on interaction in multivariable process control, Manuscript Ctr., 28 Aug. 1978, Foxboro Co., MASS. 02035, USA

[5] Brosilow, C.B., The structure and design of Smith predictors from the view point of inferential control, Joint Autom. Control Conf., Denver, CO, 1979.

[6] Doyle, J., Analysis of feedback systems with structured uncertainties, IEE Proc., 129, pt. D, 6, 242-250, Nov. 1982.

[7] Edgar, T.F., Current problems in process control, IEEE Control System Mag., 13–15, Apr. 1987.

[8] Fan, M.K.H. and Tits, A.L., Characterization and efficient computation of the structured singular value, Tech. Rep. No. TR–85–2, Systems Res. Centre, Univ. of Maryland, College Park, Maryland 20742, 1985.

[9] Fan, M.K.H. and Tits, A.L., Geometric aspects in the computation of the structured singular value, Proc. ACC, Seattle, Wash., 437–441, June 18-20,1986.

[10] Foo, Y.K., Robust analysis with structured uncertainties, Int. Rep., School of Elect. and Electronic Eng., Nanyang Tech. Inst., Nanyang Ave., Singapore, Sept. 1986.

[11] Foss, A.S., Critique of chemical process control theory, IEEE Trans. Autom. Control, 646–652, Dec. 1973.

[12] Gagnepain, J–P and Seborg, D.E., Analysis of process interactions with applications to multiloop control system design, Ind. Chem. Process Des. Dev., 21, 5–11, 1982.

[13] Garcia, C.E. and Morari, M., Internal model control. 1. A unifying review and some new results, Ind. Eng. Chem. Process Des. Dev., 21, 308–323,1982.

[14] Garcia, C.E. and Morari, M., Internal model control. 3. Multivariable control law computation and tuning guidelines, Ind. Eng. Chem. Process Des. Dev.,24, 484–494, 1985.

[15] Garcia, C.E. and Morari, M., Internal model control. 2. Design procedure for multivariable systems, Ind. Eng. Chem. Process Des. Dev.,24, 472–484,1985.

[16] Grosdidier, P., Morari, M. and Holt, B.R., Internal controllability, failure tolerance, robustness and the Relative Gain Array, Proc. ACC, 1290–1295, San Diego, USA, June 6–8,1984.

[17] Grosdidier, P., Morari, M. and Holt, B.R., Closed–loop properties from steady–state gain information, Ind. Eng. Chem. Fundam., 24, 221–235, 1985.

[18] Grosdidier, P., Morari, M., Interaction measures for systems under decentralized control, Automatica, 22,3, 309–319, 1986.

[19] Johnson,R.D. and Barton, G.W., Structural interaction analysis. Int. J. Control, 41, 4, 1005–1013, 1985.

[20] Lau, H., Alvarez, J. and Jensen, K.F., Synthesis of control structures by singular value analysis: Dynamic measures of sensitivity and interaction, AIChE J., 31, 3, 427–239, Mar. 1985.

[21] Levien , K.L. and Morari, M., Internal Model Control of coupled distillation columns, AIChE J., 33,1, 83-98, Jan. 1987.

[22] Manousiouthakis, V. and Arkun, Y., Selection of decentralized process control structures. Proc. 24th CDC, Ft. Louderdale, F1., USA, 299-304, Dec. 1985.

[23] Morari, M., Skogestad, S. and Rivera, D.F., Implications of Internal Model Control for PID controllers, Proc. ACC, 661–666, San Diego, USA, June 6–8, 1984.

[24] Witcher, M.F. and McAvoy, T.J., "Interacting control system: Steady–state and dynamic measurement of interaction", Instr. Soc. Amer. Trans., 16, 3, 35–41, 1977.

[25] Zafiriou, E. and Morari, M., Robust H_2–type IMC controller design via structured singular value, Proc. 10th IFAC World Congress, 275–280, July 1987.

A general result on adaptive stabilization by switching function controllers

Bengt Mårtensson

Abstract

In the present work the concept of *switching function adaptive controllers* is presented in a slightly more general setting than usual. A general result is proved on adaptive stabilization of "large" classes of plants under very weak a priori assumptions. As a corollary, it follows that for the special case of linear, time-invariant multivariable systems, the order of any stabilizing controller is sufficient a priori information for adaptive stabilization.

Keywords: Adaptive Stabilization, Adaptive Control, Non-linear Systems, Time-Variable Systems

1 Introduction

Faced with the problem of controlling an unknown plant one may naively suggest: "Try a number of different 'standard' controllers, one at a time, selecting between them according to some reasonable algorithm, that will eventually settle on some satisfying controller!" This, in a nutshell, is what "adaptive control by switching functions is about.

Adaptive stabilization by switching function controllers was introduced in [12], where the switching served the purpose of finding the right sign of the gain of a minimum phase, single-input, single-output, linear, time-invariant plant of relative degree 1. The switching took place between two different controllers. Later, this result was generalized in [2] to square multivariable minimum phase plants with invertible instantaneous gain. Here the switching took place in a countably infinite set; however it was later shown [8, Section 6.2] that in this case, the switching can be taken finite. In [8, Section 3.5] a more general result was proved, which as a special case (Section 4.2) gave a switching function adaptive controller capable of stabilizing any linear, time-invariant plant for which the order of *some* linear time-invariant controller was known. This was achieved by switching among the countable set of linear time-invariant controllers with rational coefficients.

A paper similar in spirit is [3], which, by restricting the attention to compact set of plants, can restrict the switching to a finite set. Another similar approach is presented in [11] which also switches among all rational linear time-invariant controllers. It should be noted that the algorithm presented therein is not differential equation based (as described in e.g. [1], see also (2.2) below), but rather "event-alarm" based. (It can, however, be interpreted as a distributional limit of (2.2).)

During the last years, several papers on adaptive stabilization have used switching in order to determine the sign of the gain of scalar systems. The papers [4] and [6] generalize [12] to certain classes of infinite dimensional plants, the latter also allows for certain cone-bounded static non-linearities in the input and output. In both, the algorithm switches between two different controllers in order to overcome the problem with unknown sign of the gain. In [5] and [9], high-gain adaptive stabilization of scalar linear time-invariant plants of higher relative degree is considered, and switching function controllers are used to turn up the gain parameter in a stepwise fashion. The relative degree was assumed to be known exactly in [5], while only an upper bound was required for the algorithm presented in [9]. The work [8, Chapter 6] contains a similar solution to the multivariable problem.

The present paper can be considered as an extension of the above-mentioned results in [8, Section 3.5], in particular in the non-linear direction.

In the next section, we define the problem and introduce the concepts formally. Section 3 presents three pieces of a priori information and proves the main theorem, stating that these pieces of a priori information are in fact sufficient a priori information for adaptive stabilization. In Section 4, the main theorem is specialized to the special case of linear, time-invariant plants. It is used to construct an adaptive switching function algorithm, capable of stabilizing any linear, time-invariant plant under the only a priori assumption of known order on any stabilizing linear time-invariant controller. The last section rounds off with some conclusions. The proof of a technical lemma is given in the appendix.

Due to the general nature of the main result, Sections 2 and 3 are by necessity fairly abstract. The reader is advised to read the hopefully more familiar material in Section 4 concurrently, to see more concretely what the general terms can mean.

2 Problems and machinery

In this paper we will treat sets $\mathcal{G}$ of time-varying nonlinear systems. For simplicity, we will assume their state-, input-, and output spaces to be $\mathbf{R}\,n$, $\mathbf{R}\,m$, and $\mathbf{R}\,p$ respectively. That is, systems which can locally be described by

$$\begin{array}{lll} \dot{x} & = f(x,u,t) & x \in \mathbf{R}\,n \quad u \in \mathbf{R}\,m \\ y & = h(x,t) & y \in \mathbf{R}\,p \end{array} \tag{2.1}$$

where f and h are smooth functions. Note that m and p, but not necessarily n, are the same for all plants $G \in \mathcal{G}$. We let a controller for (2.1) be a system of the form

$$\begin{array}{lll} \dot{z} & = f(z,y,t) & z \in \mathbf{R}\,l \\ u & = h(z,y,t) & u \in \mathbf{R}\,p \end{array} \tag{2.2}$$

We shall also write this on operator form as $u = Ky$. It is assumed that $0 \in \mathbf{R}\,n$ is an equilibrium point (2.1), (i.e. $f(0,0,t) = 0$), and that the control objective is stabilization to 0, as defined below. The definition is completely standard (cf. e.g. [1]):

Definition 2.1 *By saying that (2.2) (globally) stabilizes (2.1) (to 0) we shall mean that for all $x(0), z(0)$ it holds that*

$$(x, z) \rightarrow (0, z_\infty) \quad \text{as } t \rightarrow \infty$$

where $z_\infty < \infty$.

This is what we shall mean by a *switching function controller:*

Definition 2.2 *Let $\mathcal{K} = \{K_i\}_{i=1}^{\infty}$ be a set of controllers. Let $\sigma : \mathbf{R} \longrightarrow \mathbf{Z}_+$ be a piecewise constant function with discontinuity points $\{\tau_i\}_{i=1}^{\infty}$ such that for all $a \in \mathbf{R}$, it holds that*

$$\sigma(\{k \in \mathbf{R} : k \geq a\}) = \mathbf{Z}_+ \tag{2.3}$$

For notational convenience we require σ to be right continuous. Then σ is called a switching function, *and the controller*

$$\begin{aligned} u &= K_{\sigma(k)} y \\ \dot{k} &= \varphi(y, u, k, t) \end{aligned}$$

(where $\varphi : \mathbf{R}^p \times \mathbf{R}^m \times \mathbf{R} \times \mathbf{R} \longrightarrow \mathbf{R}_+$ is a smooth function) is called a switching function controller.

The quantity k is to be thought of as an increasing parameter. This parameter is "acted upon" by σ, which picks out controller $j = \sigma(k)$ when $k \in [\tau_i, \tau_{i+1})$. Controller j is then used to generate the input to the plant as long as $k \in [\tau_i, \tau_{i+1})$. The condition (2.3) ensures that the switching will use any particular controller infinitely often as $k \rightarrow \infty$.

3 The main theorem

Let $\mathcal{G}$ be a set of plants of the type (2.1). Assume the following quantities to be known

1. A countable set of controllers of the type (2.2): $\mathcal{K} = \{K_i\}_{i=1}^{\infty}$
2. A function $\varphi : \mathbf{R}^p \times \mathbf{R}^m \times \mathbf{R} \times \mathbf{R} \longrightarrow \mathbf{R}_+$
3. An integer $q \geq 1$

For a particular choice of $y(\cdot)$ and $u(\cdot)$ define $k(\cdot)$ by

$$\dot{k} = \varphi(y, u, k, t); \quad k(0) = k_0 \tag{3.1}$$

for k_0 some initial condition. We shall assume that the quantities (i)–(iii) are chosen so that the following assumptions hold.

(A) For all (G, K_i) such that K_i does ***not*** stabilize G, it holds that

$$\int^{\infty} \varphi(y, K_i y, k, t)\, dt = \infty$$

(If the solution x of the equation (2.1) ceases to exist at time $t\dagger < \infty$ the integral should be interpreted as $\int^{t\dagger} \varphi\, dt$.)

(B) There exists a q-th order polynomial p_q so that the following holds: For all $G \in \mathcal{G}$ there exists $K_i \in \mathcal{K}$ such that

(B1) K_i stabilizes G.

(B2) For all x_0, k_0 there exists T such that

$$\int_{t_0}^{\infty} \varphi(y, K_i y, k, t)\, dt \leq p_q\left(\int_0^{t_0} \varphi(y, u, k, t)\, dt\right) \tag{3.2}$$

for all $t_0 \geq T$, all $u(t)$; $0 \leq t \leq t_0$.

(C) If the solution x of the equation (2.1) ceases to exist at time $t\dagger < \infty$, then $k(t) \nearrow \infty$ as $t \to t\dagger$.

More informally stated, these conditions mean the following: The function φ is to be thought of as a parameter updating law, as in (3.1). Assumption (A) states that the parameter k cannot converge unless a stabilizing controller is used asymptotically. Assumption (B1) requires that $\mathcal{K}$ is so rich that for any plant $G \in \mathcal{G}$, there is a controller $K \in \mathcal{K}$ stabilizing it. Further, assumption (B2) quantifies this stabilization, so that (after long time) the future increase in the parameter, providing that we stay with K_i to the end of time, ($= \int_{t_0}^{\infty} \varphi\, dt$) can be estimated by a polynomial in $k(t_0) = \int_0^{t_0} \varphi\, dt$. Finally, assumption (C) captures finite escape time; whenever $x(t)$ explodes, $k(t)$ also explodes, thus giving the controller a chance to act fast, before the solution ceases to exist. It is easy to see that this is a necessary condition for global adaptive stabilization. Compare [10].

We are now ready for the main theorem. To emphasize its character as an existence result rather than a practically useful algorithm, two formulations are given, one constructive and one non-constructive.

Theorem 3.1 (Non-constructive version) *Let G be a plant of the type (2.1). Then $\mathcal{K}$, φ, and q in (i)–(iii), subject to assumptions (A)–(C) are sufficient a priori information for adaptive stabilization of G.*

Theorem 3.1 (Constructive version) *Let G be a plant of the type (2.1). Assume $\mathcal{K}$, φ, and q in (i)–(iii) known, and that assumptions (A)–(C) hold. Then any controller of the type*

$$\begin{aligned} u &= K_{\sigma(k)} y \\ \dot{k} &= \varphi(y, u, k, t) \end{aligned}$$

with σ any switching function with switching points $\tau_j = \alpha^{(q+1)^j}$ for $j = 1, 2, \ldots$ and some $\alpha > 1$ will stabilize G.

Of course, we prove the constructive version. The proof is basically elementary: first we show that it is enough to show that the controller switches only finitely many times. To this end, we assume the contrary. Using (3.2) this leads to a contradiction provided the sequence $\{\tau_j\}$ increases sufficiently fast. A lemma shows that the gives sequence is in fact sufficiently fast.

Proof Assumption (C) states that if k stays bounded, then the solution to the system equations is indeed defined for all t. By assumption (A), finitely many switchings as $t \to \infty$ also implies that G will be stabilized. Therefore, it is enough to show that $\sigma(k(t))$ switches only a finite number of times as $t \to \infty$.

Now assume that $\sigma(k(t))$ switches infinitely many times, i.e. $k(t) \nearrow \infty$ as $t \to \infty$. Say that K_i is one stabilizing controller for G. Assumption (B1) guarantees that such exist. The assumption of infinitely many switchings means that whenever K_i is used (i.e. $\sigma(k) = i$) k will increase sufficiently for the next switching to take place. By (2.3), this happen infinitely many times, and for arbitrarily large k. Let t_0 denote one of the times when the adaptive controller starts using K_i. That is:

$$k(t_0) = \int t_{00}\varphi \, dt + k(0) =: \tau_j$$

and $\sigma(\tau_j) = i$. To say that we will reach the next switching point is to say that

$$\tau_{j+1} - \tau_j \leq \int_{t_0} \infty\varphi(y, K_i y, k, t) \, dt$$

The right hand side denotes how much k will increase if no more switchings occurs. Using (3.2) we get

$$\tau_{j+1} - \tau_j \leq p_q \left(\int_0 t_0\varphi \, dt \right) = p_q \left(\tau_j - k(0) \right)$$

that is, with p'_q some q-th order polynomial

$$\tau_{j+1} \leq p'_q(\tau_j) \tag{3.3}$$

So the assumption on infinitely many switchings implies that (3.3) holds for arbitrarily large j. To complete the proof we shall show that with $\{\tau_j\}$ as above, (3.3) will be violated for large j. This follows from the next lemma, the proof of which is given in the Appendix. □

Lemma 3.2 *Let $\alpha > 1$ and put $\xi_j := \alpha \ (q+1) \ j$. Then, for p_q any polynomial of degree q it holds that*

$$\xi_{j+1} > p_q(\xi_j)$$

for large j.

4 The linear time-invariant case

We next specialize Theorem 1 to the case of $\mathcal{G}$ being a set of linear, time-invariant plants for which the order l of a stabilizing linear controller is known. The theorem is used to construct an adaptive switching function algorithm, capable of stabilizing any linear, time-invariant plant under the only a priori assumption that for any plant $G \in \mathcal{G}$ there is a linear, time-invariant controller of order l which stabilizes G. (It was shown in [1] that this is in fact also necessary a priori knowledge for adaptive stabilization by the class of controllers of the type (2.2).) Since the result is—at least in principle—well known, the exposition will be slightly more terse.

The plant can be described by

$$\begin{aligned} \dot{x} &= Ax + Bu & x \in \mathbf{R}\ n \quad u \in \mathbf{R}\ m \\ y &= Cx & y \in \mathbf{R}\ p \end{aligned} \qquad (4.1)$$

We will consider dynamic feedback in the spirit of [7], namely as static feedback from a plant augmented with l integrators, each one with its own input and output. Thus, a controller of order l can simply be identified with a matrix of order $(m + l) \times (p + l)$. Let $\tilde{u}$ and $\tilde{y}$ denote the input and output of the augmented plant, respectively. We claim that, if we put

$$\begin{aligned} \mathcal{K} &= \{K_i\}_{i=1}\ \infty = \mathbb{Q}\ (m+l) \times (p+l) \\ \varphi &= \|\tilde{y}\|\ 2 + \|\tilde{u}\|\ 2 \\ q &= 1 \end{aligned}$$

(for some enumeration of $\mathbb{Q}\ (m + l) \times (p + l)$), Theorem 1 will apply. That is, equivalently, we shall prove the following result, where—in the interest of brevity—only minimal plants are considered:

Corollary L *et l be a non-negative integer and $\mathcal{G}$ a set of minimal plants of the type (4.1), with the property that, for any plant $G \in \mathcal{G}$, there exists an l-th order linear time-invariant controller stabilizing G. Let $\{K_i\}_{i=1}\ \infty$ be an enumeration of $\mathbb{Q}\ (m + l) \times (p + l)$. Then the controller*

$$\begin{aligned} \tilde{u} &= K_{\sigma(k)}\tilde{y} \\ \dot{k} &= \|\tilde{y}\|\ 2 + \|\tilde{u}\|\ 2 \end{aligned}$$

with σ any switching function with switching points

$$\tau_j = \alpha\ 2\ j \ \ for\ j = 1, 2, \ldots \qquad (4.2)$$

and some $\alpha > 1$ will stabilize any $G \in \mathcal{G}$.

Remark 1. The switching points given by (4.2) can also be described as $\tau_{j+1} = \tau_j 2; \tau_1 > 1$. This sequence can be found in [2], [6], and [8]. It can, by Lemma 3.2, be described as a sequence that grows faster than any exponentially growing sequence.

The proof consists of showing that (A)–(C) are fulfilled:

Proof For notational simplicity, we drop the "~" in the proof. It is trivial that (A) and (B1) are fulfilled. For the proof of (B2) and (C) we need the following lemma, which is a slight simplification of the lemma in [7].

Lemma 4.1 *Assume that the system (4.1) is observable. Then, for all* $T > 0$ *there is a constant* c_1 *so that*

$$\|x(t)\|^2 \leq c_1 \int_0^t (\|y(\tau)\|^2 + \|u(\tau)\|^2)\, d\tau$$

for all $x(0)$, $u(\cdot)$, *and* $t \geq T$.

From this, (C) follows immediately. It is a standard exercise in basic linear system theory to show that, if u is generated by a stabilizing controller for (4.1), then there is a constant c_2, such that for all $x(t_0)$

$$\int_{t_0}^{\infty} (\|y(\tau)\|^2 + \|u(\tau)\|^2)\, d\tau \leq c_2 \|x(t_0)\|^2$$

Together with Lemma 4.1 (with t replaced by t_0), this proves (B2) for $q = 1$, thus concluding the proof of Corollary 4. □

5 Conclusion

A very general result on adaptive control has been proved. As a special case, it contains the result, originally shown in [7], that the order of a stabilizing controller is sufficient a priori information for adaptive stabilization of linear time-invariant multivariable systems. It is natural to use the result on new classes of plants, e.g. infinite-dimensional. (Essentially for notational convenience, we here made the assumption of finite dimension of the plants.)

The character of existence result rather than practical algorithm should be emphasized. For this reason, the present work contains no simulations, since there is absolutely no reason to believe that e.g. the sledge-hammer attitude of dense search in controller space is a reasonable way of controlling a more-or-less real plant.

Appendix. Proof of Lemma 3.2

Proof For some c and large positive x it holds that $p_q(x) < c x^q$. By elementary algebra

$$\begin{aligned} \xi_{j+1} &= \alpha^{(q+1)^{(j+1)}} = \alpha^{(q+1)^j (q+1)} \\ &= \alpha^{q(q+1)^j} \alpha^{(q+1)^j} = \xi_j^q \alpha^{(q+1)^j} \end{aligned}$$

Together,

$$\xi_{j+1} > \frac{1}{c}\alpha\,(q+1)\,jp_q(\xi_j) > p_q(\xi_j)$$

for large j. □

References

[1] Christopher I. Byrnes, Uwe Helmke, and A. Stephen Morse. Necessary conditions in adaptive control. In Christopher I. Byrnes and Anders Lindquist, editors, *Modelling, Identification and Robust Control*, North-Holland, Amsterdam, 1986.

[2] Christopher I. Byrnes and Jan C. Willems. Adaptive stabilization of multivariable linear systems. In *Proceedings of the 23rd IEEE Conference on Decision & Control*, pages 1574–1577, Las Vegas, NV, December 1984.

[3] Minyue Fu and B. Ross Barmish. Adaptive stabilization of linear systems via switching control. *IEEE Transactions on Automatic Control*, 31(12):1097–1103, December 1986.

[4] T. Kobayashi. Global adaptive stabilization of infinite dimensional systems. *Systems & Control Letters*, 9:215–223, 1987.

[5] Hassan Khalil and Ali Saberi. Adaptive stabilization of SISO systems with unknown high-frequency gain. *IEEE Transactions on Automatic Control*, 1986.

[6] Hartmut Logemann and David H. Owens. Input-output theory of high-gain adaptive stabilization of infinite-dimensional systems with non-linearities. *International Journal of Adaptive Control and Signal Processing*, 2:193–216, 1988.

[7] Bengt Mårtensson. The order of any stabilizing regulator is sufficient a priori information for adaptive stabilization. *Systems & Control Letters*, 6(2):87–91, July 1985.

[8] Bengt Mårtensson. *Adaptive Stabilization*. Ph.D. dissertation, Lund Institute of Technology, Department of Automatic Control, Lund, Sweden, February 1986.

[9] Bengt Mårtensson. Dynamic high-gain stabilization of multivariable linear systems, with application to adaptive control. In *Linear Algebra in Signals, Systems, and Control*, pages 384–400, Society for Industrial and Applied Mathematics, 1988.

[10] Bengt Mårtensson. Remarks on adaptive stabilization of first order non-linear systems. In *Proceedings "Mathematical Theory of Networks and Systems 1987"*, 1988. Submitted.

[11] D. E. Miller and E. J. Davidson. An adaptive controller which can stabilize any stabilizable and detectable LTI system. In *Procedings of the MTNS-conference 1987*, 1988. To Appear.

[12] Jan C. Willems and Christopher I. Byrnes. Global adaptive stabilization in the absence of information on the sign of the high frequency gain. In *Proc. INRIA Conf. on Analysis and Optimization of Systems*, pages 49–57, Lecture Notes in Control and Information Sciences 62, Springer Verlag, Berlin, 1984.

Some aspects of the mathematical theory of evidence in multi-sensor data fusion

C J Harris

Abstract

This paper shows a method, the so-called Dempster-Shafer consensus or evidential theory, for integrating or combining evidence derived from a variety of disparate knowledge sources. These knowledge sources may include sensors, intelligence data, encyclopaedic or look-up tables, algorithms or even expert views. It is more general than Boolean or the more usual Bayesian inferencing technique for uncertain events, since it combines information from disparate sources at different levels of abstraction with a non-arbitrary representation of ignorance. The Dempster-Shafer theory is based upon the concept of upper and lower probabilities to represent belief (or support) and disbelief (plausibility) in a proposition; the evidential interval — the difference between such limits, is of prime importance in this theory since it effectively represents ignorance in the support for the proposition. The resulting theory is a set of theorems for the combination (or decomposition) of evidential intervals for propositions from a set of independent knowledge sources.

1 Introduction

There are a large number of complex real time systems that require human based supervisory control eg: nuclear power, military command and control, air traffic control and satellite and spacecraft mission control centres. Such systems are inevitably hierarchical, with the highest levels of supervisory control functions demanding the most intellectually arduous cognitive skills such as (i) anomaly and fault diagnostics (ii) stochastic plausible reasoning associated with problem solving, learning and planning. The large databases (including on line sensor information, look-up tables, algorithms, operational rules and plans etc), the need for real time signal processing and control, and the need for fusing together incomplete, uncertain and/or conflicting inputs into decision algorithms, suggest the requirement for real time autonomous intelligent data or sensor integration or fusion techniques.

Multi-Sensor Data Fusion (MSDF) or sensor integration [1, 3, 4, 7], involves the integration of multiple and disparate sources of information about the same entity or object. The resulting sensor fusion or reinforcement provides fault tolerance and redundancy in safety critical applications such as nuclear power stations, air traffic control, aircraft fighter control etc. Historically, Bayesian methods have been utilized [2, 3, 11]. The Bayesian approach requires the evolution of the probability of every proposition in the space of possibilities, irrespective of the quality of evidence.

Given that there is a wealth of statistical data, this methodology is appropriate, however when it is impoverished, or even unobtainable, or at different levels of abstraction (corresponding to each and every type of data or knowledge source), an alternative methodology is required. Similarly, reasoning based upon subjective elements of these probabilities is intractable. Also what is the interpretation of low probabilities, have we not observed an object and have no reason to believe its presence, or have strong reasons to disbelieve its presence? Intuition fails at this level of detail leading to frequently inconsistent, unjustifiable and unverifiable stochastic models of the system. Here point estimates of the probabilities are incompatible with the required level of precision for decision or hypothesis making.

2 Evidential reasoning

Reasoning from uncertain, incomplete, and sometimes inaccurate information is necessary whenever the system has to interact in an intelligent manner (through say situation assessment or decision making) with the environment. The perception of the world is achieved through the set of knowledge sources of data fusion produced through partially processed sensory information and intelligence. This information is limited by sensor capabilities (temporal and spatial coverage, resolution, ambiguity rejection, portion of EM spectrum utilized, diversity of features, response times etc) and is inherently *evidential.* This is particularly the case with knowledge based expert systems which have to reason on the basis of evidence generated by multi-sensor integration according to the degree of partial belief associated with each sensor. Even with prior knowledge of a situation in which a knowledge source is observing a portion of the environment, to locate and classify objects within it, by multi-sensor integration is not a simple problem, since this presupposes that each knowledge source partially understands what observables/characteristics can be linked with each type of object.

There are essentially three types of evidence: —

(i) **Uncertain evidence**. A knowledge source probing an environment does not usually reveal exactly what the environment contains, instead it provides or attributes various degrees of belief to several environmental possibilities. The attributed degrees of belief to a proposition reflect the relative strength of the contributing evidence. For example it may have determined from ESM and radar tracks that a sensed aircraft target is unlikely to be a flighter, but can't determine if it is a bomber or a transport (EW or AWACS or other) without other information (such as EO or ESM).

(ii) **Incomplete evidence**. The precise degree of belief accorded to *every* environmental possibility generally cannot be determined on the basis of a single body of evidence, nor indeed is it possible to know all the possibilities or eventualities (yet this is the basis for Bayesian statistics). Any single body of evidence can determine the degree of belief in a possible outcome, but should remain uncommitted to another. That is *ignorance* (as a measured bounded quantity) is an important component of evidence (or the lack of). Understanding what remains unknown (and therefore requires unveiling) is just as important a component of evidential information as what is known. For example, an incoming unobserved sea skimming missile over the radar horizon is of greater significance than the observed fighter at 60km that deployed it!

(iii) **Incorrect evidence**. If a knowledge source's information is largely correct, except for sensor resolution or numerical 'noise' errors, then it contains some 'measurement' errors within its conclusions. However if the evidence is incorrect (say detecting an erroneous target), through some qualitative error or hypothesis, then it contains gross errors. This is further complicated by bodies of evidence being correct about some things, yet incorrect, to varying degrees, about others.

2.1 Representation of uncertainty of evidence

Since evidence is typically uncertain, something beyond the purely logical approach is necessary. Boolean expressions of propositional truth and falsity are inadequate (so called crisp logic) as they fail to represent the various degrees of partial beliefs. Nevertheless, evidence in the context of intelligent aids in decision making is well expressed in terms of truths and falsities, therefore whilst Boolean expressions are inadequate they must be represented, or included as a subset of a more representative probabilistic description. Partial beliefs are usually represented by probability measures and as such have formed the basis of much work in expert systems.

For a given situation suppose that there are fixed sets of environmental possibilities: —

$$F = \{q_1, q_2, \cdots, q_n\} \tag{2.1}$$

with every proposition of interest either true or false relative to each possibility. Then every proposition is completely defined by the subsets A of F ($A \subseteq F$) containing exactly those q_i for which the proposition is true. For example $F = \{$ Aircraft Target Classes $\} = \{$ air superior, multi-role, bomber, transport, helicopter, drone, missile, other $\}.A = \{$ Fighter $\} = \{$ air superior, multi-role $\} \subseteq F$. The possibility measures, a partial belief in each q_i, can be expressed through a Bayesian distribution over F. This is achieved by distributing a *unit* of belief amongst the elements, q_i, of F based upon their relative occurrance or likelihood. That is

$$\begin{aligned} &dist : F \rightarrow [0, 1] \\ &\sum_{q_i \in F} dist(q_i) = 1 \end{aligned} \tag{2.2}$$

Thus, for all $A \subseteq F$, the probability of the proposition $A, P(A)$ defined on the space F is

$$P(A) = \sum_{q_i \in A} dist(q_i) \qquad \forall A \subseteq F \tag{2.3}$$

from which it follows that $P(\bar{A}) = 1 - P(A)$ ($\bar{A}$ signifies the complement or *not* of A).

The fundamental problem with the Bayesian approach is the failure to represent the *incompleteness* of the evidence. However, the principle of indifference is frequently utilized to overcome this shortcoming, whereby *equal* probabilities are ascribed to the various propositions when the available evidence provides no reason for preferring one proposition to another. For example if we have evidence to support the disjunction of mutually exclusive observations A_1 (= fighter), A_2 (= transport A/C), A_3 (= helicopter), A_4 (= missile), as $P(A)$ = prob ($A_1 \cup A_2 \cup A_3 \cup A_4$) = 0.8, clearly $P(A)$ = 0.2 (= probability of drone or other aircraft class). The principle of indifference suggests that $P(\bar{A}_i) = 0.2$, therefore $P(A_i \cup A_j) = 0.4$ for $i, j = 1, 2, 3, 4.\ \ i \neq j$, which implies that there is a distinct preference for the disjunction of any two propositions over the third. If however the evidence does not provide a preference for these disjunctions over the singletons, there are several incompatible ways in which the principle could be used. Further what if these propositions are themselves disjunctions of other more primitive propositions?

This confusion is avoided in the Dempster-Shafer(DS) theory [8], in which the belief in a proposition A is represented by an *evidential interval* $EI(A) = \{S(A)p\ell(A)\}$ of the closed real interval [0, 1]. The lower bound $S(A)$ represents the degree to which the evidence *supports* the proposition, $p\ell(A)$ represents the degree to which the evidence fails to refute the proposition (or for which it is *plausible*). In Bayes theory prob $(A) = S(A) = p\ell(A)$, that is the evidential interval has collapsed to a point – when we have precise likelihood assignment. The principle of indifference applies in D-S theory, but now over intervals instead of probabilities, with singleton propositions and their disjunctions being assigned identical intervals, thus avoiding the dilemma of Bayes approach.

3 The Dempster-Shafer approach to sensor integration

Let F be a set of propositions about the exhaustive and exclusive possibilities of some numerical or symbolic variable x. F is called the *frame of discernment* and 2^F is the set of all subsets of F. Let $m(q)$ (so called basic probability assignment or mass) denote the degree of belief that the true value of x is in a subset $q \subseteq F$, and no smaller subset of q. It is useful to think of x as a mass of weight $m(q)$, which is constrained to stay in q, but is otherwise free to move inside q. This freedom represents the noncommittal nature of our belief or ignorance.

Assume that F has subsets $q_1, \cdots, q_n$ (called *focal elements*) – representing the set of knowledge sources propositions such that

$$q_i \neq \emptyset \text{ (the empty or null set), and } m(q_i) \neq 0$$

$$\begin{aligned} & m(\emptyset) = 0 \\ & 0 < m(q_i) \leq 1 \\ & \sum_{i=1}^{n} m(q_i) = 1 \end{aligned} \tag{3.1}$$

That is a knowledge source distributions a unit of belief across the set of propositions of focal elements, q_i, in proportion to its weight (or amount of evidential support). The so called *partial belief* $m(q_i)$ represents the proportion of belief committed to the proposition q_i. $m(q)$ can be interpreted as a segmented/partitioned line of unit length, with the length of each sub segment corresponding to the mass attributed to its focal elements q_i.

A *simple* probability mass function $m(q)$, $q \neq \Phi$, and $q \neq 0$ is such that

$$m(q) + m(\theta) = 1 \tag{3.2}$$

A special case is when $m(F) = 1$, and $m(q_i) = 0$ for all $q_i \neq F_j$. This function represents total ignorance, since no proportion of belief is committed to a proposition $q_i \subseteq F$. The definition of a partial belief function establishes an isomorphism between subsets of F and logical propositions. Hence the concepts of conjunction, disjunction, negation and implication are equivalent to the logic forms of intersection, union, complementation and inclusion.

We define the *supports* or total belief function $S : 2^F \to [0, 1]$ as: —

$$S(q_i) = \sum_{q_j \subseteq q_i} m(q_i), \quad 1 \leq i, j \leq n \tag{3.3}$$

this represents the degree or total belief to which the evidence *supports* the proposition q_i (and all its subsets). Note in general:

$$S(q_i) + S(\bar{q}_i) \neq 1$$

$$\sum_{i=1}^{n} S(q_i) \neq 1 \tag{3.4}$$

(compare with equations (3.1) and (3.2)) – this is because some mass may be attributed to propositions that imply neither q_i nor $\bar{q}_i$.

The ***plausibility*** function $p\ell : 2^F \rightarrow [0,\ 1]$ is defined by

$$p\ell(q_i) = 1 - S(\bar{q}_i) = 1 - \sum_{q_j \subseteq \bar{q}_i} m(q_j) \tag{3.5}$$

which represents the degree to which the evidence fails to refute the proposition q_i, that is the degree to which it remains ***plausible***; $S(q_i)$ is called the ***disbelief function***. The following propositions are possible but not necessary: —

(i) $q_i = F$ for some i, $1 \leq i \leq n$

that is one of the focal elements of F can be the entire frame of discernment

(ii) $q_i \cap q_j \neq 0;$ $1 \leq i,\ j \leq n$

that is the focal elements need not be disjoint (cf probability distributions).

(iii) $\bigcup_{i=1} ..q_i \neq F$

that is the union of all the focal elements need not cover the entire frame of discernment, that is there is a domain of ignorance or uncertainty.

The ***evidential interval***, $EI(q_i)$ or uncertainty of q_i is given by

$$EI(q_i) = p\ell(q_i) - S(q_i) = 1 - S(\bar{q}_i) - S(q_i) \tag{3.6}$$

represents the residual ignorance associated with q_i. Hence complete ignorance is the unit interval $EI(\cdot) = [0,\ 1]$, that is the proposition is plausible, but there is no evidence to support it. A zero interval represents the Bayes probability assignment to q_i (that is Bayesian probability is a special case of D-S theory); in general

$$S(q_i) \leq P(q_i) \leq p\ell(q_i) \tag{3.7}$$

Example 1

$F = \{q_1, q_2, q_3\}$; $A = \{q_1\}$, $A \cup B = \{q_1, q_2\}$, $\neg A = \{q_2, q_3\}$, with $m(< A, \neg A, A \cup$

$B, F >) =< 0.4, 0.2, 0.3, 0.1 >$ then the evidential intervals or ignorance associated with their propositions are respectively $EI(A) = (0.4, 0.8)$, $EI(A \cup B) = (0.7, 1)$, $EI(\neg A) = (0.2, 0.6)$ and $EI(F) = (1, 1)$.

The last evidential interval is a general rule, in that the proposition corresponding to the frame of discernment F, always receives full support since $S(F) = p\ell(F) = 1$. Also any support committed to a proposition is by definition supportive to any other proposition it implies. Given that a knowledge source is represented in terms of $\{S(q_i), p\ell(q_i)\}$ for a set of propositions from the frame of discernment F, then a set of inference rules will enable these estimates to be refined, producing reduced uncertainty, through a set of translation propositions. The following set of evidential intervals are obvious from the above definitions

(i) Given proposition A with evidential intervals $EI_1(A) = \{S_1(A), p\ell_1(A)\}$ and $EI_2 = \{S_2(A), p\ell_2(A)\}$ then

$$EI(A) = \{S(A), p\ell(A)\} = \{\max(S_1(A), S_2(A)),\ \min(p\ell_1(A), p\ell_2(A))\} \tag{3.8}$$

Also

$$EI(\neg A) = \{S(\neg A), p\ell(\neg A)\} = \{(1 - p\ell(A)),\ (1 - S(A))\} \tag{3.9}$$

(ii) Given propositions A and B with evidential intervals $EI(A), EI(B)$ then

$$\begin{aligned} EI(A \cup B) &= \{S(A \cup B), p\ell(A \cup B)\} \\ &= \{\max\,(S(A),\ S(B)),\ \min\,(1, p\ell(A) + p\ell(B))\} \end{aligned} \tag{3.10}$$

Also given $EI(A \cup B)$ and $EI(A)$ defined above then:

$$\begin{aligned} EI(B) &= \{S(A),\ p\ell(A)\} \\ &= \{\max\,(0, S(A \cup B) - p\ell(A), p\ell(A \cup B)\} \end{aligned} \tag{3.11}$$

Similarly given $EI(A)$ and $EI(B)$

$$\begin{aligned} EI(A \cup B) &= \{S(A \cap B),\ p\ell(A \cap B)\} \\ &= \{\max\,(0, S(A) + S(B) - 1),\ \min\,(p\ell(A),\ p\ell(B))\} \end{aligned} \tag{3.12}$$

(iii) Given $EI(A)$ and $EI(A \cap B)$, then

$$\begin{aligned} EI(B) &= \{S(B),\ p\ell(B)\} \\ &= \{S(A \cap B), \min\,(1, 1 + p\ell(A \cap B) - S(B))\} \end{aligned} \tag{3.13}$$

Should the evidential intervals become either $\{0,0\}$ or $\{1,1\}$ the corresponding propositions become false and true, and the rules of combining evidential intervals reduce to the corresponding rules of propositional calculus. Therefore belief mass functions and the associated propositional inference rules enable a natural transition between probabilistic and deterministic propositional calculus. The primary advantage of the partial belief method of representing knowledge, is that each knowledge source can express itself at the most appropriate level, allowing decisions over one proposition or another to be suspended until further data is available. This representation of ignorance reduces the possibility of false knowledge source reports. Each knowledge source can represent exactly what it believes without eliciting unsupported statements or speculating about outcomes for which it has little or no supportive data.

4 Consensus of evidence-data fusion

Dempster's Consensus rule [8], combines multiple bodies of information such as to increase the belief in propositions that are supported by the various bodies of evidence and away from all others. Bayes rule for probabilities is just a special case of Dempster's rule. To derive Dempster's rule, consider two knowledge sources, KS1 and KS2 with respective partial belief functions $\{m_1(q_{1i})\}$ and $\{m_2(q_{2j})\}$; $j,\ i = 1, 2, \cdots, n$. Construct a square of unit area (representing certainty), with each side representing the partial belief functions of KS1, KS2 respectively.

The horizontal strip of weight $m_1(q_{1i})$ represents the committment by KS1 to q_{1i} whereas the vertical strip $m_2(q_{2j})$ represents the committment by KS2 to q_{2j}. The intersection of these strips commits a belief $m_1(q_{1i})\ m_2(q_{2j})$ to the combination of propositions $q_r = q_{1i} \cap q_{2j}$; there may of course be other squares (such as $q_{2j} \cap q_{1i}$, $q_r = q_{2j} \cap \theta_1 = \theta_2 \cap q_{1i}$ for $\theta_1, \theta_2 =$ ignorance in KS1, KS2 respectively) which contributes to this proposition. Therefore the total belief committed to the proposition $q_r = q_{1i} \cap q_{2j} \subseteq F$ is

$$m(q_r) = \sum_{q_r = q_{1i} \cap q_{2j}} m_1(q_{1i}) m_2(q_{2j}) \tag{4.1}$$

Clearly some belief in the *orthogonal* combination of KS1, KS2 ($m = m_1 \oplus m_2$) is committed to the proposition $q_{1i} \cap q_{2j} = \emptyset$, the null set. Since the total belief in the combined knowledge source must be unity, then equation (4.1) must be normalised by disregarding all squares for which $q_{1i} \cap q_{2j} = \emptyset$, or equivalently multiply (4.1) by the correlation factor

$$k = (1 - \sum_{q_{1i} \cap q_{2j} \neq \emptyset} m_1(q_{1i}) m_2(q_{2j}))^{-1} \tag{4.2}$$

Hence we have Dempster's consensus rule for combining evidence from two orthogonal or *independent* knowledge sources as

$$m(q_r) = k \sum_{q_r = q_{1i} \cap q_{2j}} m_1(q_{1i}) m_2(q_{2j}) \qquad (4.3)$$

If $k = 0$, then $m = m_1 \oplus m_2$ does not exist and m_1 and m_2 are totally contradictory source of evidence, and as such k is a correlation measure between data sources. Dempster's rule (4.3) can be generalised to n-sources of evidence [6] and can also be used to generate the evidential intervals (measures of ignorance) of combined data or knowledge sources. Given propositions $A,\ B \subseteq F$, with evidential intervals $EI(A) = \{S(A), p\ell(A)\}$, $EI(B) = \{S(B), p\ell(B)\}$, then evidential interval of their combined knowledge/evidence is given by

$$EI(A \oplus B) = \{1 - k(1 - S(A))(1 - S(B)),\ \ kp\ell(A)p\ell(B)) \qquad (4.4)$$

where $k = (1 - (S(A)S(B) + S(A)\ .\ S(B)))^{-1}$. This formation of the consensus evidential interval is useful since it demonstrates that Dempster's rule is both commutative and associative, that is the grouping and ordering of combinations of evidence, inferencing etc is immaterial. Also if we denote $\ominus$ as the inverse operator to $\oplus$ in Dempster's rule, then we can *decompose* or retract earlier combinations of evidence or inferences by

$$EI(A) \ominus EI(B) =$$

$$\{k^{-1}(1 - S(B))(S(A)p\ell(B) - p\ell(A)S(B)), \qquad (4.5)$$

$$k^{-1}(1 - S(B)p\ell(A)p\ell(B) - S(B))\}$$

where
$k = (1 - S(B))(p\ell(B) - p\ell(A)S(B)) - (1 - S(A)p\ell(B))(1 - p\ell(B))$.

This inversion result enables a conclusion of an earlier consensus to be withdrawn without influencing conclusions reached by other mechanisms. Similarly the *conditional* support for a proposition A given that proposition B is true $(A, B \subseteq F)$ can be generated as

$$S(A/B) = \frac{S(A \cup \bar{B}) - S(\bar{B})}{p\ell(B)} \qquad (4.6)$$

Note:

(i) this result should be compared with the Bayes result

$$p(A/B) = \frac{p(A \cap B)}{p(B)}$$

(ii) also this consensus rule reduces to a singleton probability for the mass function

$m_2(A)$ for a proposition A. Hence for all $q_r \in F$ the equivalent Bayes rules are

$$m(q_r) = k \sum_{A \cap q_{1i} = A} m_1(q_{1i}) \tag{4.7}$$

where

$$k = (1 - \sum_{A \cap q_{1i} \neq A} m_1(q_{1i})), \text{ and}$$

$$p(B/A) = \sum_{q_r \in B} m(q_r), \ \forall B \subseteq F \tag{4.8}$$

(iii) The Dempster-Shafer rule can be generalised to the orthogonal sum of a series of belief functions $m = m_1 \oplus m_2 \oplus m_3 \oplus \cdots \oplus m_n$ for $q_r = \bigcap_i q_{ij} \subseteq F$ as

$$m(q_r) = k \sum_{\bigcap_i q_{ij} = q_r} \prod_{i \leq j \leq n} m_j(q)_{ji} \tag{4.9}$$

where

$$k = (1 - \sum_{\bigcap_i q_{ij} = \emptyset} \prod_{i \leq j \leq n} m_j(q_{ji}))^{-1}$$

with $q_r \neq \emptyset$

(iv) Other special cases of (4.4) follow: —

a.

$$\{S(A), p\ell(A)\} \oplus \{0, 1\} = \{S(A), p\ell(A)\} \tag{4.10}$$

the evidential range $\{0, 1\}$ represents total uncertainty or ignorance and results from the application of an inapplicable rule.

b.

$$\{S(A), 1\} \oplus \{S(B), 1\} = \{S(A) + S(B) - S(A)S(B), 1\} \tag{4.11}$$

Note in this case the support functions combine in the usual manner for *independent* propositions, and $S(\neg A) = S(\neg B) = 0$, or we have no disbelief in the propositions A and B.

c.

$$\{1,1\} \oplus \{S(B), p\ell(B)\} = \{1,1\} \tag{4.12}$$

for $\{S(B), p\ell(B)\} \neq \{0,0\}$, and

$$\{0,0\} \oplus \{S(B), p\ell(B)\} = \{0,0\} \tag{4.13}$$

for $\{S(B), p\ell(B)\} \neq \{1,1\}$.
These combinations imply that there is no application of Dempster's rule that can overrule a logical certainty.

5 Discussion

The Dempster-Shafer theory of evidence has found wide recent applications including Command and Control data fusion systems [7, 1], target identification systems [3, 7], supervisory control [9], and fault diagnosis [10], and is currently under study for application in air traffic control, intelligent sensor management, and intelligent autonomous land vehicles [University of Southampton research programmes].

The Dempster-Shafer method moves belief in the direction of propositions that are supported by multiple bodies of evidence and away from those which are unsupported. The consensus rule is both commutative and associative, therefore the grouping and order of evidence combinations are immaterial. Essentially the Dempster-Shafer method combines the Bayesian notion of probabilities with the Boolean concept of sets. The Dempster-Shafer method deals with evidential conflict through the frame of discernment F and the associated measure of ignorance or uncertainty, since F covers the set of all possible propositions, a non-zero contribution to F ensures that no two sensor or evidential sources are in complete conflict. Bayesian methods handle this dilemma by artificially distributing various (usually equal) likelihoods where they are not known. There are however obvious difficulties with the Dempster-Shafer approach, independence of knowledge or evidence sources, computational explosion associated with evaluating plausibility functions etc as the number of focal elements increase, and the requirement that $(m_1 \oplus m_2)(\emptyset) = 0$ (see eqn (4.3)). This latter problem can be overcome [11] by adding an additional element m_0 to the space:

$$\begin{aligned} &m_0(q_r) = 0 \quad \text{for } q_r \neq 0 \\ &m_0(\emptyset) = 1 \end{aligned} \tag{5.1}$$

such that $m_1 \oplus m_2 = m_0$ if $k = 0$ or

$$\sum_{q_r = \emptyset} m_1(q_{ij}) m_2(q_{2j}) = 1 \tag{5.2}$$

The binary combination rule is then defined for all m_1, m_2 with m_0 an absorbent state since $m_0 \oplus m = m \oplus m_0 = m_0$ for all m. The interpretation of m_0 is that it represents total ignorance in the combination of evidence, since its combination yields no change in belief or mass functions.

References

[1] Bogler, P., *Shafer-Dempster reasoning with applications to multi-sensor target identification systems*, IEEE Trans. S.M.C., vol **17**, No **6**, pp 968-77, 1987.

[2] Chair, Z. and Varshrey, P. K., *Optimal data fusion in multi-sensor detection systems*, IEEE, Trans. A.E.S., Vol **22**, pp 98-101, 1986.

[3] Harris, C. J., *Reducing uncertainty and ignorance in C^2 systems.* Ch 2.3 in *Application of Artificial Intelligence to Command and Control Systems* ed C. J. Harris, Peter Peregrinus, 1988.

[4] Harris, C. J. (ed). *Application of Artificial Intelligence to Command and Control Systems*, Peter Peregrinus, 1988.

[5] Hummel R. A., and Landy, M. S., *A statistical viewpoint on the theory of evidence*, IEEE Trans. P.A.M. I, Vol **10**, No **2**, pp 235-47, 1988.

[6] Lu, S. Y., and Stephanou, H. E., *A set theoretical framework for the processing of uncertain knowledge.* Proc. Nat Conf on AI. August 6-10, Texas University, USA, 1984.

[7] Miles, J. A. H., *Artificial Intelligence applied to Data Fusion and Situation Assessment for Command & Control*, PhD, Southampton University, 1988.

[8] Shafer, G., *A mathematical theory of evidence*, Princeton University Press, 1976.

[9] Shenoy, P. P., and Shafer, G., *Propagating belief functions with local computations*, IEEE Expert. Fall issue, pp 43-51, 1986.

[10] Silverman, B., *Distributed inference and fusion algorithms for real time supervisory controller positions*, IEEE Trans. S.M.C. Vol **17**, No **2**, pp 230-39, 1987.

[11] Tenney, R. R., and Sandell, N. R., *Detection with distributed sensors*, IEEE Trans. A.E.S. Vol **17**, No **4**, pp 501-509, 1981.